JUL 17 '85

Strength and Fracture
of Engineering Solids

Strength and Fracture of Engineering Solids

DAVID K. FELBECK
University of Michigan, Ann Arbor

ANTHONY G. ATKINS
University of Reading, United Kingdom

PRENTICE-HALL, INC., *Englewood Cliffs, NJ 07632*

Library of Congress Cataloging in Publication Data

Felbeck, David K.
    Strength and fracture of engineering solids.

    Includes bibliographies and index.
    1. Strength of materials.   2. Fracture mechanics.
I. Atkins, Anthony G. (date).   II. Title.
TA405.F418   1984   620.1'12   83-9488
ISBN 0-13-851709-6

Editorial/production supervision: Lori Opre
Cover design: Edsal Enterprises
Manufacturing buyer: Anthony Caruso

© 1984 by Prentice-Hall, Inc., Englewood Cliffs, New Jersey 07632

Printed in the United States of America

10 9 8 7 6 5 4 3 2 1

ISBN 0-13-851709-6

PRENTICE-HALL INTERNATIONAL, INC., *London*
PRENTICE-HALL OF AUSTRALIA PTY. LIMITED, *Sydney*
EDITORA PRENTICE-HALL DO BRASIL, LTDA., *Rio de Janeiro*
PRENTICE-HALL CANADA INC., *Toronto*
PRENTICE-HALL OF INDIA PRIVATE LIMITED, *New Delhi*
PRENTICE-HALL OF JAPAN, INC., *Tokyo*
PRENTICE-HALL OF SOUTHEAST ASIA PTE. LTD., *Singapore*
WHITEHALL BOOKS LIMITED, *Wellington, New Zealand*

To
George Atkins
and
Egon Orowan

Contents

4 STRUCTURE OF SOLIDS 74

5 CRYSTAL IMPERFECTIONS AND SLIP 84

6 PURE ELEMENTS 132

7 SINGLE PHASES 151

8 PHASE DIAGRAMS 162

15 IRONS   363

16 CERAMICS AND GLASSES   378

17 COMPOSITES   403

18 FRACTURE BY GRADUAL CRACK GROWTH: FATIGUE AND STRESS-CORROSION CRACKING   440

19 FAILURE ANALYSIS   471

experts, Drawings and specifications, Applicable standards 492
Misapplication 499   Analysis 501   Operational Tests 502
Assessment 506   References 507   Problems 507

Preface

Strength and fracture of solids are important topics to most practicing engineers, many technicians, some lawyers, and students preparing for these fields. Theories of strength and practical developments of stronger materials have been maturing and improving rapidly during the past two decades, to the extent that engineers and teachers not specializing in this area are often hard-pressed to keep up with the field. This book is intended to serve these groups as a self-contained presentation of information on the fundamentals of strength, with plentiful data on real materials, presented in the simplest possible way.

This book is written for the second- or third-year college or technical school student who has had a good grounding in chemistry, physics, and a basic course in materials science. We review here material from each of these three subject areas, particularly materials science. Since the topic of strength and fracture is much narrower than materials science, we are here able to include much new information beyond the usual materials science coverage. The topics covered here and the method of presentation have been developed through use in a third-year course at the University of Michigan, through notes, a prior short text, and laboratory experiments.

Three subjects are incorporated here that are new and developing subjects not usually found in fundamental texts: (1) a detailed presentation of the fundamentals and applications of high-performance composite materials (Chapter 17); (2) discussion of brittle fracture from the perspective of both the classic Griffith–Orowan–Irwin analysis and the more general fracture toughness approach of Gurney (Chapter 14); and (3) presentation of the fundamentals of failure analysis (Chapter 19), which is emerging as a separate and growing field of specialization as a result of stricter product and worker safety laws. We have had some experience in these areas and may succeed in providing new perspectives to them.

The book is divided into five general areas: (1) a kind of preamble, or *raison d'être*, Chapters 1–3; (2) fundamentals of strength of ductile metals, Chapters 4–11; (3) polymers and glasses, Chapters 12–13; (4) fracture of solids, Chapters 14–17; and (5) fatigue and failure analysis, Chapters 18–19. Beginning with Chapter 4, after the introductory material, we have tried to develop the subject from the simple to the complex, with a minimum of reference to subsequent material. We have in most cases selected materials for discussion as examples of certain classes of properties and have made no attempt to provide a comprehensive description of any class of engineering material. We do believe strongly in the value of repetition in learning, so the reader should not be annoyed when a term or symbol is defined a second or third time, perhaps with slightly different words. We believe this practice is better than forcing the reader to search back for the original reference to the term or symbol.

We have incorporated exercises and solutions into the body of the text, in order to allow the reader to check on progress and to illustrate particular points or paradoxes. The reader should work all of these exercises as they are presented, for they form an integral part of the presentation.

Many of the problems at the ends of the chapters have been selected from the large collection used in our third-year course. We are indebted to our close colleagues Robert M. Caddell, Julian R. Frederick, and Kenneth C. Ludema for many of these problems and for their other substantial contributions over the years in developing this subject with us. We have been influenced in our approach, consciously and subconsciously, by the work of many other authors, as our list of references shows. We are grateful for their guidance and to the many authors who permitted diagrams and photographs to be reproduced. For numerous original photographs and test results, we are indebted to W. H. Durrant.

Ann Arbor, Michigan   DAVID K. FELBECK
                                                              ANTHONY G. ATKINS

1

Engineering Design with Materials, and the International System of Units

The engineering designer requires the device he is creating to meet certain design criteria and be manufactured at the lowest possible cost. He therefore wants to select those materials that will best meet these requirements. That so many different materials are in common use is the result of their diversity of mechanical and physical properties, their differences in cost and geographical distribution, and the propensity of a society to continue to use materials with which it has experience. This chapter considers in a very general way the procedure followed by the designer in selecting materials for engineering applications.

In its simplest form, the task the designer faces is shown in Fig. 1-1, which represents a portion of an engineering machine or structure. A force P must be transmitted from point B to point C, which are located a distance L apart. The part to be designed, shown with dashed lines, must be no higher or thicker than H so that it will not interfere with other parts. The distance between B and C must not increase when under load by more than a length δL_{max}; *failure* occurs in this case when, through stretching or fracture, $\delta L > \delta L_{max}$. (Other limitations may be imposed as well, such as minimum weight or resistance to a corrosive environment or to repeated loadings, but these are omitted now for simplicity.) The finished product must also be as inexpensive as possible.

The designer's approach to this problem is essentially iterative. He can first select a material that is low in cost and easily fabricated. He could make a part that would just fit in the space provided, then test it to find whether it supports the specified force. If it is not strong enough, the designer selects a

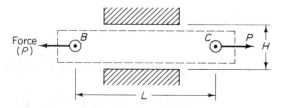

Figure 1-1   The essential design problem.

stronger material and repeats the process. But testing is very expensive, so the designer would prefer to make use of a handbook in which earlier test results are reported and tabulated. Although the designer is interested only in finding material for the part that will resist the applied force P, he is unlikely to find test results for a load of magnitude P on the precise material he has selected, for a part of exactly the size he has in mind. (For more complex design problems, the likelihood of finding the exact test results desired becomes virtually zero.)

STRESS

To overcome this problem of various test specimen sizes, the concept of stress was developed. It is obvious that a larger specimen cross section will support a larger load before fracture, and a little experimentation demonstrates that the maximum load a specimen will support is directly proportional to its cross-sectional area (except for very large or very small areas, which will be discussed in later chapters). Thus, for a failure load P_F and initial cross-sectional area A_0,

$$P_F = \text{constant} \cdot A_0 \qquad (1\text{-}1)$$

The constant is called failure stress, given the symbol S_F, and can thus be written

$$S_F = \frac{P_F}{A_0} \qquad (1\text{-}2)$$

The units of stress are thus force/area, either newtons/square meter (N/m^2), also called pascal (Pa), or pounds/square inch, $lb/in.^2$ (psi), or kilopounds per square inch (1000 psi = 1 ksi). (See the discussion of units later in this chapter.)

For the discussion of design procedure in this chapter, the general expression "failure" is used and denoted by the subscript "F." However, several explicit terms that will be used in this book should also be defined. The failure load P_F described above may or may not describe the conditions at the instant of fracture. Specifically, the *fracture strength* $S_f = P_f/A_0$, where P_f is the load at fracture. The maximum load may occur at or prior to fracture, and the stress at maximum load is called *ultimate tensile strength*, $S_u = P_{max}/A_0$. (Note that when $P_f = P_{max}$, $S_f = S_u$.) *Yielding* is the onset of substantial permanent deformation, so yield strength $S_y = P_y/A_0$, where P_y is the load at yielding. Failure by fracture or by excessive elongation may occur at any stage.

The relationships among the various definitions of "failure" may be summarized as follows: Failure to perform the desired function will occur at the lowest value of the failure stress S_F that is determined when one of the following happens:

1. Maximum allowable deflection δL_{max} is exceeded. This may occur at either low or high values of S_F, depending on the magnitude assigned to δL_{max}.
2. Discernible permanent (plastic) deformation occurs. For many applications, permanent deformation constitutes failure of a part, even though $\delta L < \delta L_{max}$. Then *yield strength* $S_y = P_y/A_0 = S_F$.
3. Fracture occurs. Then $S_F = S_u = P_{max}/A_0$. This can occur either:
   a. At maximum load, in which case $S_F = S_u = S_f$; or
   b. After maximum load is reached and the load drops before fracture, in which case $S_F = S_u > S_f$.

Now the designer needs to know only the numerical value of S_F for the selected material to determine whether the material is adequate. (There are a number of factors in selecting S_F; these will be discussed in detail as the subjects arise.) If the maximum cross-sectional area of the part in Fig. 1-1 is H^2, then the required value of S_F must be at least P/H^2 or the part will fail.

FACTOR OF SAFETY

In order to allow for the possibility of a range of the strength of the material used, variations in dimensions resulting from manufacturing variables, and service loads that may exceed the design load, a *factor of safety* is employed. Defined as the ratio of the failure stress (the smallest S_F as defined above) to the maximum calculated service stress, the factor of safety takes into account such variables as the precision of the estimate of loads and calculation of maximum stress, and whether human life would be jeopardized by a failure. Thus the maximum calculated service stress s_{max} will be less than the failure stress S_F of the material by a factor equal to the factor of safety (f.s.).

$$s_{max} = \frac{S_F}{f.s.} \tag{1-3}$$

In Fig. 1-1 the maximum service stress must be s_{max} when the load is P, so

$$s_{max} = \frac{P}{A_0} \tag{1-4}$$

Then combining Eq. (1-3) with Eq. (1-4) gives the required area A_0:

$$A_0 = \frac{(P)(\text{f.s.})}{S_F} \tag{1-5}$$

In Eq. (1-5), A_0 is the design size to be calculated, P is the load to be supported by the device, f.s. is the factor of safety, and S_F is the property (some kind of strength) of the material to be used. If the A_0 required is too large to satisfy the dimensional constraint H, then another material must be selected. If the first material selected already has the largest available value of S_F, then either the factor of safety must be shaved down (a risky procedure) or the original design must be changed so as to alter the strength and area requirements; otherwise, the device cannot be built.

STRAIN

The second requirement for the design is that the distance BC in Fig. 1-1 does not increase by more than a distance δL_{max} when the load P is applied. In a manner similar to that in defining stress, the elongation under tensile load is found to be related to the length under stress. The nondimensional term *strain* (e) is thus defined as

$$e = \frac{\delta L}{L_0} \tag{1-6}$$

where L_0 is the unloaded length BC in Fig. 1-1.

After preliminary selection of the material with failure strength S_F based on S_y or S_u, calculations must be made to determine whether the part will extend more than the maximum allowable distance δL_{max}. The value of strain e for each value of stress s will then be available, for example as in Fig. 1-2, and the value of strain e_{max} corresponding to the failure strength S_F can be

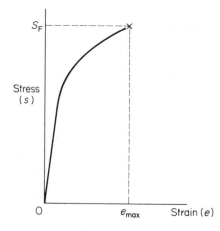

Figure 1-2 Schematic stress–strain relationship for a moderately ductile solid.

established. (Note that this strain is not necessarily the failure strain for the part in Fig. 1-1.) Then, from Eq. (1-6), the δL corresponding to S_f can be calculated:

$$\delta L = e_{max} L_0$$

If this δL exceeds the allowable δL_{max}, then the material and/or design must be changed. For example, if the cross-sectional area can be increased without exceeding the limiting dimension H in Fig. 1-1, the stress can be reduced and thus the maximum strain will be reduced.

MANUFACTURING AND ECONOMIC CONSIDERATIONS

As an integral part of the material selection process, the designer must determine the processes required to manufacture the part. He must estimate the total cost of the part, including the cost and availability of the raw material or mill product (e.g., rolled plate, extruded beams) and the cost of manufacturing the final part. At this stage the decision-making process becomes quite complicated in considering interactions among material choices, cost of material, and the number of parts to be made by the manufacturing method selected. There are seldom any easy answers, but the extremes of choices in a simple case can illustrate the point.

The design of a single novel automobile carburetor for a research project might specify that the complex shape of this part might best be made by machining it from a small number of solid pieces of free-machining brass, to be brazed together prior to the final machining. On the other hand, the manufacturer of a million carburetors per year currently makes them of aluminum–zinc alloys cast directly from molten metal under pressure in a metal mold called a die. The mold can be very expensive, but this cost is spread among all the parts, so that the unit cost will be much less than for the brass machined part. Improvements in the properties of polymers (plastics) allow the manufacturer an even better choice, because the polymers melt at a relatively low temperature and thus the dies will be cheaper.

ADDITIONAL MATERIAL PROPERTIES

The general requirements for strength, size, and deformation discussed earlier can result in severe constraints on the range of material selection available to the designer. But the situation is made much more difficult by the existence of several different modes of failure of materials, any one of which can lead to the smallest failure stress:

1. Failure from the excessive deformation associated with the onset of permanent (plastic) deformation, called *yielding*.

2. Failure from fracture that follows after substantial plastic deformation (*ductile fracture*).

3. Failure from fracture that occurs with little or no plastic deformation (*brittle fracture*) and hence usually without warning.

4. Failure from fracture that results from a large number of repeated stress cycles (*fatigue*) at stresses below the stress for yielding.

5. Excessive deformation and/or fracture under conditions of elevated temperature (*creep*) at stresses below the stress for yielding.

6. Rate-dependent deformation (*viscoelasticity*), as in polymers, where load-elongation behavior depends on speed.

7. Crack growth and fracture that occur with time under tensile stresses in certain environments (*hydrogen embrittlement* or *stress-corrosion cracking*).

8. Combinations of the preceding, of which the most conspicuous are fatigue at high temperature and fatigue associated with stress-corrosion environments.

STRESS CONCENTRATION

The stress considered so far is the average stress over an area, such as in Fig. 1-1. But since a notch, hole, defect, or sudden change in cross section can create a region of locally high stress, the designer must take care to minimize the magnitude of these local stresses. For example, if the part in Fig. 1-1 is a flat plate containing a central hole with diameter one tenth the width of the part, it turns out that the maximum stress will be 3.04 times the average stress in the part without the hole; thus the *stress-concentration factor* is 3.04. This value applies only if no plastic deformation occurs. The effective stress-concentration factor under fatigue loading conditions will be smaller than the facture under single-tension loading.

Thus the designer must cope with the additional complexity of calculating the stress concentrations that would result from the use of each material considered, and for the design configuration and manufacturing method used. Since all these variables can influence the stress concentration (e.g., because of differences in gross dimensions, surface finish, inside radii), the maximum stress may occur at different locations in the part for different materials and processes. Before he can consider his design complete, the designer must therefore calculate the maximum stress at every location where a maximum might occur.

FRACTURE TOUGHNESS

Brittle fracture was mentioned earlier as one of the possible failure modes, where a part fractures with little or no prior plastic deformation. In many applications failure in an otherwise strong structure has occurred by brittle fracture as the result of an unanticipated flaw or notch in a material whose resistance under stress to propagation of a flaw (called *fracture toughness*) is too low. The lower the fracture toughness, the smaller is the size of a flaw that can grow to failure under ordinary service loads. Since all practical structures contain flaws or stress-concentrating notches of some size, high fracture toughness should be an important criterion for material selection.

Failure of the designer to recognize the importance of fracture toughness of materials and to control his design accordingly has resulted in failures in spacecraft (the Apollo 13 moon mission fracture of titanium alloy spheres containing liquid oxygen), aircraft (the first commercial jet, Comet I, suffered complete fracture of three aluminum alloy fuselages in service), merchant ships (60 merchant ships with steel hulls have broken completely in two during the past 50 years), several megameters of oil and gas pipe lines, and many oil storage tanks, bridges, and lesser structures.

In the balance of this book, the many facets of the interaction among material behavior, design, and manufacturing will be considered in detail, with the objective of providing the prospective engineering designer with the information needed to design safe structures and components that can be reliably manufactured and assembled.

THE INTERNATIONAL SYSTEM OF UNITS

The International System of Units, usually called SI (for Système International), will be used throughout this book. This system is a modification of the metric system (cgs, MKS), which has been simplified through elimination of many superfluous units. Of particular interest for the subject area of this book are the following changes: (1) The unit of force, 1 newton (abbreviated N), is derived as the force required to accelerate a mass of 1 kilogram (kg) 1 meter per second per second (m/s^2), where meter and second are base units; (2) a system of prefixes is used to indicate the magnitude of a unit, with each larger prefix increasing by a factor of 1000; and (3) several commonly used combinations of units have been assigned the names of famous people associated with the early use of the unit. The newton has been mentioned; the unit of pressure, or stress, N/m^2, is the pascal (abbreviated Pa).

The six basic units are: meter (m) for length; kilogram (kg) for mass; second (s) for time; ampere (A) for electric current; degree Kelvin (K) for thermodynamic temperature; candela (cd) for luminous intensity. In addition, there are two supplementary units: radian (rad) for plane angle, and steradian (sr) for solid angle. From these all other units are derived.

Some of the derived units pertaining to the materials area are: density (kg/m^3 or Mg/m^3); energy, work, and quantity of heat (N-m, called the joule,

J); force (kg-m/s^2, called the newton, N); frequency (cycle/s, called the hertz, Hz); power (J/s, the watt, W); pressure and stress (N/m^2, the pascal, Pa); dynamic viscosity (N-s/m^2, which is Pa-s); degree Celsius ($°C = K - 273.15$), which is the common temperature unit used outside of thermodynamics and studies at very low temperatures.

The prefixes used to indicate multiples of 10^3 are given in Appendix 1. Note that the prefixes for all positive exponents are capitalized, except k, and the prefixes for all negative exponents are lower case.

Several rules that apply to SI style and usage in written form serve to keep units as simple and clear as possible. Prefixes are to be used in their simplest forms; for example, use GW (gigawatt), not kMW. Prefixes are to be used only in the numerator (except for kg, see below). For example, use MN/m^2 (MPa) for stress instead of N/mm^2. Use of cm and all other units with prefixes not included in Appendix 1 is to be avoided, although some residual use of these units will, no doubt, survive.

Because the kilogram is a base unit of SI, its use in the denominator is not a violation and it should be used in place of the gram (g); for example, use MJ/kg, not kJ/g.

Because prefixes are also raised to the power of the unit, mm^3 is 10^{-9} m^3, not 10^{-3} m^3. The convenience of having use of a different prefix for every step by a factor of 10^3 is lost thereby: The next unit smaller than m^2 is mm^2, which is 10^{-6} smaller, and for volume the situation is even worse, with a step of 10^{-9} from m^3 to mm^3. One solution to this is to use the liter (L), where 1 liter is $(100 \text{ mm})^3 = 10^{-3}$ m^3. Then 1 kL = 1 m^3 and 1 μL = 1 mm^3. The symbol L for liter has been adopted by the Society of Automotive Engineers and avoids the confusion that arises from the lower-case L, which is the same as the number one on an ordinary typewriter. (For area, there should be a separate name for an area of 10^{-3} m^2, which if in the shape of a square would be 31.62 mm on a side, so we could then apply all the prefixes to this unit in the same way as with the liter.)

Symbols for SI units are usually not capitalized unless the unit is derived from a proper name. One common symbol, m, might mean either "milli" or "meter," except that when two or more units are written together they should be separated by a raised dot or a hyphen. Thus ms means 10^{-3} second, while m-s means meter-second.

Because of opposite use of decimal points (periods) and commas in different countries, the comma is no longer used with numbers. Numbers are simply separated by a space between every group of three on either side of the decimal point, except that common practice usually runs four digits together without a space. For example, use:

1 532 or 1532 instead of 1,532; and 0.692 4 (correct but confusing) or, better, 0.6924

98 769.816 782 instead of 98,769.816782

SI symbols are always written in singular form. For example, 50 newtons is 50 N.

One space is left between the number and the first symbol; 200 GPa, not 200GPa. Periods should not be used after SI unit symbols except at the end of a sentence.

The SI nomenclature is particularly well suited to use with the hand calculator or computer having exponential notation. With the exception of kg, each prefix in a calculation simply represents its appropriate exponent and is so processed. The exponent in the units of the answer is thus read directly upon completion of the calculation, since all conversions of units within SI are made by multiplying by one! For example, consider the calculation of the plastic surface energy in Orowan's modification of the Griffith equation as the explanation for the brittle fracture of steel (see Chap. 14). The relation is

$$\sigma \approx \left(\frac{Ep}{c}\right)^{1/2}$$

where σ is the fracture stress, say 180 MPa; E is the elastic modulus of steel, 200 GPa; p is the plastic surface energy to be calculated; and c is half the crack length in the steel, say 16/2 mm. The calculation of p is thus

$$p = \frac{\sigma^2 c}{E} = \frac{(180 \text{ MPa})^2 (16 \text{ mm}/2)}{200 \text{ GPa}} = 1.3 \text{ kJ/m}^2$$

The evaluation involves squaring (180×10^6), then multiplying by 16×10^{-3}, then dividing by 2, and finally dividing by (200×10^9). The units of the answer can be determined by writing out in full (if necessary) all units abbreviated by names, such as pascal. The units are thus $(N^2/m^4)(m)/(N/m^2) = N/m$ which is usually written as $N\text{-}m/m^2 = J/m^2$. The exponent of the answer, in this case $+3$, dictates the prefix of the answer, k.

Treatment of the kilogram in problems involving mass and force requires a 10^3 shift, since the kilogram is a base unit and $N = \text{kg-m/s}^2$. The procedure is to reduce all prefixes for grams by one step of 10^3, to units of kilograms. For example, the force to accelerate a mass of 4 Gg by 7 m/s^2 is

$$F = (4 \text{ Gg})(7 \text{ m/s}^2)$$

$$= (4 \times 10^6 \text{ kg})(7 \text{ m/s}^2) = 2.8 \times 10^7 \text{ kg-m/s}^2 = 28 \times 10^6 \text{ kg-m/s}^2$$

$$= 28 \text{ MN}$$

Throughout this book, most graphical data will be given in both SI units and the common form of English system of units. If measurements are taken in SI, there is no reason for converting to or from any other system of units. But since we recognize that we must communicate in two or three systems of units for many years to come, some of the more common conversions are listed in Appendix 2. All problems in this book will be given in SI units, so the only conversion factor that the reader is forced to learn is the number one.

REFERENCES

Several references that cover substantial portions of the background and scope of this book are listed below, as well as at the ends of many chapters. The *Metals Handbook* series of the American Society for Metals is surely the principal source in the United States for data and practice. The reader should also consult those *Metals Handbook* volumes published after the publication of this book.

1-1. American Society for Metals, *Metals Handbook*, 8th ed. Metals Park, Ohio: American Society for Metals:
Vol. 1, *Properties and Selection of Metals*, 1961.
Vol. 2, *Heat Treating, Cleaning, and Finishing*, 1964.
Vol. 7, *Atlas of Microstructures of Industrial Alloys*, 1972.
Vol. 9, *Fractography and Atlas of Fractography*, 1974.
Vol. 10, *Failure Analysis and Prevention*, 1975.

1-2. American Society for Metals, *Metals Handbook*, 9th ed. Metals Park, Ohio: American Society for Metals:
Vol. 1, *Properties and Selection: Irons and Steels*, 1978.
Vol. 2, *Properties and Selection: Nonferrous Alloys and Pure Metals*, 1979.
Vol. 3, *Properties and Selection: Stainless Steels, Tool Materials and Special-Purpose Metals*, 1980.

1-3. Alexander, J. M., and R. C. Brewer, *Manufacturing Properties of Materials*. London: Van Nostrand, 1963.

1-4. Caddell, Robert M., *Deformation and Fracture of Solids*. Englewood Cliffs, N.J.: Prentice-Hall, 1980.

1-5. Cottrell, A. H., *The Mechanical Properties of Matter*. New York: John Wiley & Sons, 1964.

1-6. Gordon, J. E., *The New Science of Strong Materials—or Why You Don't Fall Through the Floor*. London: Penguin Books, 1968.

1-7. Gordon, J. E., *Structures—or Why Things Don't Fall Down*. London: Penguin Books, 1978.

1-8. Hertzberg, Richard W., *Deformation and Fracture Mechanics of Engineering Materials*. New York: John Wiley & Sons, 1976.

1-9. Kelly, A., *Strong Solids*. Oxford: Clarendon Press, 1964.

1-10. Knott, J. F., *Fundamentals of Fracture Mechanics*. London: Butterworths, 1973.

1-11. LeMay, Iain, *Principles of Mechanical Metallurgy*. New York: Elsevier North-Holland, 1981.

1-12. McClintock, F. A., and A. S. Argon, eds., *Mechanical Behavior of Materials*. Reading, Mass.: Addison-Wesley, 1966.

1-13. Rolfe, Stanley T., and John M. Barsom, *Fracture and Fatigue Control in Structures: Applications of Fracture Mechanics*. Englewood Cliffs, N.J.: Prentice-Hall, 1977.

1-14. Rollason, E. C., *Metallurgy for Engineers*. London: Edward Arnold, 1956.

1-15. Ruoff, Arthur L., *Introduction to Materials Science.* Englewood Cliffs, N.J.: Prentice-Hall, 1972.

1-16. Smith, Charles O., *The Science of Engineering Materials.* Englewood Cliffs, N.J.: Prentice-Hall, 1977.

1-17. VanVlack, Lawrence H., *Elements of Materials Science and Engineering,* 4th ed. Reading, Mass.: Addison-Wesley, 1980.

2 | Traditional Strength Tests and Mechanical Processing

As preparation for a discussion of the mechanisms that influence the strength of solids, this chapter defines strength and some of the common related mechanical properties. Although metals will often be used as examples, the definitions and descriptions of terms apply equally to nonmetallic solids, such as polymers and ceramics. Specific differences in behavior of polymers and ceramics will be brought out in later chapters. Furthermore, because differences among the standard tests used for measurement of static mechanical properties seldom alter the comparative ranking of different solids, only a few of the many tests will be mentioned here. A separate list of references on standard tests is given at the end of this chapter. We will show that mechanical working is one method of sometimes improving the strength of solids. Some illustrations of the technique, with the increases in strength that may be obtained, are given toward the end of the chapter.

GENERAL FEATURES OF THE TENSION TEST

When selecting a solid for a structural application, the design engineer's chief concern is that the final part fabricated from this material will be able to support the loads it will experience during its service life. Although these loads may be in the form of compression, tension, torsion, bending, internal or external pressure, or a combination of any of these, the *tension test* (also known as a *tensile test*) usually provides a good measure of the relative ability of materials to resist statically applied stresses at moderate temperatures. Thus tension-test data are particularly useful to the designer.

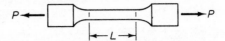

Figure 2-1 Simple tensile test specimen.

The results of tensile tests on low-carbon steel and brass given presently will show most of the forms of behavior commonly observed, and we will see how the strength-related properties of principal interest have evolved.

A tensile test can be made on a simple cylindrical or flat specimen such as that shown in Fig. 2-1. The portion of the specimen being tested lies within the gage length, which is the distance L between two marks on the specimen. The larger-diameter portions at the ends fit the grips of the *testing machine*, which applies opposing tensile loads P. Simultaneous measurement of P and L can be made a number of times during a tensile test to produce a plot of load vs. extension, that is, P vs. ΔL to fracture. Here $\Delta L = L - L_0$, the change in length, where L_0 is the initial length.

A plot of P vs. ΔL for a 0.20% carbon steel is shown in Fig. 2-2, together with a plot for a 70% Cu–30% Zn brass. For both metals the straight line *oa* exhibits *elastic* behavior; Hooke observed that in this region deformation is proportional to load. The curve from *a* to *b* shows *yielding* in mild steel, where the behavior departs suddenly from the linearity of the elastic region to give large deformation with little or no increase in load. Mild steel is *not* typical of most metals in its form of yielding, as seen in comparison with brass, where the region *ab* is absent. There, a smooth transition occurs between the elastic region, *oa*, and the region of *uniform strain hardening*, *bu*, where *u* is the

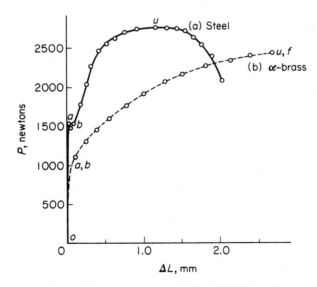

Figure 2-2 Tensile load P vs. deformation ΔL for (a) 0.20% carbon steel and (b) 70% Cu–30% Zn brass of initial cross sections 135 mm^2 and gage lengths 50 mm.

Figure 2-3 Fractured tensile specimen of 0.20% carbon steel of initial diameter 13 mm. (Photograph courtesy of W. H. Durrant.)

ultimate point of maximum load. In steel, *ab* appears as a discontinuity in the curve up to *u*. Between *b* and *u*, the specimen elongates and reduces in diameter uniformly, but beyond *u*, *nonuniform strain hardening* occurs. The deformation is concentrated in a *neck*, ending with fracture at *f*. Note that for most ductile metals the load at fracture P_f is less than the maximum load P_u. Figure 2-3 is a photograph of a fractured specimen of mild steel. If the broken parts are fitted back together, it is seen that the specimen has been *permanently elongated* and *reduced in diameter*. *Plasticity* is the name given to the regions *buf* in Fig. 2-2, which produce these permanently deformed changes.

NOMINAL STRESS AND STRAIN

The data given in Fig. 2-2 apply only to a specimen of the dimensions given. What would be the result of doubling *L* or doubling the cross-sectional area, etc.? Clearly a multitude of plots for *P* vs. Δ*L* are possible, *for the same material*, depending on the size of the test piece.

Because the designer wants to be able to apply the results of such tests to parts ranging widely in size, the test results should be presented in more general form. Furthermore, precise definition of terms that are less ambiguous and more generally useful than "load" and "elongation" will make possible a reasonably rigorous description of the conditions for deformation and fracture in solids.

Figure 2-4 is a sketch of the test length of the tensile specimens in Fig.

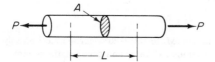

Figure 2-4 The gage section of a circular tensile test piece.

2-1 showing the cross-sectional area A. The *nominal* or *engineering stress s* is defined as

$$s \equiv \frac{P}{A_0} \tag{2-1}$$

where the subscript 0 refers to conditions before the load is applied.

The *nominal strain e* is defined as

$$e \equiv \frac{L - L_0}{L_0} \tag{2-2}$$

Since for any given specimen the values of A_0 and L_0 are constant, a plot of s vs. e would have exactly the same *shape* as the curve for P vs. ΔL in Fig. 2-2. However, *all* separate curves for P vs. ΔL would reduce to *one* plot for s vs. e. For example, if the area A_0 of a specimen is doubled, the load P at any time during the test would be doubled, but the nominal stress s would be unchanged.

Some traditional definitions of mechanical properties that have arisen from the tension test are as follows:

The *Young's modulus* is the *slope* of the initial reversible straight line portion *oa* of the plot for s vs. e.

The *yield strength* is the load at yield divided by the initial area. For annealed mild steel, the yield load is given by P at a or b. For materials whose P vs. ΔL plot smoothly curves over, an arbitrary definition of yield load has to be used. One method, when the elastic region is clearly defined, is to use the load at the point of departure from linearity. Another method, when the transition from elasticity to plasticity is imperceptible, is to define an *offset* or *proof yield stress*, given by the intersection of an arbitrary line parallel to *oa* with the curve for s vs. e.

The *tensile strength*, or *ultimate tensile strength*,[*] is the nominal stress at maximum load, that is,

$$S_u \equiv \frac{P_{max}}{A_0} \tag{2-3}$$

P_{max} is the maximum load during the entire tensile test, irrespective of the strain at which the load reaches a maximum. In Fig. 2-2 for ductile steel, for example, the load reaches a maximum, with $dP/dL = 0$, then drops to a lower load at fracture. If fracture had occurred between a and u, the material would be considered to have moderate to poor ductility (see later); if between o and a, the material would be very brittle (see later). Note that any fracture prior to u in Fig. 2-2 would not have zero slope (excepting the special use during yielding in the region ab). Because the definition of the *ultimate* point of Eq.

---

[*]We use here the capital letter S to indicate that strength is a *property* of the material, in accordance with conventional design notation. The lower case s will continue to be used for any value of nominal stress.

(2-3) does not account for these important differences in the maximum load, we designate the ultimate that occurs when $dP/dL = 0$ (and $ds/de = 0$) as the *zero-slope ultimate*.

The *percent elongation* is the value of e at fracture (converted from a decimal).

The *reduction in area to fracture* (often merely called the *percentage reduction in area* in the tensile test) is given by $(A_0 - A_f)/A_0$ where A_f is the cross-sectional area of the specimen at fracture.

The latter two quantities give some indication of the *ductility* of the solid, that is, of how much deformation can take place before fracture.

For reasons that are explained later, the use of some of the foregoing definitions is to be discouraged as they can lead to confusion. Nevertheless, they are quantities that are often quoted in handbooks of materials properties.

Exercise 2-1

A tensile specimen of gage length 50 mm and initial cross-sectional area 10 mm^2 yields when the load is 4.2 kN, at which instant it has stretched to 50.1 mm. Find the yield stress and Young's modulus.

Solution. $S_y = (4.2 \times 10^3)/(10 \times 10^{-6}) = 420$ MPa

$e_y = (50.1 - 50)/50 = 0.002; \qquad E = S_y/e_y = (420 \times 10^6)/(2 \times 10^{-3}) = 210$ GPa.

Exercise 2-2

The specimen in Exercise 2-1 withstood a maximum load of 9 kN when the length was 65 mm and the cross-sectional area 7.7 mm^2. It subsequently fractured at a load of 6.5 kN, when the length was 90 mm and the cross-sectional area 5 mm^2. Find the ultimate tensile strength, the percent elongation at fracture, and the percent reduction in area to fracture.

Solution. Ultimate tensile strength $S_u = (9 \times 10^3)/(10 \times 10^{-6}) = 900$ MPa; the percent elongation at fracture is $(90 - 50)/50 = 80\%$; percent reduction in area to fracture is $(10 - 5)/10 = 50\%$.

LOADING/UNLOADING AND WORK HARDENING

The tension tests described so far have taken place under continuous loading, that is, pulled continuously until fracture occurs. What happens if we unload at some stage and reload later? Unloading during the initial elastic loading (*oa* in Fig. 2-2) simply restores the specimen to its initial condition, like unloading a spring. There is no permanent deformation, because the yield point was not reached, and there would be no evidence at all that the test piece had been pulled. This lack of permanent deformation is the definitive feature of elasticity. However, specimens that are unloaded after the yield stress has been exceeded do display some permanent stretch. If the load and shrinking length are measured during unloading, we obtain curves such as those in Fig. 2-5,

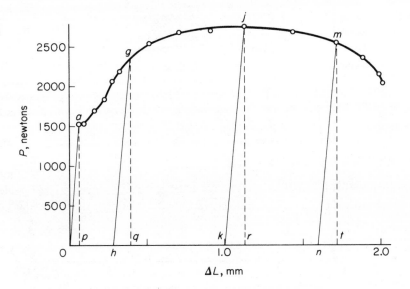

Figure 2-5 The effect of loading, unloading, and reloading on P vs. ΔL tensile plots of a 0.20% carbon steel of initial diameter 13 mm and gage length 50 mm.

where three specimens are represented being loaded (1) from o to g and back to h, (2) o to j to k, and (3) o to m to n. The unloading lines gh, jk, and mn are all essentially parallel to the initial elastic portion oa, even when unloading after passing the zero-slope ultimate load. The extension in the samples just before unloading is given by the values of ΔL at q, r, and t, and after unloading oh, ok, and on represent the permanent (plastic) extension remaining in the bars. The (*small*) values of hq, kr, and nt are called the elastic recovery of the bars resulting from unloading. Clearly all these ΔL values may be converted to permanent strains and elastic recovery strains. Note that extensions up to op in magnitude are possible before yielding takes place, and typically these represent the working deflections (and strains) of elastically deforming bridges, structures, machine parts, and so on.

What happens if a test piece that has been loaded beyond yield and unloaded is subsequently reloaded? Consider a test piece following loading along path $oagh$ in Fig. 2-5. A new plot of P vs. ΔL would look as in Fig. 2-6. Here $\Delta L'$ is measured from the new starting length. The slope $o'a'$ is exactly the same as the slope hg in Fig. 2-5, and the yield load, $P_{a'}$, is equal to P_g and thus is greater than the initial yield load of the virgin specimen given by P_a in Fig. 2-5. Evidently the process of loading beyond yield and unloading has *strengthened* the material. The process is variously called strain hardening, *work hardening*, or *cold working*. The reasons for the effect are discussed in subsequent chapters, but for the moment let us simply acknowledge that *mechanical working* can be a means of strengthening a material.

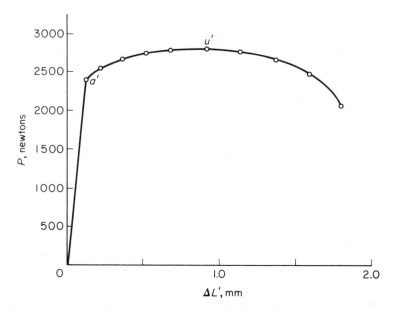

Figure 2-6 P vs. $\Delta L'$ plot for subsequent testing of specimen prestrained through the loading path *oagh* in Fig. 2-5, where $\Delta L'$ is measured from the (longer) starting length of the prestrained specimen. The yield strength of the prestrained specimen is increased.

The as-received P vs. ΔL curve is an envelope for all the other curves. In particular, the yield loads of the cold-worked samples are the loads at which the earlier deformation stopped, and from where unloading took place. This means that if we know the continuous P vs. ΔL curve for the annealed material, we have a pretty good idea of the increases in yield load obtainable by cold working. Note that the ductility of the worked material is reduced. Note also that bars cold-worked beyond zero-slope ultimate (i.e., unloaded from a point such as m in Fig. 2-5) neck immediately upon yielding when reloaded. Thus reloading a bar cold-worked to m results in simultaneous yield and ultimate, but the ultimate does not exhibit zero slope. Bars of this metal, when cold-worked to strains less than ultimate, strain harden after yielding and then display a zero-slope ultimate load.

As explained earlier, it is preferable to normalize loads and extensions into stresses and strains. What does work hardening do to s and e? We have said that the continuous P vs. ΔL curve of the "as-received" material is the same as the equivalent s vs. e curve, since A_0 and L_0 are fixed starting values. However, a piece of material work-hardened less than ultimate and unloaded has a smaller cross-sectional area than A_0, from which it came. Thus values of s for cold-worked bars will be *greater* than those based on A_0. Again, since there has been permanent stretch, the gage length defined by the original marks on the bar will be greater than before, meaning that values of e are less.

Thus, although the P vs. ΔL curves may be made to coincide by horizontal shifting, the s vs. e curve of the unworked material would be an envelope of the s vs. e curves of the strain-hardened materials *only* if s were defined in terms of the first A_0, and e were defined in terms of the first L_0.

We shall see later that these difficulties can be resolved in terms of so-called *true stress* and *true strain*, but whatever the definition of stress, the fact remains that cold working increases the strength of solids and most often reduces the ducitility.

Exercise 2-3

A 10-mm diameter bar of an annealed aluminum alloy yields in tension at a load of 6 kN. When the diameter is reduced to 9 mm, the load is 8 kN, at which point the bar is unloaded. Find the nominal yield strength of the metal on first loading, and the value upon reloading.

Solution. S_y at first yield is $6 \times 10^3/(\pi/4)(10 \times 10^{-3})^2 = 76.4$ MPa; S_y on reloading after cold work is $8 \times 10^3/(\pi/4)(9 \times 10^{-3})^2 = 125.8$ MPa.

TORSION AND SHEAR

The tension test involves tensile or *normal* stresses, which are perpendicular to the plane over which they act, that is, the load P is normal to A_0 in the definition of s.

Consider a round bar that is twisted, rather than pulled. A straight line scribed along the length of the bar would be distorted as shown in Fig. 2-7(a), the amount of movement depending on the magnitude of the torque. Radial lines on the end of the bar would remain straight, and the twisting deformation process may be viewed as if the bar were made up of a series of thin

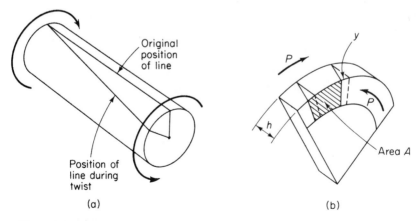

Figure 2-7 (a) Shear distortion of round bar in torsion; (b) section of "disk" along the length of the bar rotated incrementally relative to its neighbors.

disks, each one being rotated incrementally relative to its neighbors. Such action is called *shear*, a section of one of the disks being shown in Fig. 2-7(b). Consider a small element. We can define *shear stress τ* as

$$\tau \equiv \frac{P}{A} \tag{2-4}$$

and the shear strain γ as

$$\gamma \equiv \frac{y}{h} \tag{2-5}$$

τ and γ are counterparts of s and e (and of σ and ϵ defined later) when the deformation is by relative sliding, rather than pulling apart.

Yielding can occur in the twisted section, if the moment or torque is great enough. However, for small deformations before yield (i.e., at small angles of twist), the behavior is reversibly elastic. Then τ and γ are related by

$$\tau = G\gamma \tag{2-6}$$

where G is called the shear modulus. G for shear is obviously analogous to E for tension. They are related, and it can be shown that $G \approx E/3$.

Another illustration of shearing is in the processes of *punching, cropping,* or *blanking*, where a disk, for example, is pushed out from a sheet. Simplistically this is accomplished by shearing along OA and $O'A'$ (Fig. 2-8) until fracture ensues.

Exercise 2-4

A thin-walled tube of mean diameter 12 mm, wall thickness 1 mm, and length 200 mm is twisted by means of a torque wrench attachment, the lever arm of which is effectively 300 mm long. Find the shear stress in the wall of the tube when the arm of the wrench is moved 5 mm ($G = 100$ GPa).

Solution.   Viewing the whole length of the tube as shown in Fig. 2.7, we have $y = (5 \times 6)/300$, and h is now the length of the tube, so

$$\gamma = \frac{y}{h} = \frac{5 \times 6}{300 \times 200} = 0.0005$$

$$\tau = G\gamma = 100 \times 10^9 \times 5 \times 10^{-4} = 50 \text{ MPa}$$

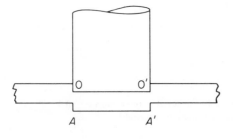

Figure 2-8 The punching, cropping, or blanking of materials to produce holes or disks. Shear followed by fracture occurs in zones OA and $O'A'$.

COMPRESSION AND COMBINED STRESS STATES

A compression test is the reverse of a tension test, that is, the material is squashed instead of pulled (Fig. 2-9). A plot of compression load vs. decrease in height of the specimen shows the same general features as a P vs. ΔL plot in tension. There is a reversible elastic region, followed by a plastic region, unloading from which gives a permanently deformed (shorter) cylinder. Upon reloading, the specimen has a greater yield stress than the original starting material. The same difficulty about "starting area" in definitions of s in tension arises with compression, except that since specimens become greater in cross section as the height reduces, s defined in terms of A_0 is always *greater* than s defined in terms of the real area. Also, strict application of Eq. (2-2) for strain gives *negative* values in compression, since the deformation is in the "opposite direction" to tension.

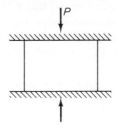

Figure 2-9 A simple compression test piece.

Although simple in principle, compression tests are complicated by friction between the specimen and the testing machine faces (platens). During deformation, material has to slide in relation to these faces, so the load required to cause a given amount of deformation depends on the friction. Stresses are thus always greater than in the absence of friction, although there are means of compensating for friction.

Since material adjacent to the compression faces is restrained by frictional effects, a test piece often becomes barrel-shaped (Fig. 2-10). The defor-

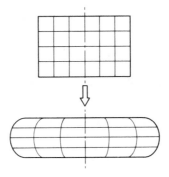

Figure 2-10 "Barreling" distortion of compression test piece caused by frictional restraint at the platens. Nonhomogeneity of deformation is shown by previously marked rectangular grids becoming skewed.

mation is nonuniform, or nonhomogeneous, with the strain patterns varying through the height of the test piece. Tension tests and frictionless compression tests, on the other hand, produce homogeneous deformation. A useful technique to check homogeneity of deformation is to scribe grids on the material and compare the patterns before and after loading (Fig. 2-10). Nonhomogeneous deformation is characterized by originally orthogonal (perpendicular) grids becoming skewed.

It is sometimes convenient to regard a skewed pattern such as shown in Fig. 2-10 as being produced by a combination of squashing or pulling (maintaining homogeneously the right-angled grids) together with shearing, which skews the grids. In this way the compression test with friction is an example of *combined stresses*. The frictional stresses are unwanted but must be tolerated. In other combined stress cases, the geometry produces loads acting in different directions. For example, a state of equal biaxial tension exists in the skin of a balloon when blown up, which also exists in a metal sheet when bulged into a dome by fluid pressure; see Fig. 2-11(a). Calculations show that a capped thin cylinder under internal pressure has both longitudinal and circumferential stresses in the wall, where the longitudinal stress is half the circumferential stress; see Fig. 2-11(b). Thus the walls of a beer can have unequal (2 : 1) biaxial tensions.

It is important, then, to remember that although we shall emphasize the simple tension test as a means of characterizing materials, there are many situations where materials are subjected to combined stresses. We must therefore have means for predicting the elastic and plastic behavior of solids under combined stresses, and means of interpreting data on stress vs. strain obtained from tests. Details are not within the scope of this book, but we shall touch on some points (such as the Tresca and Maxwell–von Mises criteria for yielding under combined stresses) in later sections.

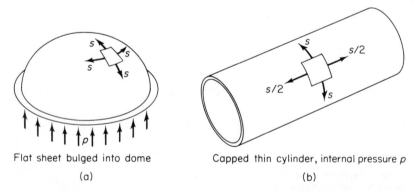

Flat sheet bulged into dome   Capped thin cylinder, internal pressure p

(a)   (b)

Figure 2-11 Combined stress states in (a) a sheet of material bulged into a dome (equal biaxial tensions) and (b) an internally pressurized thin cylinder (unequal biaxial tensions, where the longitudinal stress is one half the circumferential stress).

One test that occurs under combined stresses is the hardness test, which concerns the resistance of a solid to penetration by some sort of indenter. Such tests are described later in this chapter, as they are a means of quickly establishing strength levels without making, for example, specially shaped tensile bars.

MOHR'S CIRCLE TRANSFORMATIONS

In combined-stress situations, it is often required to establish the magnitude of stresses and strains in directions different from those along which stresses and strains are known by virtue of the external loading. Two reasons for so doing are that brittle fracture is often connected with the maximum tensile stress in a body and that plastic flow is often connected with the maximum shear stress in a body. Given some set of loads on a body, what are the absolutely greatest tensile and shear stresses? On what planes do they occur, and what are the orientations of those planes relative to some known directions in the body? In other words, we must know how to "resolve" stresses and strains in different directions.

Stress (whether normal or shear) is defined as (force $\div$ area). Both force *and* area are vector quantities, that is, they both require magnitude and direction to be fully described. The idea of a force being a vector is, no doubt, familiar to the reader, but perhaps not so for area. Think, however, of the cross-sectional area of a tensile test bar; we usually talk in terms of the size of the normal cross-section, for convenience, but we could just as well think of the size of an oblique cross section, the orientation of which would be known; hence "magnitude" and "direction" are required for area. Stress, which is evidently the quotient of two vectors, is *not* a vector itself, but rather something called a second-order tensor. The rules for resolving vectors do *not* apply to stress, and we must establish what the correct rules are. We shall learn that a simple graphical technique (Mohr's circle) aids the process.

Consider a tension test of a flat sheet of material. If the specimen were saw cut in two along an inclined direction (Fig. 2-12), we might ask what distribution of normal and shear stresses would have to be applied to the cut plane in order to maintain the body in equilibrium. This is the same as asking what the distribution of stresses is on such an inclined plane in an *uncut* specimen, since the force induced by the applied stresses on a cut specimen must be equivalent to the axial force applied to the uncut tension specimen by the testing machine.

Let the stresses on the plane, which is inclined at θ to the horizontal, be s_θ and τ_θ (see Fig. 2-12, where positive rotations are counterclockwise). If the nominal cross-sectional area of the specimen is A, the area of the inclined plane is $(A/\cos \theta)$. We cannot resolve stresses, but we may resolve forces. Thus

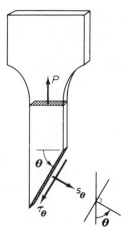

Figure 2-12 A flat tensile specimen cut along an inclined direction, where the cut plane is identified by the angle θ to the horizontal (this is the same as defining the plane by its normal). s_θ and τ_θ are the normal and shear stresses on the inclined plane. Positive rotations are counterclockwise.

along the axis of the bar

$$s_\theta\left(\frac{A}{\cos\theta}\right)\cos\theta + \tau_\theta\left(\frac{A}{\cos\theta}\right)\sin\theta = P \left.\vphantom{\begin{array}{c}1\\1\\1\\1\end{array}}\right\}$$

and across the bar   (2-7)

$$s_\theta\left(\frac{A}{\cos\theta}\right)\sin\theta = \tau_\theta\left(\frac{A}{\cos\theta}\right)\cos\theta$$

Thus

$$s_\theta = \frac{P}{A}\cos^2\theta \left.\vphantom{\begin{array}{c}1\\1\\1\end{array}}\right\}$$

and   (2-8)

$$\tau_\theta = \frac{P}{A}\sin\theta\cos\theta = \left(\frac{1}{2}\right)\left(\frac{P}{A}\right)\sin 2\theta$$

The quantity P/A is the normal stress induced in the bar on cross sections perpendicular to the direction of loading. Equations (2-8) are the resolved or *transformed* stresses on the inclined plane. Note that they are not merely the simple $\sin\theta$ or $\cos\theta$ resolution that would apply to forces.

With regard to maximum values of s or τ, the form of the trigonometric functions says that s is greatest when $\theta = 0$ (i.e., the usual normal stress) and τ is greatest when $\theta = 45°$ (from differentiation of $\sin 2\theta$). The magnitude of the greatest τ is $P/(2A)$, or one half of the maximum tensile stress in this case.

The foregoing treatment may be generalized to include normal stresses acting at right angles on a cube (Fig. 2-13), and also to include shear stresses. A pressurized beer can being twisted is one example of such a two-dimensional stress state.

The horizontal shear stresses τ_{yx} on opposite faces of the cube are equal in magnitude, as we see from considerations of force equilibrium (they act on

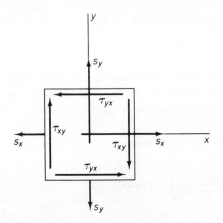

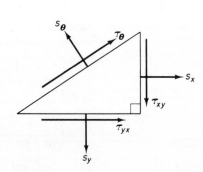

Figure 2-13 General two-dimensional stress state with both normal (s_x and s_y) and shear (τ_{xy} and τ_{yx}) stresses.

Figure 2-14 Stresses s_θ and τ_θ acting on an inclined plane in a general two-dimensional stress system having s_x, s_y, and τ_{xy}.

areas of equal magnitude); so are the vertical stresses, τ_{xy}. Equilibrium of moments, which is necessary as the body does not spin in space, shows furthermore that the pairs of orthogonal shear stresses are themselves equal, that is, $\tau_{xy} = \tau_{yx}$. For this reason they are called complementary shear stresses. The pairs are given double suffixes in order to signify both the *plane* in which the shear stress acts (the first suffix) and the *direction* of the shear stress (second suffix). The definition of planes follows the mathematical convention of employing the normal to the plane; thus τ_{yx} is the shear stress acting in the plane whose normal is the y-axis, with the direction of stress along the x-axis. Similarly for τ_{xy}. Strictly there should be double suffixes on the normal stresses to define the plane, such as s_{xx}, but usually only a single suffix is used, since different symbols (s, and later σ, rather than τ) are being employed for the two types of stress.

Given some combination of s_x, s_y, and τ_{xy}, what are the expressions for some s_θ and τ_θ (Fig. 2-14)? Proceeding as before, we may resolve forces. If we take, this time, the *inclined* area to be of size A, the plane defined by the y-axis has area ($A \cos \theta$) and that of the x-plane is ($A \sin \theta$). Then resolving horizontally gives

$$s_\theta A \sin \theta = \tau_\theta A \cos \theta + s_x A \sin \theta + \tau_{yx} A \cos \theta$$

and resolving vertically gives

$$s_\theta A \cos \theta + \tau_\theta A \sin \theta = s_y A \cos \theta + \tau_{xy} A \sin \theta$$

$$(2\text{-}9)$$

Manipulation of these equations gives

$$s_\theta = \tfrac{1}{2}(s_y + s_x) + \tfrac{1}{2}(s_y - s_x) \cos 2\theta + \tau_{xy} \sin 2\theta$$

$$\tau_\theta = \tfrac{1}{2}(s_y - s_x) \sin 2\theta - \tau_{xy} \cos 2\theta$$

$$(2\text{-}10)$$

For $s_x = 0 = \tau_{xy}$ these equations reduce to Eq. (2-8) with $P/A \equiv s_y$, as in that case A is defined differently.

Exercise 2-5

A tensile bar such as shown in Fig. 2-12 has cross section 5 mm × 2 mm and is under a load of 3 N. What are the stresses on the plane inclined at 30° to the horizontal?

Solution. $A = 5 \times 2 \times 10^{-6} = 10 \times 10^{-6}$ m². Using Eq. (2-8), we have

$$s_{30} = \left(\frac{3}{10 \times 10^{-6}}\right)\left(\frac{\sqrt{3}}{2}\right)^2 = 225 \text{ kPa}$$

$$\tau_{30} = \left(\frac{3}{10 \times 10^{-6}}\right)\left(\frac{\sqrt{3}}{2}\right)\left(\frac{1}{2}\right) = 130 \text{ kPa}$$

Exercise 2-6

A stress state on two perpendicular planes consists of a normal tension stress on one, a normal compressive stress of equal magnitude on the other, and a pair of complementary shear stresses. On a plane at 60° (counterclockwise) to the plane subjected to the compressive stress, the stress state is 40 MPa tension and 10 MPa shear. What are the values of the unknown orthogonal stresses?

Solution. If the magnitude of the unknown tension stress (corresponding to s_x in Fig. 2-14) is s, the orthogonal compressive stress (equivalent to s_y) is given by $(-s)$. Then we have in Eq. (2-10)

$$s_\theta = 40 \times 10^6 = \tfrac{1}{2}(s - s) + \tfrac{1}{2}(-s - s) \cos 120° + \tau_{xy} \sin 120°$$

$$\tau_\theta = 10 \times 10^6 = \tfrac{1}{2}(-s - s) \sin 120° - \tau_{xy} \cos 120°$$

that is,

$$40 \times 10^6 = -s\left(-\frac{1}{2}\right) + \tau_{xy}\frac{\sqrt{3}}{2}$$

and,

$$10 \times 10^6 = -s\left(\frac{\sqrt{3}}{2}\right) - \tau_{xy}\left(-\frac{1}{2}\right)$$

From this, $s = 11.34$ MPa and $\tau_{xy} = 39.6$ MPa. Therefore the unknown stress state is 11.34 MPa tension, 11.34 MPa compression, and complementary shears of magnitude 39.6 MPa (with senses as in Fig. 2-14).

Maximum and minimum values of s_θ and τ_θ may be found by first differentiating Eq. (2-10) with respect to θ to establish stationary values and then inserting those values of θ back in the expressions for s_θ and τ_θ. Alternatively, we may note that according to Eq. (2-10), $\tau_\theta = 0$ when $\tfrac{1}{2}(s_y - s_x) \sin 2\theta = \tau_{xy} \cos 2\theta$, that is,

$$2\theta = \tan^{-1} \frac{2\tau_{xy}}{s_y - s_x} \tag{2-11}$$

for maximum or minimum s_θ. There will be *two* values of (2θ) satisfying this relation, which are 180° apart. In other words, whatever the (s_x, s_y, τ_{xy}) system, there will always be a pair of planes at right angles to each other $(\theta = 90°)$ on which there is no shear stress τ_θ, merely tensile or compressive s_θ. It may be shown (e.g., by differentiation) that the pair of s_θ are the algebraically greatest and least values of normal stress. These values are called the *principal stresses* of the given (s_x, s_y, τ_{xy}) system, and the planes on which they act are called the *principal planes*.

Substitution of θ given by relation (2-11) in expression (2-10) for s_θ gives

$$(s_\theta)_{\text{principal}} = \tfrac{1}{2}(s_y + s_x) \pm \tfrac{1}{2}[(s_y - s_x)^2 + 4\tau_{xy}^2]^{1/2} \qquad (2\text{-}12)$$

The planes for maximum shear stress may be found by differentiation of Eq. (2-10). We obtain

$$(2\theta) \text{ for maximum } \tau_\theta = \tan^{-1}\left(\frac{-(s_y - s_x)}{2\tau_{xy}}\right) \qquad (2\text{-}13)$$

which again gives another (different) pair of planes at 90° with respect to each other, on which

$$(\tau_\theta)_{\text{max}} = \tfrac{1}{2}[(s_y - s_x)^2 + 4\tau_{xy}^2]^{1/2} \qquad (2\text{-}14)$$

Note, however, that planes having $(\tau_\theta)_{\text{max}}$ do have some s_θ, unlike the principal planes of $(s_\theta)_{\text{max}}$, which have no τ_θ.

Comparison of Eqs. (2-11) and (2-13) for the θ values shows that the pairs of planes are at 45° to one another (Fig. 2-15).

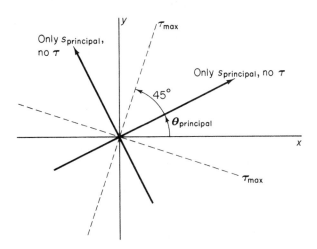

Figure 2-15 Relative orientations of the principal planes on which the normal stresses are greatest and least (and on which there are *no* shear stresses associated with them) and those planes on which the shear stresses are greatest (where there *are* associated normal stresses).

Exercise 2-7

Determine the magnitude of the maximum shear stress in a system with a normal stress of 10 MPa and a shear stress of 5 MPa on one plane, and a compressive stress of 5 MPa and a complementary shear stress of 5 MPa on an orthogonal plane.

Solution. In Eq. (2-14), $s_y = +10$ MPa, $s_x = -5$ MPa, and $\tau_{xy} = 5$ MPa; thus $(\tau_\theta)_{\text{maximum}} = \frac{1}{2}\{[10-(-5)]^2 + 4.5^2\}^{1/2} = 9$ MPa.

Equations (2-10) give the general (s, τ) combinations on an arbitrary plane orientated at some θ from the y-plane given an orthogonal set of (s_x, s_y, τ_{xy}). If the orthogonal set of stresses happened to coincide with the principal stress state (i.e., $s_x =$ one principal stress, $s_y =$ the other, and $\tau_{xy} = 0$), Eq. (2-10) would be the special case of giving the general (s, τ) combinations on an arbitrary plane, orientated at some θ from a principal plane, in terms of the principal stresses only. If the principal stresses are called s_1 and s_2 (where $s_1 > s_2$ algebraically), we have

$$\left.\begin{aligned} s_\theta &= \tfrac{1}{2}(s_1 + s_2) + \tfrac{1}{2}(s_1 - s_2)\cos 2\theta \\ \tau_\theta &= \tfrac{1}{2}(s_1 - s_2)\sin 2\theta \end{aligned}\right\} \tag{2-15}$$

In a plot of τ vs. s, Eq. (2-15) describes a circle whose center is at $[(\tfrac{1}{2})(s_1 + s_2), 0]$ with radius $\tfrac{1}{2}(s_1 - s_2)$, as shown in Fig. 2-16 (where s_1 and s_2 are arbitrarily shown as being both positive; they could be both negative or a mixture, of course, which would shift the circle along the abscissa). The coordinates of point Q on the circle are $s_\theta = OT = [\tfrac{1}{2}(s_1 + s_2) + \tfrac{1}{2}(s_1 - s_2)\cos 2\theta]$ along the s-axis and $\tau_\theta = TQ = [\tfrac{1}{2}(s_1 - s_2)\sin 2\theta]$ along the τ-axis. Thus *all* stress states in the body, as θ is varied, lie on the circle. The circle is known as a *Mohr's stress circle* and is equivalent to a graphical tensor transformation.

As angles are in terms of 2θ in the *stress plane*, a rotation of 180° in the

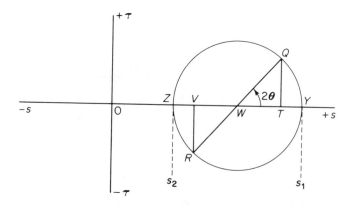

Figure 2-16 The basic Mohr's circle representation of stress at a point. The relation between τ and s given by Eq. (2-15) is a circle with center at $(\tfrac{1}{2}(s_1 + s_2), 0)$ and radius $\tfrac{1}{2}(s_1 - s_2)$.

physical plane of the stressed body (which comes back to the stresses in line before rotation) corresponds with a 360° rotation along the Mohr's circle. To use the circle, an arbitrary sign convention must be employed for shear stresses. In this book we use the following convention: If τ would produce *clockwise* rotation of the body, it is *positive*. Thus in Fig. 2-14, τ_{xy} is positive and τ_{yx} negative. In Fig. 2-13, τ_{xy} is positive and τ_{yx} negative. The shear stress at point Q on the circle (Fig. 2-16) is TQ and is positive; the corresponding complementary shear is $VR = TQ$ corresponding to the point R diagonally opposite Q (i.e., 180° in the stress plane, 90° in the physical plane) and is negative. The normal stress on that orthogonal plane is OV.

The plane displaying the (OT, TQ) stress state is located at θ counterclockwise from the plane in which s_1 acts; that displaying (OV, VR) is at $(\theta + 90)°$ from the plane of s_1, and so on.

The circle has been constructed on the basis of knowing the principal stresses, which thus gave the pair of points Y and Z on the abscissca and on the circle. The circle could equally well have been constructed from knowledge of a pair of points such as Q and R, by joining them, thus locating W the center of the circle. In fact, if the end τ_{xy} terms are eliminated in Eq. (2-10) by squaring and adding, the equation of the circle in terms of $[s_x, s_y, \tau_{xy}]$ is given, rather than the equation in terms of s_1 and s_2. An important point to remember in such constructions is that the orientation of other planes is always measured from a known plane. Thus the zero of rotation 2θ would be the WQ radius (Fig. 2-16) if the circle were drawn on the basis of points Q and R; the zero of rotation is the axis *only* when one starts with the principal stresses.

Exercise 2-7 can be solved through use of a roughly drawn Mohr's circle, as follows: In Fig. 2-17(a), the applied stresses are shown. For the x-plane (that plane perpendicular to the x-direction), $s_x = -5$ MPa and $\tau = 5$ MPa; the point representing this stress condition is thus labeled "x" on the τ vs. s plot in Fig. 2-17(b). Likewise, the point representing $s_y = 10$ MPa, $\tau = -5$ MPa is labeled "y." (The subscripts for the shear stresses are dropped here, as the Mohr convention for shear differs from the general sign convention for shear, and no confusion results from omitting the subscripts.)

Points x and y in Fig. 2-17(b) are the opposite ends of a diameter of the Mohr's circle for stress. The maximum shear stress τ_{max} is the radius r of the circle:

$$\tau_{max} = r = \left\{ \left(\frac{s_y - s_x}{2} \right)^2 + \tau^2 \right\}^{1/2} = \left\{ \left(\frac{10 + 5}{2} \right)^2 + 5^2 \right\}^{1/2} = 9.01 \text{ MPa}$$

The maximum principal stress s_1 is thus $\overline{OC} + r$, where

$$\overline{OC} = \frac{s_y + s_x}{2} = \frac{10 - 5}{2} = 2.50$$

$$s_1 = 9.01 + 2.50 = 11.5 \text{ MPa}$$

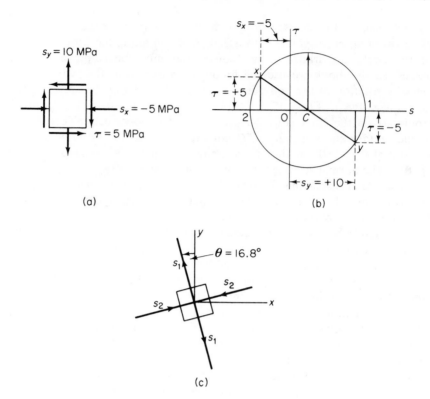

Figure 2-17   The Mohr's circle solution to Exercise 2-7.

Likewise, the minimum principal stress s_2 is

$$s_2 = \overline{OC} - r = 2.50 - 9.01 = -6.51 \text{ MPa}$$

If the angle between, say, the y-direction and the maximum principal stress were needed, the Mohr's circle provides it easily, since all angles on the circle are double those in real space. Thus $2\theta = \sin^{-1}(\tau/r) = \sin^{-1}(5/9.01) = 33.7°$ and $\theta = 16.8°$, counterclockwise from the y-direction. Figure 2-17(c) shows the plane of maximum principal stress s_1; since the minimum principal stress s_2 is 180° from s_1 in the Mohr's circle, it is 90° from s_1 in Fig. 2-17(c).

If, instead of a two-dimensional stress system, we were to consider three-dimensional stress systems, Fig. 2-13 would have to be redrawn as a "cube" rather than a "square" and stresses such as $(s_z, \tau_{xz}, \tau_{yz})$ would enter the picture. The complete treatment of Mohr's circles for three-dimensional stress systems is beyond the needs of this book—and there are certain pitfalls in constructing them—but it suffices to say that the problem can be viewed as three separate two-dimensional circles. Because of the existence now of three principal stresses, the three circles are defined by s_1, s_2, and s_3 on the abscissca, and they touch at those points (see Exercise 2-8, following).

Exercise 2-8

Draw the Mohr's circles for the following cases: (a) uniaxial tension prior to necking, (b) direct compression before barreling, (c) simple torsion, (d) a pressurized beer can. Determine the magnitude of the maximum shear stresses in each case.

Solution. The circles are shown in Fig. 2-18(a), (b), (c), (d), located in the first instance by points YZ. In cases (a) and (b), the systems are uniaxial, with one principal stress being the applied stress and the other two being zero. Because of the convention that $s_1 > s_2 > s_3$ algebraically, the zero stresses are (s_2, s_3) in Fig. 2-18(a), but (s_1, s_2) in Fig. 2-18(b). In case (d) the stress in the hoop direction is $s_1 = pD/2t$, and in the axial direction it is $s_2 = pD/4t$ where p is the internal pressure, D the cylinder mean diameter, and t the wall thickness. For all practical purposes $s_3 = p \approx 0$ as $D \gg t$.

The magnitude of the maximum shear stress in cases (a)–(c) is straightforward to calculate and is equal to $YZ/2 = (s_1 - s_3)/2$. In case (d), however, although the maximum shear stress in the plane of the beer can wall is given by $YZ/2$ ($= pD/8t$), there is a *greater* shear stress, $(s_1 - s_3)/2 = s_1/2 = pD/4t$, in the body, which acts in a plane that is normal to a line which lies in the transverse plane of the cylinder and is 45° to the radius.

Note that Fig. 2-18(c) shows that pure shear is equivalent to equal tension and compression at right angles.

Much confusion arises with sign conventions for angles and for shear stresses in Mohr's circles. For merely the magnitudes of s and τ it is possible to do without a sign convention because of the symmetry of the circle, but for

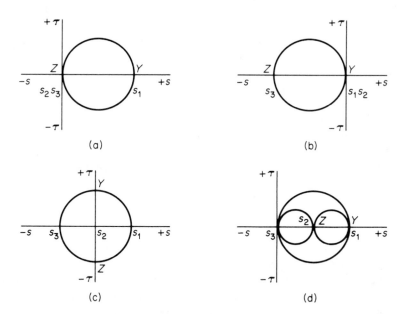

Figure 2-18 Solution to Exercise 2-8, showing the Mohr's circles for (a) uniaxial tension prior to necking, (b) homogeneous compression, (c) simple torsion, (d) a pressurized beer can.

orientations of planes, signs are required. The convention used here is that *counterclockwise angles* in both the stress *and* physical planes are positive but that positive shear stresses are those producing *clockwise rotations*. This system makes sense physically when considering the distortion of elements. Other conventions sometimes require different senses of rotation in the physical and stress planes, which is a needless complication.

Exercise 2-9

Find the magnitude of the principal stresses and their orientation for the orthogonal two-dimensional stress system having $s_x = 25$, $s_y = 20$, $\tau_{xy} = 5$ (MPa).

Solution. Either the Mohr's circle may be constructed from the coordinate points $(+25, +5)$ and $(20, -5)$, or Eq. (2-12) may be solved, to give $s_1 = 28.1$ MPa and $s_2 = 18.9$ MPa.

The Mohr's circle construction shows that the radius joining the center of the circle to the point (25, 5) lies at $+63.4°$ (i.e., $63.4°$ counterclockwise) from the abscissca. Thus the s_1 plane lies at $63.4/2 = 31.7°$ *clockwise, relative to* the plane in which the 25-MPa tension stress acts. Of course, $63.4°$ is also the solution of Eq. (2-11).

TRUE STRESS AND STRAIN

As mentioned in connection with loading and unloading, the significance of nominal stress as load per original area becomes more and more distorted as the real area over which the load acts changes. For example, in a tension test of ductile metal, the fracture area in the neck A_f may be of the order of half the initial area A_0. To provide a more meaningful description of instantaneous load per unit area, the *true stress σ* is defined as

$$\sigma \equiv \frac{P}{A} \tag{2-16}$$

Thus σ will be twice s when A is half A_0, and so on.

In the same way, the meaning of nominal strain as deformation per unit length is distorted when the length L becomes appreciably larger than L_0 during the test. For example, when $e = 1$, the extended length is equal to $2L_0$; that is, we have 100% extension. To what height would we have to compress a cylinder to get compressive strain of 100%? We would have

$$e = -1 = \frac{L_f - L_0}{L_0}$$

Hence $L_f = 0$, which is impossible.

A better measure of strain is given by referencing changes in length, height, etc., to the *current* length or height, rather than to some original dimension. Then an increment of this *true strain* is given by

$$\Delta\epsilon = \frac{\Delta L}{L}$$

so that
$$\epsilon = \ln \frac{L}{L_0} \qquad (2\text{-}17)$$

Relation (2-17) is valid only when strain is uniform all along L and is thus invalid for strains greater than the necking strain.

Exercise 2-10

Show that doubling the length of a tensile specimen is equivalent, in terms of true strain, to halving the height.

Solution. In halving the height, $L = \frac{1}{2}L_0$, so $\epsilon = \ln (L/L_0) = \ln \frac{1}{2}$. But $\ln \frac{1}{2} = -\ln 2$ and $\ln 2 = \ln (2L_0/L_0)$, which is the true strain when the length is doubled. Thus, with due change of sign for tension and compression, the processes are equivalent. Compare in the text what happens for e under identical conditions.

Another useful form for ϵ in the plastic range of strains comes from invoking *volume constancy*. Measurements of diameter and length in the regions bu (Fig. 2-2) show that the volume of the gage length is essentially constant, that is,

$$A_0 L_0 = AL$$

so
$$\frac{\pi d_0^2 L_0}{4} = \frac{\pi d^2 L}{4} \qquad (2\text{-}18)$$

for a round test piece, where A, d, and L are the values anywhere between b and u, either during deformation or after unloading beyond yield. Strictly, there are very small changes in volume, but these are associated with the elastic component of strains and are negligible when compared with the plastic strains. Then, from Eq. (2-18),

$$\frac{L}{L_0} = \frac{A_0}{A} = \left(\frac{d_0}{d}\right)^2$$

Thus

$$\left.\begin{aligned}
\epsilon &= \ln \left(\frac{L}{L_0}\right) \\[6pt]
&= \ln \left(\frac{A_0}{A}\right) \\[6pt]
&= \ln \left(\frac{d_0}{d}\right)^2 = 2 \ln \left(\frac{d_0}{d}\right)
\end{aligned}\right\} \qquad (2\text{-}19)$$

Beyond zero-slope ultimate, u in Fig. 2-2, the deformation, which is concentrated in a neck, is nonuniform, both along and across the bar, although the volume is still constant. Between u and f in Fig. 2-2, the length over which the deformation occurs is much smaller than L_0, and Eq. (2-17) for ϵ cannot be used.

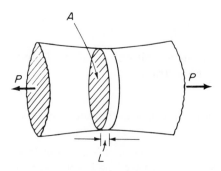

Figure 2-19   Force equilibrium of a thin disk within the neck of a tensile specimen over which the strain may be assumed to be uniform.

However, Eq. (2-19) *can* still be used, by imagining a very short cylinder located right in the neck (Fig. 2-19) over which the strain may be assumed uniform. If we really knew such a cylindrical length, we could use Eq. (2-17), but it is much easier to measure *diameters* and to calculate ϵ from Eq. (2-19). In fact, for the plastic range of deformation, measuring diameter d in the range *buf* (Fig. 2-2) is both easier and more accurate than measuring lengths from b to u and then having to change.

Thus, $\epsilon = \ln (A_0/A)$ is valid from b to f, but $\epsilon = \ln (L/L_0)$ is valid only from o to u, the zero-slope ultimate. The elastic strain oa is negligible compared with the plastic strains beyond a, so $\epsilon = \ln (A_0/A)$ can be used for the full range of plastic strains.

Exercise 2-11

Find the true stresses and strains at ultimate load and fracture for the test piece quoted in Exercises 2-1 and 2-2.

Solution. $\sigma_u = \dfrac{9 \times 10^3}{7.7 \times 10^{-6}} = 1.17 \text{ GPa.}$

$\sigma_f = \dfrac{6.5 \times 10^3}{5 \times 10^{-6}} = 1.3 \text{ GPa.}$

$\epsilon_u = \ln (10/7.7) = 0.26.$

$\epsilon_f = \ln (10/5) = 0.69.$

RELATIONSHIPS BETWEEN s AND σ AND BETWEEN e AND ϵ

Relationships between e and ϵ and between s and σ may be obtained as follows: Since

$$e = \frac{L - L_0}{L_0} = \frac{L}{L_0} - 1$$

it follows that

$$1 + e = \frac{L}{L_0}$$

So

$$\ln \frac{L}{L_0} = \ln (1 + e) = \epsilon \tag{2-20}$$

Also, since $\ln (1 + e) = e - e^2/2 + e^3/3$, we note that ϵ and e are essentially the same at small strains, but that up to zero-slope ultimate $\epsilon < e$ when differences become significant at large (plastic) strains.

For a relation between s and σ we have

$$s = \frac{P}{A_0} \quad \text{and} \quad \sigma = \frac{P}{A}$$

So we require a relation between A and A_0. If elastic volume changes are neglected, it has been shown that

$$\left. \begin{array}{c} \epsilon = \ln \dfrac{A_0}{A} \\[2em] \exp \epsilon = \dfrac{A_0}{A} \end{array} \right\} \tag{2-21}$$

Hence

So

$$s = \frac{P}{A_0} = \frac{P}{A \exp \epsilon} = \frac{\sigma}{\exp \epsilon} \tag{2-22}$$

Thus $\sigma = s \exp \epsilon$. Again, for small ϵ, $\sigma \approx s$, since $\exp \epsilon = 1 + \epsilon + \epsilon^2/2!$, but generally $\sigma > s$.

Exercise 2-12

A round wire of initial diameter 2 mm and length 300 mm yields at a tensile load of 314 N. Further loading to 400 N increased the length to 320 mm. The maximum load was 450 N when the length was 380 mm. After necking the wire fractured at a load of 390 N, when the final length and diameter were 430 mm and 1.5 mm, respectively. Compare s and σ, and e and ϵ at various stages in the loading. What is the true stress in the region of the specimen away from the neck at the instant fracture occurs in the neck?

Solution. At initial yield $s_y = \sigma_y = 314/(\pi/4)(2 \times 10^{-3})^2 = 100$ MPa; $s_y = \sigma_y$, as there is only imperceptible change in wire diameter. However, by the time the 400-N load is reached, the diameter is $(300/320)^{1/2} \cdot 2 = 1.94$ mm. The true stress is then $400/(\pi/4)(1.94 \times 10^{-3})^2 = 136$ MPa; the corresponding nominal stress is $400/(\pi/4)(2 \times 10^{-3})^2 = 127$ MPa. The associated strains are $\epsilon = \ln (320/300) = 0.064$ and

$e = (320 - 300)/300 = 0.067$. At maximum load the diameter has become $(300/380)^{1/2} = 1.78$ mm. The true stress is

$$\sigma_u = \frac{450}{(\pi/4)(1.78 \times 10^{-3})^2} = 1.81 \text{ MPa}$$

$$s_u = \frac{450}{(\pi/4)(2 \times 10^{-3})^2} = 143 \text{ MPa}$$

Again,

$$\epsilon_u = \ln \frac{380}{300} = 0.23 \quad \text{and} \quad e_u = \frac{380 - 300}{300} = 0.27$$

At fracture,

$$\sigma = \frac{390}{(\pi/4)(1.5 \times 10^{-3})^2} = 221 \text{ MPa}$$

and

$$s = \frac{390}{(\pi/4)(2 \times 10^{-3})^2} = 124 \text{ MPa}$$

Again, the associated strains are $\epsilon = 2 \ln (2/1.5) = 0.57$ and $e = (430 - 300)/300 = 0.43$. (This is now less than ϵ instead of being greater than ϵ, as the deformation has become nonuniform.) Note that we cannot use lengths for evaluation of ϵ after necking, and so we use diameters. The regions away from the neck remain at the uniform diameter that the whole wire had just prior to necking, that is, 1.78 mm. Thus the true stress at fracture in the "shoulder" regions away from the neck is $390/(\pi/4)(1.78 \times 10^{-3})^2 = 157$ MPa. We see that these regions elastically unload after necking sets in elsewhere, since 157 MPa < 181 MPa, which is the highest true stress experienced by them. The true stress in the neck, of course, goes on increasing right up to fracture.

The P vs. ΔL curve for steel in Fig. 2-2 may be reinterpreted as s vs. e and σ vs. ϵ curves, as shown in Fig. 2-20.

The numerical differences between the nominal and true stresses and strains for ultimate u and fracture f in Fig. 2-20 can be explained from the definitions of the terms and from the behavior shown in Fig. 2-21. The true stress at ultimate σ_u is larger than the nominal stress s_u, which is the tensile strength, because the area A_u is smaller than A_0. At fracture σ_f is much larger than s_f, since the instantaneous area A_f is much smaller than A_0.

True strain is the integral of the incremental change in length divided by the instantaneous length [Eq. (2-17)], while the nominal strain is the total change in length divided by the initial length [Eq. (2-2)]. Throughout the uniform-strain portion ou of the tensile test (with any discontinuous yield behavior in the ab region neglected), the instaneous length L is always greater than the initial length L_o; hence the true strain is always smaller, in the ou region, than the nominal strain. Figure 2-20 shows, for example, that $\epsilon_u < e_u$.

The most striking difference between the two methods of presentation is

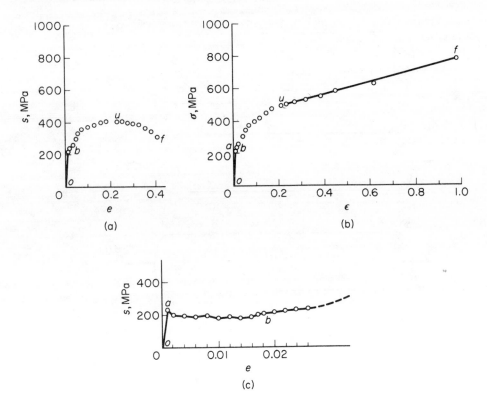

Figure 2-20 The P vs. ΔL data of Fig. 2-2 reinterpreted: (a) as nominal stress s vs. nominal strain e; (b) as true stress σ vs. true strain ϵ; (c) in a magnified view of the s vs. e curve in the yield region.

that there is no maximum in the σ/ϵ curve. Thus, the curious feature of Fig. 2-2, that there are apparently two compatible e values for one s value (i.e., $e < e_u$ and $e > e_u$), is eliminated. It is of particular importance that the zero-slope ultimate point (σ_u, ϵ_u) is located in an unobvious place somewhere on the σ vs. ϵ curve.

When loading/unloading/reloading P vs. ΔL curves are expressed in terms of σ and ϵ, it is found that the concept of "horizontal shifting" works quite well. That is, when a bar is strained to some value ϵ^* and then unloaded, the σ vs. ϵ curve produced by reloading blends into the original σ vs. ϵ curve if the strain origin is shifted to ϵ^* (Fig. 2-22). This suggests that deformation produced in more than one stage is equivalent to the same *total* deformation produced in one step, *only* when the deformations are given in terms of ϵ. That is, deformation of magnitude ϵ_1, followed by unloading, followed by deformation to ϵ_2 beyond that, is the same as deformation of magnitude $(\epsilon_1 + \epsilon_2)$ in one step. This may be illustrated as follows: let $\epsilon_1 = A_0/A_1$ and $\epsilon_2 = A_1/A_2$

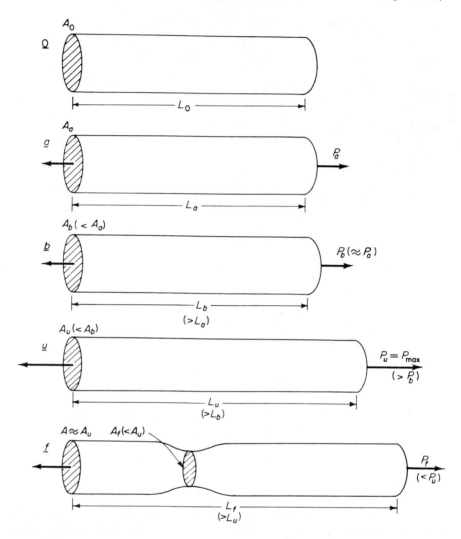

Figure 2-21 Five stages of deformation during a tensile test, corresponding with the lettered points in Fig. 2-20.

(remember that ϵ_2 is measured relative to *its own* starting area). Then

$$\epsilon_1 + \epsilon_2 = \ln \frac{A_0}{A_1} + \ln \frac{A_1}{A_2} = \ln \left[\frac{A_0}{A_1} \cdot \frac{A_1}{A_2} \right] = \ln \frac{A_0}{A_2}$$

which is the same strain reached by deforming from A_0 to A_2 in one step.

These arguments do *not* hold when written in terms of e. Thus, true strains are *additive*, but nominal strains are not.

In the above example, if we wished to find the yield strength of the

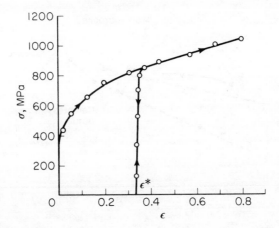

Figure 2-22 Horizontal shifting of σ vs. ϵ curves: material prestrained by ϵ^* gives a σ vs ϵ curve on reloading, which blends in with the curve of the unstrained material if the origin of strain is moved to ϵ^*.

work-hardened bars after the first, second, or total deformations, we would have to use the σ vs. ϵ plot and find the σ at the appropriate ϵ. Matters would be simplified if there were algebraic relationships between σ and ϵ in the plastic region. Logarithmic plotting of σ vs. ϵ for ductile metals shows that over the plastic range ab to u a reasonable linear relation is followed, that is,

$$\log \sigma = (\text{slope}) \log \epsilon + \log (\text{constant}) \left.\vphantom{\begin{matrix}a\\b\end{matrix}}\right\}$$

or $$\sigma = K \epsilon^n \qquad\qquad\qquad\qquad (2\text{-}23)$$

where n is the slope (called the work-hardening index) and K is the constant, which is evaluated by extrapolating the data to $\log \epsilon = 0$, that is, to $\epsilon = 1$. Experiments in compression suggest that relationship (2-23) is valid beyond the necking strain in tension, although data from a tensile test usually show an upward trend after ϵ_u.

In the elastic region at small strains less than the yield strain, $\sigma = E\epsilon$. Logarithmic plotting of elastic data gives

$$\log \sigma = \log \epsilon + \log E \qquad\qquad\qquad (2\text{-}24)$$

that is, a straight line of slope (45° on equal scales), with intercept equal to the Young's modulus at $\epsilon = 1$. Hence plotting of both elastic and plastic data gives two linear segments. The transition from one to the other is not necessarily sharp with real-life data, and the yield stress is not necessarily given by the intersection of the lines. It must always be remembered that algebraic σ vs. ϵ relationships are only approximations of real behavior (Fig. 2-23).

When reloading data are plotted on logarithmic paper, where the plastic strains are measured relative to the bar's new starting size, the applicability of Eq. (2-23) comes into question, as the prestrained σ/ϵ data tend to follow

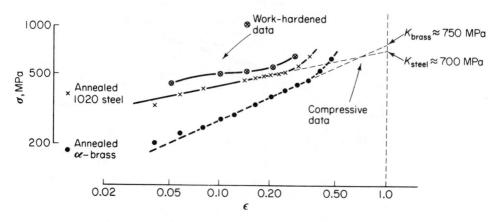

Figure 2-23 Logarithmic plotting of σ vs. ϵ data showing reasonable agreement with a relationship of the form $\sigma = K\epsilon^n$ in the plastic range.

shallow concave-up curves (Fig. 2-23). Even so, some workers do draw straight lines through the data points and quote different pairs of (K, n) values for each prestrain value. However, in the spirit of horizontal shifting, it is found that all data fit on the one envelope curve, if ϵ_{cw} (prior strain from cold work) is added to the true strains measured in the reloading test before the data are plotted.

Thus, the full version of Eq. (2-23), allowing for any prestrain ϵ_{cw}, is

$$\sigma = K(\epsilon_{cw} + \epsilon)^n \qquad (2\text{-}25)$$

where ϵ is the strain measured beyond the condition of prior cold work. Now K and n take the same values for *all* degrees of prestrain.

If we think of initial yielding of a cold-worked bar as being expressed by $\epsilon = 0$ in Eq. (2-25), we have, for the yield strength σ_y of a material prestrained an amount ϵ_{cw},

$$\sigma_y = K\epsilon_{cw}^n \qquad (2\text{-}26)$$

Thus, solids with large n give greater changes in strength when cold-worked. Representative values of K and n for many metals are given in Appendix 3; Chap. 12 discusses the effects of strain rate and temperature on these parameters.

TENSILE INSTABILITY, POLYMER "DRAWING," AND LÜDERS BANDS

During the early portion of the strain-hardening region of the stress–strain curve, the load increases because the rate of increase of true stress resulting from strain hardening is greater than the rate of decrease of area. With further strain the rate of increase of stress [i.e., the slope ds/de in Fig. 2-20(a)] gradually decreases. The zero-slope ultimate condition represents a turning point where the increase in load resulting from strain hardening is exactly equal to the decrease in load associated with the area reduction. Beyond ultimate the decrease in load from decrease in area exceeds the increase in load from increase in true stress. The reduction of area that results from the first increment of strain beyond ultimate thus lowers the load-carrying capacity of the region where this strain occurs below that of the other regions along the length of the specimen. Since the load required to cause additional strain is lower in this region than elsewhere, further strain beyond ultimate is confined to this region of instability. Characteristic necking results.

For a dead-load tensile test, in which masses are hung on a tensile bar, the reduction of load beyond ultimate would never be observed, because the specimen would fracture when the total mass reached a value corresponding to s_u. In the usual laboratory hydraulic or constant-speed electric tensile-testing machine, however, the machine is designed to pull with force P just enough to maintain the existing strain and cause a small increment of additional strain, and thereby is capable of picking up the decreasing load beyond the zero-slope ultimate condition.

Figure 2-21 shows schematically the test section of the specimen at each point labeled on the stress–strain curve in Fig. 2-20. Beyond u, associated with the drop in load P is the local reduction in the cross-sectional area already mentioned. The last sketch in Fig. 2-21 shows the localization of strain and resultant necking as it appears just prior to fracture at f.

Because the calculation of true stress accommodates the effect of reduction of area, the σ/ϵ curve in Fig. 2-20(b) shows a continuous increase of true stress (although at a reduced rate) with strain from b to u to f. The true-stress–true-strain curve beyond u indicates the average conditions at the minimum diameter of the neck, where most of the deformation takes place. Because the load beyond ultimate is less than the ultimate load, no substantial deformation occurs in regions away from the neck, which in fact elastically unload beyond u.

The load maximum occurs at the onset of necking, and the incremental change in load is zero, that is, $\Delta P = 0$. Since $P = \sigma A$ at all points on the σ/ϵ curve,

$$\Delta P = \sigma \cdot \Delta A + A \cdot \Delta \sigma$$

Hence at the point of tensile instability (zero-slope ultimate)

$$\sigma_u \cdot \Delta A_u + A_u \cdot \Delta \sigma = 0$$

that is,
$$d\sigma/\sigma_u = -dA/A_u \qquad (2\text{-}27)$$

If we again invoke plastic volume constancy, we can show that $-dA/A = d\epsilon$. This follows from the volume being equal to AL for all A and L, whence

$$\Delta(volume) = A\Delta L + L\Delta A$$

and if $\Delta(\text{volume}) = 0$,

$$\frac{+dL}{L} = \frac{-dA}{A} = +d\epsilon$$

Thus, back in the equation describing the changes at necking, we have

$$d\sigma/\sigma_u = d\epsilon \quad \text{or} \quad (d\sigma/d\epsilon)_u = \sigma_u \qquad (2\text{-}28)$$

Instability points are thus given when the *slope* of the σ vs. ϵ curve is numerically equal to the *magnitude* of the stress at that point. Hence, the point u in Fig. 2-20(a) is the same point as point u in Fig. 2-20(b).

Exercise 2-13

Show that if $\sigma = K(\epsilon_{cw} + \epsilon)^n$, the strain at zero slope ultimate is given by $\epsilon_u = n - \epsilon_{cw}$. Hence determine the length at neck initiation of a tensile bar, initially 50 mm long, and initially prestrained by $\epsilon_{cw} = 0.05$ before testing ($n = 0.2$).

Solution. From Eq. (2-28), $d\sigma/d\epsilon = \sigma_u$, so

$$d(K(\epsilon_{cw} + \epsilon)^n)/d\epsilon = K(\epsilon_{cw} + \epsilon_u)^n$$

that is,

$$nK(\epsilon_{cw} + \epsilon_u)^{n-1} = K(\epsilon_{cw} + \epsilon_u)^n \quad \text{or} \quad (\epsilon_{cw} + \epsilon_u) = n \quad \text{or} \quad \epsilon_u = (n - \epsilon_{cw})$$

If the bar is prestrained by $\epsilon_{cw} = 0.05$, $\epsilon_u = 0.2 - 0.05 = 0.15$. Hence $\ln (L_u/50) = 0.15$ or $L_u = 58$ mm. See Exercise 2-18, later.

Exercise 2-14

A tensile specimen, initially of 11.3 mm diameter, has a diameter at zero-slope ultimate of 10 mm, and a diameter at fracture of 8.5 mm. If the maximum load during the test is 75 kN, find the load at fracture. Assume that $\sigma = K\epsilon^n$.

Solution. The true strain at ultimate is $2 \ln (11.3/10) = 0.24$. Therefore $n = 0.24$, as there is no prestrain. The true stress at ultimate is $75 \times 10^3/(\pi/4)(10 \times 10^{-3})^2 = 955$ MPa. If $\sigma = K\epsilon^n$, $K = 955 \times 10^6/(0.24)^{0.24} = 1.35$ GPa. The true strain at fracture is $2 \ln (11.3/8.5) = 0.57$. The true stress at fracture is $1.35 \times 10^9 (0.57)^{0.24} = 1.18$ GPa. The load at fracture is $1.18 \times 10^9 (\pi/4)(8.5 \times 10^{-3})^2 = 67$ kN.

In the tensile test of most ductile metals, the necked region continues to thin down until fracture takes place. With some polymers and superplastic alloys in tension, a neck forms but it grows stably, material being fed into the necked region from the thicker adjacent areas. This behavior is called "drawing" in the polymer literature and is discussed in Chap. 12, where it is contrasted with the unstable propagation of necks which is the subject of this section. Also in Chap. 12 the similarity between regions of discontinuous yielding [such as *ab* in Fig. 2-20(a), (c)], and the growth of stable necks is described. Discontinuous yielding is often accompanied by a series of striations called Lüders bands, which ripple along the gage section in the manner of stable necks; however, Lüders bands occur at much smaller strain levels than the strains encountered in stable neck propagation in polymers. It must be noted that not all polymers display stable necks, and likewise only some metals display discontinuous yielding behavior.

Exercise 2-15

A bar of initial area 25 mm^2 is reduced to area 22 mm^2 and unloaded. At what area and load will it neck if reloaded? Assume $\sigma = 1500\epsilon^{0.4}$ MPa.

Solution. $\epsilon_{cw} = \ln (25/22) = 0.13$. $\epsilon_u = (n - \epsilon_{cw}) = 0.4 - 0.13 = 0.27$. The area at ultimate is obtained from $\ln (25/A_u) = 0.27$, so $A_u = 19$ mm^2. The true stress at ultimate is $1500 (0.4)^{0.4}$, where the true strain at ultimate is $0.4 = 0.27 + 0.13$, which becomes 1.04 GPa. Therefore the necking load at ultimate is $1.04 \times 10^9 \times 19 \times 10^{-6} = 19.7$ kN.

HARDNESS

Although the tensile test is usually a good measure of the ability of a material to resist plastic deformation and failure, it is expensive and destroys the specimen. Hardness tests, of which several kinds are in use, can overcome both of these disadvantages.

Hardness is a measure of the resistance to penetration under load by some form of indenter. In the *Brinell* hardness test, for example, hardness is determined by the diameter of the permanent deformation left by a hard, spherical indenter: The larger the diameter, the lower the Brinell hardness number (Bhn), which is calculated by dividing the load by the spherical area of the indentation. The *Vickers* hardness number is measured in terms of indentation by a diamond pyramid. Neither of these is a "direct-reading" test, and the indentation (less than a few mm in size) has to be measured in an eyepiece and the hardness number looked up in tables.

Rockwell hardness machines give hardness numbers directly on a dial or digital display on the machine, as they rely on the *in situ* depth of indentation. They are thus quicker to use. Various different indenters and loads are used for various ranges of hardness: Thus Rockwell C is used for the harder steels (the indenter being a cone (brale) under 150 kgf load), and Rockwell B (a ball of 1/16 in. diameter under 100 kgf) for softer steels. Since indentations recover more in their depth than in their cross-sectional area, Rockwell tests tend to give higher hardnesses for materials with larger Young's modulus. Nevertheless, there are approximate relationships between the various scales, and tables of equivalent numbers are available (Appendix 4).

Although these and other hardness standards may differ in the means of producing and measuring plastic formation, they all depend on the resistance of the metal to plastic deformation over a wide range of strain. The strain field under an indentation will in general be complicated and vary from point to point, but it may be expected that there will exist a "representative" value of the yield stress, characteristic of some average strain produced by the indentation.

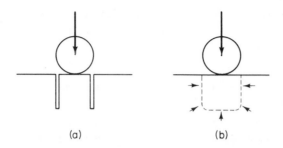

(a)   (b)

Figure 2-24 The origin of plastic constraint in a hardness test: (a) a hypothetical unconstrained indentation test on a free column of material where the applied stress (i.e., hardness) would be about the same as the uniaxial yield stress; (b) the actual constrained conditions of a practical hardness test where the applied stress is some $2\frac{1}{2}$ to 3 times the uniaxial yield stress.

Since a hardness measurement is a "constrained" compression test (Fig. 2-24), Brinell hardness and Vickers hardness numbers turn out to be between 2.5 and 3 times this "representative" yield stress, and not merely the same as the yield stress; for these tests the units of hardness developed historically as kgf/mm^2. Consequently measurements of hardness in the laboratory are good indicators of changes in strength produced by cold working as described in this chapter, or changes in strength produced by alloying or heat treatment described in later chapters. Hardness measurements are also useful for the determination of forces required to produce any given observed indentation that resulted from contact between two metal parts. This technique is discussed in Chap. 19.

Since for many *steels* (but not other materials) the representative yield stress of material deformed by a Brinell indenter is approximately numerically equal to S_u, we have

$$\left. \begin{aligned} S_u &\approx \frac{Bhn}{2.85} \ (kgf/mm^2) \\ &\approx 3.45 \ Bhn \ (MPa) \end{aligned} \right\} \tag{2-29}$$

where Bhn is the Brinell hardness and 2.85 is the constraint factor relating hardness with uniaxial tensile or compressive yield strength. Figure 2-25 shows this relation for steels. For other materials with different work-hardening indices, this S_u/Bhn relation is not followed, nor does the relation hold for materials that fracture in tension with little plastic deformation (compare data points for some of the martensitic steels in the figure and see Chap. 10).

The effect of creep and time-dependent mechanical behavior on hardness is described in Chap. 12.

Exercise 2-16

The Brinell hardness of a certain steel is 207. The work-hardening index of the same steel is 0.15. Determine the values of K in $\sigma = K\epsilon^n$ for the steel.

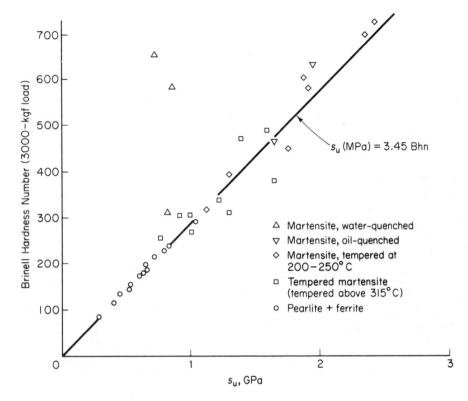

Figure 2-25 Brinell hardness number vs. tensile strength for steels of a wide range of carbon content and heat treatment.

Solution. $S_u = 3.45 \times \text{Bhn} = 714$ MPa. The true strain at ultimate is $n = 0.15$. From Eq. (2-22), $\sigma_u = s_u \exp n = 714 \exp 0.15 = 830$ MPa. But $\sigma_u = Kn^n$, so $K = 830/0.15^{0.15} = 1.1$ GPa.

DYNAMIC AND NOTCHED-IMPACT TESTING; CREEP TESTING; FATIGUE TESTING

The tension, compression, and hardness tests described in this chapter are quasi-static, that is, they are slow and take place under low rates of straining or stressing. There is also need for mechanical property information under *dynamic* or *high strain rate* conditions where, in broad terms, data are obtained from "fast" tensile tests. Such experiments are described in Chap. 12, where we also discuss *creep testing*, which concerns the behavior of materials under *very low* rates of strain. Again, materials can respond differently if subjected to many repeated applications of the load instead of only one or a small number of applications; such *fatigue* behavior is the subject of Chap. 18.

In *notched-impact testing* the energy of a pendulum (swinging under standardized conditions) that is absorbed in fracturing a notched test piece is measured. In a *Charpy* test the specimen is broken in three-point bending by being struck behind the V-notch; in an *Izod* test the specimen is broken as a notched cantilever with the notch facing the direction of the blow from the pendulum. As shown in Chap. 14, the energies thereby determined are a rough measure of the area under the σ vs. ϵ curve, as distorted by the complex nature of the deformation at the notch. It is often said that Charpy and Izod data reflect the *toughness* (or resistance to cracking) of materials under dynamic conditions. However, the results of the test are determined more by the notch than by the rate of straining, and although notched-impact data give some qualitative indication of crack resistance, it is preferable to use the quantitative concepts of *fracture toughness* and *fracture mechanics* discussed in Chap. 14 to assess the possibility of cracking leading to failure in a given situation.

MECHANICAL PROCESSING

We have seen that cold working increases the strength of ductile materials. However, simple stretching as a means of strengthening is limited by necking, which means that cold work no greater than $\epsilon_{cw} = n$ may be imparted. Thus, in a medium-carbon steel for which $\sigma = 1500 \, \epsilon^{0.15}$ MPa, the maximum increased yield strength would be

$$\sigma = 1500 \, (0.15)^{0.15} \approx 1130 \text{ MPa} = 1.13 \text{ GPa}$$

In compression, however, tensile instabilities are absent, so it would seem that greater amounts of cold work could be imparted. This is so, and the widely used process of squeezing material through *rollers* may be thought of as continuous compression (Fig. 2-26). In this way, true strains as high as unity can be enforced, and appreciable improvements in strength thus gained, along with desirable changes in shape.

In other types of mechanical processing, shown in Fig. 2-26, material is pushed or pulled through a converging channel (die) to effect a reduction in cross section. Pulling through a die, which is called *drawing*, is the means by which wire is made. The difference between simple tension and wire drawing is the presence of sideways compressive stresses in the die. They aid the plastic flow, and the state of combined stress within the die requires a *smaller* external stress than the equivalent reduction in cross-sectional area would require in simple tension. True strains greater than n can be readily imparted in wire drawing, the limit of which occurs when the drawing stress becomes equal to the strength of the drawn material (see Exercise 2-20). Pushing through a die, which is called *extrusion*, is the same as the action in a toothpaste tube. There is no obvious limiting strain in this process, but friction in particular does effectively limit the available reductions. Complicated sections may be made

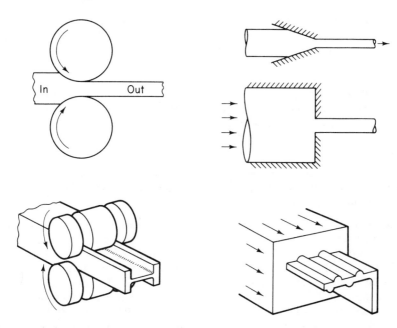

Figure 2-26   Various types of mechanical processing, such as rolling, drawing, and extrusion, showing also production of profiled sections.

from round or rectangular starting materials, if the rollers or dies are profiled (compare the action of tubes for cake frosting or icing decoration.)

In mechanical processing it is common to use the percentage reduction of area, r, as a measure of strain, given by $r = (A_0 - A)/A_0$. Clearly, this expression has the same basis as the percentage reduction of area to fracture in the tensile test, but here A represents the new cross-sectional area produced from the initial area A_0 by the process of rolling, drawing, extrusion, and so on. Since $(1 - r) = A/A_0$, the relationship between ϵ and r is

$$\epsilon = \ln \frac{1}{1 - r} \qquad (2\text{-}30)$$

Note that percentage reductions, unlike true strains, are *not* additive.

Exercise 2-17

A strip of initial thickness 10 mm is rolled down in successive stages to 9 mm and 5 mm. Find the true strains imparted and demonstrate that percentage reductions in area are not additive. Assume that there is no sideways spread in rolling.

Solution.   The cross-sectional area of the strip is tw where t is the thickness and w the width. Thus from Eq. (2-10), $\epsilon = \ln(A_0/A) = \ln(t_0 w_0/tw)$. But if there is no sideways spread in the rolling process, w is constant, so $\epsilon = \ln(t_0/t)$ alone. Hence for rolling from 10 mm to 9 mm (abbreviated 10 mm → 9 mm), and using subscripts before and after ϵ to indicate the thickness before and after rolling, for 10 mm → 9 mm the strain is

$_{10}\epsilon_9 = \ln(10/9) = 0.105$ and for 9 mm$\rightarrow$5 mm the strain is $_9\epsilon_5 = \ln(9/5) = 0.588$. Had the reduction been performed in a single rolling pass from 10 mm to 5 mm, the strain would be $_{10}\epsilon_5 = \ln(10/5) = 0.693$. Note that $_{10}\epsilon_5 = _{10}\epsilon_9 + _9\epsilon_5$.

In terms of reductions of area, again since w is constant, $r = (t_0 - t)/t_0$. Therefore (using the same subscript notation as for ϵ) $_{10}r_9 = (10 - 9)/10 = 0.1 = 10\%$; $_9r_5 = (9 - 5)/9 = 0.44 = 44\%$; $_{10}r_5 = (10 - 5)/10 = 0.5 = 50\%$. We see that $10\% + 44\% \neq 50\%$. The true strains above could, of course, have been evaluated alternatively using Eq. (2-30), that is, $_{10}\epsilon_9 = \ln(1/(1 - 0.1))$, etc.

Exercise 2-18

If the material in Exercise 2-17 obeys $\sigma = 2000\ \epsilon^{0.2}$ MPa, determine the yield strengths that the rolled material would have after the two stages of rolling, and sketch the tensile stress–strain curves that each would display.

Solution.   After a 10 mm$\rightarrow$9 mm rolling pass $\sigma_y = 2000(0.105)^{0.2} = 1.27$ GPa. S_y will be the same referenced to the 9-mm thickness of the rolled sheet from which tensile test pieces would be made. Again after the 9 mm$\rightarrow$5 mm rolling pass $\sigma_y = 2000\ (0.105 + 0.588)^{0.2} = 1.86$ GPa and $S_y = 1.86$ GPa referenced to a 5-mm thick test piece. As far as σ/ϵ curves are concerned, the two samples of cold-rolled sheet would behave as shown in Fig. 2-22, with the prestrained curves blending into the σ/ϵ of the unrolled sheet, the ϵ^* values being 0.105 and 0.693 (i.e., $0.105 + 0.588$). However, the details of the response of test pieces made from the two samples of cold-rolled sheet would be quite different. According to Exercise 2-9, the true strain at ultimate of a sample prestrained by ϵ_{cw} is $\epsilon_u = (n - \epsilon_{cw})$. The prestrains in the two sheets here are 0.105 and 0.693, and $n = 0.2$. This suggests that $\epsilon_n = (0.2 - 0.105) = 0.095$ and $\epsilon_u = (0.2 - 0.693)$, a negative value. This brings out a very important point, which is that samples for which $\epsilon_{cw} < n$, when tested in tension, display uniform plastic flow over the strain interval from initial yield to $(n - \epsilon_{cw})$ beyond, at which point the specimen necks. However, specimens already cold-worked greater than the strain at zero-slope ultimate $(\epsilon_{cw} > n)$ show *no uniform reduction* in cross section after initial yield and instead *neck immediately*; yield and ultimate coincide. Thus load/extension (or s/e) curves for the two samples of cold-rolled sheet will be as shown in Fig. 2-27. Compare the difference with the σ vs. ϵ curves (Fig. 2-22). Clearly, when $\epsilon_{cw} > n$, the method of prestraining cannot be direct tension.

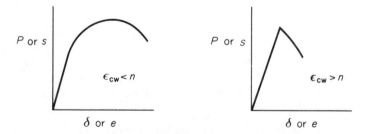

Figure 2-27 Solution of Exercise 2-18, showing how annealed materials prestrained less than n (the strain at zero-slope ultimate) display uniform plastic flow prior to necking in subsequent tensile tests, but how materials cold-worked to a strain greater than n neck immediately on yielding.

WORKING LOADS

Approximate values for the loads required in mechanical processing may be obtained by considering work. Consider first the work done in a simple tension test, which we shall use as a reference process. The force on the specimen at some instant during deformation is given by $(\sigma\ A)$, where σ and A are the current values of true stress and cross-sectional area; in mechanical processing we often call the current value of true stress during deformation the *flow stress*. The increment (ΔW) of work done in extending the specimen a further ΔL is given by force × distance moved, so

$$\Delta W = (\sigma\ A)\Delta L$$

The work done per unit volume is

$$\frac{\Delta W}{V} = \sigma\ \frac{\Delta L}{L} = \sigma\ \Delta\epsilon$$

where the volume of the test piece is $V = AL$. If we assume constancy of volume, the total work done per unit volume in extending the test piece from an initial length L_0 to a final length L^* [i.e., from $\epsilon = 0$ to $\epsilon^* = \ln(L^*/L_0)$] is

$$\frac{W}{V} = \int_0^{\epsilon^*} \sigma\ \Delta\epsilon \tag{2-31}$$

which is the area under the σ vs. ϵ curve up to ϵ^*, as shown in Fig. 2-28.

If we have a highly work-hardened material, with a flat σ vs. ϵ curve or if we assume some average yield stress $\bar{\sigma}$, we may write

$$\frac{W}{V} = \bar{\sigma}\epsilon^* = \bar{\sigma}\ \ln\frac{L^*}{L_0} \tag{2-32}$$

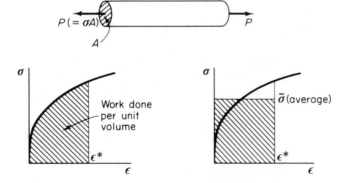

Figure 2-28 Work done in a simple tension test. The area under the σ/ϵ curve out to some strain ϵ^* is the work done per unit volume of gage section in attaining that strain ϵ^*. The area in a work-hardening case may be viewed as an equivalent rectangle, over the same strain base, but with average (constant) flow stress $\bar{\sigma}$.

with units of J/m^3, for example. In the case of a material that work-hardens according to $\sigma = K\epsilon^n$,

$$\bar{\sigma} = \frac{1}{\epsilon^*} \int_0^{\epsilon^*} K\epsilon^n \, d\epsilon = \frac{K(\epsilon^*)^n}{n+1} \tag{2-33}$$

Equation (2-32) gives the work required for *homogeneous deformation* (discussed in the section on compression and combined stress states earlier in this chapter; see Fig. 2-10). If we are prepared to neglect the distortion effects of friction in the die, and so-called redundant deformation (which concerns work of "skewing the grids" and which, although inherent in a given process, is not necessary as far as the external shape change is concerned), we can use this work expression to estimate extrusion and drawing loads.

Consider extrusion, where a "slug" originally of length L_0 and cross-sectional area A_0 becomes a length L of extrudate of area A (Fig. 2-29). The work done by the extrusion ram force P in pushing out the slug is (force) × (distance moved) $= PL_0$. From Eq. (2-32)

$$PL_0 = V\bar{\sigma} \ln \frac{L}{L_0}$$

So

$$P = \left(\frac{V}{L_0}\right) \bar{\sigma} \ln \frac{L}{L_0}$$

or

$$\left.\begin{array}{l} \sigma_{\text{extrusion}} = \dfrac{P}{A_0} = \bar{\sigma} \ln \dfrac{L}{L_0} \\[2ex] \qquad = \bar{\sigma} \ln \dfrac{A_0}{A} \\[2ex] \qquad = \bar{\sigma} \ln \dfrac{1}{1-r} \end{array}\right\} \tag{2-34}$$

Thus the work done per unit volume, the area under the σ vs. ϵ curve between the strains in question, and the magnitude of the extrusion stress are all identical.

The same arguments apply for working stresses in wire drawing and other processes, where the assumption of homogeneous deformation is made. Where friction and redundant deformation are significant, and thus homoge-

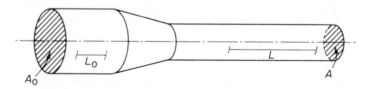

Figure 2-29 The extrusion of a billet where a length L_0 of area A_0 becomes a length L of extrudate with area A.

neous work underestimates the working loads, an efficiency factor η can be introduced, so that

$$\sigma_{\text{extrusion}} \text{ or } \sigma_{\text{drawing}}, \text{ etc., } \text{ is given by } \left(\frac{1}{\eta}\right) \bar{\sigma} \epsilon^* \qquad (2\text{-}35)$$

The effect of strain rate in altering $\bar{\sigma}$ in working processes is discussed in Chap. 12.

Exercise 2-19

Find the load necessary to extrude a round bar 100 mm in diameter that is made of an aluminum alloy of average yield strength 300 MPa through a hole of 20 mm diameter.

Solution. The extrusion pressure, given by Eq. (2-33), is $(300 \times 10^6)\, 2 \ln(100/20) = 966$ MPa. The cross-sectional area at ram is

$$\frac{\pi}{4}(100 \times 10^{-3})^2 = 7.85 \times 10^{-3} \text{ m}^2$$

So the extrusion load is $966 \times 10^6 \times 7.85 \times 10^{-3} = 7.6$ MN.

Exercise 2-20

Determine the maximum attainable reduction in wire drawing.

Solution. The drawn wire will yield, and flow will take place downstream of the die, when the drawing stress is as great as the yield stress of the drawn material. From Eq. (2-34) the drawing stress is $\bar{\sigma}\epsilon$, where $\bar{\sigma}$ is the average flow stress within the die. If the material has a constant yield stress $\bar{\sigma}$, the yield strength of the drawn (and, indeed, the undrawn) material will be the same, so in these circumstances the maximum strain attainable in wire drawing will be set by $\bar{\sigma} = \bar{\sigma}\epsilon$ or $\epsilon = 1$ or $r = 63\%$. If the material work-hardens through the die according to $\sigma = K\epsilon^n$, then $\sigma = K\epsilon^n/(n + 1)$, but the yield strength of the wire as it emerges from the die will be $\sigma_y = K\epsilon^n$. The limit of drawing is then set by $K\epsilon^n = (K\epsilon^n/(n + 1))\epsilon$ or $\epsilon = (1 + n)$, which is greater than before; for example, with $n = 0.1$, $r = 70\%$, and with $n = 0.5$, $r = 78\%$. Note that since ϵ (equal to 1 or to $1 + n$) is greater than n, yielding and necking would coincide at the exit to the die (see Exercise 2-18). In practice, because of friction and redundant deformation, practical limiting reductions in wire drawing are smaller than these values.

ANISOTROPY

The discussions of strength so far have not suggested that the properties of materials might vary with direction. That is, we assume that materials are *isotropic* and if test pieces are taken at various directions from a large piece of material, they all give essentially the same σ vs. ϵ curve. Almost no materials are truly uniform and isotropic. An obvious example is wood, which has different properties along and perpendicular to the grain. Rope has little strength across its width. Figure 2-30(a) shows the results of tension tests of specimens cut, at different angles, from a sheet of a cold-rolled aluminum

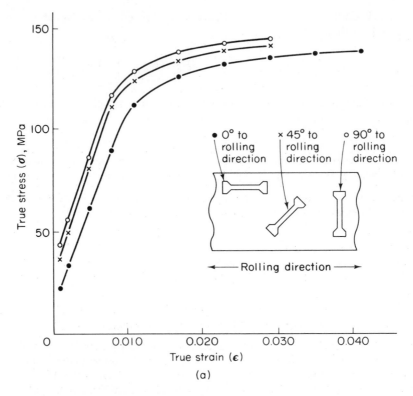

Figure 2-30 Anisotropy in (a) commercial-purity aluminum (99.0% Al) in the "half-hard" condition (i.e., cold-rolled to a reduction of some 21% in area after the last anneal), and (b) Ti6Al4V titanium aluminum vanadium alloy sheet (8 mm wide by 500 μm thick).

alloy. Figure 2-30(b) gives similar data for Ti-6Al-4V sheet material. It is seen that there are considerable variations in strength with orientation.

These effects are called *anisotropy*. The magnitude of the differences in strength and ductility depend on the degree of cold work, and anisotropy is commonly encountered in metals and polymers that have been rolled, drawn, extruded, forged, and so on. Directional variations may be present in solids for other reasons, the bases of which are covered in later chapters. For example, single-phase crystalline materials have some degree of variation in grain size associated with point-to-point variations in cooling rates. Two-phase materials show in addition compositional variations that result from finite cooling rates during solidification. Some structural reasons for anisotropic effects are mentioned in Chap. 5, and fiber-reinforced composites, which are inherently anisotropic, are discussed in Chap. 17. In general terms, amorphous materials in bulk can be more nearly isotropic than crystalline materials, except for fabrication effects. It is important to be aware that *preferred orientation* or *texture*

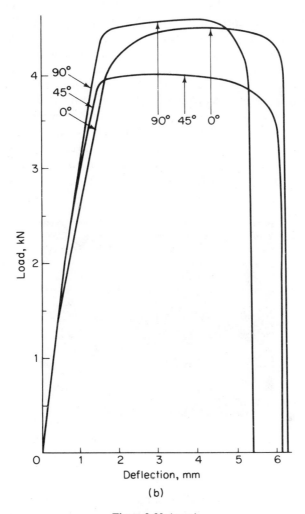

Figure 2-30 (cont.)

effects (other names associated with anisotropy) can persist to some extent
even after so-called thorough homogenizing or annealing heat treatments, as
mentioned in later chapters, such as Chap. 6.

An important indicator of anisotropy in metals is the r factor (not to be
confused with cold work, which uses the same symbol), which can be deduced
from measurements of the strains in a tension test. As a specimen extends in
one direction, it shrinks in the other directions; from considerations of constant plastic volume we have

$$\epsilon_L + \epsilon_w + \epsilon_t = 0 \tag{2-36}$$

where ϵ_w and ϵ_t are (negative) width and thickness strains. In an isotropic material, with no preferential orientations, $\epsilon_w = \epsilon_t$. In a round bar, for example, the gage section remains circular and reduces uniformly. In an anisotropic material, the gage section may become elliptical, with some directions shrinking preferentially; again, ball hardness indentations are often noncircular in anisotropic solids. This type of effect is shown in a tensile test specimen in Fig. 2-31. Then, and in sheet specimens for which "width" and "thickness" have obvious meanings, $\epsilon_w \neq \epsilon_t$. The degree of anisotropy may be quantified in terms of the ratio $r = \epsilon_w/\epsilon_t$. This can be obtained by mounting strain sensors on tensile specimens in the width and thickness directions. Sometimes, with very thin sheet material, ϵ_t is too small to measure accurately. In that case, ϵ_t can be inferred from ϵ_w and ϵ_L by means of Eq. (2-36).

Figure 2-31   Elliptical cross section of neck in tensile bar of 1100 aluminum. (Photograph courtesy of W. H. Durrant.)

Anisotropy can be beneficial in some situations if the major component of load is always in one direction. Furthermore, in sheet metal working certain sorts of anisotropy help to prevent premature failure of the metal during forming processes.

Exercise 2-21

A strip of low-carbon steel, 1 mm thick by 10 mm wide by 75 mm long, is observed to become 9 mm wide when it is stretched to 80 mm in length. Find the r factor for this steel.

Solution. $\epsilon_w = \ln(9.5/10) = -0.051$. $\epsilon_L = \ln(80/75) = 0.064$. $\epsilon_t = -(-0.051 + 0.064) = -0.013$. Therefore

$$r = \frac{-0.051}{-0.013} = 3.9$$

REFERENCES

2-1. American Society for Metals, *Metals Handbook*, 8th ed., vol. 1. Metals Park, Ohio: American Society for Metals, 1961.

2-2. American Society for Testing and Materials, *Index to Standards*. Philadelphia: American Society for Testing and Materials (use latest edition). The Index will give the location in the *Book of ASTM Standards* of any ASTM test.

2-3. American Welding Society, *Welding Handbook*, 5th ed., sec. 1. New York: American Welding Society, 1963.

2-4. Crane, F. A. A., *Mechanical Working of Metals*. London: Macmillan, 1964.

2-5. Dieter, George E., Jr., *Mechanical Metallurgy*. New York: McGraw-Hill Book Co., 1961.

2-6. Fenner, A. J., *Mechanical Testing of Materials*. London: George Newnes, 1965.

2-7. Gordon, J. E., *The New Science of Strong Materials—or Why You Don't Fall Through the Floor*. London: Penguin Books, 1968.

2-8. Gordon, J. E., *Structures—or Why Things Don't Fall Down*. London: Penguin Books, 1978.

2-9. Marin, Joseph, *Mechanical Behavior of Engineering Materials*. Englewood Cliffs, N.J.: Prentice-Hall, 1962.

2-10. McClintock, Frank A., and Ali S. Argon, eds., *Mechanical Behavior of Materials*. Reading, Mass.: Addison-Wesley, 1966.

2-11. O'Neill, H., *Hardness Measurement of Metals and Alloys*, 2nd ed. London: Chapman and Hall, 1967.

2-12. Polakowski, N. H., and E. J. Ripling, *Strength and Structure of Engineering Materials*. Englewood Cliffs, N.J.: Prentice-Hall, 1966.

2-13. Tabor, D., *The Hardness of Metals*. Oxford: Clarendon Press, 1951.

PROBLEMS

2-1. A tensile test on 1020 steel gives the following results:

	Load, kN	Diameter, mm	Length, mm
0	0	12.8	50.800
1	22.2	—	50.848
2	28.5	—	— (yielding begins)
3	51.2	12.2	56.1
4	51.2	10.4	67.3
5	43.6	—	69.8 (fracture)

No data were taken at maximum load.
(a) Calculate the elastic modulus.
(b) Calculate the maximum load during the test.
(c) Calculate the maximum nominal strain.
(d) Calculate the strength of this steel.

2-2. A tensile test of 0.40% carbon steel produced the following data:

	Load, kN	Diameter, mm	Length, mm
Initial	0	12.8	35.6
Max. load	88.1	11.9	40.6
Fracture	75.0	9.22	46.7

(a) Calculate the strength of this specimen.
(b) Calculate the maximum nominal strain during the test.

2-3. The following data were taken during a tensile test of a 0.20% carbon steel:

Reading	0	1	2	3	F
Load (kN)	0	40.0	49.8	50.8	46.3
Length (mm)	50.0	52.8	56.9	65.3	69.1
Diameter (mm)	12.8	12.6	12.1	11.0	9.65

None of these readings was taken at maximum load. The reading at "F" was at fracture.
(a) Calculate the maximum nominal strain during the test.
(b) Calculate the nominal stress at reading number 2.
(c) Calculate the maximum true stress during the test.
(d) Calculte the strength of this specimen. (Show clearly the data you have used.)

2-4. A tensile test of 70-30 brass produced the following data (*none* of these readings was taken at maximum load):

	Load, kN	Diameter, mm	Length, mm
Initial	0	12.8	50.8
Reading 1	53.8	11.4	63.5
Reading 2	55.1	10.7	68.6
Fracture	—	8.26	74.9

(a) Calculate the strength of this specimen.
(b) Calculate the maximum reduction of area during the test.
(c) Calculate the maximum nominal strain during the test.
(d) Calculate the diameter of the specimen at the instant of maximum load.

2-5. Calculate the magnitude of the maximum shear stress in the stressed element in Fig. 2-32. All stresses are in MPa.

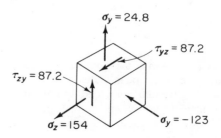

Figure 2-32

2-6. Calculate the maximum normal stress and the maximum shear stress in the cube shown in Fig. 2-33.

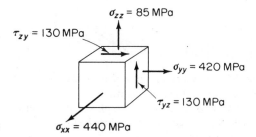

Figure 2-33

2-7. Calculate the maximum normal stress and the maximum shear stress in the cube shown in Fig. 2-34.

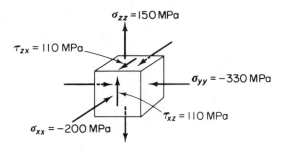

Figure 2-34

2-8. (a) Calculate the magnitude of the maximum tensile stress in the cube shown in Fig. 2-35.
   (b) Calculate the angles (degrees) between the normal to the plane of maximum stress and the x-, y-, and z-axes.
   (c) Calculate the maximum shear stress in the cube.

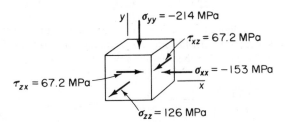

Figure 2-35

2-9. An element of a solid exposed to the stress system is shown in Fig. 2-36. Normal stress fracture will occur when the maximum normal stress is 323 MPa; shear fracture occurs when the maximum shear stress is 192 MPa.
   (a) If A is gradually increased from zero, calculate A (in MPa) when fracture occurs.
   (b) By which mode does fracture occur, normal stress or shear stress?

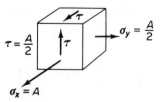

Figure 2-36

2-10. The graph in Fig. 2-37 shows results of a tensile test of 1020 steel. Specific values at several points are given in the table below.
  (a) Calculate the constants K and n for the plastic strain-hardening region.
  (b) Calculate the modulus of elasticity for this steel.
  (c) Estimate the yield strength.
  (d) Calculate the maximum load during the test.

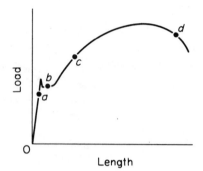

Length   **Figure 2-37**

	Load, kN	Gage Length, mm	Diameter, mm
Init.	0	50.800	12.800
a	23.6	50.848	12.796
b	27.1	51.900	—
c	46.4	55.6	12.4
d	52.5	62.5	10.9

2-11. You are given one half of a fractured brass tensile test piece (Fig. 2-38) and told that the initial diameter was 12 mm and the maximum load was 36 kN. Calculate the numerical value of K in the relation $\sigma = K\epsilon^n$.

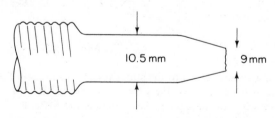

10.5 mm   9 mm

Figure 2-38

2-12. Annealed steel wire and cold-worked aluminum wire kink when bent; cold-worked steel wire and annealed aluminum wire do not kink but bend "smoothly." Explain this seeming paradox.

2-13. A laboratory experiment involved a tensile specimen that had been pulled beyond necking, unloaded, and machined down to a diameter slightly larger than the neck. In the subsequent test, why did the specimen not break where the neck had been?

2-14. Plot separately the nominal stress–strain and true stress–strain curves to the same scale (but without numbers) for the situations indicated below. Assume that elastic deformation is negligible compared to plastic deformation. Make your graphs for parts (a), (b), and (c) consistent with each other with respect to relative magnitude. Label each curve *a, b, c,* or *d.* Plot *s–e* and σ–ϵ curves for:

   (a) A specimen of ductile metal that necks; indicate the ultimate point on both curves with *u.*

   (b) A specimen machined from a bar of the same kind of metal as used in (a). The bar had been subjected earlier to a stress about halfway between yield and ultimate.

   (c) A specimen machined from a bar of the same kind of metal as used in (a). The bar had earlier been cold-rolled to a longitudinal true strain well beyond the ultimate true strain but less than the fracture strain.

   (d) A ductile material that does not strain-harden.

2-15. A tensile test of a metal was run in the laboratory, but only the maximum load was measured during the test (it was 59 kN). All other data were taken before and after the test as follows:

Initial diameter:	13.0 mm
Minimum diameter after test (at point of fracture):	10.0 mm
Maximum diameter after test (away from fracture):	11.6 mm

Find:

   (a) The strain-hardening exponent, *n,* in the equation $\sigma = K\epsilon^n$ for this metal.

   (b) The load on the specimen at the instant just prior to fracture.

2-16. The following data were obtained in a tensile test of naval brass:

	0	1	Fracture
Load, kN	0	50.7	48.0
Diameter, mm	12.8	11.9	8.5

Condition 1 as measured was *not* the maximum load condition. Calculate the strength of this brass.

2-17. For the plastic strain-hardening graph for metals A and B shown in Fig. 2-39:

   (a) Find the value of *K* for metal A.

   (b) Find the value of *n* for metal B.

   (c) Determine which is stronger, A or B.

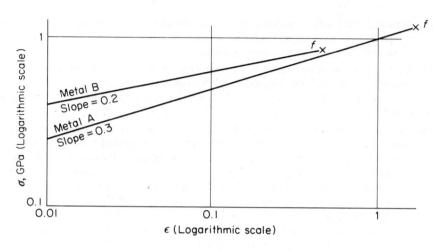

Figure 2-39

2-18. For the log σ vs. log ϵ plots shown in Fig. 2-40, which specimen has
   (a) The largest K?
   (b) The largest true strain at ultimate?
   (c) The largest strength?
   (d) The largest elastic modulus?

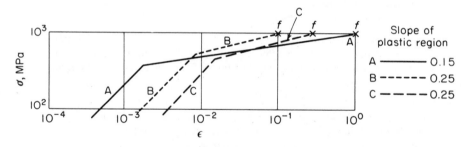

Figure 2-40

2-19. Why are strength and hardness usually related? For what general category of metals are strength and hardness less likely to be related and why?

2-20. If a steel bar undergoing a Rockwell B test has a burr on it that contacts the anvil of the testing machine, will the hardness reading obtained be too high, unchanged, or too low?

2-21. When the R_B hardness at the bottom of a Brinell indentation was measured, these data resulted:

Material	Annealed	500-kgf Load	1000-kgf Load
1020 steel	46	72	77
Brass	-35	66	84

(a) Estimate the R_B that you would expect to measure at the bottom of a 1500-kgf Brinell indentation for 1020 steel and for brass.

(b) Which material has the higher value of n?

(c) Which material has the higher value of K? Describe exactly the line of reasoning you used to reach this answer.

2-22. Two separate Brinell hardness tests using a ball 10 mm in diameter are made at different locations on the same piece of 70-30 brass, (a) at 500 kgf (4.90 kN) and (b) at 1500 kgf (14.7 kN). Which test gives the higher hardness, (a) or (b)? Explain why this happens.

2-23. An annealed metal follows $\sigma = (700 \text{ MPa}) \times \epsilon^{0.3}$. To what final thickness must a piece, initially 15 mm thick, be cold-rolled to increase its hardness to 160 Bhn? (Assume $S_u = 3.45 \times$ Bhn MPa.)

2-24. A low-carbon steel that follows $\sigma = (630 \text{ MPa}) \times \epsilon^{0.18}$ is annealed and then cold-worked 55%.

(a) Find the Brinell hardness before the cold-working and after the cold-working, if $S_u = 3.45 \times$ Bhn.

(b) The reduction of area to fracture from the annealed condition is 70%. If you give the 55% cold-worked specimen to a stranger who does not know its history, what value should he obtain for reduction of area (cold work) to fracture?

2-25. An annealed 70–30 brass bar yields the following when rolled:

Condition	Thickness, mm	R_B
A (initial)	12.7	−12
B	11.5	57
C	7.5	83

(a) Calculate the longitudinal (that is, rolling direction) true strain from A to B.

(b) Calculate the longitudinal true strain from B to C.

(c) Contrast the true strain increments in (a) and (b) with the changes in hardness, and explain why the changes are not proportional.

2-26. Data from the cold rolling of annealed brass, initially 12.7 mm thick, are as follows:

Thickness, mm	12.7	11.5	10.0	9.0	7.9	6.3
R_B	8	51	75	82	88	95

If it is known from tensile data that this brass follows the relationship $\sigma = 700\epsilon^{0.5}$ MPa, find the R_B hardness of the broken tensile bar at a point midway between the fracture surface and the shoulder at the end of the test length.

2-27. Three hardness readings during the cold rolling of 70-30 brass are as follows: initially 13 mm thick, 13 R_B; 12 mm thick, 40 R_B; 6.4 mm thick, 85 R_B. Calculate n in the relation $R_B = (\text{constant})\epsilon^n$, where ϵ is the strain in the rolling direction.

2-28. A square bar of brass that fractures at a true tensile strain of 0.6 is annealed, then cold-rolled so that the reduction in thickness is 35%, and given to you. If you were unaware that the bar had already been cold-rolled, what additional percentage of reduction in thickness would you obtain if you cold-rolled the bar until fracture occurred, assuming that the true tensile strain at fracture applies to rolling?

2-29. A fabricator wishes to use some aluminum sheet in a rolling operation. He conducts a tensile test on a sample initially 6 mm × 6 mm in cross section and determines that this material follows $\sigma = (210) \times \epsilon^{0.21}$ MPa and has a final area at fracture of 19 mm². The cold-rolled sheet must have a yield strength of 170 MPa and be 10 mm thick. Find the *required initial thickness* of the sheet before rolling.

2-30. You are given several identical metal bars of 25 mm square cross section and about 300 mm long. You conduct a tensile test on one bar and observe that the maximum load during the test was 200 kN and that the cross-sectional area was 500 mm² when maximum load was reached. The minimum cross-sectional area at fracture was 280 mm². You wish to produce a bar of 12 mm by 25 mm cross section that will support a maximum load of 140 kN; such a bar could be made of one of the 25 mm square bars by cold rolling, to get the strength, and then machining to get the correct dimensions. If the final bar is to have the highest possible ductility (i.e., do not cold-roll any more than necessary), to what thickness (before final machining) should the 25 mm square bar be rolled?

2-31. Sketch a typical work-hardening stress–strain curve, and superimpose the "drawing stress"–"drawing strain" relationship *on the same axes*, using Eq. (2-33) for the stress in wire drawing. Hence show how a maximum attainable reduction comes about in drawing, and evaluate it for $\sigma = K\epsilon^n$.

What factors does the simple theory above neglect? What effect do they have on the value of maximum reduction? What effect do they have on the mechanical properties of the final product?

2-32. Find the extrusion stress necessary to extrude iron, already cold-worked by 10%, through a reduction in area of 30%. Assume that $\sigma = 560(\epsilon_{cw} + \epsilon)^{0.32}$ MPa.

2-33. Experiments were performed to determine the "best efficiency" in drawing through some fixed-angle dies. Best efficiency means the ratio of the theoretical drawing stress for homogeneous deformation to the *actual* drawing stress. The

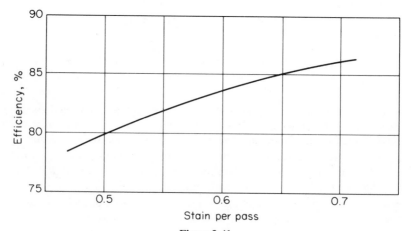

Figure 2-41

results for copper wire are shown in Fig. 2-41 for reductions between $\epsilon = 0.45$ and $\epsilon = 0.70$. Copper wire is to be drawn in the factory from a diameter of 6.5 mm to 2.5 mm. It has been decided that the drawing stress should be no more than 75% of the wire flow stress in order to prevent fracture. If the copper being drawn has an almost constant flow stress (i.e., displays only minor workhardening), how many drawing passes are required to end up with the correct size wire?

2-34. What is the r-value of an isotropic plate?

2-35. Dimensions of the gage section of a piece of cold-rolled aluminum strip were measured before and after a tensile test which was halted before necking occurred. The results were as follows:

	Length, mm	Width, mm	Thickness, mm
Initial	159	27.2	5.36
Final	200	24.4	4.71

Calculate the r-value for this aluminum.

2-36. Two flat strip tensile specimens, having gage section dimensions of $L_0 = 50$ mm, $w_0 = 19$ mm, and $t_0 = 1.2$ mm, are made from a sheet of metal 1 m by 2 m whose thickness is 1.2 mm. One specimen is made parallel to the 2-m length, while the second is at right angles to the first. The following tensile results were obtained:

	Specimen 1			Specimen 2		
	Load, N	L, mm	w, mm	Load, N	L, mm	w, mm
(i) Yield	720	50.1	19.0	(1) 720	50.1	19.0
(ii)	1100	58.0	17.7	(2) 1000	58.0	17.4
(iii) Max	1400	64.0	17.0	(3) 1280	63.0	16.5

As a means of studying anisotropy:
(a) Determine the r-values for these specimens.
(b) What is the thickness of specimen 2 at the maximum load?
(c) What is the true stress at maximum load for specimen 1?
(d) Compare the values of the tensile strengths of these specimens.

2-37. An internally pressurized aluminum sphere, of radius 73 mm and initially 1.65 mm thick, fractures at a pressure of 5 MPa. The thickness at fracture is 1.07 mm. Calculate the equivalent through-thickness compressive stress at fracture. If the sphere is isotropic, calculate the tensile circumferential strain at fracture.

2-38. The following data were taken on a specimen of 0.20% carbon steel (1020 steel):

Load, kN	Length, mm	Dia., mm	Load, kN	Length, mm	Dia., mm
0	50.8	12.8	48.9	55.6	12.3
32.9	52.1	12.7	51.9 (max.)	64.0	11.4
37.4	52.8	12.6	48.9	68.6	10.1
40.0	53.1	12.6	41.8 (fract.)	71.1	9.07
44.5	53.8	12.5			

   (a) Calculate the strength of this steel.
   (b) Calculate the maximum nominal strain during the test.
   (c) Calculate the maximum true stress during the test.

2-39. A tensile test on a specimen of annealed metal X results in the following data:

Condition	Load, kN	Minimum Diameter, mm
Initial	0	12.83
Yield	48.9	12.80
Maximum load	75.6	9.45
Fracture	—	7.98

   (a) Find the maximum nominal stress during the test.
   (b) Find the maximum true stress during the test.
   (c) You are given a plate 25 mm thick of annealed metal X. What is the minimum thickness to which this plate can be cold-rolled without further heat treatment?

2-40. A standard tensile test is run on a piece of steel, and the following data are obtained:

	Load, kN	Minimum Diameter, mm
Initial	0	12.8
Maximum load	103	12.2
Fracture	—	8.94

   (a) Calculate the exponent n in $\sigma = K\epsilon^n$.
   (b) Calculate K in the same equation.
   (c) Find the true strain at fracture.
   (d) Calculate the true stress at fracture.
   (e) This was an actual test. The load at fracture was measured as 81.8 kN. What value does this give for true stress at fracture? Explain any discrepancy with the answer in part (d).

2-41. A low-carbon steel is found to follow $\sigma = (717 \text{ MPa})\epsilon^{0.234}$. If its initial diameter is 12.8 mm and its diameter at fracture is 9.27 mm, calculate the maximum tensile load (kN) that is reached during a tensile test of this steel.

2-42. An element of material is subjected to the following stresses: $\sigma_x = 60$ MPa, $\sigma_y = 140$ MPa, $\sigma_z = 220$ MPa, $\tau_{xz} = 50$ MPa. Calculate the maximum shear stress in the element.

2-43. For a cube stressed by $\sigma_1 = 110$ MPa, $\sigma_2 = -70$ MPa, $\sigma_3 = -150$ MPa, what value of compressive stress must σ_3 reach in order for τ_{max} to be equal to 150 MPa?

Classes of Properties
of Engineering Materials

The variety of materials used for engineering purposes is the result of differences in mechanical and physical properties, cost, availability, and prior experience by the designer or user. Eskimoes build igloos of ice because it is handy (cheap), an adequate insulator, strong enough in compression, and easily fabricated, and they know how to build an ice igloo. The Pueblo Indians used adobe (clay) for exactly the same reasons. Each designer has obviously made a wise application of the materials available, although his choices are quite limited.

In contrast, many more factors influence the choice of material for a component of a modern engineering structure, to the extent that often the better choice between two materials can only be made after trial of both. For example, the choice between gray cast iron, aluminum or aluminum–silicon alloy, and magnesium alloy for automobile engine blocks remains unresolved to the present day. Attempts have been made in the past to produce large engine blocks [with displacement of the order 5 L (305 cu in.)] of aluminum alloy, but today most aluminum and magnesium alloy engine blocks are found in smaller engines. The better thermal conductivity and lighter weight of the aluminum and magnesium alloys must be evaluated against the lower cost and greater experience with gray cast iron.

In this chapter we examine the mechanical properties exhibited by the different classes of metals and polymers of engineering interest. The properties of particular interest are given in Appendix 5. Structural materials such as concrete, brick, asphalt, soil, and wood (and ice and adobe, as well) will not be included here, as all except wood lack sufficient tensile strength for general mechanical use, and wood, while a magnificently balanced structural material, is relatively immune to changes in its properties by humans. The mechanical properties of greatest importance are described in the next three paragraphs.

Stiffness, which is another word for Young's (elastic) modulus, is the ratio of stress to strain in the linear elastic portion of the stress–strain graph, line *oa* in Fig. 2-2. Stiffness controls the deflection under load and is thus important in retaining close tolerances of parts at stresses lower than yield strength. In parts exposed to bending, failure by buckling or excessive deflection can occur at low stresses if the elastic modulus is too low.

The *strength* of any engineering material is the maximum nominal stress in a tensile test [Eq. (2-3)]. For a solid to be able to support the forces that it will experience during service, it must at least have sufficient strength.

Toughness is the ability of a material to resist cracking; it will be described more explicitly in Chap. 14. Green wood is tough, and window glass is not tough (brittle); a wide range of materials have toughness between these extremes. One rough indicator of toughness is the nominal strain to fracture, e_f. Another approximate measure of toughness is S_u/S_y, since large values of this ratio indicate large plastic deformation after yield and, hence, high toughness.

The properties of metals and alloys depend in part on the properties of the pure elements from which they are made. Appendix 6 lists some approximate properties of pure metals that will be useful for reference.

DUCTILE STEELS

Ductile steel, usually steel that is not heat-treated, is used more than any other single class of metal in the world today, because of its high stiffness, high toughness, low cost, and adequate strength. The microstructural characteristics of ductile steel that lead to its good ductility will be developed in Chaps. 4–9.

Steel, both ductile and high-strength, has the highest elastic modulus (about 200 GPa) of any ordinary structural material except the high-performance composites (Chap. 17). The strength of ductile steels usually exceeds about 350 MPa, with an upper limit of the order 700–800 MPa. This range of strength for the cheapest steels corresponds approximately to the strength range of the highest-strength heat-treated aluminum alloys and greatly exceeds the maximum strength of wood (which is of the order 100 MPa) and polymers (which range from 5 to 190 MPa).

Ductile steel is one of the toughest metals, giving it ease of fabrication by rolling, for example, and high reliability in service because of its relative insensitivity to notches and cracks. The nominal strain to fracture, e_f, is of the order 0.2–0.5 for ductile steels.

The widespread distribution and availability of iron in the earth's crust is a happy accident that has profoundly influenced technological development.

CAST IRONS

Cast irons were discovered by ancient man as a consequence of the high carbon content of his iron and the resulting lower melting temperature (see Chaps. 8 and 15). For thousands of years, gray cast iron has found applications in civilization even though it is very brittle. The nominal fracture strain of most gray and white irons is virtually zero. Nevertheless, this castable metal has found many applications because it is relatively easy to fabricate (compared with steel, which has higher melting temperatures), its elastic modulus is about half that of steel and pure iron (about 100 GPa), it has moderate strength, and it can be made very hard and wear-resistant (white iron). Furthermore, modern man has discovered how to produce cast iron of greatly improved toughness and thus has widened the range of its applications.

Fabrication includes pouring molten metal into a mold, where it freezes into the desired shape, and afterward machining the casting to produce very flat surfaces or precise holes. Most cast irons are easier to machine than ductile steels simply because of their low ductility and tensile strength. In many applications, stiffness and low cost of fabrication are more important than high strength and ductility, so cast iron is used.

Strengths of gray cast irons range from 140 to 320 MPa, depending on chemical composition (see Chap. 15), and are thus close to the range of strength of ductile steel. Higher-ductility cast irons, called *nodular, ductile,* or *malleable,* can be made very ductile ($e_f = 0.20$, $S_u = 380$ MPa) or very strong ($e_f = 0.04$, $S_u = 650$ MPa), but they cost more.

Hardness of cast iron ranges from about 150 to 260 Bhn for gray irons, 110 to 240 Bhn for the ductile irons, and up to 670 Bhn for specially treated alloy white iron.

ALUMINUM ALLOYS

The mass density of pure aluminum is 2.70 Mg/m^3, compared with 7.87 Mg/m^3 for pure iron. This nearly threefold weight advantage of aluminum over iron alloys accounts for many of its applications. Pure aluminum is much more resistant to atmospheric corrosion than steel or pure iron, a quality that leads to additional applications. But with an elastic modulus of 70 GPa, aluminum has about 35% of the stiffness of iron-base alloys.

Aluminum alloys usually include a few percent of magnesium, silicon, or copper to increase strength and hardness through mechanisms described in Chaps. 8 and 9. These alloys can be cast or wrought. Casting involves melting a bar of the alloy and then pouring the molten metal into a mold of the desired shape. Pure aluminum melts at 660°C, and these alloys melt at even lower temperatures, so high-production parts are frequently cast under pressure in molds made of steel; this is called *diecasting.* Wrought aluminum alloys are hot- or cold-rolled or extruded as parts of constant cross section, such as bars, sheets, ell or tee sections, or tubes.

The complexity of cast aluminum parts does not permit the same control over microstructure that is possible for wrought shapes, so cast parts generally have lower strengths. Characteristic tensile strengths and fracture strains are given in Appendix 5; the maximum strength of cast aluminum alloys is of the order 300 MPa.

Wrought aluminum alloys have maximum strengths of 650 MPa and generally much better ductility than the cast alloys. The opportunity for cold-forming these alloys also provides an additional strengthening mechanism (see Chap. 6). The maximum *specific strength*, or strength/density, of the wrought aluminum alloys is 0.2 $(km/s)^2$, or about twice that of ductile steel. The higher specific strength thus accounts for the wide application of aluminum alloys to subsonic aircraft skin, where good ductility is essential for forming the product. Even if ductile steel possessed the same specific strength as wrought aluminum alloy, the better corrosion resistance of aluminum would probably dictate its use in many applications despite the higher elastic modulus of steel.

POLYMERS

The past 40 years have produced enormous expansion in the engineering applications of man-made polymeric materials. Detailed discussion of the properties of polymers is presented in Chap. 13. Polymers exhibit mechanical behavior that is different from that of the metals already described in this chapter. The most obvious difference is that most polymers continue to deform for a long time after a stress is applied; that is, their behavior is *time dependent*. This characteristic of polymers is called *viscoelasticity*. Continued deformation over a long time thus limits stresses to values much lower than the stresses allowable for a short-term loading. Such behavior requires the designer of a part made of a polymer to consider the stress–time history of the part in addition to the factors already described for metals.

In conjunction with their time-dependent behavior, most polymers exhibit strong sensitivity of mechanical properties to temperature. Many polymers exhibit marked changes in properties in the range between $-50°C$ and $150°C$. For example, with modest increases in temperature in this range the deformation rate of polymers can increase markedly, to the point where the material is a viscous liquid. With reduction of temperature, many rubbery polymers become brittle and unusable at $-50°C$. Such behavior severely limits the upper and lower temperatures for polymers, and often this is the chief reason that polymers cannot be used in many applications.

When a polymer changes during cooling from a flexible rubbery material to a brittle, glassy state, this transition is called the *glass transition*. The highest-strength polymers are generally brittle at 20°C, a reflection of their having high *glass-transition temperatures*, well above 20°C. Even low-strength polymers will become brittle at sufficiently low temperatures. Thus the glass transition temperature, T_g, is an important property of each polymer. Applications requiring rubbery behavior, such as for flexible tubing, demand polymers with T_g well below the operating temperature; anyone who has tried to uncoil a plastic garden hose outdoors in winter knows how stiff and useless the hose is at temperatures approaching T_g.

SPECIAL APPLICATIONS

The categories of materials described above include most of the common structural materials. In addition, special requirements dictate the use of materials having special mechanical properties. Some of these materials are described in this section. In achieving some special property, we usually pay a higher price for the material and often must also accept inferiority in some of its other properties.

Low-alloy steels.   These steels contain 0.30–0.50% carbon (see Chap. 10), have small (0–5%) amounts of other elements in iron, and can be heat-treated to have strengths up to about 2 GPa. However, their toughness is lower than that of plain carbon steels, and nominal fracture strain can be as low as 0.05. It is possible to manufacture low-alloy steels of still higher strength, but the resulting loss in ductility causes these steels to be of only limited value. The maximum specific strength of 0.25 $(km/s)^2$ of low-alloy steel leads to its use for many structural parts in aircraft, as it has slightly higher specific strength than the wrought aluminum alloys. The better fatigue properties (long-term strength under varying stresses) of low-alloy steels give them a further advantage over aluminum alloys in many applications. The low-alloy steels are widely used in vehicles and machinery wherever a combination of high strength and good hardness is desired, and where moderate ductility is tolerable.

Tool steels.   These steels have high hardness and wear resistance, often at elevated temperatures. These properties are achieved through high carbon, high alloy content (manganese, silicon, nickel, chromium, vanadium, tungsten, molybdenum, and cobalt), and heat treatment. Hardness up to 68 Rockwell C and good stability and hardness at temperatures to 600°C are attainable. Tool steels are also used in a limited number of aerospace structural applications where their high hardness, strength, and temperature resistance provide the only material available, but low toughness is usually the penalty for their use.

Sintered carbides.   These are usually mixtures of very hard intermetallic compounds that are formed by being pressed together at high temperature, usually with cobalt as a binder. Typical compounds in use are carbides of tungsten, titanium, and tantalum. The hardness is above 70 R_C at room temperature, and hardness up to 67 R_C may be retained at 760°C. Carbides thus provide hardness at high temperature that cannot be attained by any steel, as well as extraordinary wear resistance. Carbide-tipped tools generally have longer lives than tool-steel tools in the same application, but of course the carbide tool costs more. The good corrosion resistance of the sintered carbides in many environments provides an added advantage for many applications, such as for mining machinery and oil well drilling bits.

High-temperature materials.   These are intended for service at temperatures above 650°C. The most common are alloys of iron with chromium, nickel, and cobalt, often called *superalloys*. The alloy content is generally high, up to 30% chromium, for example, in contrast to the low-alloy steels, which generally contain no more than 5% of alloying elements. Typical good high-temperature materials can support a tensile stress of 140 MPa at 650°C, 50 MPa at 750°C, and 25 MPa at 900°C for acceptable rates of *creep deformation*, deformation that continues with time. (High-temperature creep is similar to viscoelastic deformation of polymers; see Chap. 12.) High-temperature materials usually have good corrosion resistance at high temperature. However, in many cases the corrosion rate rather than the creep rate can limit the operating temperature of a material. Applications of these materials are in the high-temperature components of engines, turbines, furnaces, and manufacturing equipment.

Ceramics.   These are compounds of metals and nonmetals, often of very complex structure, that are used for a wide range of structural and nonstructural applications. Brick and concrete are structural materials because of their low cost and good compressive strength. Glass is used because it is corrosion-resistant, transparent, easily formed to complex shapes, and cheap. For applications at very high temperatures (to 2500°C), refractory materials such as MgO are used. Ceramics generally have very low ductility and associated low tensile strength.

Stainless steels.   These are designed to resist corrosion. Since corrosion resistance and high-temperature resistance are found in similar alloys, many materials serve both purposes. The stainless steels are iron alloys with 12–30% chromium, low carbon (usually less than 0.30%), and often substantial amounts of nickel (3–22%). Strengths range up to 1.4 GPa for heat-treated or severely cold-worked grades, with nominal fracture strain of 0.10–0.30, and acceptable levels of toughness. In spite of their much higher cost over iron and ductile steel, the stainless steels have numerous applications such as in chemical processing (often combined with high temperatures), food-handling equipment, and where good resistance to environmental corrosion is required.

Copper.   Copper is used in the nearly pure form (at least 99.9% Cu) because it is an excellent electrical conductor, combined with excellent corrosion resistance and ease of mechanical forming. Alloys of copper retain much of its corrosion resistance and exhibit good strength and ease of fabrication. Although beryllium-copper gives an acceptable specific strength of 0.17 $(km/s)^2$, most of the copper alloys, usually called brasses or bronzes, fall below 0.1 $(km/s)^2$. Thus the high cost of copper alloys and their low specific strength restrict their use in structural applications to corrosive environments (especially fresh and salt water) in which no other material will serve as well.

Titanium alloys.   These provide a strong, lightweight metal capable of withstanding operating temperatures considerably higher than aluminum alloys can. Pure titanium has a mass density of 4.51 Mg/m^3, between those of iron and aluminum, a melting temperature of 1668°C, which is 130°C above that of iron, and an elastic modulus of 116 GPa. Alloys of titanium involve aluminum and vanadium up to about 10% in many alloys, plus smaller quantities of up to about eight other elements. The highest-strength alloys have maximum specific strengths of 0.35 $(km/s)^2$, the highest of any material discussed so far in this chapter. Titanium alloys retain useful ductility to very low temperatures, say −200°C. Furthermore, commercially pure titanium has excellent corrosion resistance in many environments, while titanium alloys have somewhat poorer but still adequate corrosion resistance. Production and fabrication costs for titanium are high, so its use is restricted to those applications for which it is uniquely suited.

Retention of useful strength (1.7 GPa maximum at 430°C for one alloy) up to 500°C allows titanium alloys to be used in supersonic aircraft, where structural components may be subjected to high continuous operating temperatures. For example, the temperature of the air at the leading edge of the wing of an aircraft flying at three times the speed of sound at an altitude of 20 km is about 340°C. As will be discussed in Chap. 9, aluminum alloys rapidly lose strength above 200°C, with the best aluminum alloys useful at reduced stresses up to a maximum of about 300°C. For this reason, aircraft made of aluminum alloys are limited to a little more than twice the speed of sound in air, for which the leading-edge air temperature is 120°C at 20 km altitude. The first commercial supersonic aircraft, the Concorde, was limited to such speeds because it used long-established aluminum alloys. Titanium alloys provide the highest specific strength of any commercially available metal suitable for high-speed aircraft.

Magnesium alloys.   These have the lowest density of any structural metal; pure magnesium has a mass density of 1.74 Mg/m^3, and the common alloying elements of magnesium, such as aluminum, zinc, manganese, and thorium, raise the density to a maximum of about 1.84 Mg/m^3. Thus magnesium is used in many applications where low mass is important, such as aerospace structures, truck and trailer body components, small portable power tools, and ladders. Magnesium alloys can be cast and can be wrought by processes such as extrusion and rolling, in many ways similar to aluminum alloys. In certain applications the magnesium alloys offer better performance than the aluminum alloys. On the basis of a structure having the same mass, the lower mass density of magnesium permits, for example, the use of sheet of about 50% greater thickness and a consequent reduction of maximum stress in bending by a factor of about 2.3. In a design in which bending stresses are critical it is possible to fabricate from magnesium alloys a structure that has a lower mass than the same structure made of aluminum alloys. By the same reasoning, with the lower elastic modulus of magnesium taken into account (see Appendix 5), deflections from bending stresses in sheet material are lower in magnesium by a factor of about 2.2, for a given mass.

In simple tension and compression, magnesium and aluminum alloys provide nearly equal load-carrying capacity for the same mass, so that the lower cost of aluminum (per kilogram) dictates its widespread use in such applications. Magnesium alloys are more resistant to ordinary atmospheric corrosion than steel but less so than aluminum alloys. Magnesium alloys require surface protection when used in salt water environments.

Composite materials.   These are man-made combinations of materials on a very fine scale (1–100 μm) that provide properties unattainable in any one component material alone. Very fine filaments usually have much greater strength than the same material in bulk (see Chap. 17, in which composites are discussed in detail), so surrounding an array of strong filaments with a low-density matrix material can produce a composite material of high specific strength. By proper selection, special desired properties such as corrosion resistance or electrical resistance can be obtained.

Typical properties of two composite materials are given in Appendix 5. Moldable glass fiber-resin is a commercial mixture of very fine glass fibers, either in short lengths or as continuous filaments, which can be produced in complex shapes by high-production equipment. Although still expensive compared with ductile steel, this class of materials provides specific strength up to twice that of ductile steel. Low ductility and elastic modulus will restrict uses of this material and require careful design. Applications include molded boats, automobile panels, and equipment housings.

Graphite–epoxy is one of the newest commercially available composites. Because of the extraordinary strength and stiffness of the graphite filaments, the composite exhibits by far the highest specific strength of any commercial material, combined with the stiffness of steel. Low ductility and high cost will continue to limit its applications. Graphite–epoxy has been used for components of orbiting space vehicles, helicopter rotors, aircraft frames, bicycles, and fishing rods.

This chapter has examined the wide range of properties that govern the selection of different materials for different applications. In many cases materials are selected because they have good nonmechanical properties, such as conductivity or corrosion resistance. Usually such physical properties cannot be changed for a given material, so the designer must change materials in order to change physical properties. However, most mechanical properties can be changed. The balance of this book is directed toward the fundamentals that control mechanical properties. Once the fundamentals are understood, methods for improving mechanical properties can be determined. We will start with the basic structure of simple solids and then advance to more complex materials.

REFERENCES

3-1. American Society for Metals, *Metals Handbook*, 8th ed., vol. 1, Properties and Selection of Metals. Metals Park, Ohio: American Society for Metals, 1961.

3-2. American Society for Metals, *Metals Handbook*, 9th ed., vol. 1, Properties and Selection: Irons and Steels. Metals Park, Ohio: American Society for Metals, 1978.

3-3. American Society for Metals, *Metals Handbook*, 9th ed., vol. 2, Properties and Selection: Nonferrous Alloys and Pure Metals. Metals Park, Ohio: American Society for Metals, 1979.

3-4. American Society for Metals, *Metals Handbook*, 9th ed., vol. 3, Properties and Selection: Stainless Steels, Tool Materials and Special-Purpose Metals. Metals Park, Ohio: American Society for Metals, 1980.

3-5. Brick, R. M., R. B. Gordon, and A. Phillips, *Structure and Properties of Alloys*, 3rd ed. New York: McGraw-Hill, 1975.

3-6. DeGarmo, E. Paul, *Materials and Processes in Manufacturing*, 4th ed. New York: Macmillan Pub. Co., 1974.

3-7. Ruoff, Arthur L., *Introduction to Materials Science*. Englewood Cliffs, N.J.: Prentice-Hall, 1972.

3-8. Smith, Charles O., *The Science of Engineering Materials*, 2nd ed. Englewood Cliffs, N.J.: Prentice-Hall, 1977.

3-9. Suh, N. P. and A. P. L. Turner, *Elements of the Mechanical Behavior of Solids*. New York: McGraw-Hill, 1975.

Structure of Solids

Chemical bonds hold matter together. We may distinguish between those forces that hold atoms together within a molecule and those that attach molecules to one another. The strong primary valency forces associated with covalent, ionic, and metallic bonds are in the former category, while van der Waals forces are weak secondary valency forces. The state of the material (solid, liquid, or gas) depends on whether the internal energy of the material (which is increased by increasing the temperature) is capable of overcoming its internal cohesion, which is a manifestation of the chemical bonds. The commonly experienced strength properties of solids, discussed throughout this book, are also manifestations of solid-state bonding forces that resist deformation.

CRYSTALLINE AND AMORPHOUS SOLIDS

At absolute zero all atoms and molecules are at rest and everything must be a hard solid. Increase of temperature produces random vibrations of atoms and molecules, which is another way of describing heat. Eventually the internal cohesion of a solid (where the atoms and molecules have little freedom) is lost and the molecules acquire the sort of "shuffling freedom" that characterizes a liquid. Further increase of temperature eventually leads to boiling, where the molecules are released from any kind of mutual cohesion. It is possible under some conditions for the liquid stage to be missed in this sequence (i.e., sublimation), but that need not concern us here for common engineering solids. Cooling reverses the process, gases condensing to liquids and liquids freezing to solids.

In the solid state atoms and molecules are arranged in either of two ways:

1. In a *crystalline* form, where the atoms or molecules are packed together in a highly regular, definite geometric form;
2. In an *amorphous* manner, with no specific form.

A crystal is analogous to bricks laid regularly in a wall, and the same number of bricks heaped in a pile is the equivalent of an amorphous solid. Clearly, amorphous solids are porous and less dense than the corresponding crystalline solid.

All materials have the tendency to become crystalline on freezing from a melted state, but not all achieve it. Metals have highly regular arrangements of molecules, with systematic three-dimensional geometric packing. Other substances may be crystalline to a greater or lesser extent. Some engineering solids have no clear packing of atoms or molecules. It all depends on the magnitude of the attractive forces between the atoms and molecules in the solid state and on how successful these cohesive forces are in drawing them together, that is, whether the "bricks" are drawn together into a regular pattern or whether they end up in a higgledy-piggledy fashion. Some materials can exist readily in either form; others exist with mixtures of both forms in the same solid; in others again the crystalline or the amorphous state predominates. For example, amorphous metals *can* be produced by tremendously fast cooling, but for most practical purposes metals are crystalline.

GLASSES

The condition of the liquid just before freezing, and the rate at which it is cooled, play important roles in determining the form of the solid. If a liquid just before freezing is very viscous (where the molecules "interfere" with one another) or if the cooling rate at freezing is very rapid, the molecules will not be able to choose the closely packed sites of a crystalline arrangement but will rather end up as an irregular, amorphous array. The general term for such an amorphous solid is a *glass*. Sugar is normally crystalline, but if cooled very quickly it becomes toffee, which is a glass. Sometimes a bit of crystallization does take place; thus fudge is partially crystallized toffee. Sand (silica, SiO_2) melts as a viscous liquid and on cooling gives one of the types of glassy, amorphous solids used for windows, beer bottles, and the like.* Hence the common use of the word "glass"; but the term really has the wider, precise scientific meaning given above. Silica does also exist in a crystalline form, which is the naturally occurring mineral quartz; this had plenty of time to

*As the melting point of pure silica is some 1600°C, practical glasses have lime and soda added to allow easier fusing at lower temperatures.

arrange its atoms into crystals as it was formed during slow cooling in the earth's crust.

Exercise 4-1

Give a possible reason for windows in medieval churches cracking and falling out in recent times.

Solution. The transition from the glassy state to the crystalline state, which is called devitrification, can occur for some materials over very long periods of time at ambient temperature. Since solids would prefer to be in the crystalline form, the church window glass has devitrified over the centuries. The crystalline form is more dense than the amorphous form, so the glass has shrunk, leading to cracks and misfits. (See Exercise 16-3 for a general discussion of this matter.)

Exercise 4-2

Expensive glass articles are often called "crystal glass," but glass is amorphous. Explain.

Solution. Scientifically, this is a silly use of the word. It arises because the type of glass designated "crystal" and used for crystal chandeliers, decanters, etc., is much better than common glass and in the early days was considered to "sparkle" and reflect as well as rock crystal (another name for quartz). But "crystal" glass is still amorphous.

VISCOSITY

The actual behavior of the thick, slugglish liquid before freezing depends on the form of the molecules and the way they associate above the melting point. Although we use the words "melting" and "freezing," it may be impossible with the thickest liquids to know precisely what we mean by those terms. Viscosity is one measure of the amount by which the liquid molecules, instead of remaining independent, form torpidly mobile networks or chains. The coefficient of viscosity (η) relates the applied stress to the strain rate (see Chap. 12). Changes in η with temperature may be used to distinguish between, for example, the clear-cut transformation from ice to water at one end of the spectrum of melting/freezing behavior and the behavior of treacly molasses or pitch at the other.

The distinction between fluids and solids is arbitrarily drawn at a viscosity of 100 TPa-s. At this value, a cube of 25 mm sides could support a man's mass for a year without sinking more than a couple of millimeters. Insofar as the mechanical properties of crystalline solids are time-dependent at temperatures just below their melting points, the choice of 100 TPa-s attempts to indicate how "thick" a treacly liquid should be to have similar properties, more or less, to crystalline solids.

GLASS-TRANSITION TEMPERATURE

Figure 4-1 shows how η changes typically with temperature, for the two extreme conditions of solidification down to crystalline and to amorphous solids (and, of course, in reverse for melting). Approximate values for the viscosity of gases at boiling/condensing points and of liquids at melting/freezing points are 10 μPa-s and 1000 μPa-s, respectively. If crystallization occurs, the viscosity changes immensely at the melting/freezing point, up to perhaps 100 PPa-s (recall the SI prefixes from Appendix 1: P = peta = 10^{15}), and goes on rising as the crystalline solid cools down. If, on the other hand, the liquid fails to crystallize, although η does increase as the temperature drops, the increase is continuous and gradual, with no sharp rise in η at a melting/freezing point. When such a "supercooled liquid" does attain the critical 100 TPa-s, the structure is such that we have effectively an amorphous solid, thereafter called

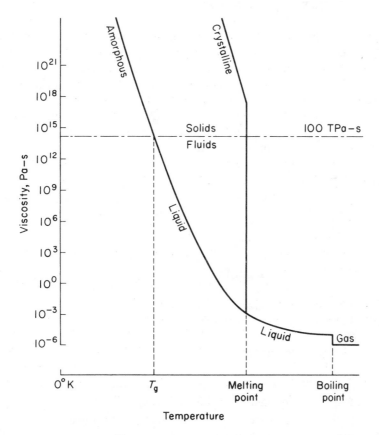

Figure 4-1 Changes in viscosity with absolute temperature for crystalline and amorphous solids.

a *glass*. The temperature at which 100 TPa-s is reached is called the *glass-transition temperature*, T_g. For many materials it is about one third of the boiling temperature. The glass-transition temperature also reflects changes in other properties, such as the so-called secondary thermodynamic quantities (for example, the coefficient of expansion); elastic moduli also change at T_g, glasses stiffening as the temperature drops (see Chap. 13). Different materials behave in different ways in the transition regions from 1 mPa-s to 100 TPa-s; some get stiff gradually, whereas others are more "liquidy" for a longer temperature (or time) span. Others display rubbery behavior, at temperatures just above T_g, with enormous reversible mechanical extensions. These differences are related to the different molecular structure and arrangements in the cooling range of temperatures down to T_g.

TYPES OF CRYSTAL STRUCTURE

Metal alloys freeze over a *range* of temperature (between the liquidus and solidus on the phase diagram, Chap. 8), during which the liquid is transformed through a slushy, partly liquid and partly solid state completely to a solid, and the almost vertical line marked "crystalline" in Fig. 4-1 has a clear backward slope over this range of temperature. Unlike the curved line marked "amorphous" in Fig. 4-1, however, the alloy line still has obvious kinks and almost discontinuous changes in viscosity with temperature.

Metals and simple inorganic substances readily form *crystals*. Rocks, gemstones, and the like had eons to cool in their formation and so are crystalline. Indeed, the first postulates about crystal structures were made from models of spheres (atoms) packed in different ways to give the external shapes of naturally occurring inorganic crystals, such as quartz, calcite, and salt. There are basically six patterns of crystal structure, of which only two are important in traditional engineering solids. These patterns are the shapes taken by the building bricks of the crystal. In Fig. 4-2 the spheres at the corners of the bricks represent atoms; in a pure element they will all be the same, but in a metal alloy or crystalline chemical compound they will be of more than one sort. The edges of the bricks represent the directions of the bonding forces, and the whole network of edges in three dimensions is called a *space lattice*.

The types of space lattice are:

1. *Cubic*, where all edges are mutually at right angles and the edges are of equal lengths (many metals fit this category);
2. *Tetragonal*, where all edges are mutually at right angles and two of the edges are equal in length, but the third is different;
3. *Orthorhombic*, where all edges are mutually at right angles and all the edges have different lengths (a brick is of this sort);

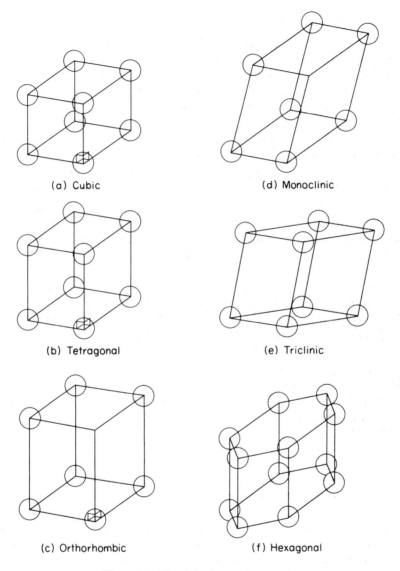

(a) Cubic

(d) Monoclinic

(b) Tetragonal

(e) Triclinic

(c) Orthorhombic

(f) Hexagonal

Figure 4-2 The six basic crystal structures.

4. *Monoclinic*, where only two of the edges are mutually perpendicular and the other edge is at some other angle, and where all edges have unequal lengths;

5. *Triclinic*, where none of the edges are mutually perpendicular and where none of the edges is equal to another (a special case of this, where the edges do have equal lengths, is called *rhombohedral*);

6. *Hexagonal*, where the base of the building brick is a regular hexagon.

A given lump of a solid may be a *single crystal* (perhaps a few millimeters in extent), but common engineering solids are *polycrystalline*, having many small, individual crystals (or *grains*)—some only a micrometer in extent— which have grown from the melt in different directions, leading to *grain boundaries* between the individual crystals. These topics are dealt with in succeeding chapters.

TYPES OF POLYMER STRUCTURE

On the other hand, easy glass formers are substances with molecules of complex shapes (which find it difficult to pack together and which interfere in the liquid state) and/or substances whose molecules link together strongly in the liquid to form long chains or networks. Thus many oxides, silicates, borates, and phosphates readily form glasses. Again, many molecules have the propensity to form long chains, where thousands of repeat units link up. The repeat unit (or *monomer*) can be a single atom, such as sulfur or selenium, but commonly it is understood that the monomer is a molecule. Furthermore, these can be inorganic molecules, such as chrysolite, $3MgO \cdot 2SiO_2 \cdot 2H_2O$, in asbestos and silicone oil, $Si(CH_3)_3O$, in the silicones; or organic, such as ethylene in polyethylene, isoprene in natural rubber, urethane in polyurethane, the amides in nylon, saccharides in cellulose (wood, for example) and formaldehydes in bakelite. It is the latter carbon-based organic materials that are familiarly called *polymers* or *plastics*, through strictly "polymer" means "many mers," whatever their nature. Most animal and vegetable living matter is in the form

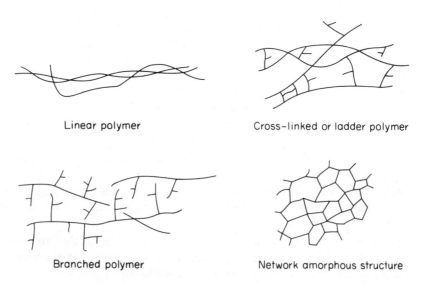

Linear polymer   Cross-linked or ladder polymer

Branched polymer   Network amorphous structure

Figure 4-3   Various types of amorphous polymer structure.

of organic polymers. Some of the mers bring atoms other than carbon into the main chain; for example, there is oxygen in cellulose and nitrogen in nylon. Again, structural units with excess valency or spare bonds in the main chain allow side chains to grow out, like branches on a tree. These may link up with side groups on other chains, giving a *cross-linked* polymer; if this process occurs many times, *network* polymers are produced. Figure 4-3 illustrates some of these amorphous structures. Polymers are often visualized as tangled masses of spaghetti or string; polymers can therefore be quite simple or extremely complicated in their makeup and behavior, as discussed in Chap. 13. A characteristic of some polymers is a range of temperature over which the mechanical response is rubberlike, with tremendous reversible strains. When that rubbery range encompasses room temperature, the polymers are called *elastomers*.

COMPARISON OF TYPES OF MATERIALS

The schematic in Fig. 4-4 attempts to interrelate most engineering solids. It should be viewed in conjunction with the viscosity–temperature plot of Fig. 4-1. Three distinct types of materials are identified:

1. *Crystals* (with clear-cut melting/freezing points);
2. *Glasses* (with no sharp transition to the solid state);
3. *Polymers* (also with no sharp transition to the solid state, but also with the feature of rubbery mechanical behavior)

It is seen that (2) and (3) are subdivision of amorphous solids that do not have specific melting/freezing points. Each type of material is represented by a cylindrical column that forms the edges of a triangular prism. Temperature is the vertical scaling, $0°K$ being represented by the base plane, temperatures being to an arbitrary scale where boiling and melting temperatures are normalized to the same level.

Let us trace what happens when materials are cooled from the gaseous state, that is, follow down through the temperature levels A, B, C, and D. Vapors condense to liquids at the boiling point. Then for crystalline materials, liquids freeze solid at the melting point, with marked changes in viscosity; for metal alloys, freezing may occur over a small range of temperature around point B. Such a solid between levels B and C is a "creepy" material, whose mechanical properties are time dependent (Chap. 12). Time dependency progressively diminishes below about $\frac{1}{2}$ of the absolute melting temperature (below level C), until the solid is not markedly time dependent (Chaps. 2 and 5). Glasses, on the other hand, remain liquidy as they cool and get thicker and thicker, until at about $\frac{1}{3}$ of the absolute boiling temperature (which corresponds, more or less, with T_g) they reach the borderline viscosity of 100 TPa-s. Below the temperature at which the viscosity attains this level, they are con-

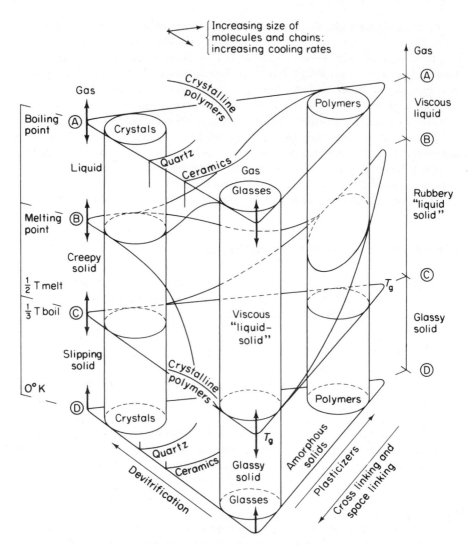

Figure 4-4 The interrelation between crystalline materials, glasses, and polymers.

sidered glassy solids, whose mechanical behavior is not markedly affected by
time and may be related merely to stress, like slipping crystals (Chap. 5).
Polymers too are liquidy above T_g, but they also can exhibit a range of
rubbery behavior between the viscous liquid and glassy regimes.

Many solids in practice show characteristics between the foregoing distinct groupings and could be located vertically along the sides of the schematic
prism in Fig. 4-4. For example, many ceramics consist of small crystals of
metallic silicates and oxides kept together by a thin layer of a glassy material
(Chap. 16). Similarly, there are crystalline polymers that have regions within

the bulk of the tangled mass of polymer chains, in which the chains are oriented in a regular, repeatable manner. Some short-chain polymers with simple mers can crystallize almost completely on cooling (e.g., polyethylene), but small oriented regions called *crystallites*, if they occur in commercial polymers, exist in isolated regions, with the same molecular chain often extending through both crystalline and amorphous regions. Again, between the polymer column and the column representing glasses in Fig. 4-4 an immense range of properties can be produced by *crosslinking* and/or *space linking*, tending to eliminate the rubbery regime from the polymer starting position; the vulcanization of natural rubber (where sulfur is added to cross-link the long chains of the natural rubber molecules) is that very thing. Indeed, it is convenient sometimes to think of silica glass as a densely cross-linked polymer. These ideas are further developed, and other methods described of altering the properties of polymers (e.g., by *plasticization*), in Chap. 13.

The overall pattern of events depicted in the three-dimensional schematic drawing in Fig. 4-4 can be markedly dependent on cooling rate. Increased rates of cooling, where less time is allowed for molecular sorting out, tend to move things away from the crystalline behavior edge of the prism.

REFERENCES

4-1. Cottrell, A. H., *The Mechanical Properties of Matter*. New York: John Wiley & Sons, 1964.

4-2. Gordon, J. E., *The New Science of Strong Materials—or Why You Don't Fall Through The Floor*. London: Penguin Books, 1968.

4-3. Harris, B. and A. R. Bunsell, *Structure and Properties of Engineering Materials*. London: Longmans, 1977.

5 | Crystal Imperfections and Slip

Strength of a solid is one of its most important properties, and an examination of the atomic-scale behavior of materials is essential to an understanding of the reasons for strength. This chapter provides (1) an introduction to the rudiments of crystallography, as many solids of engineering importance are crystalline in nature, (2) the classical development of the theoretical shear stress for slip in a perfect crystal, and (3) a description of the nature of imperfections in real crystals and how they affect strength.

EQUILIBRIUM SPACING OF ATOMS AND CRYSTAL STRUCTURE

Atoms can be regarded as relatively hard spheres, since the outer electron shell (or the electron cloud in metals) acts as a strong and very short-range repelling force to other atoms. Whereas the positively charged nucleus, or center, of an atom is in force equilibrium with its own surrounding electrons, the balance of forces is altered by the negative charge of the electrons of a nearby atom. The classical representation of these forces, usually neglecting the small repelling force between the two positive nuclei, is shown in Fig. 5-1. The spacing a_0 where the net force between atoms is zero represents the equilibrium spacing, that is, the spacing of the atoms in the absence of other forces.

The idea of two isolated atoms as hard spheres resting a distance a_0 apart is valid only at absolute zero temperature ($0°$ K). At any other temperature, atoms are in motion and thus may or may not have equilibrium spacing a_0 as their average position.

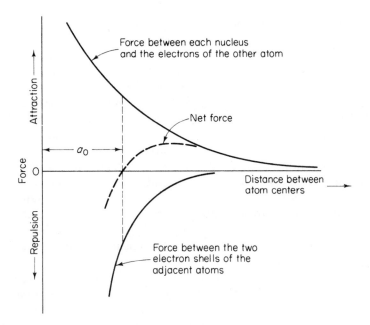

Figure 5-1 Schematic representation of forces between two atoms. Equilibrium spacing is a_0.

Let us first consider a large number of like atoms (that is, atoms of the same element) together at $0°$ K, and then later gradually increase the temperature. At $0°$ K we could expect the atoms to assume some equilibrium spacing, since they are completely at rest. The simplest and commonest model is what we get when we carefully fill a box with ping pong balls, a layer at a time. The resulting spacing and positions of the ping pong balls are analogous to one kind of *crystal structure*. In real atoms, the rules that hold for equilibrium spacing of only two atoms become much more complicated when each atom is close to many other atoms, and the equilibrium spacing is different, but the principle is the same. Thus, at $0°$ K a large number of like atoms will assemble together in some equilibrium geometrical pattern that is the characteristic *crystal structure* for that element. This three-dimensional structure, with imaginary lines connecting the atom centers, is called a *crystal lattice*. Different elements are found to prefer different crystal structures; after a brief excursion into the subjects of melting and vaporization, we will consider the three most common crystal structures in metals.

SOLIDS, LIQUIDS, AND VAPORS

Everyone knows that solids retain their shape, liquids fill a container up to a level, and vapors (gases) expand to fill a container no matter how big it is. The enormous mechanical differences among these three usual forms, or *phases*, tend to cloud the simple changes that occur with increasing temperature. But what about solids that ooze, or liquids that do not flow, or the processes of evaporation (liquid to vapor) or sublimation (solid to vapor)? Our simple atomic model, if realistic, should help to explain some of these phenomena.

Return now to the solid crystal at $0°$ K, with the hard-sphere atoms at rest and in a fixed crystal structure. If heated, the individual atoms will vibrate at their positions, maintaining nearly the same average spacing as before. The spacing does increase a little with temperature, a process called *thermal expansion*. The vibration is, essentially, the nature of increased temperature, so we are actually using a kind of circular argument in saying increased temperature leads to increased atomic motion. Nevertheless, as temperature (vibration) increases further, atoms eventually vibrate so much that some actually switch positions at random, a process called *diffusion*; at still higher temperatures some atoms may vibrate out of their positions and not be replaced, leaving holes or *vacancies*.

The *melting temperature* represents a transition in atomic structure, on heating, from an ordered crystalline structure (*solid*) to a substantially disordered, or amorphous, structure (*liquid*) of enormously lower *viscosity*. Viscosity, which is the ratio of shear stress to the resulting shear strain rate, has already been discussed in Chap. 4 and will be discussed more in Chaps. 12 and 13. The transition from solid to liquid in pure elements is not gradual, and the melting point is uniquely defined; in the case of alloys (mixtures of elements, described later) melting occurs over a range of temperatures. As heat flows to the solid, individual atoms of the solid acquire much more energy and thus vibrate so much that they break away from their crystal lattice sites. The *heat of fusion* of a material is simply the total energy necessary to allow all atoms of a given mass to acquire enough energy to break away from the ordered crystalline structure. The atoms are still close enough together in the new liquid state to experience the bonding forces in Fig. 5-1, but the mobility of the atoms is so great that they have no fixed equilibrium positions with respect to each other.

Vaporization occurs when an atom receives enough energy so it can break loose from the local forces in Fig. 5-1. This can happen either from the liquid state, in which case the energy required is the *heat of vaporization*; or from the solid state, a process called *sublimation*, which requires the sum of the energies required for melting and vaporization. Atoms or molecules (groups of atoms) of a vapor, or gas, are usually widely spaced and move at high velocities with occasional collisions with each other.

Thus, in a rather oversimplified way, the essential changes from solid to liquid to vapor can be understood in terms of whether the balance between bonding forces and atomic vibration constrains the atoms to a closely spaced regular structure (solid), a closely spaced near-random structure (liquid), or a widely spaced random structure (vapor). As we shall see, although the transformation from solid to liquid is very sharp and occurs at a fixed temperature for pure elements, the atoms in solids just below their melting temperature have sufficiently high mobility to take on some of the characteristics of liquids.

One more point before we return to the crystal structure of solids. It is possible to consider the phase changes described above in terms of *surface free energy*, which is the energy required to create a free surface, such as exists at the interface between a solid and its vapor. An atom on the surface of a solid is only partially surrounded by adjacent atoms, in contrast with an atom in the interior of the solid. Clearly, there are fewer attached bonds associated with surface atoms, and energy is required to break free some of the bonds of an interior atom when the atom becomes (by diffusion, say, or fracture [see Chap. 14]) an atom on the surface of the crystal. An atom that has no fixed bonds, such as in the liquid state, has thus broken even more bonds and requires still higher energy. And an atom in the vapor state has virtually no bonds, so it requires the highest energy to attain that state. Thus for an interior atom in a solid to sublime requires first that it diffuse to the free surface through accidentally acquiring energy from the vibrations of its neighbors. Then the atom must be given a relatively large "kick" to impart the energy required for complete separation from the solid crystal and into the vapor state.

The *vapor pressure* is the equilibrium pressure of the vapor atoms in the vicinity of the free surface (between either solid or liquid and the vapor), for which as many atoms drop back into the solid crystal (or liquid) as are ejected. It is fortunate that the vapor pressure for solids is usually extremely small, or the earth's atmosphere would be filled with atoms of metals, say, and no solid would last very long. For example, molybdenum at 2000 K, which is 69% of its melting temperature, has a vapor pressure of 105 μPa. This is a rather low pressure (atmospheric pressure being some 101 kPa): At 2000 K, 96 g (1 mol) of Mo would just fill, with vapor at a pressure of 105 μPa, a cubic box roughly 500 m on each side. As the melting temperature is approached, vapor pressure increases rapidly; the vapor pressure of Mo at 2566 K (88.5% of melting temperature) is 133 mPa, about 1300 times larger than at 2000 K.

CRYSTAL STRUCTURE OF METALS

Of the metallic elements, 44 solidify in one or more of three simple crystal structures. Of these three structures, two are versions of the cubic system [Fig. 4-2(a)] and the other is a version of the hexagonal system [Fig. 4-2(f)]. The common metals that are exceptions are manganese (complex cubic), mercury [rhombohedral; Fig. 4-2(e)], and tin [a body-centered version of the tetragonal system; Fig. 4-2(b)]. References to thorough discussions of all crystal systems are given at the end of this chapter. Appendix 6 lists the crystal structures of many metals of engineering interest.

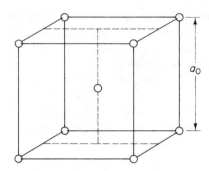

Figure 5-2 Body-centered cubic (bcc) unit cell, with small circles representing the centers of atoms. Dashed lines are only to show the position of the center atom.

Body-centered cubic. The geometry of the *body-centered cubic* (bcc) crystal lattice consists of a simple cube with atom centers at each corner, plus an atom positioned at the cube center, as shown in Fig. 5-2. Such a characterization of the crystal is useful for remembering the spacial relationships of the various atoms, but the lines in Fig. 5-2 are solely for this purpose and do not necessarily represent the most important atomic bonds. For example, the distance between corner atoms and the center atom is less than the distance a_0 between corner atoms. The cube formed by the solid lines is called the *unit cell*, and the complete crystal is formed of many repetitions of this unit cell. Appendix 6 gives the size of the *lattice constant* a_0 for the common bcc metals.

There is no difference at all between the corner and center atoms in Fig. 5-2, although a superficial examination of the unit cell might suggest that there

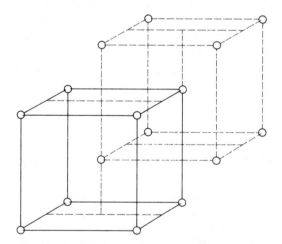

Figure 5-3 A bcc crystal regarded as interlocking simple cubes, one set shown with solid lines and one set with dashed lines. The atom at the center of the solid set is thus the corner atom of the dashed set, and vice versa.

are differences. We could just as well have drawn a unit cell using the center atom as one corner. In fact, we could even think of the bcc structure as a set of interlocking simple cubes (Fig. 5-3), but this is a more complicated model to treat mathematically and is therefore replaced by the simpler bcc model.

For clarity we have so far shown the bcc crystal lattice by locating only the center points of each atom. However, since the crystal is really (in our thinking, anyway) made up of hard spheres that are touching, the bcc unit cell really looks as in Fig. 5-4. Only one eighth of each corner atom is located within each unit cell cube, while the entire body atom is located within the unit cell. Thus the requirements that the atoms be touching and be situated in the bcc geometry lead to a fixed relationship between the lattice constant a_0 and the size of the atoms.

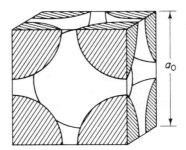

Figure 5-4 The bcc unit cell, with atoms shown as full-sized spheres.

Exercise 5-1

Calculate the ratio a_0/r for a bcc crystal, where r is the radius of the atoms.

Solution. Cut the unit cell along a diagonal plane through the center atom (Fig. 5-5). The line of atoms touching is thus from one corner atom through the body atom to the opposite corner atom, shown dashed in Fig. 5-5. This diagonal is $a_0 \sqrt{3}$, equal to 4 atomic radii. Thus $4r = a_0 \sqrt{3}$, and so $a_0/r = 4/\sqrt{3} = 2.31$.

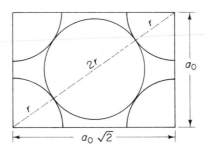

Figure 5-5 Solution to Exercise 5-1.

Face-centered cubic.   The other common cubic crystal structure in metals is *face-centered cubic* (fcc), shown in Fig. 5-6. Here, in addition to atoms at each corner of the cube, atoms are located at the center of every face of the imaginary cube surface. The differences between bcc and fcc are much greater than might appear from a comparison between Figs. 5-1 and 5-6. Figure 5-7 shows all the atoms, full sized, in a unit fcc cell.

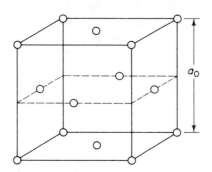

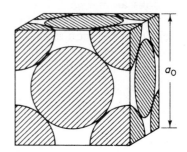

Figure 5-6 Face-centered cubic (fcc) crystal unit cell, with small circles representing centers of atoms. Dashed lines are only to show the positions of the face atoms.

Figure 5-7 The fcc unit cell, with atoms shown as full-sized spheres.

Exercise 5-2

How many atoms are in the bcc unit cell? How many in the fcc unit cell?

Solution.   From Fig. 5-4, for bcc, there is one center atom and 8 corner parts, each $\frac{1}{8}$ atom, so the total is $1 + 8 \times (\frac{1}{8}) = 2$ atoms.

From Fig. 5-7, for fcc, there are $8 \times (\frac{1}{8})$ corner atoms, and $6 \times (\frac{1}{2})$ face atoms, so the total is $8(\frac{1}{8}) + 6(\frac{1}{2}) = 4$ atoms. We would thus conclude that for about the same density of atoms, the fcc cube would have about twice the volume.

Exercise 5-3

Calculate a_0/r for fcc.

Solution.   From Fig. 5-7, the atoms touch along the face diagonal, which is thus $4r$ long. Thus $a_0 \sqrt{2} = 4r$, and so $a_0/r = 4/\sqrt{2} = 2.83$.

Hexagonal close-packed.   The third and last common crystal structure of metals is *hexagonal close-packed* (hcp), shown in Fig. 5-8. The atoms of the top and bottom planes in this sketch are centered on the outlines of hexagons and at the hexagon centers. Six spheres will perfectly surround and touch a seventh sphere if all are on a plane; these *close-packed* planes are called *basal planes*. Between the top and bottom planes lies an identical plane of atoms (also a basal plane), shifted so that the spherical atoms rest in the valleys between the spheres in the top and bottom planes. Figure 5-9 shows a

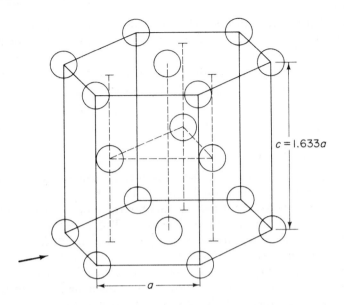

Figure 5-8 The hexagonal close-packed (hcp) unit cell, with small circles representing centers of atoms. Solid lines outline the unit cell, and dashed lines show the relative positions of atoms. The arrow indicates the viewing direction of Fig. 5-9.

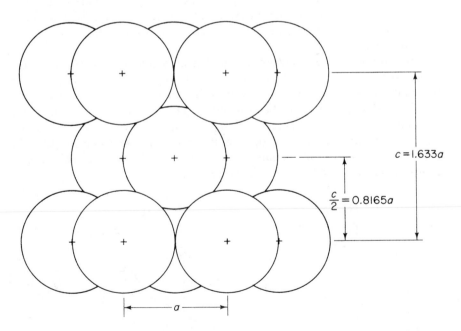

Figure 5-9 Hexagonal close-packed crystal, as viewed from the direction of the arrow in Fig. 5-8.

side view of the hcp structure. The ratio of the unit cell height c to the hexagonal side dimension a would be $c/a = 1.633$ if the spacing of the planes depended solely upon the size of the atoms as hard spheres. In the hcp metals c/a varies from the theoretical value of 1.633 by as much as 15%, so we may conclude that factors not incorporated into our model influence the spacing of the basal planes.

MILLER INDICES

Strength of crystalline materials (such as metals) can usually be related to events that take place in individual crystals. To help in describing the geometry of events on a crystalline scale, a spatial nomenclature called *Miller indices*, akin to vector notation, has been developed. The principles of vectors are incorporated in the use of Miller indices, but there are differences in application, such as the method of defining planes, and the symbols used are different from conventional vector notation. In order to understand strength-related processes such as slip, the reader will find it desirable to study and work with Miller notation until he can use it easily and quickly.

Directions in cubic crystals.   A direction vector in a cubic crystal *always starts at the origin* of the three axes that lie parallel to the three edges of the unit cubic cell. The point in space where the vector ends is given in *square brackets* by the coordinates $[xyz]$, where x, y, and z are the number of unit cell dimensions (a_0) along each respective axis. If the coordinate is negative, a bar is placed above it.

An example for the face-centered cubic (fcc) system is the $[110]$ direction. In Fig. 5-10, a single cell of an fcc metal is shown, with small circles to represent each atom. The $[110]$ direction is thus one unit cell dimension (a_0) in the x-direction and one a_0 in the y-direction. The vector always starts at the origin and here ends at the point $(x = 1a_0, y = 1a_0, z = 0)$. If we wish to talk about the same direction on the top face of the unit cell, we can also call it the $[110]$, for once the vector is defined, it can be moved anywhere in space, as long as it remains parallel to its original direction.

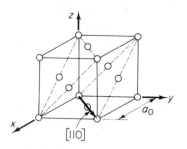

Figure 5-10   Directions in fcc.

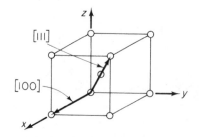

Figure 5-11   Directions in bcc.

Sometimes for clarity the unit cell dimension is given as well, for example, $a_0[110]$. This procedure then allows us to vary the length of the direction vector. For example, the vector from the center of the atom at the origin in Fig. 5-10 to the center of the atom at the center of the bottom face is thus $a_0/2[110]$; this is often stated as $1/2[110]$. The family of directions crystallographically equivalent to $[110]$ is designated as $\langle 110 \rangle$. This includes $[101]$, $[01\bar{1}]$, $[\bar{1}10]$, $[\bar{1}\bar{1}0]$, etc.

For the body-centered cubic system (bcc) the same rules apply. Figure 5-11 shows the $[111]$ direction and the $[100]$ direction.

Directions in hexagonal close-packed crystals. The hexagonal close-packed (hcp) system appears geometrically more complicated, and it has four indices, one of which is redundant. Here the first three coordinates (a_1, a_2, a_3) are at $120°$ to each other, and the fourth coordinate (c) is perpendicular to the plane of the three coordinates. Figure 5-12 shows the hcp system, with the intermediate (identical) layer of atoms positioned by dashed lines. Note that a_3 is positive toward the back. In order that directions be perpendicular to planes with the same indices (a fact that is probably not apparent at this stage), the directional indices must be selected so that the a_3 index is equal to the negative of the sum of the a_1 and a_2 indices. For example, the $[11\bar{2}0]$ direction is shown; this is one unit-cell (here atom-to-atom) dimension in the $+a_1$ direction, one unit-cell dimension in the $+a_2$ direction, 2 unit-cell dimensions in the $-a_3$ direction, and zero in the c direction. The vector connecting the origin atom with the first atom in the $-a_3$ direction is thus $1/3[11\bar{2}0]$. Selection of indices to describe a desired direction is not easy, because of the constraining rule that $a_1 + a_2 = -a_3$.

When the c-direction component is not zero, the directions get even more complicated; for example, the $1/3[\bar{2}203]$, displaced here from the origin, is shown in Fig. 5-12. It passes through the atom in the intermediate plane. (See Exercise 5-4.)

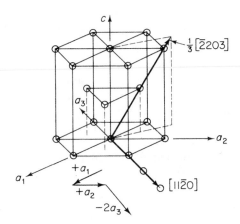

Figure 5-12 Directions in hcp.

Exercise 5-4

Draw enough unit cells of hcp to show that the tip of the vector $1/3[\bar{2}203]$ lies at the point shown in Fig. 5-12.

Solution.   The construction is given in Fig. 5-13. Start from the origin in the bottom left cell, move $-2a_1$, then $+2a_2$, zero a_3, then $+3c$. Connect the origin to the terminal point for vector $[\bar{2}203]$. This vector lies in the plane described by dashed lines parallel to the basal direction $[\bar{2}200]$. Thus, $1/3[\bar{2}203]$ ends one unit cell up from the starting point.

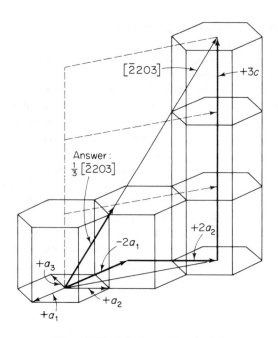

Figure 5-13   Solution to Exercise 5-4.

Planes in cubic crystals.   Planes are described by the reciprocal of their intersections with the perpendicular coordinate axes, x, y, z. To be described, a plane cannot pass through the origin. (Although this obscure procedure may seem to be a mad plot against the reader's sanity, this nomenclature is useful and mathematically the easiest to handle, once it is mastered.) As will be evident later, to be of significance in crystallography, planes must pass through significant numbers of atoms; this restriction greatly limits the number of interesting planes.

Planes are indicated by Miller indices in curved brackets (parentheses). Figure 5-14 shows the atoms in the (110) plane, shown shaded, in an fcc crystal, with most of the face atoms omitted here for clarity. The indices are developed as follows: x-intercept $= 1a_0$; y-intercept $= 1a_0$; z-intercept $= \infty$.

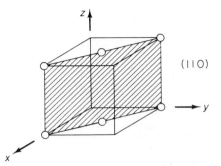

Figure 5-14 The (110) plane in fcc.

Thus the plane is $(\frac{1}{1}\frac{1}{1}\frac{1}{\infty})$, or (110). Indices are normally reduced to the smallest possible set of integers. The family of planes equivalent to (110) is denoted {110}. One memory aid to distinguish the indices for directions and planes is to recall that directions are straight lines, and Miller indices for directions are enclosed by square brackets, which are made up of straight lines; curved and curly brackets then apply to planes.

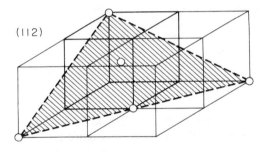

Figure 5-15 The (112) plane in bcc.

The same procedure applies to the bcc system. Figure 5-15 shows four unit bcc cells, which are needed in order to describe the intercepts for the (112) plane, with only one body atom shown, that for the cell that contains the origin. The atoms shown lie in this plane. Note that the intercepts are $x = 2a_0$, $y = 2a_0$, $z = 1a_0$, giving $(\frac{1}{2}\frac{1}{2}\frac{1}{1})$ or (112) to the smallest integer.

Planes in hexagonal close-packed crystals. The procedure for hcp planes is the same as for the cubic system, except that the redundant a_3 must be accommodated. Since numerical values of indices for a plane are the same as for the direction normal to the plane (which is why reciprocals are used to define a plane), the rule that $a_1 + a_2 = -a_3$ must also hold for planes. Figure 5-16 thus shows the (11$\bar{2}$0), for intercepts of $a_1 = 1$, $a_2 = 1$, $a_3 = -\frac{1}{2}$, $c = \infty$ to give $(\frac{1}{1}\frac{1}{1}\frac{1}{-1/2}\frac{1}{\infty})$ or (11$\bar{2}$0). Note that the a_3 index is fixed as soon as a_1 and a_2 are set.

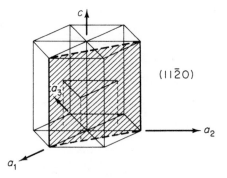

Figure 5-16   The $(11\bar{2}0)$ plane in hcp.

Manipulations and tricks with Miller indices.   For a cubic system of lattice constant a_0, the spacing $d_{(hkl)}$ between the set of planes defined by (hkl) can be found from

$$d_{(hkl)} = \frac{a_0}{\sqrt{h^2 + k^2 + l^2}}$$

It is essential that (hkl) include *all* atoms in the crystal.

Exercise 5-5

Calculate the spacing of the (110) in bcc.

Solution. $d_{(110)} = a_0/(1^2 + 1^2 + 0^2)^{1/2} = a_0/\sqrt{2}$. The set of planes of this spacing includes all the atoms in the crystal.

An example where the planes as defined by the lowest common integers do not include all atoms is the (110) in fcc. Figure 5-17 shows how this family of planes excludes, for example, the atoms marked with x that lie between the two planes of open circles. The above equation gives $d_{(110)} = a_0/\sqrt{2}$ for the spacing of planes, but this includes only half the atoms. To avoid this problem, the integer rule must be violated and the planes containing the complete set of atoms must be designated (220). This gives the correct spacing as $a_0/(2\sqrt{2})$.

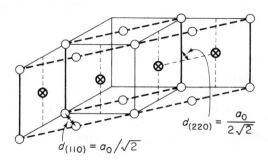

Figure 5-17   Spacing of (110) in fcc.

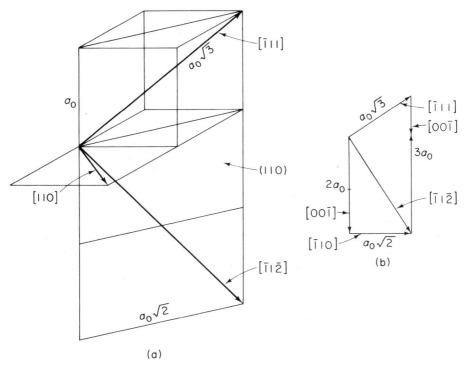

Figure 5-18   To find the direction in (110) that is perpendicular to $[\bar{1}11]$ in bcc.

A useful result of Miller indices is that the normal has the same indices as its plane. (Miller indices were chosen with this in mind.) Consider the problem shown in Fig. 5-18(a), where a bcc slip plane, the (110), contains a slip direction $[\bar{1}11]$. It is desired to find the direction lying in (110) that is perpendicular to $[\bar{1}11]$. This is solved as a geometry problem in Fig. 5-18(b), and the direction is found to be $[\bar{1}1\bar{2}]$. One must demonstrate that $[\bar{1}1\bar{2}]$ is perpendicular to $[\bar{1}11]$. Even for a simple case this is not easy. For more complex cases the geometry rapidly becomes very messy.

Using Miller indices, consider the same problem, restated in Fig. 5-19,

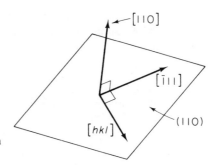

Figure 5-19   The same problem as in Fig. 5-18, solved using Miller indices.

without regard to the geometry of the cubic unit cell. (The reader who has not studied vector manipulation can skip the next two paragraphs.) The slip direction $[\bar{1}11]$ lies in (110). The problem is to find (hkl), lying in (110), normal to $[\bar{1}11]$. Construct the normal [110] to the plane (110). Then $[hkl] = [\bar{1}11] \times [110]$; for **x**, **y**, **z** as unit vectors of length a_0 in the three directions,

$$[hkl] = \begin{vmatrix} \mathbf{x} & \mathbf{y} & \mathbf{z} \\ \bar{1} & 1 & 1 \\ 1 & 1 & 0 \end{vmatrix}$$

$$[hkl] = \mathbf{x}(0 - 1) - \mathbf{y}(0 - 1) + \mathbf{z}(-1 - 1)$$

$$[hkl] = [\bar{1}1\bar{2}]$$

This quick solution also tells us that the plane containing $[\bar{1}11]$ that is normal to the (110) is the $(\bar{1}1\bar{2})$.

Miller indices also give us a quick test to determine whether a direction lies in a plane, since the dot product of this direction and the direction normal to the plane must be zero. Does a direction $[ghj]$ lie in a plane (def)? Yes, if $gd + he + jf = 0$. Thus, in Fig. 5-18, we see that $[\bar{1}1\bar{2}]$ does lie in the (110), since $-1 + 1 + 0 = 0$.

SPACING OF ATOMS

William H. Bragg and his son William L. Bragg first demonstrated quantitatively with X-rays that the atoms in a solid metal crystal exist in a regular, repeated pattern with remarkably constant spacing between atoms. Using X-rays of wavelength λ striking a metal crystal at an angle θ with a set of planes of atoms spaced d apart, they showed by simple geometry that the X-rays were diffracted at the angle θ to the planes if

$$\lambda = 2d \sin \theta \tag{5-1}$$

Diffraction will occur, in fact, if the left side of the equation is equal to any integral number of wavelengths, although the higher orders are not usually important. The significance here of the Braggs' work is that it shows the high precision and regularity of atomic spacing within a metal, thus confirming the hypothesis that atoms are found in a regular three-dimensional array, with each atom at a fixed mean position relative to each other atom, in a characteristic *crystal structure*.

Exercise 5-6

A single crystal of nickel (face-centered cubic, atomic radius = 124.3 pm) gives first-order diffraction with X-rays from a copper target (wavelength = 154 pm) at an angle $\theta = 26.0°$ (0.454 rad). Find which set of planes is producing this diffraction.

Solution. $n\lambda = 2d \sin \theta$; $d = n\lambda/(2 \sin \theta) = 1 \times 154$ pm/(2 sin 26°) = 176 pm. For Ni, $a_0\sqrt{2} = 4r$; $a_0 = 4 \times 124.3$ pm/$\sqrt{2}$ = 352 pm. Thus $d/a_0 = 176/352 = 0.5$, and diffraction would be from {200} planes. If this problem is still not clear, a review of simple crystal structures and Miller indices discussed earlier in this chapter may be in order.

If the crystal structure contains a relatively small number of instances where the atomic spacing is not uniform, diffraction will still occur. Thus X-ray diffraction cannot readily discriminate between crystals that are perfect and those that have a few instances of nonuniform spacing and are thus slightly imperfect. It will be shown later in this chapter that imperfections do exist. If the number of imperfections grows to a relatively high value in a crystal, the X-ray diffraction will be blurred by the irregularities in crystal spacing and orientation.

SLIP IN PERFECT CRYSTALS

Assume that a metal crystal is perfect. The yield strength of this crystal must be related to the way in which atoms move relative to each other to cause strain and to the magnitude of forces required to produce this motion. Observation of a large number of deformed crystals reveals that the most common motion is sliding of one plane of atoms relative to an adjacent plane of atoms. This is called *slip*.

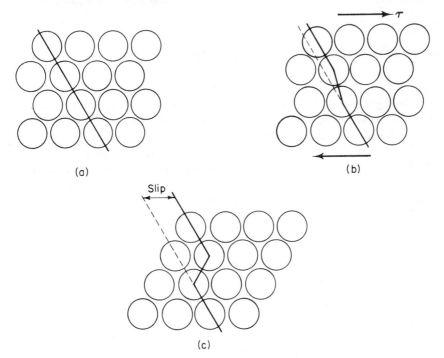

Figure 5-20 Schematic sketches of slip in a perfect crystal: (a) initial position; (b) an intermediate position, with shear stress τ; and (c) final position, no stress.

The essential aspects of the process of slip can be described with a two-dimensional sketch. Although some observed behavior is attributable to the three-dimensional geometry of a real crystal, the plane sketches in Fig. 5-20 show how slip might occur in a perfect crystal. Figure 5-20(a) shows the initial position of 16 atoms lying in a plane and having a uniform crystal structure. The diagonal line will be used to keep track of the relative positions of atoms.

A simple tensile or compressive stress acting on the crystal in any direction except vertical or horizontal will have some shear-stress component τ acting on the horizontal and vertical faces of the crystal; for simplicity the vertical stresses are not shown in Fig. 5-20(b). When τ becomes large enough, the interatomic bonds will be overcome on some plane of atoms, shown in Fig. 5-20(b) as the middle plane, and sliding will begin.

If τ is further increased, the upper eight atoms can move to the right to a new equilibrium position, shown in Fig. 5-20(c). The magnitude of this relative motion is the amount of *slip* during the complete process. Because the atoms are in new equilibrium positions, no stress is required to hold them there; thus the deformation, or slip, that has occurred represents permanent *plastic* deformation that remains after the stress is removed. The crystal in Fig. 5-20(c) is just as perfect as in Fig. 5-20(a), so it would not be possible to say that plastic deformation had occurred if one did not know the history of the crystal.

MAXIMUM THEORETICAL STRESS FOR SLIP

When the relative sliding of one layer of atoms over another has reached the midpoint between the two equilibrium positions, the shear stress is zero. By symmetry it can be seen that the upper two layers of atoms in Fig. 5-21, which represents such a condition, could fall with equal ease to the right or to the left. Thus the shear stress required to move the upper group of atoms to the right from the initial equilibrium position must reach a maximum before the middle position shown in Fig. 5-21. A plot of the variation of shear stress with sliding distance x from the initial equilibrium position might be roughly ap-

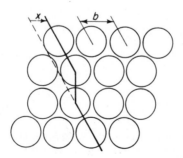

Figure 5-21 Unstable midpoint of slip process; $\tau = 0$ when $x = b/2$.

proximated by a sine function

$$\tau = \tau_{max} \sin \frac{2\pi x}{b} \qquad (5\text{-}2)$$

where τ_{max} is the maximum shear stress, b is the distance between equilibrium positions, and x is the relative slip between the two planes at the instant the applied shear stress is τ. Figure 5-21 thus shows the case where $x = b/2$.

If strength is related to the shear stress required to cause slip to occur, then the highest value of shear stress, τ_{max}, should give an indication of the strength of a perfect crystal. If τ_{max} is not reached, extensive slip will not occur. A rough approximate calculation of τ_{max} should therefore provide an order-of-magnitude estimate of the strength to be expected. This is done in the following paragraphs.

Exercise 5-7

At what value of x will $\tau = \tau_{max}$ in a perfect crystal?

Solution.   Since $\tau = \tau_{max} \sin(2\pi x/b)$ from Eq. (5-2), $\tau = \tau_{max}$ when $\sin(2\pi x/b) = 1$; thus $2\pi x/b = \pi/2$, and $x = b/4$. Other information suggests that the maximum stress occurs at a smaller value of x, since the stress does not follow an exact sine relationship, but this does not alter the order of magnitude of τ_{max}.

The slope of the τ-x curve can be calculated for small values of x and equated to the measured slope of the elastic stress–strain curve. This step can be taken because elastic behavior occurs at relatively small reversible displacements of atoms from the equilibrium position, a condition represented by the τ-x curve at $x \ll b$. From Eq. (5-2),

$$\frac{d\tau}{dx} = \tau_{max} \cdot \frac{2\pi}{b} \cdot \cos \frac{2\pi x}{b}$$

For $x \ll b$, $\cos(2\pi x/b) \approx 1$:

$$\frac{d\tau}{dx} = \tau_{max} \cdot \frac{2\pi}{b}$$

Elastic deformation resulting from shear stress can be represented by Fig. 5-22, where the dashed lines show the position of the upper two atoms after

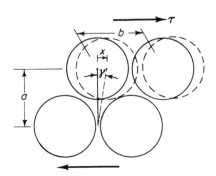

Figure 5-22   Elastic displacement under low stress τ.

the shear stress τ is applied. The resultant angular shear strain γ on the entire crystal can be measured, and γ is found to be directly related to the shear stress τ.

$$\frac{\tau}{\gamma} = G \tag{5-3}$$

G, the *shear modulus of elasticity*, is a constant for each elastic solid.

By geometry in Fig. 5-22, for $x \ll a$, $\gamma = x/a$. Combining with Eq. (5-3) gives $x/a = \tau/G$. In the straight-line elastic range, the slope is $d\tau/dx = \tau/x = G/a$. Assuming that $a \approx b$, as can be seen in Fig. 5-22, and combining with Eq. (5-2), we obtain

$$\tau_{max}\left(\frac{2\pi}{b}\right) = \frac{G}{a}$$

$$\tau_{max} \approx \frac{G}{2\pi} \tag{5-4}$$

In Chap. 14 relating to fracture, a similar type of calculation for *pulling* atoms apart (as opposed to *sliding* over one another as here) gives $E/2\pi$ (where E is the tensile modulus of elasticity) for the theoretical tensile fracture strength.

MEASURED STRESSES FOR SLIP

Theoretical values [from Eq. (5-4)] and average measured values of shear stress for slip for three metals of fairly high purity are given in the following table:

	Theoretical τ_{max}, MPa	Measured τ_{slip}, MPa
Zone-refined iron	14 000	35
99.996% aluminum	3 500	7
Electrolytic copper	7 000	17

The theoretical strengths are thus 400 or 500 times larger than the measured strengths of high-purity metals. Similar results are obtained for a wide range of metals.

Even with the inaccuracies in the assumptions involved in the derivation of Eq. (5-4), it is unlikely that a discrepancy of the order of 500 could be explained away. Therefore the basic premise, that these crystals are perfect and that slip occurs as described in Fig. 5-20, must be invalid. If slip occurs in some other way not involving simultaneous movement between two planes of atoms, then the crystal cannot remain perfect. This conclusion has far-reaching implications for the nature of strengthening mechanisms, as will be shown in the next section.

One exception to the relatively low strength of real metals has been observed. In some metal crystals in the shape of a very fine filament or whisker, stresses of the order 10 GPa have been measured. Since the theoretical maximum stress is of this same magnitude, we may conclude that these fine crystals either are perfect or contain imperfections that do not substantially influence strength. Applications of these very-high-strength filaments are discussed in Chap. 17.

DIRECT OBSERVATION OF IMPERFECTIONS

The tension test reveals a wide range of behavior among different metals and alloys. For example, at the same temperature, tensile strengths of different metals and alloys can differ by a factor of about 100, and there is a wider range for true strain at fracture, which can vary by a factor of about 1000. There is no consistent correlation between these two mechanical properties, although high fracture strain is more often associated with low strength than with high strength. Furthermore, a wide range of mechanical properties may be produced in a single alloy of constant chemical composition through heat treatment and mechanical working. The fundamental reasons behind these variations in properties must be understood if the selection, heat treatment, and properties of metals and alloys are not to appear as a mysterious and inconsistent sort of alchemy.

Since the assumption of perfectly regular crystal structure in metals leads to calculated stresses for slip that are several orders of magnitude larger than the measured stresses for slip, the search for strengthening mechanisms leads to a study of the nature of possible imperfections. These imperfections must provide a mechanism for permitting slip to occur at relatively low stresses. We now turn to the current evidence for plausible imperfections that may exist in crystals, with the objective of finding general rules governing strength.

It is possible to observe the shape and distribution of imperfections in many solids. The scale of observation necessary to view imperfections directly must be much finer than the wavelength of light, which is of the order 500 nm, since the interatomic spacing of most metals falls in the range 0.2 to 0.4 nm. For observation of imperfections on this size scale, an electron beam accelerated through, say, 100 kV is well suited, as it has a wavelength of 3.7 pm and it can penetrate a specimen of metal 10 to 100 nm thick. An electron microscope arrangement for observing atomic-scale defects in crystals is shown in Fig. 5-23. An electron beam traveling through a thin crystal specimen may find some defects oriented so as to cause local Bragg diffraction, when this condition occurs, the defects are effectively opaque to these electrons. The electrons that do not pass near any defect are not diffracted, and after enlargement

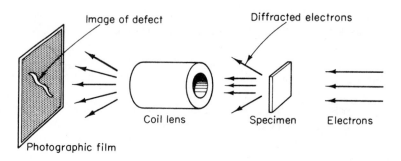

Figure 5-23   Schematic arrangement for thin-foil electron microscopy.

by the lens system the electrons darken all of the photographic film except in the region of the defect.

It is important to note that the study of thin foils by the method of Fig. 5-23 is not the same as a shadow picture. The conditions for Bragg diffraction are quite critical, and very few of the defects present in a crystal will be visible at any one time. Tilting the specimen to change its angle with the electron beam can cause the first defects to disappear and other defects to appear.

Figure 5-24 shows a typical electron micrograph of a thin specimen of stainless steel. The dark lines are imperfections in the crystal; these imperfections appear as light lines in the negative that is exposed to the electron beam, and the reversal occurs in the print made from the negative. A series of such pictures made of a single thin specimen, but in different regions and with different orientation to the electron beam, will reveal similar lines throughout the specimen. If the specimen is stressed so that slip occurs, the pattern of lines changes. Thus on two counts imperfections must be associated with slip: (1) the observed imperfection configuration is altered with slip, and (2) the strength of a real metal is much lower than the calculated strength of a perfect crystal.

Figure 5-24   Dislocations in an electron micrograph of a thin foil of 18% chromium–8% nickel stainless steel. Magnification is given by the 1-μm scale marking, which represents the size a 1-μm dimension would have if enlarged the same amount as the specimen. (Photograph courtesy of P. D. Goodell.)

Exercise 5-8

What two characteristics must imperfections have in order to explain the behavior of real metals?

Solution. These imperfections must (1) provide a mechanism by which slip can occur, and (2) be capable of reducing the strength by several orders of magnitude below the strength of a perfect crystal. Many kinds of imperfections could be imagined that would not provide both of these features, so the number of kinds of imperfections of interest here is reduced substantially.

Other experimental methods have been applied successfully to the observation of imperfection patterns. These include etching (accelerated corrosion) of a polished surface to reveal pits where imperfections intersect the surface; accumulation of opaque foreign atoms in the locally distorted region surrounding an imperfection in a transparent crystalline solid; and formation of an image by emission of ions from a very small spherical surface, which reveals the local surface configuration of atoms. Although these methods reveal some additional information, they will not be further discussed here. The important result of the several independent methods of imperfection observation is that they agree on the existence of the imperfections described in the next section.

THE EDGE DISLOCATION

Figure 5-25 shows a plane of atoms of a simple cubic crystal arranged in a perfect matrix. (The cubic matrix is used here instead of the 60° matrix in Fig. 5-20 because the cubic structure is simpler to view. No metals have this simple cubic structure, but the argument that follows applies equally well to the more complex crystal structures.) Note in Fig. 5-25 that all atoms, here represented by small circles, follow a regular, repeated pattern, the necessary requirement for a perfect crystal. The lines between circles represent the principal bonds between each atom and its neighbors. In three dimensions, lines would also be

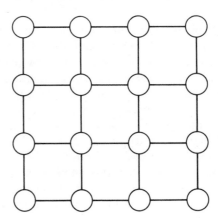

Figure 5-25 Plane of a simple cubic crystal without imperfections.

drawn normal to the plane of this figure to connect these atoms with atoms lying in the adjacent parallel planes.

Exercise 5-9

Would the next parallel plane of atoms lying in back of those in Fig. 5-25 have atoms (a) directly in back of the atoms in the sketch, (b) directly in back of the squares in the sketch, or (c) elsewhere?

Solution. (a) Directly in back of the atoms in the sketch. If the next atoms were behind the squares, this would be a body-centered cubic structure. For the simple cubic case, then, a line of atoms in the two-dimensional sketch represents a whole plane of atoms perpendicular to the plane of the sketch and lying in back of (and in front of) the line. The equivalent three-dimensional sketch is shown in Fig. 5-26.

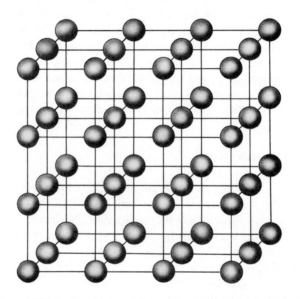

Figure 5-26   Perfect simple cubic crystal, in three-dimensional sketch.

A plane of atoms containing an imperfection, called a *dislocation*, is shown in Fig. 5-27. This sketch differs from Fig. 5-25 in that two extra atoms are present. These represent an extra plane of atoms that ends within the crystal. Figure 5-28 shows this configuration in three dimensions. In Fig. 5-28 a line along the lower edge of the plane that ends within the crystal defines the region of atomic discontinuity; such a line describes an *edge dislocation*. The interatomic spacing cannot be exactly the same both above and below the dislocation line, since in Fig. 5-28 there are four spacings across the top and three spacings across the bottom. (Real dislocations distort a crystal over a much wider region than this simple sketch shows.) Thus an electron beam may be diffracted in the region of the dislocation line without being diffracted in the bulk of the crystal, and so the dislocation is visible on the photographic film

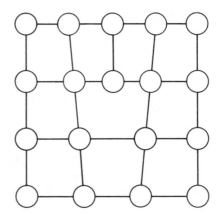

Figure 5-27 Plane of a simple cubic crystal that contains an edge dislocation.

because some electrons did not travel in a straight path through the aperture (see Fig. 5-23).

The edge dislocation was first hypothesized in 1934 by Orowan, Polanyi, and Taylor, each working independently, to explain the observed low strength of real crystals. This occurred long before the development of the electron microscope; since other techniques for dislocation observation had not yet been developed, many years were to pass before the first direct observation of dislocations and confirmation of their existence.

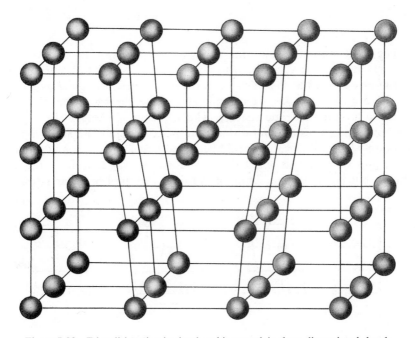

Figure 5-28 Edge dislocation in simple cubic crystal, in three-dimensional sketch.

SLIP BY DISLOCATION MOTION

If the edge dislocation is to qualify as an imperfection of interest here, it must provide a mechanism for slip that differs from the way slip occurs in a perfect crystal. The edge dislocation is the line where the extra plane of atoms ends. If this extra half-plane is able to shift slightly so as to form interatomic bonds with a single half-plane, rather than being situated between two planes, then the dislocation line is moved laterally to a new extra half-plane. Figure 5-29 shows one plane of an edge dislocation, with atom A initially representing the bottom row of atoms in the extra plane. By a very slight motion of the upper four atoms to the right, the bond between atoms B and C can be broken and replaced by a bond between A and C, shown dashed in Fig. 5-29. Atom B would then be at the lower end of the extra plane, and the edge dislocation would have shifted one interatomic spacing to the right. For the full edge-dislocation *line* to move to the right, all the bonds between the atoms under B and C must likewise be broken and replaced by bonds between the atoms under A and C.

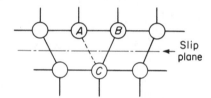

Figure 5-29   Motion of edge dislocation.

The total motion of the four upper atoms in Fig. 5-29 relative to the three lower atoms is very small during this process. If slip is to occur, that is, if all the atoms above the *slip plane* (a horizontal plane through the middle of Fig. 5-29) are to move one interatomic spacing to the right, the dislocation must move further than one spacing. The sequence of sketches in Fig. 5-30 shows how slip can result from dislocation motion. Figure 5-30(a) is a perfect crystal into which a dislocation could be forced by an indentation at the upper left edge (not shown). This is not the most common way that dislocations are produced, but it serves as a simple illustration. An applied shear stress then

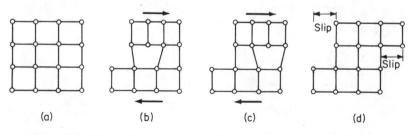

Figure 5-30   Slip by full traverse of edge dislocation moving from left to right.

causes the dislocation to move by shifting of interatomic bonds [Fig. 5-30(b) and (c)], until the extra plane emerges from the crystal and appears as a step on the right edge [Fig. 5-30(d)]. Note that while the dislocation moves the full width of the crystal, each of the upper atoms moves only one interatomic space to the right. In this case slip can be measured at the left or right edge and is equal to one interatomic spacing in the direction parallel to the direction in which the dislocation moves. To produce a slip step large enough to be observed by an optical microscope, many dislocations must move along or near the same slip plane.

Exercise 5-10

Slip in nickel commonly occurs in a ⟨110⟩ direction, as is generally true of fcc (face-centered cubic) metals. If the atomic radius of Ni is 124.3 pm, how many edge dislocations must emerge from a crystal of Ni to produce a step that is just visible in the optical microscope, say, a step 1 μm high?

Solution.   Interatomic spacing in ⟨110⟩ in fcc is $2r = 248.6$ pm. 1 μm/248.6 pm = 4020 dislocations.

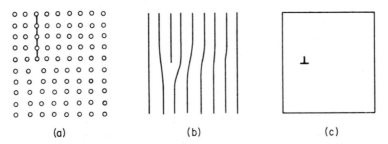

Figure 5-31   Three ways to represent the same edge dislocation.

   A simple symbol for the edge dislocation can be introduced here, to replace the cumbersome sketches of individual atoms. Figure 5-31(a) shows an edge dislocation of the form already discussed, but it shows lines connecting atoms only in the extra plane. In Fig. 5-31(b) the individual atoms have been replaced by lines representing planes perpendicular to the sketch. The symbol used in Fig. 5-31(c) shows the location of the edge dislocation by its position, the direction of the extra plane by the single vertical leg of the symbol, and the horizontal slip direction of the dislocation by the horizontal line of the symbol. The edge-dislocation symbol can be used in any orientation in a plane sketch.

STRAIN FROM DISLOCATION MOTION

It is possible to calculate the shear strain that results from the motion of one or more edge dislocations moving on parallel planes in a single crystal. Figure 5-32(a) shows the shear strain γ resulting when one dislocation moves all the

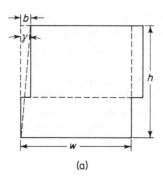 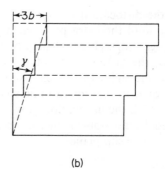

(a)   (b)

Figure 5-32 Shear strain associated with traverse of a crystal by a dislocation: (a) one dislocation; (b) three parallel dislocations.

way across the crystal. If γ is always measured by a line from the atom in the lower left corner to the one in the upper left corner, then no matter where horizontal slip occurs, the shear strain γ will be the same for the same total slip. For Fig. 5-32(a), $\gamma = b/h$, where b is the slip caused by the one dislocation and h is the height of the crystal. If n dislocations traverse the crystal, the total slip will be nb, and if $nb \ll h$, $\gamma = nb/h$. These n dislocations need not occur on the same plane, as each parallel slip displacement b will move the atom in the upper left corner a distance b to the right. Figure 5-32(b) shows the result of complete traverse by three dislocations; here $\gamma = 3b/h$.

     In real crystals the number of dislocations involved in a simple slip process is very large, as the calculation for Exercise 5-10 demonstrates. Thus the average motion on a gross scale resulting from many small dislocation motions will be measured as shear strain. Consider Fig. 5-33, which shows the influence on the atom in the upper left corner (any point on the upper surface will serve, in fact) of motion of an edge dislocation under a shear stress from position A to position B, where the distance moved is $w/2$. In the vicinity of the dislocation, individual atom motion will range from zero to b, but at a

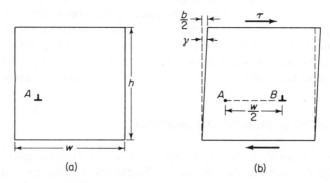

Figure 5-33 Limited motion of a dislocation from position A to position B.

large distance from the slip plane, the average displacement during this process must be $b/2$. Thus for this case, $\gamma = b/2h$, or for the general case of n parallel edge dislocations of magnitude b moving an average glide distance g in a crystal of height h and width w, the shear strain γ is

$$\gamma = \left(\frac{nb}{h}\right)\left(\frac{g}{w}\right) \qquad (5\text{-}5)$$

Such highly simplified parallel-dislocation motion seldom occurs in real metals, so Eq. (5-5) must usually be further modified to accommodate complexities of geometry and dislocation configuration.

Exercise 5-11

A crystal face 15 mm high and 20 mm wide intersects 10^8 parallel edge dislocations on parallel slip planes. Interatomic distance in the direction of dislocation motion is 0.30 nm in this crystal. How far must each dislocation move, on the average, to give a shear strain of 0.001?

Solution. Equation (5-5) gives $\gamma = nbg/hw$; $g = \gamma hw/nb$, $g = (0.001 \times 15 \text{ mm} \times 20 \text{ mm})/(10^8 \times 0.30 \text{ nm}) = 10 \ \mu\text{m}$. In the real case it is more likely that new dislocations will be generated and some of the existing dislocations will not move, but the effect is the same.

STRESS FOR DISLOCATION MOTION

The second requirement for a dislocation is that, in addition to providing a means for producing slip and shear strain, a dislocation must lower the strength of a perfect crystal by several orders of magnitude.

The maximum shear stress required for slip results from the summation of the forces required at any one time to break all atomic bonds involved in the slip process. Consider, for example, a perfect crystal cube with 30-mm edges. Each edge will have about 10^8 atoms, and a slip plane would thus have 10^{16} atoms. Dividing the total force by 10^{16} should give an estimate of the force f required to break one interatomic bond. For a cross-sectional area of roughly 1000 mm^2,

$$f \approx \frac{(\tau_{\text{max}})(1000 \text{ mm}^2)}{10^{16}} = 10^{-19} \ \tau_{\text{max}} \text{ m}^2$$

τ_{max} is the maximum theoretical shear stress in a perfect crystal, which was calculated earlier, in Eq. (5-4).

If slip occurs by dislocation motion, then the total shear force required will be f times the number of interatomic bonds being broken at one time. It is possible to calculate the maximum number of atoms that can be involved in a dislocation at any one instant in order to account for the observed strength-reduction factor of about 500. The requirement for this is that the average shear force (stress τ acting on an area of 1000 mm^2) that will move the dislocation must be $\frac{1}{500}$ times the theoretical shear force for a perfect crystal:

$$\tau(1000 \text{ mm}^2) = \frac{1}{500} (\tau_{max})(1000 \text{ mm}^2)$$

But the average shear force can be considered as acting only to move the dislocation, and the force f required to break the bonds in a dislocation has been estimated above. The total force on the dislocation must then be f times the number of atoms involved in the dislocation, N_{dis}:

$$\tau(1000 \text{ mm}^2) = f N_{dis}$$

$$\frac{\tau_{max}}{500} (1000 \text{ mm}^2) = f N_{dis}$$

$$2 \tau_{max} \text{ mm}^2 = (10^{-19} \tau_{max} \text{ m}^2)(N_{dis})$$

$$N_{dis} = 2 \times 10^{13}$$

Since the length of each dislocation in the simple case of a 30-mm cube contains about 10^8 atoms, the width of the region involving breaking of interatomic bonds at one time would be less than or equal to $2 \times 10^{13}/10^8 = 2 \times 10^5$ atomic spacings, or about $(2 \times 10^5)(0.3 \text{ nm}) = 60$ μm wide. Dislocations appear to be much narrower in influence than 60 μm, so the maximum width requirement is amply met.

Thus we see that the essential reason for the low stress for slip is that the dislocation is narrow compared to the width of the slip plane, so at any one time only a relatively small number of the atomic bonds across the slip plane are being broken. A number of analogies have been noted that exhibit this effect. For example, an earthworm moves by first extending its head (front end) so as to create a short tensile extension, or dislocation, just behind its head. This dislocation then travels backwards the length of its body and eventually causes the tail to move forward the same distance that the head moved at the outset, with a net forward motion of the whole worm. Another analogy is in the local vertical displacement, or hump, of a carpet that can be easily moved from one edge of the carpet to the other so as to shift the whole carpet slightly. [Note that the net motion of the worm was opposite to the direction of motion of the extensional dislocation, but the net motion of the carpet is in the same direction as the motion of the compressive (actually in the third dimension) dislocation.] These two analogies exhibit one feature in common with dislocations in crystals: The forces resisting motion are being overcome over only a relatively small fraction of their total area at any one time, so the maximum force, or stress, is much less than would be required in the absence of a dislocation.

The conclusion reached through the foregoing arguments is that dislocations can provide a mechanism for slip at stresses much lower than the theoretical stress required for a perfect crystal, and that the observed widths of dislocations are consistent with the magnitude of strength reduction required.

The essential mechanism for reduction of strength by a dislocation is in the breaking of a few interatomic bonds at one time rather than breaking all bonds at once, as would be required in the perfect crystal.

DISLOCATION DENSITY IN REAL SOLIDS

A common way to measure dislocation density is to cut a crystal and count the number of dislocations that intersect the cut face. The *areal density* ρ is then the number of dislocations divided by the area. Using the dimensions as defined in Fig. 5-33(a), and assuming that n is the total number of dislocations present, we have

$$\rho = \frac{n}{hw} \tag{5-6}$$

Measurements of real metals give dislocation densities of 10^{10} to $10^{11}/m^2$ for annealed metals and of the order of $10^{16}/m^2$ for severely cold-worked metals. The causes of increase in dislocation density and the resulting changes in strength are discussed in Chap. 6.

STRENGTH AND DISLOCATION MOTION

For those materials in which slip is significant, strength is the stress required to produce a certain strain. As it is a quantity which is easily determined, a commonly used value for strength is nominal stress at maximum load. A perhaps more important measure of strength is the yield stress, the stress to produce the onset of plastic deformation; lower stresses produce merely reversible elastic deformation. Sometimes the departure from elastic to irreversible plastic deformation is difficult to identify, in which case the strain for the determination of yield stress is assigned an arbitrary value such as 0.002 (see Chap. 2). Whichever value of strain is used for specifying strength—the strain for ultimate or the strain for initial plastic deformation—comparison of strengths for two different solids should be made at the same strain. The following discussion will show how such a comparison leads to a correlation between strength and dislocation glide distance.

The general relationship between normal and shear strain follows a Mohr's circle form of representation (see Chap. 2), so the maximum shear strain can be used as a general measure of strain. Equation (5-5) can be simplified by Eq. (5-6) to give

$$\gamma = \rho b g \tag{5-7}$$

where γ is the plastic shear strain, ρ the density of *moving* dislocations, b the local slip resulting from movement of the parallel edge dislocations, and g

their average glide distance. Equation (5-7) is not of general quantitative application because of the specialized geometric restrictions in its derivation, but qualitatively it shows that the plastic shear strain is directly related to the density of active dislocations and the average glide distance. The value of b is normally a constant for a given solid and is called the *Burgers vector* of a dislocation. The Burgers vector is thus the slip that would result from motion of a dislocation across the full width of a crystal, as shown in Fig. 5-32(a). Numerically, minimum and maximum values of b differ by a factor of about 2 for all common metals. For applications covered in this book, b is equal to twice the atomic radius.

Geometric complexities of dislocations also make the behavior of real solids more difficult to predict than the above case would indicate. If slip in a metal crystal involved only slip of parallel dislocations completely traversing the crystal, all specimens of this metal would have the same strength. Tests of single crystals substantially fulfilling these requirements do in fact show nearly constant low stress over moderate ranges of strain. But, in fact, dislocations usually move on more than one plane, do not remain in straight lines, and interact strongly with each other; so the average dislocation glide distance can be greatly altered by the structure of the solid. Furthermore, the geometry of real dislocations is seldom as simple as for the special case of the edge dislocation. If an edge-dislocation line turns 90° in the slip plane within a crystal, the discontinuity that results has the form of the axis of a spiral staircase, where it is impossible to match up planes; this is called a *screw dislocation* and is discussed in the next section. Most of the strength-related properties of edge dislocations are valid for screw dislocations and for dislocations that are part edge and part screw.

Dislocations often occur as loops that grow in size rather than as straight lines. (Loops are made possible by the presence of the screw dislocation.) The shear strain that results from the motion of straight parallel dislocations [Eq. (5-5)] may be further modified by the ratio of the length t of the loop perpendicular to the glide distance g, in a crystal of depth d.

$$\gamma = \left(\frac{nb}{h}\right)\left(\frac{g}{w}\right)\left(\frac{t}{d}\right)$$

This situation is shown in Fig. 5-34. As the dislocation has here moved a distance g, and thus has swept out the shaded area gt, the above equation is geometrically equivalent to the strain from a dislocation loop that has grown from zero area to the present area gt. (Since it is geometrically impossible for a dislocation to end within a crystal, only a dislocation loop can in fact exist entirely within a crystal.) If the swept area gt of one dislocation is replaced by A_{swept} and the denominator hwd is replaced by the crystal volume V, the shear strain becomes

$$\gamma = \frac{nbA_{swept}}{V} \tag{5-8}$$

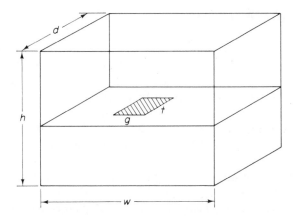

Figure 5-34 Rectangular dislocation loop of swept area gt in a crystal of volume hwd. (Real loops are more nearly circular.)

This equation is valid for the real case of a very large number of nearly circular dislocation loops that grow from point sources in crystals.

To obtain an estimate of the number of dislocations involved in any slip process of a real polycrystalline metal, some modification of the assumption of parallel slip planes must be made to suit the individual case.

Equation (5-7) links strain with active dislocation density and glide distance. To show how strength, or stress at a given strain, is related to dislocation behavior, compare two crystals identical in active dislocation density but with crystal A having a greater average dislocation glide length than crystal B. Assume further that the second dislocation in each crystal moves at a stress slightly higher than the stress for the dislocation that moves first, and so on. At each stress level a certain number of dislocations will have moved. Thus, if some minimum stress τ_y is required for motion of the first dislocation in each crystal, a plot of τ vs. γ for crystals A and B would look like Fig. 5-35. Consider some stress $\tau_1 > \tau_y$. If the same number of dislocations in crystals A and B have moved, then from Eq. (5-7), $\gamma_{A1} > \gamma_{B1}$, since $g_A > g_B$. For comparison of strengths at the same strain, Fig. 5-35 shows that for any fixed strain, say γ_{B1}, $\tau_B > \tau_A$. This leads to the conclusion that strength can be increased by a reduction in average dislocation glide distance.

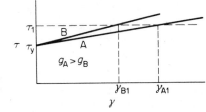

Figure 5-35 Shear stress vs. shear strain, where crystal A has greater average dislocation glide length than crystal B.

THE SCREW DISLOCATION

The edge dislocation must, by definition, be a straight line along the edge of an atomic-scale discontinuity, which is an extra plane of atoms. In a similar way we can describe a *screw dislocation*, for it is also a straight line along an atomic discontinuity, but the discontinuity is the center of a helix, or spiral, of atoms that are not properly matched. Figure 5-36 shows the essential geometry. Note that if we follow a single plane of atoms in a path around the screw dislocation line, we will then be on a higher or lower plane than at the start. The closest physical reality to this geometry is found in an automobile parking structure that consists of a continuous flat spiral ramp of the general shape shown in Fig. 5-36(a). Just as there can be no atoms at the exact center of a screw dislocation, the spiral ramp must have a hole or step along its axis.

The manner by which the screw dislocation permits slip at low average

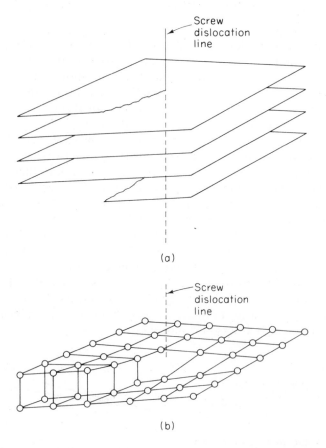

(a)

(b)

Figure 5-36 A screw dislocation in a simple cubic crystal, (a) showing only the atomic planes, and (b) showing some of the atoms in the top two planes.

stress is shown in Fig. 5-37. The presence of the dislocation permits the bond between atoms A and B to be broken; when this happens all along the dislocation line, the dislocation moves into the position shown in Fig. 5-37(b). The *Burgers vector b* for this screw dislocation is thus the slip that results from the motion of the dislocation, which can be seen to be the height of the shaded step in Fig. 5-37. The Burgers vector b is shown here with arrowheads on both ends, as we have here not assigned any sign convention to slip. For more complicated situations it would be necessary to have a system of sign conventions, including b.

The progress of a screw dislocation across a crystal is shown in Fig. 5-38. As with the edge dislocation, slip is always in the direction of the shear stress, but the motion of the screw dislocation is *perpendicular* to the shear stress and

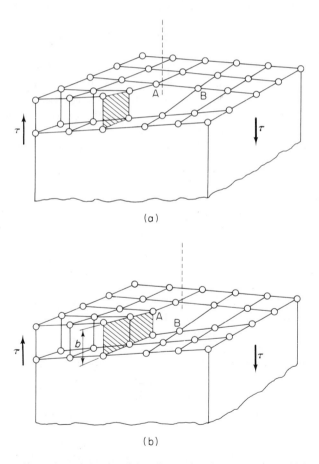

(a)

(b)

Figure 5-37 Motion of a screw dislocation under shear stress. (a) Initial position, with dislocation indicated by dashed line. (b) When right side moves down and left side moves up under stress, dislocation has moved one interatomic spacing. Note the new positions of atoms A and B.

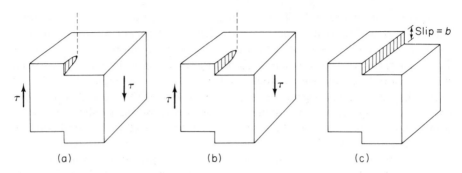

Figure 5-38 Slip caused by motion of a screw dislocation. (a) Dislocation line shown dashed, near front edge. (b) Dislocation has moved to center of crystal. (c) Dislocation has emerged from the far side of the crystal, and permanent slip results.

slip. This characteristic makes the geometry of the screw dislocation more difficult to follow.

Screw dislocations have some properties that are different from edge dislocations. The most obvious is that the slip direction is *parallel* to the screw dislocation line but *perpendicular* to the edge dislocation line [see Fig. 5-30(d)]. The slip plane must contain both the dislocation line and the Burgers vector, so an edge dislocation is always restricted to one plane. But any plane that cuts a screw dislocation will also contain the Burgers vector and is thus a potential slip plane; however, crystallographic considerations limit the number of possible slip planes for a given screw dislocation to a maximum of three.

MIXED DISLOCATIONS AND FORCES

Real dislocations are usually curved; thus they are a mixture of edge and screw dislocations and vary from point to point depending on the orientation of the dislocation line segment with respect to the Burgers vector. Figure 5-39 shows such a dislocation, lying in a horizontal slip plane. The vertical lines along the right face indicate atomic planes and illustrate the edge dislocation. As the stress τ is applied, point S will move to the left, point E will move parallel to τ toward the back of the crystal, and point M will be a mixture of the two motions. Thus the dislocation arc will expand and eventually cause slip in the entire crystal [Fig. 5-39(b)]. At this final stage it is impossible to tell what kind of dislocation caused the slip; in fact, if the end steps are cut off, it is impossible to know even that slip has occurred, since the crystal is perfect in Fig. 5-39(b).

Screw dislocations of the same sign (e.g., if both are right-handed, as in Fig. 5-36) will repel each other when they are parallel and on the same slip plane, with the repulsive force varying with $1/x$, where x is the distance between them. Screw dislocations of opposite sign will attract each other and if

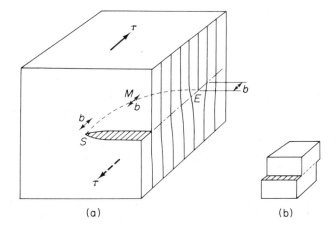

(a)   (b)

Figure 5-39 (a) Dislocation line SME, shown dashed, varies from screw (S), where it is parallel to b, to mixed (M), where it is neither parallel nor perpendicular to b, to edge (E), where the dislocation line is perpendicular to b. (b) A smaller view of this crystal after the dislocation has run fully through it.

allowed to come fully together will annihilate each other. Both statements also hold for interactions between edge dislocations. But a screw dislocation and an edge dislocation parallel to it have no interaction, as their Burgers vectors are mutually perpendicular. Two mixed dislocations interact to the extent that their screw components interact and their edge components interact.

In general, dislocations tend to repel each other, so in addition to permitting slip at low stresses, dislocations are self-inhibiting. We will see in Chap. 6 how this behavior is in itself an important strengthening mechanism.

CRYSTALLOGRAPHIC CONSTRAINTS TO DISLOCATION MOTION IN METALS

It is observed that dislocations usually move on those planes that are the ones most widely spaced apart: the {111} in fcc, the {110} in bcc, and the {0001} in hcp. The most widely spaced planes must thus also be the most densely packed planes, for the average number of atoms in any volume must be constant. The {111} in fcc and the {0001} in hcp are of maximum packing density, called *close-packed*, and both have an identical hexagonal matrix. Figure 5-40 shows the (111) in fcc, and Fig. 5-8 shows the hexagonal matrix of the basal planes of hcp.

The planes in bcc of maximum packing density are the {110}, but because the atoms in these planes are not in the most closely packed configuration, which is the hexagonal pattern of the {111} in fcc and {0001} in hcp, the {110} planes in bcc do not achieve the maximum theoretical packing density.

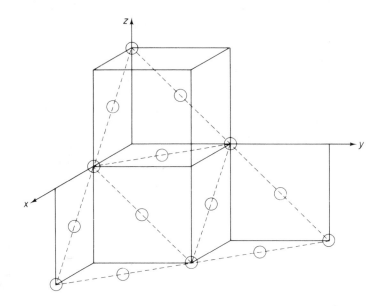

Figure 5-40 Atoms on the (111) in fcc. If the atoms were drawn full-sized, they would be touching along ⟨110⟩.

Figure 5-41 shows these differences. No other planes in bcc are even as densely packed at the {110}. Thus at least in part because of the absence of a plane of maximum theoretical density, slip in bcc iron, for example, occurs on {110}, {112}, and {123}. These planes are shown in Fig. 5-41(c)–(e).

Exercise 5-12

Calculate the ratio of spacing d of planes to lattice constant a_0 for each of the planes in Fig. 5-41.

Solution.

(a) For (111) in fcc: Use $d_{(hkl)} = a_0/(h^2 + k^2 + l^2)^{1/2}$ or

$$\frac{d}{a_0} = 1/(h^2 + k^2 + l^2)^{1/2} = 1/(1 + 1 + 1)^{1/2}$$

$$= 1/\sqrt{3} = 0.577$$

(b) For (0001) in hcp: Obviously, the spacing of planes is the same (theoretically) as that of (111) in fcc, except that a for hcp is twice the atomic radius r. Thus d/a for hcp is equal to d/a_0 for fcc, corrected for the ratio a_0/a. In fcc, $a_0\sqrt{2} = 4r$, so that $a_0 = 4r/\sqrt{2}$. Thus

$$\frac{a_0}{a} = \frac{4r/\sqrt{2}}{2r} = \sqrt{2}$$

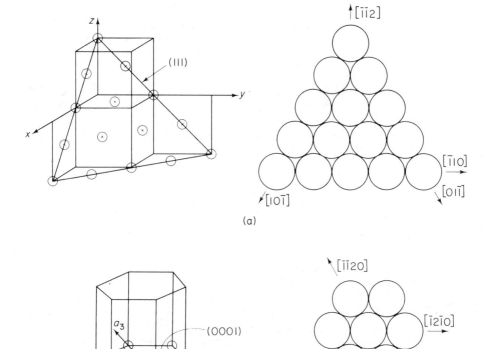

Figure 5-41 The common slip planes in metals: (a) (111) in fcc; (b) (0001) in hcp, (c) (101) in bcc; (d) (112) in bcc; (e) (213) in bcc.

and so

$$\frac{d}{a} = \frac{d\sqrt{2}}{a_0} = \sqrt{\frac{2}{3}} = 0.8165$$

(Note that this spacing is exactly half the theoretical c/a spacing of 1.633 given earlier.)

(c) For (101) in bcc:

$$\frac{d}{a_0} = \frac{1}{(1 + 0 + 1)^{1/2}} = \frac{1}{\sqrt{2}} = 0.707$$

(d) For (112) in bcc:

$$\frac{d}{a_0} = \frac{1}{(1 + 1 + 4)^{1/2}} = \frac{1}{\sqrt{6}} = 0.408$$

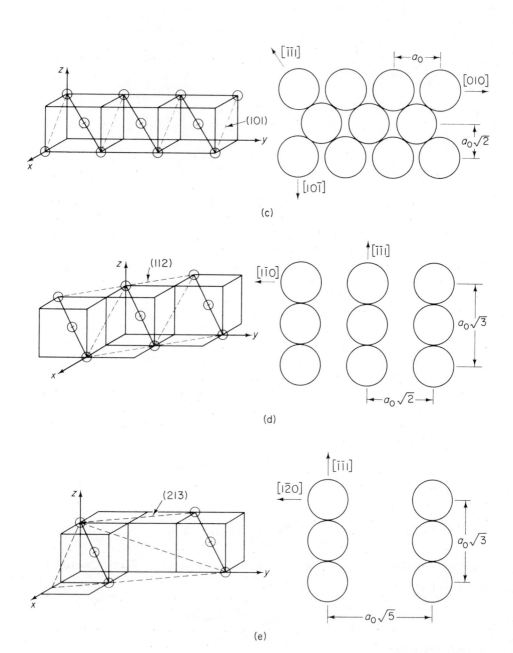

Figure 5-41 (cont.)

(e) For (213) in bcc:

$$\frac{d}{a_0} = \frac{1}{(4 + 1 + 9)^{1/2}} = \frac{1}{\sqrt{14}} = 0.267$$

It is surprising that slip occurs between such closely spaced planes as the (213), compared with the (101), which are 2.65 times as widely spaced. The reader may correctly conclude that our hard-sphere model of atoms is too simple to explain this observation.

Metals slip in the direction of closest atom-to-atom distance, which is the direction along which they touch in Fig. 5-41. Since this slip occurs by dislocation motion, the slip that occurs has the magnitude of the Burgers vector b, which is thus the distance from the center of one atom to the center of an adjacent atom that it touches. Thus $b = 2r$, as has already been noted. The detailed mechanism by which the dislocation moves is more complex than we have described here, since the dislocation often breaks up into *partial* or incomplete components. But since the net effect on slip is the same, we will consider throughout this text that $b = 2r$, without regard to the exact mechanism.

The slip directions are thus $\langle 110 \rangle$ for fcc, $\langle 11\bar{2}0 \rangle$ for hcp, and $\langle 111 \rangle$ for bcc. A *slip system* consists of a slip plane and direction. Some of the possible slip directions are shown in Fig. 5-41; thus, (111)[$\bar{1}10$] in Fig. 5-41(a) describes a slip system. Since $\langle 111 \rangle$ is the only possible slip direction in bcc, the three different slip planes for bcc must contain a $\langle 111 \rangle$. In iron (bcc), slip often leaves wavy traces, resulting from the shifting of screw dislocations from one of the three slip planes to another, and back, while maintaining the same Burgers vector. For example, the cases shown in Fig. 5-41(c)–(e) all have $\mathbf{b} = (a_0/2)[\bar{1}\bar{1}1]$, which can describe three possible slip systems having the same Burgers vector.

Slip in fcc metals is usually much easier than in hcp metals. Examples of fcc metals are Al, Ni, Cu, Au, Ag, Pb; hcp metals are Mg, Ti, Zn. This behavior can be explained in part by the uniform distribution in space of possible slip systems in fcc, in contrast with the limitation of common slip systems in hcp to the basal plane, (0001). In fcc each $\{111\}$ has three possible $\langle 110 \rangle$ lying on it; there are four possible $\{111\}$, so fcc has 12 possible slip systems. (Note that directions and planes that are the full negatives of other directions and planes are not really different. The ($\bar{1}\bar{1}\bar{1}$) is parallel to the (111), and the [$1\bar{1}0$] is the same slip direction, but opposite in sign, to the [$\bar{1}10$].) In contrast, hcp has only three common slip systems; however, when constrained by the applied shear stress or by the presence of alloying elements, hcp metals will slip on planes other than the basal plane, but the Burgers vector does not change.

Slip direction, which is always parallel to the Burgers vector, is usually stated as, for example in Fig. 5-41(a), [$\bar{1}10$]. Strictly speaking, if $b = 2r$, this value is twice too large, for [$\bar{1}10$] will reach from one corner of the unit cell across the face to the other corner. Thus we say that $\mathbf{b} = (a_0/2)[\bar{1}10]$ to be precise; or often just $\mathbf{b} = (1/2)[\bar{1}10]$, since it is clear that the basic unit cell dimension is a_0; or even more simply, $\mathbf{b} = [\bar{1}10]$, which gives direction only, the $a_0/2$ being understood.

RESOLVED SHEAR STRESS

A *polycrystalline* specimen of metal contains many crystals of more or less random orientation. When subjected to stress, the specimen will first experience slip in that crystal most favorably oriented to allow dislocation motion. Thus the critical condition is the magnitude of the net shear stress acting on a slip plane and parallel to a slip direction.

Since tensile forces are much simpler to apply than shear forces, consider the geometry of a tensile test, shown in Fig. 5-42. The area of the slip plane is $A/\cos \phi$, where A is the area perpendicular to the tensile force P and ϕ is the angle between P and the slip plane normal. Thus the average applied axial stress is $P/(A/\cos \phi)$. The component of this stress parallel to the slip direction b is the resolved shear stress τ and is the axial stress times $\cos \lambda$:

$$\tau = \frac{P \cos \phi \cos \lambda}{A}$$

Since P/A is the average applied tensile stress σ,

$$\tau = \sigma \cos \phi \cos \lambda$$

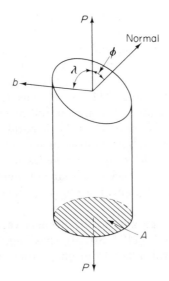

Figure 5-42 Calculation of resolved shear stress parallel to slip direction b, from tensile force P.

The maximum value of $\cos \phi \cos \lambda$ is reached when ϕ and λ are both $45°$, and $\cos \phi \cos \lambda = 0.5$.

Schmid's law, the result of measurements on real crystals, states that slip occurs in a crystal when τ reaches the *critical resolved shear stress* τ_c for that material. Schmid measured τ_c for high-purity magnesium to be 181 kPa, independent of the crystal orientation with respect to the tensile stress σ. We would indeed expect that for a given material dislocations would begin to move at a fixed level of resolved shear stress.

CROSS SLIP

It was noted earlier that a screw dislocation lies parallel to its Burgers vector and is thus not necessarily restricted to a single slip plane. We can see now that the possible slip systems in real crystals will limit the number of planes on which the screw dislocation can move. Figure 5-43 shows two planes, (111) and (11$\bar{1}$), in fcc. Dislocation AB lies parallel to its Burgers vector $b = [\bar{1}10]$ and is thus a screw dislocation. Assume that the resolved shear stress is sufficient to move this dislocation on the (111) to position CD, at which time two possible events can cause *cross slip*: (1) The resolved shear stress changes so that it is greater for this dislocation on the (11$\bar{1}$) than on the (111); or more commonly, (2) the dislocation is blocked from further motion on the (111), such as by fine particles (Chap. 9) or by other dislocations, and the resolved shear stress on the (11$\bar{1}$) is sufficient for motion on the (11$\bar{1}$). In either case, the

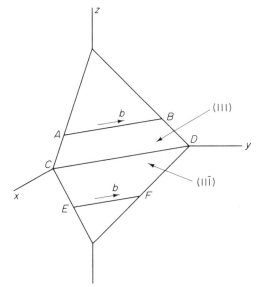

Figure 5-43 Cross slip of a screw dislocation segment in an fcc crystal. Dislocation AB on (111) moves to CD, then moves on (11$\bar{1}$) to EF.

screw dislocation will then leave the (111) and move onto the (11$\bar{1}$), a process called *cross slip*.

Cross slip is very important in ductile metals, as it permits more dislocation motion, and hence more plastic deformation, than would otherwise be possible. Cross slip is a mechanism for bypassing dislocation barriers. *Double cross slip* often occurs when a blocked screw dislocation cross slips to position *EF* in Fig. 5-43 and then returns to a new (111) that is not blocked. The net result is to avoid the barriers on the original (111) that first stopped the dislocation at *CD*.

The wavy slip associated with bcc metals is a form of cross slip, except that it appears to be more the result of the absence of a close-packed slip plane (such as {111} in fcc) rather than the result of a blocked slip plane. However, the result is the same as in fcc in that a blocked screw dislocation can easily move to another plane. In Fig. 5-41(c)–(e), note how easily a screw dislocation with $b = [\bar{1}\bar{1}1]$ could shift from (101) to (213) to (112), in contrast to the relatively large direction change required in Fig. 5-43 for the [$\bar{1}$10] screw dislocation to shift from (111) to (11$\bar{1}$).

REFERENCES

5-1. Birchenall, C. Ernest, *Physical Metallurgy*. New York: McGraw-Hill Book Co., 1959. Chapter 2 describes crystal systems, X-ray techniques, and the categories of possible dislocations.

5-2. Cottrell, A. H., *Dislocations and Plastic Flow in Crystals*. Oxford: Clarendon Press, 1953. One of the two classic early texts on dislocation theory. For advanced-level reading.

5-3. Friedel, Jacques, *Dislocations*. Reading, Mass.: Addison-Wesley, 1964. A comprehensive dislocation text on an advanced level.

5-4. Hull, Derek, *Introduction to Dislocations*. Oxford: Pergamon Press, 1965. Elementary theory, complete enough for most needs, with numerous illustrations of real materials.

5-5. LeMay, Iain, *Principles of Mechanical Metallurgy*. New York: Elsevier North-Holland, 1981. Chapter 4 discusses in detail the properties of dislocations and their interactions.

5-6. McClintock, Frank A., and Ali S. Argon, eds., *Mechanical Behavior of Materials*. Reading, Mass.: Addison-Wesley, 1966. Chapter 4, "Dislocation Mechanics," by Ali S. Argon, gives a thorough and well-illustrated introduction to the important properties of dislocations.

5-7. McLean, D., *Mechanical Properties of Metals*. New York: John Wiley & Sons, 1962. Chapter 2 provides a rapid summary of dislocation theory.

5-8. Orowan, E., "Fracture and Strength of Solids," *Reports on Progress in Physics*, **12** (1949), pp. 185–232. A classic early review of the origins of strength.

5-9. Polakowski, N. H., and E. J. Ripling, *Strength and Structure of Engineering Materials*. Englewood Cliffs, N.J.: Prentice-Hall, 1966. Chapter 4 introduces the types of forces between atoms and describes the crystallography of metals and some nonmetals. Chapter 7 describes the role of dislocations in plastic deformation.

5-10. Read, W. T., Jr., *Dislocations in Crystals*. New York: McGraw-Hill Book Co., 1953. The other classic early text (with Cottrell's) on dislocation theory; also on an advanced level.

5-11. Reed-Hill, Robert E., *Physical Metallurgy Principles*. Princeton, N.J.: D. Van Nostrand Co., 1964. Chapter 1 provides a review of crystal structures and Miller indices.

5-12. Weertman, Johannes, and Julia R. Weertman, *Elementary Dislocation Theory*. New York: Macmillan, 1964. A brief, complete, beginning-level text on the theory.

PROBLEMS

5-1. Suppose a single crystal of iron (body-centered cubic, atomic radius = 0.124 nm) of volume 10^{-6} m^3 contains 10^9 atoms that are imperfectly situated in the crystal. Calculate the percentage of atoms that are still perfectly situated.

5-2. Derive the equation for the maximum theoretical shear stress for slip in a perfect crystal, and write out the assumptions made.

5-3. What other mode of elastic deformation, in addition to the shear strain γ in Fig. 5-22, can occur in a real crystal subjected to a tensile stress?

5-4. One of the most uncertain assumptions in the derivation of Eq. (5-4) is the sinusoidal variation of τ in Eq. (5-2). What evidence suggests that this assumption is reasonable?

5-5. Develop the geometry for the Bragg equation, Eq. (5-1), relating wavelength λ of X-rays, the angle θ made with atomic planes, and the spacing d of the atomic planes.

5-6. Show how an edge dislocation can produce slip as it moves across an otherwise perfect crystal; use several sketches and indicate the dislocation by the symbol ⊥, but do not show individual atoms.

5-7. Draw a two-dimensional sketch of a dislocation in a 60° lattice (such as bubbles form).

5-8. Why cannot an edge dislocation end *within* a crystal? Sketch such a situation for a simple three-dimensional cubic lattice, and show what happens.

5-9. The strength of a specimen of iron 0.5 μm in diameter may be of the order 10 GPa, yet the strength of a bar of 10 mm diameter of the same metal is of the order 10 MPa. Briefly explain why this difference in strength occurs, in terms of the atomic mechanisms that most influence strength.

5-10. A single crystal of height y and width x contains a total of n edge dislocations of like and opposite signs lying in parallel planes, as shown in Fig. 5-44. Find an

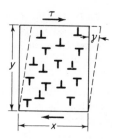

Figure 5-44

expression for the shear strain γ associated with the motion of the dislocations over an average distance L. State any additional assumptions that you make.

5-11. Calculate the ratio of the Burgers vector to the lattice constant for an edge dislocation in:

(a) Simple cubic lattice.

(b) Body-centered cubic lattice.

(c) Face-centered cubic lattice.

5-12. What must be the angle between the Burgers vector of an edge dislocation and the dislocation line? Do both the dislocation line and the Burgers vector lie in the slip plane?

5-13. A single crystal of fcc metal of $a_0 = 0.361$ nm has the dimensions given in Fig. 5-45. All dislocations are edge (as shown), lie parallel to the x-direction, and move parallel to the xy-plane. The y-direction of slip corresponds to [110], the usual slip direction in fcc. If the average dislocation glide distance is 1 μm, how many active dislocations must there be to cause a strain γ of 0.15?

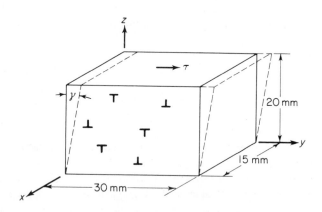

Figure 5-45

5-14. Using a three-dimensional matrix of simple cubes, for bcc iron:

(a) Draw the (312), containing at least five atoms; show all atoms that are on this plane.

(b) Draw a Burgers vector for slip that lies on the (312), being precise as to length.

(c) Label the Burgers vector with its precise Miller indices for magnitude and direction.

5-15. A single crystal of aluminum is shown in Fig. 5-46. The plane shown dashed is the (111), and the slip direction [$\bar{1}$10] is shown. One of a series of parallel dislocations lying on the (111) is shown; during slip it moves in the [$\bar{1}$10] for one fourth of the total possible distance across the crystal.

(a) Calculate the magnitude of the shear strain if 2.67×10^5 dislocations so move.

(b) Show this shear strain by drawing a sketch of the crystal after the dislocations have moved.

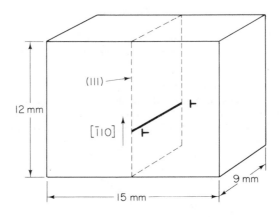

Figure 5-46

5-16. Using a simple three-dimensional cubic matrix, for an fcc crystal:

(a) Draw a (1$\bar{1}$1) that intersects at least five atoms. Show all atoms on this (1$\bar{1}$1) with small circles.

(b) Draw a Burgers vector for one of the possible slip directions that lies on this (1$\bar{1}$1); be sure its length is correct.

(c) Label this Burgers vector with its Miller indices, being precise as to direction and magnitude.

5-17. A single crystal of hcp titanium of initial dimensions 14 mm by 14 mm by 14 mm is deformed by a shear stress so as to produce a shear strain $\gamma = 0.083$. If 1.74×10^{10} dislocation loops of average diameter D are counted on all the parallel slip planes (which are parallel to one of the faces of the 14-mm cube), and if the original dislocations were infinitesimal loops, calculate the value of D.

5-18. In a crystal of copper that is 17.3 mm on each side, calculate the number of dislocations required to cause a shear strain of 0.129 if the average glide distance is 16.4 μm. State any assumptions you have made.

5-19. The copper single crystal shown in Fig. 5-47 contains parallel dislocations on

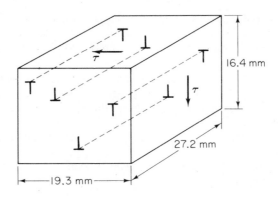

Figure 5-47

parallel slip planes, some of which are shown. If each dislocation glides an average of 2.69 μm, calculate the total dislocation line length if the shear strain is 0.127.

5-20. A single crystal of lead has a shear stress applied to its (111) in the [$\bar{1}$10] as shown in Fig. 5-48. If the average dislocation is a circular loop that is initially of diameter 24.7 μm and grows to diameter 67.1 μm, calculate the number of such dislocations that must be present to cause a plastic shear strain of 0.0736.

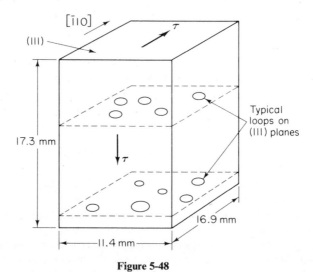

Figure 5-48

5-21. Calculate the maximum theoretical tensile strength of a perfect filament of pure magnesium.

5-22. A single crystal of iron slips on the $\{112\}\langle111\rangle$ system.
   (a) Draw a $\{112\}$, label it, then draw a Burgers vector in this plane and label it with its correct Miller indices.

    (b) Calculate the scalar magnitude of the Burgers vector, using the lattice constant a_0 from Appendix 6.

5-23. How large (in millimeters) will be the slip step in a magnesium single crystal after 3×10^6 dislocations move on the same slip plane and emerge at an edge of the crystal?

5-24. Calculate the maximum theoretical tensile load that a whisker of aluminum of 0.234 μm diameter will support. How many 1500-kg automobiles would be supported by a 0.1-m² cross section of these whiskers in tension? How many such automobiles would your pencil support if it were made of aluminum whiskers?

5-25. Using a three-dimensional simple cube representing one unit cell, draw and clearly label the following: [210], (101), (010).

5-26. Draw the (121) in bcc with the Burgers vector $(a_0/2) [\bar{1}1\bar{1}]$ lying in the (121).

5-27. Suppose that slip occurs in a nickel crystal as shown in Fig. 5-49. If 10^8 parallel dislocation segments lie on parallel (111) slip planes, oriented the same as the dislocation of length c in Fig. 5-49, and the applied shear stress τ is parallel to the Burgers vector of the dislocations, calculate the shear strain in the crystal if all the dislocation segments of length $c = 14$ μm bow out to semicircular shape (as shown dashed) and then move no more.

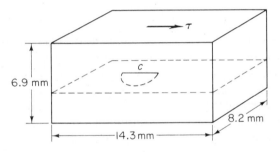

Figure 5-49

5-28. A single rectangular crystal of aluminum is oriented with its edges parallel to the coordinates x, y, z, with the x-edge 12.2 mm long, the y-edge 15.7 mm long, and the z-edge 8.5 mm long. The z-plane is (111) and the $+y$-direction is [$\bar{1}$10]. The crystal contains 10^8 parallel edge dislocations that lie parallel to the x-axis on (111) planes. If a shear stress $-\tau_{zy}$ produces a shear strain $\gamma_{zy} = -0.0500$, calculate the average distance that each dislocation has moved.

5-29. On an x, y, z coordinate system draw a cube that starts at the origin. Then draw three more identical cubes adjacent to it: one on the x-side of the first cube, one on the y-side, and one on the z-side. Assume that these cubes represent four unit cells of bcc iron.
    (a) Draw the [1$\bar{1}$1].
    (b) Draw a {110} that includes your [1$\bar{1}$1], and write its Miller indices.
    (c) Calculate the ratio of spacing of {110} to the spacing of {112} in bcc.

Pure Elements

This chapter and the five that follow apply to the study of practical metals the relationship between strength and dislocation motion that was developed in Chap. 5. The wide range of mechanical properties of metals and their alloys can be attributed largely to the range of conditions for slip. In many commercial metals the strength achieved is the result of several different mechanisms acting simultaneously to limit dislocation motion. These mechanisms can best be developed and described by considering real metals, beginning with the simplest structures. This chapter therefore examines the kinds of strengthening mechanisms possible in solids containing only one kind of atom, that is, in pure elements.

SINGLE CRYSTAL OF LOW DISLOCATION DENSITY

A pure element in the form of a single crystal containing a few dislocations will experience slip of a few interatomic spacings at a relatively low stress. There apparently exists a minimum stress for dislocation motion in a given crystal, called the *friction stress* or *Peierls–Nabarro stress*. This is the minimum stress required to move an otherwise unimpeded dislocation, and its magnitude is ordinarily much smaller than the yield stress because yielding requires the movement of a large number of dislocations. A sufficient number of dislocations usually does not exist before yielding, so they must be created and set in motion. As will be shown in the discussion that follows, this requires a substantial stress, and then the new dislocations interact and impede their own

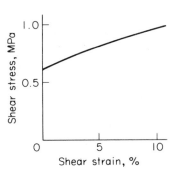

Figure 6-1 Stress–strain curve for 99.96% pure magnesium single crystal, loaded in tension at 203 K. [From H. Conrad and W. D. Robertson, "Effect of Temperature on the Flow Stress and Strain-Hardening Coefficient of Magnesium Single Crystals," *Trans. AIME*, **209** (1957), p. 503. © by AIME and reprinted with its permission.]

motion. Thus, because the strain from the unimpeded motion of one dislocation is infinitesimal, and the strain for yielding is finite and measurable, the stress for yielding, to generate and move a large number of dislocations in a real crystal with many barriers to motion, is much greater than the Peierls–Nabarro stress.

It is possible to create new dislocations by several mechanisms. For example, dislocations can be created by a force applied to a free surface, as was shown in Fig. 5-30. A similar form of this mechanism can exist at a grain boundary when a slip step from one grain acts against another grain to generate new dislocations. Cross slip within a crystal can lead to the generation of a large number of dislocation loops through a mechanism called the *Frank–Read source*. The important point here is that large numbers of new dislocations can be created. If these dislocations do not interfere with each other, then slip can continue indefinitely at this low stress. Although such a set of conditions is seldom achieved for a real crystal, careful experiments on high-purity single crystals have approached these conditions and have resulted in the expected low stress for slip. For example, Fig. 6-1 show the early part of a stress–strain curve for high-purity magnesium. Because slip in magnesium, which is hexagonal close-packed, tends to be limited to the parallel basal (0001) planes, the dislocations experience less interaction with each other than in metals that have several intersecting slip systems. Thus the stress for increasing amounts of strain in a single crystal of magnesium does not increase very rapidly until the strain is large. From similar experiments with a wide range of materials it can be concluded that very pure elements in the form of single crystals of low (but not zero) dislocation density and with little dislocation interaction have very low strength. In the previous chapter strength was shown to be inversely related to the average dislocation glide distance. This agrees with the experimental observations of pure single crystals in which the average dislocation glide distance is relatively large and the strength therefore low.

Exercise 6-1

What would happen to the stress required for slip in single-crystal magnesium if some slip system not parallel to the basal plane also came into action?

Solution.   The stress required for slip should increase. Data on magnesium show that when slip on another system occurs, such as on $\{10\bar{1}0\}$ planes in $\langle 1\bar{2}10 \rangle$ directions, the stress for slip at room temperature is more than 10 times the stress for basal slip.

INTERACTION OF DISLOCATIONS

In most crystals, only limited slip can occur before dislocations begin to inter-fere with each other. The influence of dislocation interaction on the stress to move dislocations can be understood from a simple energy analysis. In the region of a dislocation line the atoms are severely displaced from their equilib-rium positions (see, for example, Fig. 5-28). The energy of displacement of the atoms is recoverable elastic energy, which is the same kind of energy stored by the displacement of atoms in a stressed spring. When the atoms return to their equilibrium positions, they do work on the surrounding atoms; but the atoms in a dislocation cannot return to their equilibrium positions except by moving the dislocation away, and this simply shifts the site of atom displacement to a new group of atoms, resulting in almost no energy change. When a stress is applied, this shift of dislocation position results in slip. Since the stored energy in the dislocation does not depend on its general location within the crystal, there has been no change in energy of the dislocation (if its shape and length remain unchanged during slip), so the stress required is low.

To show what can happen when one dislocation approaches another, the spring analogy is particularly useful, Figure 6-2 shows a straight-line load–elongation relationship for a simple elastic spring in tension. For a given displacement of magnitude a, the energy required is represented by the area under the curve, triangle OAa. Suppose, however, that the spring is already stressed so that its initial elongation is a. What is the energy required for an additional elongation of a? Figure 6-2 shows that the additional energy to deform the spring from elongation a to b (where $b = 2a$) must be the area under the curve from A to B, or $ABba$. This area is substantially larger (in this case 3 times larger) than the area under the curve from O to A. Thus defor-mation of an already-stressed elastic system requires more energy than the same deformation in an unstressed system.

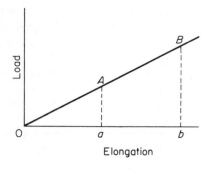

Figure 6-2   Load-elongation of an elastic spring.

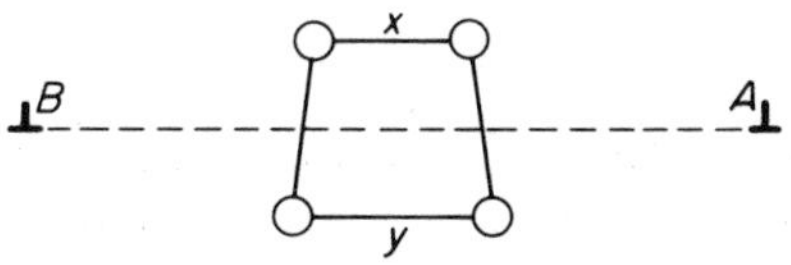

Figure 6-3 Interaction between edge dislocations of like sign.

Consider what happens in the simple case of an edge dislocation approaching another edge dislocation on the same slip plane and with the same sign (that is, with the extra planes of atoms of both dislocations on the same side of the slip plane). This situation is shown in Fig. 6-3, where A is a stationary dislocation which dislocation B is approaching. Four atoms adjacent to the slip plane are shown, already displaced from their equilibrium positions by dislocation A before dislocation B approached. Note that, even at some distance from A, the atoms above the slip plane (which is shown by a horizontal dashed line) are compressed from their equilibrium positions to accommodate the extra plane of atoms above the slip plane A, and these atoms have a spacing x. The atoms below the slip plane are correspondingly stretched apart, to a spacing y. Thus, if b is the Burgers vector of the dislocation, $x < b < y$. As dislocation B approaches, its effect will be to compress further the atoms above the slip plane and to stretch further apart the atoms below the slip plane. Then the spring analogy discussed above will be valid and the additional energy required for the displacement of these four atoms by dislocation B must be greater than the energy required for the initial displacement of these same atoms by dislocation A.

The source for this required extra energy must be the applied shear force moving through some distance. If the atoms shown in Fig. 6-3 are halfway between A and B, then it can be seen by symmetry that if A approached first from the right, then stopped, then B approached from the left, the energy required to move B into its present position must be greater than the energy that was required to move A into position. But the shear strain associated with motion of A and B is related only to the distance each has moved [see Eq. (5-5)], so the distance that the applied shear force moves would be the same for both cases. Therefore, if the energy (work done) is to be greater for case B, the shear force to move B into position after A is in position must be greater than the shear force required initially to move A into position. In other words, as dislocation B approaches dislocation A, the shear stress increases.

The shear stress to move a dislocation does not always increase as it approaches other dislocations. In a few special cases the shear stress can decrease. An obvious example would be to change the sign of dislocation B in Fig. 6-3, so that the extra plane of atoms is below the slip plane. Then interatomic spacing x would be increased and y decreased by dislocation B, resulting in a less severely strained lattice and less stored elastic energy.

The stress field surrounding an edge dislocation can be approximated from elasticity theory. For an edge dislocation lying along the z-axis, perpendicular to the xy-plane, with the extra plane of atoms in the $-y$-direction (see

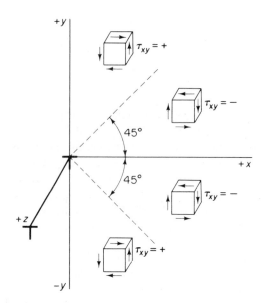

Figure 6-4 Shear stresses resulting from an edge dislocation that is parallel to the z-axis.

Fig. 6-4), the shear stress τ_{xy} at any point is

$$\tau_{xy} = \frac{-Gb}{2\pi(1-v)}\left(\frac{x(x^2-y^2)}{(x^2+y^2)^2}\right) \tag{6-1}$$

where G is the shear modulus of elasticity (see Appendix 6 for values for metals), b is the Burgers vector, v is Poisson's ratio (usually about 0.3), and x and y are the coordinates of the point in question. Note that τ_{xy} does not vary with z. The directions of the shear stresses in four regions are shown in Fig. 6-4; the shear stresses shown acting on the cubes also serve to define the directions of positive and negative shear stresses.

If the extra plane of the edge dislocation of Fig. 6-4 is in the $+y$-direction, then the sign of Eq. (6-1) is reversed. Figure 6-5(a) shows such a dislocation A located at the origin. To determine the direction of stress, and hence the force, on any dislocation B located at x, y, Eq. (6-1) is used with the sign change. If B is located as in Fig. 6-5(a), where $x > y$, then the sign of τ_{xy} from Eq. (6-1) is positive. A positive τ_{xy} will cause a force to the right on dislocation B; this will cause B to separate from A and is thus repulsive. Figure 6-5(b) shows the relative magnitude and direction of the force between A and B for any position x, y of B. For every case, the force on B is always in the $\pm x$-direction.

The most usual interaction between two parallel edge dislocations of like sign occurs when they lie on the same plane, $y = 0$. This situation cannot be shown by Fig. 6-5(b) because x is plotted in terms of y, but from Eq. (6-1), for

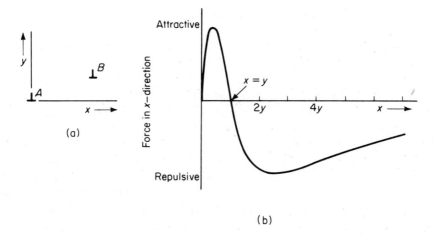

(a)

(b)

Figure 6-5 Interaction forces between parallel edge dislocations of like sign: (a) Relative positions of the dislocations, where A is at the origin and the position of B is x, y; (b) Variation of force between A and B. [From A. H. Cottrell, *Dislocations and Plastic Flow in Crystals*. (Oxford: Clarendon Press, 1953), p. 48. Reprinted by permission of the publisher.]

the dislocation in Fig. 6-4, with $y = 0$, we have

$$\tau_{xy} = \frac{-Gb}{2\pi(1 - v)} \cdot \frac{1}{x} \qquad (6\text{-}2)$$

Thus edge dislocations of like sign on the same plane will repel each other; if the signs are opposite, they will attract.

The stresses surrounding a screw dislocation are radially symmetrical and are thus most simply described in cylindrical coordinates. Figure 6-6 shows a screw dislocation having the shape of a left-handed helix lying along

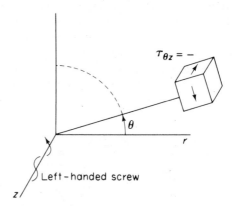

Figure 6-6 Shear stresses resulting from a left-handed screw dislocation that is parallel to the z-axis.

the x-axis. The shear stress $\tau_{\theta z}$ shown varies only with the radius r:

$$\tau_{\theta z} = \frac{-Gb}{2\pi} \cdot \frac{1}{r} \qquad (6\text{-}3)$$

where G is the shear modulus of elasticity and b the Burgers vector. Note that this equation differs from Eq. (6-2) only by the factor $(1 - v)$.

Any two parallel screw dislocations of the same sign (both right-handed or both left-handed) will repel each other. If one is a right-handed screw and one left-handed, they will attract.

Edge dislocations do not cause a $\tau_{\theta z}$ shear stress, and screw dislocations do not cause a τ_{xy} shear stress. Thus an edge dislocation and a parallel screw dislocation will not interact. The forces for the general case between non-parallel dislocations are very complex and will vary from point to point along each dislocation.

Exercise 6-2

If dislocations exert forces on each other, why do they not move in the absence of external stressses into positions where the forces are minimized?

Solution. Although a complete answer to this question is not known, loss of energy (associated with the Peierls–Nabarro force, already discussed) during dislocation motion apparently prevents free motion of dislocations. The nature of this lost energy may result in part from a "twanging" of atoms as the dislocation passes by; such vibration of atoms would be dissipated as heat, a common observation during plastic deformation. When the interaction forces between dislocations are great enough, the dislocations will move by themselves to lower their interaction forces, but dislocation motion stops before the forces vanish.

STRAIN STRENGTHENING (WORK HARDENING)

From interaction forces between dislocations it is an easy step to strain strengthening. The preceding section shows how the stress required for dislocation motion can be increased in the presence of another dislocation. A simple instance of strain strengthening is shown in Fig. 6-7, where dislocation A originating from a dislocation source at the left has been stopped by a barrier of some kind, such as a grain boundary (see below for further discussion on this point), and dislocations B and C originating from the same

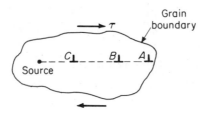

Figure 6-7 Pile-up of edge dislocations.

source are prevented from moving as far as *A* by the repulsive force of *A*. (Dislocation *C* is repelled by both *A* and *B*, so the separation of *B* and *C* is greater than the separation of *A* and *B*.) Therefore the stress required to move dislocation *C* a given distance from the source is greater than the stress that was required to move *B* the same distance from the source. Thus *C* moves a shorter distance under the same stress. Under a given stress the dislocations find equilibrium positions and move no further.

If additional strain is required, the stress must be raised. Then one or more of at least four possible events can occur: (1) the spacing between *A*, *B*, and *C* will decrease; (2) the source will generate new dislocations; (3) another set of dislocations elsewhere will begin to move; or (4) dislocation *A* will break through or move around its barrier and allow further strain. The first three of these events will promote further strain strengthening; the fourth will tend to diminish the rate of strain strengthening. Since billions of dislocations are ordinarily present in a standard test specimen, it is unrealistic to speak of just one dislocation system acting at any instant. Many thousands of such systems must be acting, but as their dislocations either pile up or are hindered in some other way (many other mechanisms have been hypothesized), the stress required for additional motion increases in the same way as described above.

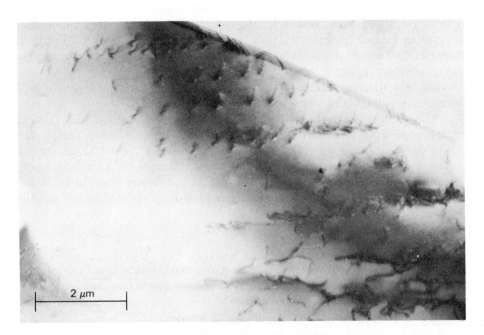

2 μm

Figure 6-8 Thin-foil transmission electron micrograph of dislocations piled up against a grain boundary (visible at upper right), in moderately cold-worked type 304 stainless steel. The short segments of dislocations run through the thickness from top to bottom of this very thin specimen. (Micrograph courtesy of Stephen J. Krause.)

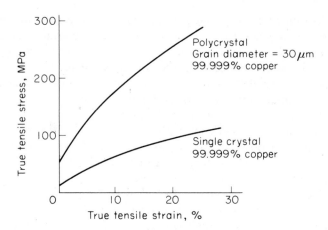

Figure 6-9 Representative stress–strain curves for single crystal and poly-crystalline copper. (Reprinted with permission from *Acta Met.*, Vol. 1, R. P. Carreker, Jr., and W. R. Hibbard, Jr., "Tensile Deformation of High-Purity Copper as a Function of Temperature, Strain Rate, and Grain Size." Copyright 1953, Pergamon Press, Ltd.)

The simple argument for one dislocation system should therefore remain valid for the real case. In this way it is possible to understand the dislocation behavior that causes strain strengthening, or, as it is often called, *strain hardening* or *work hardening*. A thin-foil electron micrograph of piled-up dislocations in type 304 stainless steel is shown in Fig. 6-8. Note the tendency toward increased spacing of dislocations to the left of the barrier.

In contrast to the easy slip of a magnesium single crystal (Fig. 6-1), most metals experience a marked increase in stress for slip as strain increases. The lower curve in Fig. 6-9 is the stress–strain curve for a copper single crystal, which is face-centered cubic. This curve is much steeper than for magnesium because many equivalent slip systems in the copper crystal begin to act, intersect each other, and provide many barriers to dislocation motion not initially present in the copper specimen.

Exercise 6-3

(a) Sketch two intersecting {111} slip planes and show their ⟨110⟩ slip directions for an fcc structure. Then sketch a third {111}⟨110⟩ system that intersects both of the others.
(b) Is it possible to have a single fcc crystal so oriented that two or more intersecting slip systems can act at the same applied stress?

Solution. (a) Figure 6-10 shows three different {111} planes and three intersecting ⟨110⟩ directions lying in these planes. The edges of the triangular portions of the planes are shown as the ⟨110⟩ directions for clarity. Note that each direction lies in two {111} planes; for example, the [10$\bar{1}$] lies in both (111) and (1$\bar{1}$1). This even further complicates the possible slip that can occur. There are 12 equivalent slip systems in fcc: three ⟨110⟩ directions on four different {111} planes.

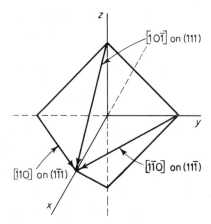

Figure 6-10   Three intersecting slip systems in fcc.

(b) By symmetry a tensile stress applied in the x-direction would exert exactly the same shear stress on all three slip systems shown in Fig. 6-10. Thus slip on intersecting systems, called *multiple glide*, could occur at a very early stage.

The upper curve in Fig. 6-9 shows the behavior of polycrystalline copper. The number of crystals capable of easy glide (i.e., slip at the low stress levels of single-crystal copper) is small and represents a very small fraction of the total number of crystals (grains) present in the specimen. Thus an easy glide portion of the curve is not seen because easy glide of the most favorably oriented individual crystals has ended in this case after an extremely small cumulative strain. The diminishing strain strengthening in Fig. 6-9 (i.e., decreasing $d\sigma/d\epsilon$) apparently results when dislocations are able to break through or go around barriers, as has already been discussed. The higher the stress, the more likely it is that dislocations will move through or around barriers, so it might be expected that the rate of strain strengthening would diminish with increasing strain.

VACANCIES

The simplest possible kind of defect is the vacancy, where a single atom is missing from the otherwise perfect crystal structure. A simple model of a vacancy in a simple cubic lattice is shown in Fig. 6-11, where the center atom

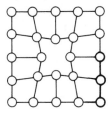

Figure 6-11   Vacancy in simple cubic lattice.

is missing. The region of displacement of the surrounding atoms from their equilibrium positions will extend much farther in a real case than is shown in the sketch.

The smooth shifting of atomic bonds that occurs when a dislocation moves in a perfect lattice will be disrupted in the region at the center of the vacancy and for many interatomic distances away from the center. The argument given above for additional energy to move a dislocation into the strain field of another dislocation will hold also in the region of a vacancy, as many of the atoms adjacent to the vacancy are already displaced from their equilibrium positions, a condition that sets up a strain field. However, in contrast to interaction between parallel dislocations, only a short length of the dislocation will be directly affected by a single vacancy. But because a dislocation cannot be broken and thus terminated within a crystal, holding up one segment of a dislocation will tend to hold up the whole dislocation. In practice, many vacancies would probably be present to influence each dislocation at a number of points.

Observation of thin metal foils containing vacancies reveals that in some cases the vacancies come together to form flat disks of vacancies, one of which is sketched in Fig. 6-12(a). Figure 6-12(b) shows how these disks can collapse

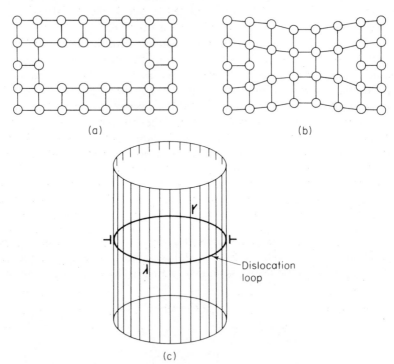

(a)   (b)

(c)

Figure 6-12 (a) Circular disk of vacancies prior to collapse. (b) Formation of circular edge dislocation loop through collapse of vacancy disk. (c) The dislocation loop is constrained to move only on the shaded slip "surface," which is unlikely.

to form a very special kind of edge-dislocation loop. This dislocation is special because the loop does not lie in the slip plane but is perpendicular to it. For this loop to move, slip would have to occur on a cylindrical or prismatic surface, so this is called a *prismatic dislocation loop*. Figure 6-12(c) shows the loop of Fig. 6-12(b) in three dimensions, with the slip "surface" shaded. Such motion is much less likely to happen than slip on a plane surface, so these loops remain fixed in position and act as barriers to ordinary dislocation motion in the same way as for other very small point defects such as individual vacancies.

Although vacancy strengthening is very effective when 0.1 to 1% of the atom sites are empty, the normal concentration of vacancies at room temperature is several orders of magnitude smaller. Thus vacancy strengthening is not now a practical means for strengthening metals except for those of very high purity and is not considered further here.

GRAIN BOUNDARIES

The boundary between two crystals is a very thin region of imperfection where the crystal orientation must shift from one crystal lattice to the other. In the region where this change in crystal orientation occurs, some atoms will be out of equilibrium position because they are attracted by atoms from both crystals. This region is the *grain boundary* (a grain is one crystal of a polycrystalline specimen), and it can be seen on a polished and etched surface because the higher energy of the grain-boundary atoms causes etching at a higher rate at the boundary than elsewhere in the grains. The resulting grooves in the polished surface can be seen in an optical microscope. Figure 6-13 shows these grooves as lines on a polished and etched specimen of iron containing very little carbon.

When some part of a dislocation approaches a grain boundary, more

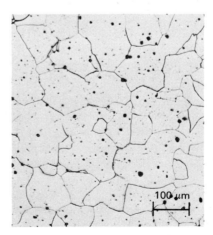

Figure 6-13 Optical micrograph of grain boundaries in Armco iron of 0.02% C.

energy, or higher stress, will be required to move the dislocation through the boundary. Even if the dislocation should get through the grain boundary, the slip plane in the next crystal will be at an angle to the original slip plane. The dislocation must then get longer in order for part of the dislocation to follow the new slip plane and part the old. Figure 6-14 illustrates a very simple case of a dislocation crossing grain boundaries. A single crystal, as in Fig. 6-14(a), contains a dislocation source S that generates one dislocation loop that moves on the slip plane, shown shaded, until it is close to the outer edges of the crystal, as in the sketch. Figure 6-14(b) shows three grains, A, B, and C, of the same total size (but not drawn to the same scale) as the single crystal in Fig. 6-14(a), with a source in grain A. When the dislocation loop reaches the boundaries between A and B, and A and C, it is forced to follow the new slip planes, shown shaded. In general, the boundary between B and C will not be a common slip plane, so the loop will be pinned at its intersection with the three grains, point a. The final dislocation length will then be greater than for the single-crystal case, and both cases have roughly the same area swept by the dislocation. Increase of the dislocation length will increase stress, by the following argument: It has already been shown that the atoms in a dislocation have greater elastic energy than if they were located in a perfect crystal lattice. Since the total energy is proportional to the number of atoms in the region of the dislocation, a longer dislocation will have greater total elastic energy because it involves more atoms. For this reason, lengthening a dislocation to allow it to lie on two intersecting slip planes requires greater total energy. This extra energy can only come from an increase in stress, as was concluded earlier in this chapter.

Grain boundaries therefore hinder dislocation motion. Because the boundary of a given grain fully envelops dislocations in that grain, the dislocations cannot move around the grain boundary in the same way that they

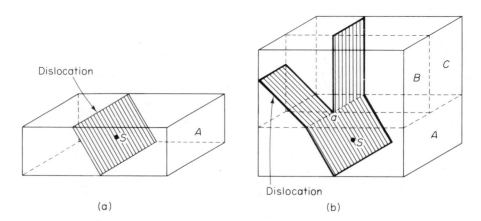

Figure 6-14   (a) Single crystal with a dislocation source at S. (b) Three crystals, A, B, and C, with a source S in grain A.

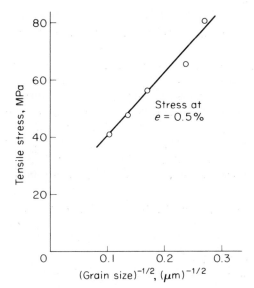

Figure 6-15 Variation of yield stress with grain diameter for copper [From D. McLean, *Mechanical Properties of Metals* (New York: John Wiley & Sons, 1962), p. 76. Reprinted by permission of the publisher.]

can sometimes bypass point defects and other dislocations. If grain boundaries are spaced closer together (i.e., for smaller grain diameters), the average dislocation glide distance is reduced and the stress must be increased for further dislocation motion. [Recall Eq. (5-5): $\gamma = nbg/hw$.] Figure 6-15 shows the marked increase of stress for initial slip (yield stress) for copper as the grain size (diameter) decreases.

Exercise 6-4

Develop the equation for the grain size relationship in Fig. 6-15.

Solution. The constants in an equation of the form $y = mx + b$ are $b = y$ at $x = 0$ and $m = $ slope. From Fig. 6-15, $b = 20$ MPa and roughly $m = (80$ MPa $- 20$ MPa$)/$ $(0.3 \ \mu\text{m}^{-1/2} - 0) = 200$ MPa-$\mu\text{m}^{1/2}$. Thus $\sigma_y = 20$ MPa $+ (200$ MPa-$\mu\text{m}^{1/2})d^{-1/2}$, where d is the grain diameter in μm. This is usually known as the *Petch equation* and written in the form

$$\sigma_y = \sigma_i + k_y d^{-1/2} \tag{6-4}$$

where σ_y is the yield stress and σ_i and k_y are constants such as developed above. One would thus expect a single crystal of this copper to have a yield strength of about $\sigma_i = 20$ MPa. (Figure 6-9 shows that a copper single crystal is close to 20 MPa at $\epsilon = 0.5\%$.)

Strain strengthening (Fig. 6-9) and reduction of grain size (Fig. 6-15) are the two most powerful and practical means for increasing the strength of pure or nearly pure metals.

SUB-BOUNDARIES AND RECRYSTALLIZATION

A ductile metal that has been plastically strained contains higher dislocation density and hence higher stored elastic energy than before straining. From thermodynamic reasoning and by observation it is known that the specimen will tend toward a minimum stored elastic energy under favorable conditions. One such condition is high temperature. As temperature is raised, the increased kinetic energy of individual atoms promotes dislocation motion at lower and lower stresses, until a temperature range is reached where dislocations move because of the stress fields of other dislocations. Thus dislocations that have attractive forces between them will move together, and dislocations that have repulsive forces will move apart. A simple example is shown in Fig. 6-16, where (a) shows the strained structure of dislocations on three parallel glide planes, and (b) shows a feasible lower energy alignment of these dislocations after some time at elevated temperature. Figure 6-5 has already shown that dislocations of like sign would tend to attract and thus to stack above each other when $x < y$. The edge dislocations must also space themselves uniformly above and below each other in order to minimize the total elastic energy. Such motion perpendicular to the slip plane is called *dislocation climb*. For climb, atoms must be added to or removed from the edge of the extra plane, a process requiring substantial diffusion, which therefore occurs only at elevated temperatures. (See Chap. 12 for a detailed discussion of high-temperature creep.) Vacancies that were created during the cold-working operation can contribute to this process by aiding diffusion and by migrating to edge dislocations. The configuration of dislocations shown in Fig. 6-16(b) has a lower elastic energy than a random distribution of dislocations. The vertical dotted lines represent two-dimensional surfaces called *low-angle grain boundaries*, since the angle between adjacent subgrains is very small. Because of the shape of the subgrains that result, the change from (a) to (b) in Fig. 6-16 is called *polygonization*.

Exercise 6-5

Calculate the angle between adjacent subgrains in Fig. 6-16(b), for a crystal of Burgers vector b and vertical spacing d as shown.

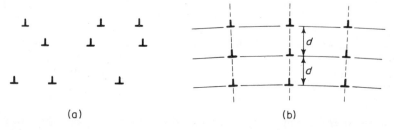

(a)   (b)

Figure 6-16 (a) Simple model of a strained (bent) crystal. (b) Formation of low-angle grain boundaries, shown as vertical dashed lines.

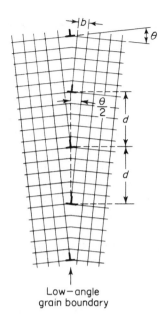

Figure 6-17 Calculation of angle between subgrains.

Low—angle
grain boundary

Solution. Figure 6-17 shows five dislocations in a sub-boundary. The left and right subgrains are at an angle θ because of the presence of the extra half-planes of atoms that end at the dislocations, acting as a wedge between the subgrains. If the distance d separating the dislocations is large compared with the Burgers vector b of each dislocation, then $\theta/2 = \tan^{-1}(b/2d) = b/2d$; thus $\theta = b/d$.

Dislocations are impeded by sub-boundaries in the same way as they would be by any set of dislocations. Since the dislocation density, and hence average spacing of dislocations, does not change during polygonization, the hardness and strength should not change very much. Figure 6-18 shows the

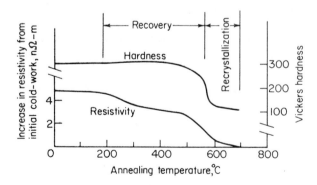

Figure 6-18 Room-temperature hardness and change in resistivity vs. annealing temperature for cold-worked 99.6% nickel. [From L. M. Clarebrough, M. E. Hargreaves, and G. W. West, "The Release of Energy During Annealing of Deformed Metals," *Proc. Royal Soc.* (London), **232 A** (1955), p. 252. Reprinted by permission of The Royal Society.]

room-temperature hardness and electrical resistivity that result after a number of specimens of identically cold-worked nickel are heated to different temperatures. In the region labeled *recovery*, no change in an initially polished and etched surface is visible in the optical microscope. Yet the resistivity drops in this range. This apparently results from the formation of low-angle grain boundaries and the absorption of vacancies by dislocations. This is the range in which polygonization occurs. The grain boundaries remain fixed in this temperature range; that is, recrystallization has not yet started.

If other identically strained specimens are heated to higher temperatures, dislocations can glide completely out of each grain so as to allow the formation of completely new grains of very low dislocation density. This is called *recrystallization* and can be observed in an optical microscope that operates at elevated temperatures. The dislocation density and elastic energy of the new crystals are much less than for crystals containing sub-boundaries. The strength and hardness of the recrystallized metal are therefore markedly reduced, as shown in the region of Fig. 6-18 labeled "recrystallization."

Exercise 6-6

Compare qualitatively the dislocation density in a pure metal after (a) slow cooling from the melt; (b) cold working; (c) recovery; and (d) recrystallization.

Solution. Dislocation density is lowest after (a) slow cooling, has virtually the same high value after (b) cold working and (c) recovery, but is again much less after (d) recrystallization.

If the purity of the nickel is increased, in this case from 99.6 to 99.85% nickel, recrystallization occurs at a lower temperature, probably because fewer impurity atoms are present to impede dislocation motion and diffusion. A mechanism by which solute atoms can interfere with dislocation motion is described in Chap. 7 and shown in Fig. 7-6.

The process of recovery and recrystallization is used to eliminate the effect of strain hardening and to control the grain size. This latter is possible because if the initial strain is increased, more elastic energy will be available to force the rate of nucleation of new grains to exceed the rate of growth of each grain, so that a larger number of smaller grains will be formed. Except by melting, it is not possible to alter the grain size of pure single-phase solid elements without first introducing some plastic strain.

REFERENCES

6-1. American Society for Metals, *Strengthening Mechanisms in Solids*. Metals Park, Ohio: American Society for Metals, 1962. Chapter 3, "Strain Hardening," by Jack Washburn, discusses strain-hardening mechanisms in detail.

6-2. Caddell, R. M., *Deformation and Fracture of Solids*. Englewood Cliffs, N.J.:

Prentice-Hall, 1980. Chapter 7 provides a thorough basic description of dislocation theory.

6-3. Guy, Albert G., *Elements of Physical Metallurgy*, 2nd ed. Reading, Mass.: Addison-Wesley, 1959. Chapter 12 gives step-by-step descriptions of recovery and recrystallization.

6-4. McLean, D., *Mechanical Properties of Metals*. New York: John Wiley & Sons, 1962. See Chap. 5, "Strain Hardening in Pure Metals."

6-5. Reed-Hill, Robert E., *Physical Metallurgy Principles*. Princeton, N.J.: D. Van Nostrand Co., 1964. Chapter 5, pp. 144–53, describes grain boundaries.

6-6. Weertman, Johannes, and Julia R. Weertman, *Elementary Dislocation Theory*. New York: Macmillan, 1964. Chapters 2 and 3 analyze the stresses and forces associated with dislocations; Chap. 5 discusses the dislocation pile-up.

6-7. Zackay, Victor F., ed., *High-Strength Materials*. New York: John Wiley & Sons, 1965. Chapter 12, "Slip Mechanism in Single Crystals of Hexagonal Close-Packed Phases," by J. E. Dorn and J. B. Mitchell, covers modes of slip and strengthening mechanisms in hcp metals.

PROBLEMS

6-1. Slip in bcc (body-centered cubic) crystals is observed to occur in $\langle 111 \rangle$, the close-packed direction. What is the most widely spaced set of planes that contain this direction? How many slip systems involving this set of planes and the $\langle 111 \rangle$ directions are possible?

6-2. (a) What is the effect of cold-rolling a bar of pure copper on its hardness? Explain why this occurs.

(b) What is the effect of cold-rolling a bar of pure copper on its load-carrying capacity? Explain why this occurs.

6-3. If you are given an annealed, coarse-grained nickel bar, describe in detail at least two different ways that you could strengthen it without changing its chemical composition.

6-4. If two parallel edge dislocations of opposite sign are on the same slip plane, they will come together (if they are not impeded) and annihilate each other. What will happen if they are of opposite sign, separated by 50 nm on the slip plane, but one dislocation is displaced to another parallel slip plane 30 nm away? If a small shear stress that tends to force the dislocations together is then applied, what happens?

6-5. Describe briefly two different atomic mechanisms for strain hardening.

6-6. If you heat a piece of cold-worked copper to a sufficiently high temperature and water-quench it, it becomes softer. Why?

6-7. Sketch the approximate curves for hardness vs. annealing temperature (omit numbers) for 20% cold-worked copper and 50% cold-worked copper. Label each curve.

(a) Explain the difference in recrystallization temperature.

(b) Which has the smaller grain size after recrystallization at the lowest possible temperature? Why?

(c) Which has the greater room-temperature strength after cold working but before annealing? Explain why, describing the basic mechanism(s) briefly.

(d) Which has the greater room-temperature strength after annealing at the lowest possible temperature? Explain why, describing the basic mechanism(s) briefly.

6-8. What is the essential difference between recovery and recrystallization in a cold-worked metal?

6-9. A long bar of pure metal that follows the relationship $\sigma = (500 \text{ MPa})\epsilon^{0.2}$ is initially 50 mm thick by 100 mm wide. It is subjected to the following rolling and heating sequence: 40% reduction of area, full annealing (recrystallization), 40% reduction of area, full annealing, then 35% reduction of area. (Each reduction is calculated on the basis of the area just before each rolling operation is started.) Assuming no change in width during thissequence, find the maximum tensile load the bar will support after this sequence.

6-10. A 30% cold-worked brass (82 R_B) is put into a furnace for 0.5 h at 350°C and water-quenched; its final hardness is 82 R_B. Another 30% cold-worked brass is held in a furnace for 0.5 h at 370°C; after water quenching, its hardness is 68 R_B. Explain why the hardness changed at 370°C but did not change at 350°C, and describe the fundamental mechanism(s) involved.

6-11. An annealed 70–30 brass bar is initially 12.70 mm thick, and its hardness is $-9 \ R_B$. It is then rolled to the thickness below and the hardness measured:

	Initial	1	2
Thickness (mm)	12.70	11.76	9.88
Rockwell B	−9	51	74

(a) Calculate the true longitudinal strains from initial to condition 1, and from condition 1 to condition 2.

(b) Explain, from a fundamental point of view, why the changes in hardness do not correspond to the changes in true strain.

6-12. Two separated Brinell hardness tests using a ball 10 mm in diameter are made on the same piece of 70–30 brass:

(a) At 500 kg_f (4.90 kN).

(b) At 1500 kg_f (14.7 kN).

Which test gives the higher hardness, (a) or (b)? Explain why this happens.

6-13. The line force acting in a direction perpendicular to a dislocation line can be expressed as

$$F = \tau b$$

where F is the line force in N/m, τ is the shear stress acting on the plane of dislocation motion, and b is the Burgers vector. Calculate the magnitude and direction of the force acting on an edge dislocation in a single crystal of pure vanadium as the result of a second edge dislocation of the same sign that lies parallel to and in the same slip plane as the first dislocation and separated by a distance of 35.7 nm.

Single Phases

Solids seldom consist of atoms of only one element, so the influence of foreign atoms is therefore an important factor in the strength of most real solids. This chapter describes those strengthening mechanisms that can be attributed to the presence of foreign atoms distributed more or less uniformly on an atomic scale, that is, it describes the relative strengths of mixtures of elements in contrast to strengths of the separate elements. Such mixtures of metals are called *alloys*. These effects on strength can act in addition to the strengthening mechanisms for pure elements that were described in the previous chapter.

PHASES

When applied to solids, the term *phase* means any collection of atoms that is homogeneous on a fine scale. When the optical microscope was still the most powerful magnifying device, any observably homogeneous portion of the structure as seen in the microscope could be conveniently termed a phase; multiple phases are described in Chaps. 8 and 9. However, with the advent of electron and ion microscopes, which can reveal details of the order of atomic dimensions, the earlier definition of a phase loses much of its meaning. For example, sometimes unlike atoms segregate themselves into a configuration that is not homogeneous on the scale of a few atoms but is homogeneous on a slightly larger scale and behaves physically and mechanically like a solid of uniform properties; such a solid can here be properly termed a phase. The distinction between one- and two-phase solids might therefore be established as homogeneity on a scale slightly larger than atomic. A hypothetical test to determine wheter a solid is a single phase could therefore be to select samples containing 100 to 1000 atoms from any two regions in the solid. If the crystal structure and atomic ratio are the same in both samples, then the solid is a single phase. This test can, of course, be applied to a pure element, which is single phase because it has no second element present. A solid chemical compound consisting of molecules with a fixed ratio of two or more elements is also a phase. Other kinds of phases are described in the following paragraphs.

INTERSTITIAL SOLID SOLUTIONS

A solid solution is a solid phase that may have a wide range of ratio of foreign (solute) atoms to solvent atoms. If the solute atoms are substantially smaller than the solvent atoms, the solute atoms may find relatively large holes between the sites of the solvent atoms. Figure 7-1 shows three solute atoms in such voids; this is called an *interstitial solid solution*. It should be made clear that a two-dimensional sketch does not show how much room is really present. For example, in a simple cubic lattice such as is shown in Fig. 7-1, if the solvent atoms are touching each other, the maximum size of an interstitial atom that can be placed between the atoms in their (100) plane without distorting the lattice is substantially smaller than the maximum size of an atom that can be placed in the voids at the centers of the cubic cells.

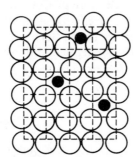

Figure 7-1 Schematic sketch of three interstitial atoms in solid solution.

Exercise 7-1

Calculate the maximum ratio of interstitial atomic radius to solvent atomic radius for the simple cubic structure for (a) the two-dimensional case, and (b) the three-dimensional case.

Solution. Let R = solvent radius and r = solute radius (see Fig. 7-2). (a) $2R\sqrt{2} = 2R + 2r$; thus $r/R = \sqrt{2} - 1$. (b) $2R\sqrt{3} = 2R + 2r$; thus $r/R = \sqrt{3} - 1$. This represents an increase of $(0.732 - 0.414)/0.414 = 77\%$ in size of atom that can fit into the space lattice. Similar calculations can be made for the other crystal structures.

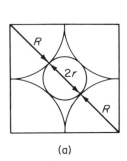
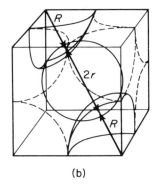

(a)   (b)

Figure 7-2   Calculation of atomic radii for Exercise 7-1.

Interstitial solid solutions do not form unless the ratio of the radius of solute atoms to the radius of solvent atoms is less than 0.59. This upper limit would involve large distortions in fcc and bcc crystals. Elements with very small atomic radii are favored as interstitial solutes, so that the most important interstitial solutes in metals are carbon, boron, nitrogen, and oxygen. For example, iron at elevated temperatures can contain up to about 2% interstitial carbon, which more then covers the range of alloys known as steels. Interstitial hydrogen is usually undesirable in most metals because it can cause cracking and embrittlement (see Chap. 18). For reasons not fully understood, these elements form interstitial solid solutions more readily with the transition metals (those with incomplete inner electron shells) than with nontransition metals.

When two elements are in contact, there is a tendency for their atoms to mix, which follows from the thermodynamic argument that systems tend toward maximum disorder. Counteracting this tendency toward mixing is a tendency toward complete separation because energy is required to distort the lattice *elastically* when the atoms are mixed, as shown in Fig. 7-1. A balance between these two effects occurs. Usually the maximum solute–solvent ratio is higher at higher temperatures; if an insufficient supply of solute atoms is available, then the ratio will be less than the maximum. For interstitial solid solutions the maximum ratio of number or mass of solute atoms to total number or mass of atoms, the *solubility*, is seldom more than a few percent.

Interstitial atoms increase strength in the same way that vacancies do (see Chap. 6), by elastically distorting the regular lattice structure of atoms and impeding dislocation motion. Even interstitial atoms smaller than the interstitial holes should cause some distortion by altering the motion of nearby solvent atoms. Although the extent of distortion shown in Fig. 7-1 is very limited, the effect of interstitials in real solids extends much farther. In a crystal containing 1% interstitial atoms, the interstitial atoms will be distributed

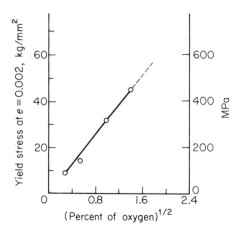

Figure 7-3 Strengthening of pure zirconium by interstitial oxygen. [From D. G. Westlake, "The Combined Effects of Oxygen and Hydrogen on the Mechanical Properties of Zirconium," *Trans. AIME*, **233** (February 1965), p. 368. © 1965 by AIME and reprinted with its permission.]

throughout the crystal so as to be less than 5 interatomic spacings apart. Thus the strain produced by the interstitial solute atoms will be general throughout the matrix, not limited to small isolated regions surrounding each solute atom. In this way, the solute atoms slightly increase the average crystal lattice spacing of the solvent atoms. The extent of strengthening of a pure metal by an interstitial element is shown in Fig. 7-3.

Exercise 7-2

A calculation of the solubility of interstitial carbon in bcc iron at room temperature gives the maximum value of 2.3×10^{-7} mass percent C. Approximately how many unit cell spacings (a_0) of Fe separate each C atom in this solid solution?

Solution. Let N represent the number of atoms. Then

$$\frac{N_{Fe} \times 55.85}{N_C \times 12} = \frac{1}{2.3 \times 10^{-9}}$$

Thus $N_{Fe}/N_C = 9.34 \times 10^7$. Fe is bcc, so each unit cell contains two Fe atoms, giving $9.34 \times 10^7/2 = 4.67 \times 10^7$ unit cells of Fe per C atom. Assume for simplicity that each C atom lies at the center of a cube containing 4.67×10^7 unit cells; then the closest distance between C atoms will be the same as the width of the large cube, which is $(4.67 \times 10^7)^{1/3} = 360$ unit-cell widths.

SUBSTITUTIONAL SOLID SOLUTIONS

Instead of placing small new atoms in the interstices between solvent atoms, as described above, the solvent atoms themselves may be replaced by other, different atoms, giving a *substitutional solid solution*. While interstitial solid solutions are limited to small amounts of solute, substitutional solid solutions usually have a wide range of solubility. The extent to which unlike atoms will form a substitutional solid solution is determined in part by their similarities

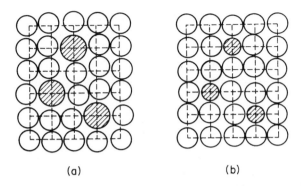

(a)   (b)

Figure 7-4 Substitutional solid solutions, viewed schematically, with (a) solute radius larger than solvent radius, and (b) solute radius smaller than solvent radius.

in atomic radius (which should not differ by more than 15%), crystal structure, and electron structure. Several pairs of elements are *completely soluble* in each other at all temperatures. Among the more common metals, the copper–nickel and cobalt–nickel systems are fcc examples, and the molybdenum–tungsten system is a bcc example. In addition, many other combinations of metals form substitutional solid solutions ranging up to 5 or 10% in temperature ranges of practical importance; thus these alloys exhibit *limited solid solubility*. Substitutional solid solutions are much more common than interstitial solid solutions in metals.

Figure 7-4 depicts schematically the distortion that is caused by solute atoms larger and smaller than the solvent atoms. The appearance of the matrix in both cases is similar to the distorted matrix of Fig. 7-1, and for the same amount of distortion the effect on dislocation motion is similar.

The effect of a substitutional solute should therefore be to strengthen the solvent, since the distorted lattice interferes with dislocation motion. More solute should distort the lattice more and thus cause further strengthening. Figure 7-5 shows that the strength of pure copper is increased by nickel in substitutional solid solution. This effect is not due to nickel being stronger

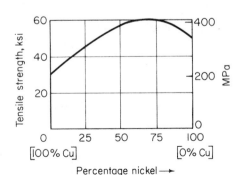

Figure 7-5 Effect of composition of substitutional solid solution on strength of Cu–Ni alloys. (Adapted from ASM data.)

than copper, since addition of copper to pure nickel (at the right side of Fig. 7-5) also increases strength. At some point between these extremes of pure elements, the strengthening effect from lattice distortion reaches a maximum. Similar results are found for solid solutions that have a limited range of solubility, although a maximum usually does not occur in this range.

Exercise 7-3

Copper forms a substitutional solid solution with zinc up to about 38% Zn. Which alloy should be stronger: (a) 95% Cu–5% Zn, or (b) 90% Cu–10% Zn?

Solution. 90% Cu–10% Zn should be stronger because the increase in this range should further distort the lattice. Typical measured values are 260 MPa for Cu–10% Zn and 240 MPa for Cu–5% Zn.

The practical utilization of solid-solution strengthening is that it provides a large number of alloys having a wide range of properties. If high strength were the only criterion, then that alloy and grain size giving the highest strength would be the only one used. However, differences in cost of metals encourage use of alloys richer in the lower-cost metal; differences in corrosion resistance promote use of the more resistant metal; and variations in toughness tend to be inversely related to strength, so that an alloy of less than maximum strength is often used in order to meet the ductility requirements of service or of the fabrication process (see Chap. 14). The single-phase copper–zinc alloys (brasses), ranging from 0 to 38% Zn, provide examples of these factors.

SUPERSATURATION

The degree of solubility may change with temperature, so that the limits to solubility may change (usually decreasing) as the temperature decreases. At a temperature approaching the melting temperature, diffusion in a solid can be very rapid. Alloys at high temperatures will thus reach stable equilibrium concentrations much more rapidly than at absolute temperatures that are small compared with the melting temperature. If a solid-solution alloy in equilibrium at high temperatre is suddenly cooled to a low temperature where the maximum equilibrium percent solute is less than that of the alloy mixture, the solid solution is *supersaturated*. More solute atoms are in solution at this temperature than would be present after slow cooling. (Note that for as long as the atoms remain uniformly distributed, such a solid solution is still a single phase.)

All supersaturated solid solutions will theoretically reach equilibrium concentrations sooner or later. But if the time required at low temperature is of the order of hundreds or thousands of years, such a solid solution may be considered stable for engineering design purposes. The strengthening mechanism that is associated with a supersaturated solid solution will be the same as for a solid solution in equilibrium, except that the presence of additional solute atoms and the resulting severe distortion will cause even greater strengthening in the supersaturated solid solution. Thus if the amount of supersaturation is diminished by a reduction in the fraction of solute atoms in solution, the strength of the solid solution will be reduced. In the next two chapters other changes that can occur in such solutions will be discussed; these changes can result in the introduction of additional strengthening mechanisms not considered here.

SOLUTES AND DISLOCATIONS

The atoms in the vicinity of a dislocation are situated at sites substantially displaced from their equilibrium sites (see Chap. 6). In an edge dislocation such as in Fig. 7-6, the atoms below the slip plane are in tension and those above in compression. Figure 7-6 shows sites in an edge dislocation where interstitial atoms and large and small substitutional atoms could reduce the lattice strain energy. That is, the strain energy is less in Fig. 7-6 than in the case where the solute atoms are located far away from dislocations. Thus when these solute atoms "fall" into sites near dislocations, there is less probability that they will leave, and if enough solute atoms are located along the line of the dislocation it is said to be *pinned*. In order to move away from the solute atom, the dislocation must return the elastic energy that was dissipated when the solute atom arrived. At normal temperatures this extra energy can only be provided by increasing the stress required to move the dislocation, thereby increasing the strength. Then either the pinned dislocations break away from these so-called *atmospheres* at the higher stress or, as appears more likely, new dislocations are first generated at other points in the structure. Once generated, these new dislocations may move and multiply at lower stresses. This probably accounts in part for the drop in stress that is often observed in annealed mild steel following initial yielding [see Fig. 2-20(c)].

Strengthening by dislocation pinning forms an important part of full solid-solution strengthening, and this effect acts in addition to the lattice-distortion effect described earlier. Solute atoms can reach dislocations in two

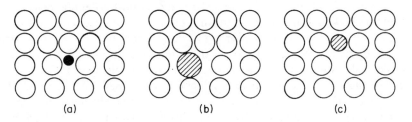

Figure 7-6 Energetically favorable lattice positions in an edge dislocation for (a) interstitial atom, (b) larger substitutional atom, and (c) smaller substitutional atom.

obvious ways: by diffusion of the solutes and by dislocation motion intersecting the solutes.

Diffusion of the solute atoms is promoted by *higher* temperatures; and interstitial solute atoms, being smaller, will diffuse more rapidly than substitutional solute atoms. For example, if a mild-steel tensile specimen is pulled slowly at 200–250°C, the carbon (and probably nitrogen) interstitial atoms can diffuse fast enough to catch the dislocations after they have broken away from their atomospheres of interstitials. In the strain-hardening region of the stress–strain curve, below ultimate, after a small amount of strain has occurred, enough interstitials will have caught and pinned dislocations to cause an increase in stress similar to initial yielding. This is followed by a drop in yield stress when the dislocations again break away from their interstitial atmospheres. This process is repeated many times to produce a sawtooth-shaped stress–strain curve of successive increases and sudden drops of stress; the phenomenon is called the *Portevin–Le Châtelier effect*. Return of the drop in stress at yielding can also occur in mild steel after room-temperature deformation, but weeks are required for the diffusion of interstitials to dislocations at room temperature. This process, whether occurring at room temperature in a week or at 150–250°C in a few minutes, is called *strain aging*.

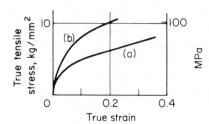

Figure 7-7 Stress–strain curves at 26°C for (a) 99.98% Al, 0.33-mm grain size, and (b) Al–0.23 atomic % Cu, 0.30-mm grain size. [From D. McLean, *Mechanical Properties of Metals* (New York: John Wiley & Sons, 1962), p. 173. Reprinted by permission of the publisher.]

The other means for bringing solute atoms and dislocations together is through dislocation motion that intersects the solute atoms. Since dislocations are generated and move at room temperature when plastic strain occurs, the increase in strength through the intersection of dislocations and solutes represents a secondary kind of strain hardening. Figure 7-7 shows the effect of a small amount of substitutional copper on the strain-hardening behavior of aluminum. (The small difference in grain size of the two specimens has only a very small influence on strain hardening.) The upward shift of the stress–strain curve cannot be attributed entirely to the pinning of dislocations by solutes; the basic solid-solution strengthening mechanism also acts. The higher strength of the solid solution forces the generation of new dislocations, which then interact with each other to increase the rate of strain hardening and to delay the onset of mechanisms by which dislocations can bypass barriers to their motion. The amount of strengthening that is due to each mechanism is not known.

ORDER–DISORDER EFFECTS

Substitutional atoms usually exhibit either a preference for atoms of the same element as adjacent neighbors—a type of behavior called *clustering*—or a preference for atoms of an unlike element as adjacent neighbors—called *short-range order*. In either case, such ordering will result in localized inhomogeneities in the atomic ratio when interatomic attractions and repulsions cause preferential positioning of atoms into ordered groups. The extreme example of clustering would thus be complete separation of the two dissimilar elements. It is clear that 1 kg of Ni and 1 kg of Cu will have lower energy when they are completely separated than when they are melted together to form a 2-kg mixture and then cooled, no matter what positions the atoms select after cooling.

Although the precise reasons for these dissimilar forces between atoms are not yet fully understood, it is clear that the equilibrium-ordered configuration has lower energy than the random solid solution. Figure 7-8(a) shows a solid solution of equal numbers of atoms of two elements that show a preference for unlike atomic bonds. After passage of a dislocation and consequent slip of one interatomic spacing, the number of unlike bonds will be reduced on the average at least to half and usually to less than half because of the initial ordering of the structure. Thus the strain energy of the crystal in Fig. 7-8(b), surface effects being ignored in this small sample of atoms, will be greater than in Fig. 7-8(a), which is nearer the equilibrium configuration. The source of this energy input must be the stress required to pass the dislocation through the region; thus the increase in the applied stress to provide the needed energy constitutes an increase in strength. If a second dislocation immediately followed behind the first, the stress required for its motion might be smaller, since the ratio of unlike atomic bonds would probably be increasing. But usually some time will pass before another dislocation traverses this same slip plane, and by then the atoms will have rearranged themselves through motion over very short distances into the lower-energy configuration, a new short-range order similar to that in Fig. 7-8(a). The process of forcing disorder by dis-

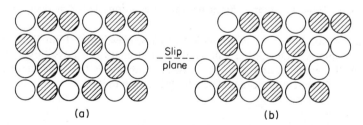

(a)   (b)

Figure 7-8   Order–disorder effect in a crystal composed of equal numbers of two kinds of atoms. (a) Ordered structure having five unlike bonds and one like bond across slip plane. (b) After slip of one interatomic space, structure has one unlike bond and four like bonds across slip plane.

location motion must then be repeated; therefore the stress for slip, and consequently the strength, will remain high.

Exercise 7-4

If a preference for unlike atomic bonds, such as in Fig. 7-8, results in strengthening, would a preference for like atomic bonds cause weakening?

Solution. No. If more like bonds existed in equilibrium before slip, then dislocation motion must decrease the number of like bonds and thus increase the local energy. The point of the argument is that dislocation motion disrupts the preferred local order of either type and thus requires a higher stress.

REFERENCES

7-1. American Society for Metals, *Metals Handbook*, 9th ed., vol. 1, Properties and Selection: Irons and Steels. Metals Park, Ohio: American Society for Metals, 1978.

7-2. American Society for Metals, *Metals Handbook*, 9th ed., vol. 2, Properties and Selection: Nonferrous Alloys and Pure Metals. Metals Park, Ohio: American Society for Metals, 1979.

7-3. American Society for Metals, *Metals Handbook*, 9th ed., vol. 3, Properties and Selection: Stainless Steels, Tool Materials and Special-Purpose Metals. Metals Park, Ohio: American Society for Metals, 1980.

7-4. American Society for Metals, *Strengthening Mechanisms in Solids*. Metals Park, Ohio: American Society for Metals, 1962. Chapter 2, "Solid Solution Strengthening," by P. A. Flinn, includes extensive theoretical treatment of this subject.

7-5. Brick, Robert M., Robert B. Gordon, and Arthur Phillips, *Structure and Properties of Alloys*, 3rd ed. New York: McGraw-Hill Book Co., 1965. Chapter 3 discusses diffusion and properties of solid solutions.

7-6. LeMay, Iain, *Principles of Mechanical Metallurgy*. New York: Elsevier North-Holland, 1981. Chapter 6 provides thorough coverage of strengthening mechanisms.

7-7. McLean, D., *Mechanical Properties of Metals*. New York: John Wiley & Sons, 1962. Chapter 6 gives a thorough analysis of the effects of solutes.

7-8. Reed-Hill, Robert E., *Physical Metallurgy Principles*. Princeton, N.J.: D. Van Nostrand Co., 1964. Chapter 8 discusses the interactions between dislocations and solute atoms.

7-9. Zackay, Victor F., ed., *High-Strength Materials*. New York: John Wiley & Sons, 1965. Chapter 9, "On Lattice Effects and the Strength of Alloys," by T. L. Johnston, A. S. Tetelman, and A. J. McEvily, Jr., emphasizes the importance in some materials of inherent lattice resistance to dislocation motion.

PROBLEMS

7-1. The largest interstitial holes in fcc crystals occur at $\frac{1}{2}, 0, 0$ and $\frac{1}{2}, \frac{1}{2}, \frac{1}{2}$ positions in the lattice, which are crystallographically equivalent. If solvent and solute atoms are assumed to be hard spheres, calculate the maximum possible ratio of solute-atom radius to solvent-atom radius without lattice distortion.

7-2. The largest interstitial holes in bcc crystals occur at $\frac{1}{2}, \frac{1}{4}$, 0 positions. If solvent and solute atoms are assumed to be hard spheres, calculate the maximum possible ratio of solute-atom radius to solvent-atom radius without lattice distortion. Why is the $\frac{1}{2}, \frac{1}{2}$, 0 position not the largest hole, since by inspection of a cube face it can be seen that there is more space between the corner atoms in the unit cell at $\frac{1}{2}, \frac{1}{2}$, 0 than there is at $\frac{1}{2}, \frac{1}{4}$, 0?

7-3. Using the results of Probs. 7-1 and 7-2, show why the bcc form of iron (atomic radius = 0.124 nm) might have lower solubility for interstitial carbon (atomic radius = 0.071 nm) than does the fcc form of iron (atomic radius = 0.128 nm).

7-4. What is a phase? Give three examples of solid phases.

7-5. Is it possible to produce a supersaturated Cu–Ni solid solution? Explain.

7-6. The *Bauschinger effect* is a reduction of the yield stress when the direction of applied stress is reversed, following initial deformation beyond yield. If solute atoms are viewed as local barriers to dislocation motion, how might this model be used to explain the Bauschinger effect?

7-7. In bcc iron, interstitial carbon can occupy a crystallographically different lattice site when the crystal is under tension or compression than when it is unstressed (this is called the *Snoek effect*). Since this shift in interstitial position requires energy, describe a mechanism by which the Snoek effect could contribute to solid solution strengthening.

7-8. Calculate the diameter of the maximum interstitial hole in aluminum.

Phase Diagrams

The presence of even a small amount of a second phase in the microstructure of a metal can markedly change the mechanical properties; in particular, strength usually increases. Control of the amount and distribution of the second phase provides substantial control over strength. This chapter will present a brief review of equilibrium-phase diagrams in order to show the origins of structure and strength of many commercial alloys and to develop the reasoning behind heat-treating operations.

COPPER–NICKEL, A SIMPLE EQUILIBRIUM PHASE DIAGRAM

Pure copper melts at 1084.5°C, and pure nickel melts at 1455°C. If 3 kg of nickel and 7 kg of copper are heated to 1500°C, all will be liquid. If the liquid nickel and copper are stirred together, they will form a liquid solution, since their atoms will be mixed together on a very fine scale. At what temperature will this mixture become solid on cooling, and in what form will the atoms be combined in the solid? These questions can be answered through use of the *equilibrium-phase diagram* (often called simply the phase diagram) for copper and nickel (Fig. 8-1).

This phase diagram tells what phases are present and their relative amounts, for any mixture of copper and nickel, at any temperature reached through a process of very slow cooling from the melt, or very slow heating from the solid form. To answer our present question, a mixture of 3 kg of

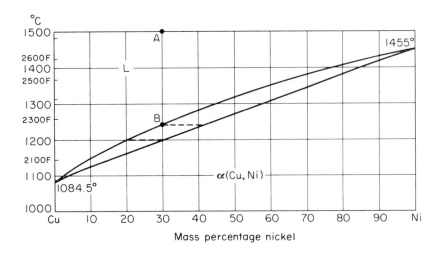

Figure 8-1 Copper–nickel equilibrium phase diagram. [From "Phase Diagrams," *Metals Handbook: Metallography, Structures and Phase Diagrams*, 8th ed., vol. 8, T. Lyman, ed. (American Society for Metals, 1973), p. 294.]

nickel and 7 kg of copper is 30% Ni. In Fig. 8-1, 30% Ni at 1500°C is point *A*, lying in a region labeled *L* (liquid), so the phase is a liquid solution and remains so on cooling until it reaches about 1240°C, point *B*.

On cooling just below 1240°C the solution enters the 2-phase region, and small amounts of a solid will precipitate, or freeze. (This solid is here designated as α; Greek letters are conventionally used to designate solid phases in equilibrium diagrams.) The 2-phase region represents combinations of composition and temperature that are impossible; that is, neither the liquid nor solid phases can have compositions of 30% Ni–70% Cu within this region. Rather, the phases present at, say, 1239°C are a solid of about 42% Ni and 58% Cu and liquid of slightly less than 30% Ni and slightly more than 70% Cu. Nothing exists with a composition between these two phases. Of course, the overall masses of Ni and Cu, taken together in the two phases, will be 3 kg of Ni and 7 kg of Cu.

Upon further slow cooling, more solid precipitates, but its nickel content gradually diminishes until at about 1200°C the solid is 30% Ni. At the same time, the content of all the solid that has precipitated earlier also changes to equal the composition of the solid currently precipitating. Thus very slow cooling is essential for true equilibrium.

At the same time, the composition of the remaining liquid at each stage changes, from its initial 30% Ni at 1240°C to about 20% Ni at 1200°C. It is obvious that this must occur because at all times the average composition of all the solid plus all the liquid must remain 30% Ni–70% Cu.

Below 1200°C, a single-phase solid solution of copper and nickel exists, with no further structural changes on cooling. The existence of a single-phase solid solution such as this one over the complete composition range is relatively rare.

SILVER–COPPER, A MORE COMPLEX SYSTEM

At a given temperature, if the amount of second element present in a two-element solid exceeds the maximum that can normally be present in solid solution with the first element (i.e., there is limited solid solubility), the excess of second element will usually be found in combination with the first element in the form of a separate and different phase. That this separate phase is different can be established by a marked difference in atomic ratio and often a difference in the crystal structure of the second phase.

The second phase forms because the energy of a single-phase solid solution of the same overall composition is greater than the total energy of the two separate phases. Each of these phases is then saturated with its respective solute element. Strain energy is required to displace solvent atoms from their equilibrium positions enough to permit solute atoms to be held in the matrix. As more solute atoms are added, the strain energy usually increases rapidly. Since the atomic-level phenomena that influence this strain energy are still not fully known, the energies cannot be calculated exactly. The atomic concentrations required for saturation are thus only observed, not calculated. With the assumption that a system will tend to change toward a minimum-energy condition, observation of real binary (two-element) combinations that are cooled very slowly usually reveals the saturation concentrations.

Thus the equilibrium-phase diagram is obtained from experiments. Data are taken with a range of concentrations of the elements involved. The discussion that follows will be limited to binary systems; since several ternary (three-element) systems are commercially important, the detailed descriptions of their use given in the references at the end of this chapter should be consulted. As an example of a binary system, Fig. 8-2 is the phase diagram for the copper–silver system. This diagram tells what phases are present under equilibrium conditions for each temperature and each possible composition, which is the percentage by mass (as here) or by ratio of the number of atoms of each element present. Each solid phase is designated by a Greek letter, in this case α and β.

At sufficiently high temperatures, all mixtures of silver and copper are liquids; the elastic strain energy associated with a distorted solid-crystal structure is therefore absent. Solidification of a pure element—of pure copper at temperatures below 1084.5°C, for example—may be looked upon as a transformation to a lower-energy configuration; otherwise, it would remain liquid. Solid copper at 1050°C thus has a lower energy than liquid copper at 1050°C, so if liquid copper is cooled quickly to 1050°C, it becomes supercooled and will soon transform to the solid form, giving off heat in the process. The

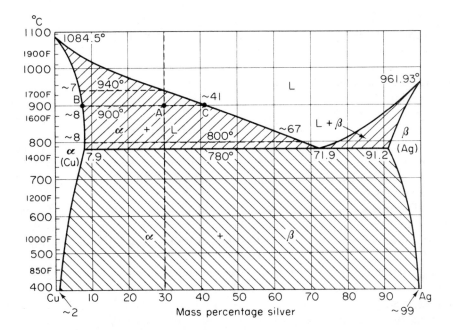

Figure 8-2　Copper–silver equilibrium phase diagram. [From "Phase Diagrams," *Metals Handbook: Metallography, Structures and Phase Diagrams*, 8th ed., vol. 8, T. Lyman, ed. (American Society for Metals, 1973), p. 253.]

freezing temperature of 1084.5°C does not mean that liquid copper cannot be cooled below that temperature and remain liquid, but rather than under equilibrium conditions (very slow cooling) pure cooper will freeze at 1084.5°C. The rest of the equilibrium-phase diagram must be interpreted in the same way.

Exercise 8-1

Name the phases present in a silver–copper alloy under the following conditions: (a) 30% Cu at 900°C; (b) 98% Cu at 800°C; (c) 3% Cu at 750°C; (d) 70% Cu at 850°C; (e) 50% Cu at 300°C.

Solution.　(a) Liquid solution; (b) α; (c) β; (d) liquid + α; (e) $\alpha + \beta$.

The unshaded areas of the phase diagram in Fig. 8-2 indicate regions where a single phase has the lowest energy and thus exists along under equilibrium conditions. The shaded areas represent temperature–composition combinations where no minimum energy single phase can exist in equilibrium. If a single phase exists at a temperature and composition that fall in a shaded area, the energy of this single phase is greater than the energy of some combination of two phases. Thus the shaded areas may be considered forbidden regions, where no one *stable* phase can exist. A mixture of silver and copper with an average composition that falls within a shaded area when it is cooled from liquid thus achieves the lowest possible energy by forming two phases from the adjacent unshaded areas. The relative amounts of these two phases are exactly those that give the overall composition of the original liquid solution.

As an example, consider slow-cooling a 30% Ag–70% Cu (by mass) liquid solution from 1100°C. This composition is shown as a vertical dashed line in Fig. 8-2. At 900°C, point A, the phase diagram indicates that this composition is forbidden, that under these conditions a solid solution rich in Cu (arbitrarily called α) of about 8% Ag (point B) and a liquid solution of about 41% Ag (point C) together provide the lowest possible energy. This forbidden region is labeled "α + L" because these are the two phases present, α and liquid, and each has a composition that falls on its boundary with the two-phase region.

The masses of α of 8% Ag and liquid of 41% Ag (we assume for now that these are exact values) must be such as to given an *average* overall composition of 30% Ag. If the average overall composition of the mix at 900°C were 8% Ag (point B in Fig. 8-2), then there would be all α and no liquid. At the other extreme, if the average composition were 41% Ag (point C), then there would be all liquid and no α present at 900°C. The fraction of liquid present varies linearly from 0 to 1.00 of the total mix, as the average composition is changed from 8% Ag to 41% Ag. Since the average composition for this problem is 30% Ag, the proportion that is liquid is the fraction BA/BC, or

$$\frac{0.30 - 0.08}{0.41 - 0.08} = 0.667$$

The proportion of α present is $1 - 0.667 = 0.333$. To check this result, assume an initial mix of 50 kg containing 15 kg of Ag and 35 kg of Cu, which would thus be 30% Ag. At 900°C, Cu will be found in the 0.333 fraction of the mix that is α (92% Cu) and in the 0.667 fraction of the mix that is liquid (59% Cu). The total mass of Cu will be distributed as follows:

Cu in α: (0.92)(0.333)(50 kg) = 15.33 kg

Cu in liquid: (0.59)(0.667)(50kg) = 19.67 kg

Total Cu = 35.00 kg

This calculation fully accounts for all the copper. A similar calculation can be made for the silver in the two phases.

The fraction of one phase of a two-phase mixture is simply the fraction of the distance across the two-phase region, moving *toward* the single-phase region of the phase in question. Thus, for the above problem as illustrated in Fig. 8-2, the *fraction* of liquid for a 30% Ag mixture at 900°C is $\overline{BA}/\overline{BC}$, or $(30 - 8)/(41 - 8) = 0.667$. Likewise, the fraction of α is $\overline{CA}/\overline{CB}$, or $(41 - 30)/(41 - 8) = 0.333$; in this case, the fraction of α varies from 0 to 1.00 *from right to left* across the α + L region, toward the α region. The same procedure applies to the calculation of the masses of liquid and solid β (silver rich) for compositions greater than 71.9% Ag.

Exercise 8-2

A mixture of 30 kg of silver and 70 kg of copper is slowly cooled from liquid to 800°C (see details in Fig. 8-2). Find (a) the total mass of α, and (b) the mass of silver in α.

Solution. From Fig. 8-2, the average composition is 30 kg/(30 + 70)kg = 0.3, or 30% Ag, the same as for the above discussion.
(a) The mass of α at 800°C is [(67 − 30)/(67 − 8)](100 kg) = 62.7 kg.
(b) The mass of silver in 62.7 kg of α, composition 8% Ag, is 0.08 × 62.7 kg = 5.02 kg.

EUTECTIC TRANSFORMATION (SILVER–COPPER)

Although the calculations in the previous section will always give the correct masses of phases in equilibrium, they tell nothing about the size, shape, and distribution of the phases. Because the configuration of the phases can have a large influence on strength (see Chap. 9), some special kinds of transformations that affect phase configuration must be studied. The Ag–Cu diagram in Fig. 8-2 is characteristic of a large number of binary systems in that it exhibits a *eutectic transformation*, where a single-phase liquid of 28.1% Cu and 71.9% Ag transforms at a constant temperature (780°C) on cooling to a solid consisting of the two phases α and β in a unique proportion. Because the entire transformation occurs at constant temperature—the *eutectic temperature*—the two solid phases that are formed are usually mixed together on a microscopic scale in a characteristic geometry that can be identified under the optical microscope and is called the *eutectic microstructure*. (Note that a microstructure may contain more than one phase.) A photomicrograph of the characteristic as-cast Ag–Cu eutectic microstructure is shown in Fig. 8-3(a). For contrast, Fig. 8-3(b) shows Ag–Cu of eutectic composition that has been cooled under controlled solidification conditions. (The effect of shape and spacing of phases is discussed in Chap. 9.) Before studying the special mechanical properties that result from the eutectic transformation, methods must be developed for calculating masses of eutectic microstructure present in compositions deviating somewhat from the eutectic.

The eutectic is not a phase; it is made up of two phases. Therefore the amount of eutectic present cannot be calculated directly from the equilibrium diagram. But since the eutectic always forms from the complete transformation of a single-phase liquid of eutectic composition at the eutectic temperature, the amount of eutectic must be the same as the amount of liquid from which the eutectic was formed. Eutectic forms *only* from this liquid (71.9% Ag at 780°C, in the Ag–Cu case), so the mass of this liquid fully accounts for the mass of eutectic formed. The horizontal constant-temperature line at eutectic temperature divides liquid-solid and solid-solid regions, so it is not possible to calculate the amounts of phases present at this exact temperature. For example, for

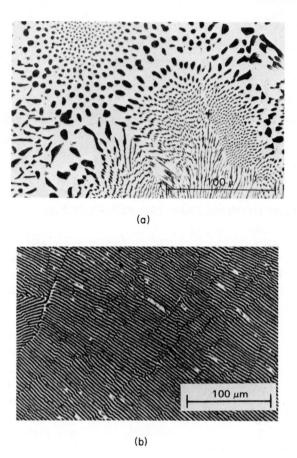

(a)

(b)

Figure 8-3 Silver–copper eutectic (71.9% Ag) in two forms, where the dark phase is α and the light phase is β (see Fig. 8-2): (a) as-cast, and (b) cooled under controlled conditions to produce directional solidification. (Photographs courtesy of Robert R. Jones.)

a eutectic composition (71.9% Ag), the equilibrium phases present at exactly the eutectic temperature (780°C) can be all liquid, liquid + (α + β), or all (α + β); how much liquid remains at exactly 780°C will depend on the cooling conditions and cannot be determined from the phase diagram. For this reason, calculation of amounts of phases must be made at temperatures slightly above or below the eutectic temperature. Instead of writing 780+°C or 780−°C, a convenient convention is to add 1°C to or substract 1°C from the eutectic temperature, a shift so small as to cause negligible changes in composition of the phases. (For example, the saturation composition of β at 779°C can be considered to be 91.2% Ag, the same as β at 780°C.) There is then no ambiguity, since for equilibrium conditions a mixture of 28.1% Cu and 71.9% Ag at 781°C is all liquid, and at 779°C it is all (α + β).

The next step, calculation of the amount of eutectic for, say, 50 kg of a 70% Cu–30% Ag mixture, involves calculation of masses of phases by the method already described. At 781°C all the liquid present is of eutectic composition (71.9% Ag), so the mass of liquid at 781°C is equal to the mass of eutectic at 779°C.

Exercise 8-3

Calculate the mass of liquid in the above case at 781°C.

Solution. The composition of liquid is 71.9% Ag, and that of α is 7.9% Ag. Thus the fraction liquid at 781°C (from left to right) is

$$\frac{30 - 7.9}{71.9 - 7.9} = 0.345$$

The mass of liquid is thus (0.345)(50 kg) = 17.3 kg.

From the solution to Exercise 8-3 we see that the mass of eutectic at 779°C is the same as the mass of liquid at 781°C, or 17.3 kg. During the eutectic transformation, the *free* α (*noneutectic*) remains unchanged. At 781°C, $50 - 17.3 = 32.7$ kg of α is present with the liquid; at 779°C, this same 32.7 kg of free α is still present with the eutectic. So α is now found in two places in the microstructure: as free α, and as part of the eutectic, mixed intimately with β in the special eutectic configuration. If the average composition were, say, 85% Ag, free β instead of free α would be present with the eutectic.

Further slow cooling of the 50 kg of 30% Ag results in no phase changes that have much influence on mechanical properties, but in some special heat treatments these changes can be important. Note that in this temperature range the solubility of both α and β decrease on cooling. The eutectic at 779°C contains $[(71.9 - 7.9)/(91.2 - 7.9)] \times 17.3$ kg $= 13.3$ kg of β of 91.2% Ag, and 17.3 kg $- 13.3$ kg $= 4.0$ kg of α of 7.9% Ag. At 400°C the eutectic contains α of about 2% Ag and β of about 99% Ag. These changes in composition as the eutectic cools do not alter the total mass (17.3 kg) or average composition (71.9% Ag) of the eutectic. The composition changes of the two phases occur by diffusion of Cu to α and of Ag to β across the boundaries between the α and β in the eutectic. The intimate mixture of the two phases provides very large surface areas of phase boundaries to accommodate this diffusion. Because the mass of metal atoms that diffuses is a small fraction of the total mass of eutectic, the shape of the eutectic microstructure changes very little in cooling from 779°C to 400°C. The mass proportions of α and β in the eutectic can also change slightly by diffusion during cooling. However, many phase diagrams have nearly symmetrical α and β *saturation lines* (separating the α region and the β region from the $\alpha + \beta$ region), so that the change in proportions of the two solid phases is small.

Exercise 8-4

Find the masses of α and β in 17.3 kg of Ag–Cu eutectic at 400°C.

Solution.   Calculating from right to left in Fig. 8-2 at 400°C, for 71.9% Ag (the eutectic composition), we find that the mass of eutectic α is

$$\left(\frac{99 - 71.9}{99 - 2}\right)(17.3 \text{ kg}) = 4.8 \text{ kg}$$

The mass of eutectic β is 17.3 kg − 4.8 kg = 12.5 kg. Notice that the relative masses of α and β at 400°C are slightly different from the amounts calculated at 779°C, already calculated above as 4.0 kg and 13.3 kg.

At 779°C the 30% Ag mixture also contains 32.7 kg of free α. During further cooling some diffusion will occur between the free α and the eutectic. Because these are not mixed on such a fine scale as the α and β in the eutectic, the phase-boundary area is much smaller, and thus diffusion between the free phase and the eutectic is assumed to be negligible. This assumption then allows a conceptual separation of the eutectic liquid and the noneutectic at 781°C, so that they can be treated as separate problems. This argument justifies the statement in the previous paragraph that the total mass of eutectic remains constant during cooling. The mixture of 70% Cu–30% Ag can now be broken up and thought of as two separate components at 781°C: 17.3 kg of eutectic liquid of average composition 71.9% Ag, and 32.7 kg of free α of 7.9% Ag. The eutectic component of the problem was covered above. The noneutectic component must maintain the same average composition as the α at 781°C. The phase diagram shows that a mixture of average composition 7.9% Ag cannot be a single phase at 400°C, and that both α and β are present.

Exercise 8-5

Find the mass of β present after 32.7 kg of free α of 7.9% Ag at 781°C is cooled to 400°C.

Solution.   Calculating at 400°C, 7.9% Ag, from left to right in the two-phase region, we find that the mass of β is

$$\left(\frac{7.9 - 2}{99 - 2}\right)(32.7 \text{ kg}) = 2.0 \text{ kg}$$

The mass of β calculated in Exercise 8-5 forms gradually as the α is slowly cooled from 781°C to 400°C. This β is observed to form at the α grain boundaries, because grain boundaries are regions of higher energy where nucleation and growth of the second phase can provide larger possible energy reduction.

Figure 8-4 shows a slowly cooled 90% Ag–10% Cu (which was used for many years for silver coins in the United States, until the high price of silver forced a change to a copper–nickel alloy), which results in a mixture of noneutectic β and eutectic. The small amount of grain boundary α that should form on slow cooling, in the manner described in the previous paragraphs, may account for the black line that can be seen along part of the β grain boundary.

The logic followed in the procedure described above for calculating the

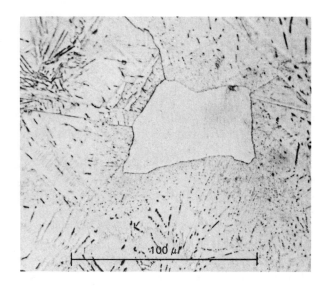

Figure 8-4 Slowly cooled 90% Ag–10% Cu (coin silver) showing a light patch of silver-rich β surrounded by eutectic.

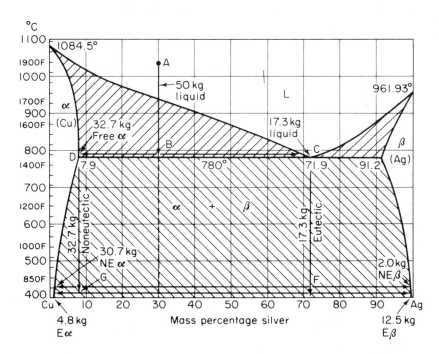

Figure 8-5 Procedure for calculating amounts of phases present during equilibrium cooling. E means eutectic, NE noneutectic. [Adapted from "Phase Diagrams," *Metals Handbook: Metallography, Structures and Phase Diagrams*, 8th ed., vol. 8, T. Lyman, ed. (American Society for Metals, 1973), p. 253.]

relative amounts of eutectic, noneutectic, free, and grain-boundary phases can be expressed graphically. Figure 8-5 shows the method of solving the above problem of 30% Ag–70% Cu cooled under equilibrium conditions from liquid. Start with 50 kg of liquid at point *A* in Fig. 8-5. At point *B*, just above the eutectic temperature, the liquid (17.3 kg) and the free α (32.7 kg) can be separated for calculation purposes by arrows from *B* to *C* and *D*. The liquid of eutectic composition at *C* first transforms to eutectic solid α + β as it passes through the eutectic temperature, then cools to 400°C. The amounts of eutectic α and β, 4.8 kg and 12.5 kg, respectively, at 400°C are shown by arrows from *F* to the saturation lines for α and β at the edges of the two-phase α + β region. Likewise, the noneutectic (free) α at point *D* cools and precipitates out grain boundary β. At point *G*, 400°C, the noneutectic can be split into 30.7 kg of noneutectic (free) α and 2.0 kg of noneutectic (grain boundary) β. Note that after the separation at point *B* into eutectic and noneutectic components, the average composition value of 30% Ag–70% Cu is not used again in any calculation, since the compositions of the eutectic and noneutectic components control the relative amounts of α and β present.

EUTECTOID TRANSFORMATION (IRON–CARBON)

A *eutectoid transformation* is the cooling transformation at a constant temperature from a single-phase *solid* to two new solid phases that are mixed together in a characteristic manner. Thus all that was said in the previous section for a eutectic transformation from liquid to solid applies to a eutectoid transformation, except that the initial phase in the eutectoid transformation is a solid, and thus this is a *solid-state transformation*. The eutectoid transformation occurs for a unique composition in a given system and at a unique equilibrium temperature. The methods for calculating amounts and compositions of phases in the eutectoid are the same as for the eutectic.

The most important of all eutectoid systems is the iron–carbon system, shown in Fig. 8-6. The forbidden two-phase regions are labeled with the two phases present. It is still necessary to observe experimentally the manner in which these phases are mixed together, which has earlier been described as the *microstructure*. The upper left region of the diagram, at 1495°C, contains a *peritectic transformation* (solid + liquid transforms on cooling to a different solid), which will not be considered further in this discussion, because in most cases further transformations on cooling obliterate any influence of the peritectic on mechanical properties at room temperature. The Fe–C system also has a eutectic transformation for 4.30% C at 1148°C, which is of importance in cast irons (Chap. 15).

Pure iron below 912°C is a bcc structure called *ferrite* (α), which can hold only very small amounts of carbon in interstitial solid solution; in calculations

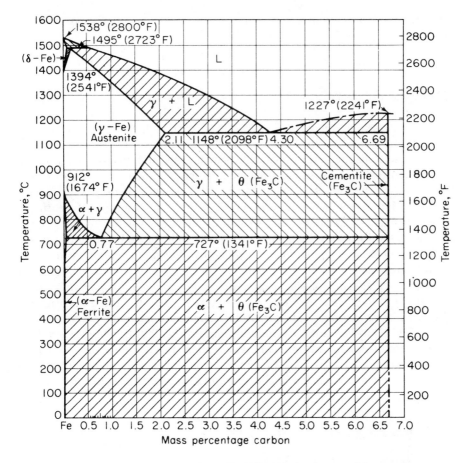

Figure 8-6 Iron–carbon equilibrium phase diagram, in absence of free graphite. [From "Phase Diagrams," *Metals Handbook: Metallography, Structures and Phase Diagrams,* 8th ed., vol. 8, T. Lyman, ed. (American Society for Metals, 1973), p. 275.]

the carbon content at 20°C is taken as zero. At higher temperatures the equilibrium structure of interstitial carbon in iron is an fcc structure called *austenite* (γ), which can have up to 2.11% carbon in solid solution (at 1148°C). Iron and carbon can also form a compound, Fe_3C, when enough carbon is present. Fe_3C has acquired several names: *iron carbide, carbide,* and *cementite* are the most common. It is usually assigned the symbol θ, the customary designation for intermetallic compounds in phase diagrams. Fe_3C is orthorhombic in crystal structure and is composed of 6.69% C by mass.

Exercise 8-6

Show that Fe_3C must have 6.69% C by mass.

Solution. The atomic ratio in Fe_3C must be $3Fe : 1C$. Using atomic masses of each element gives

$$\text{Mass fraction C} = \frac{1 \times 12.011}{(1 \times 12.011) + (3 \times 55.85)} = 0.06689 = 6.69\%$$

This is the desired result.

The *eutectoid composition* of iron and carbon is 0.77% C. When this composition is cooled slowly, at 727°C the austenite transforms at constant temperature to ferrite and iron carbide, which grow simultaneously during the transformation in the form of a layered structure called *pearlite* (because it looks, in a microscope, somewhat like mother-of-pearl). Pearlite is composed of alternating layers of ferrite and Fe_3C. A photomicrograph of pearlite is shown in Fig. 8-7. Details of the structure can be seen better in the electron micrograph in Fig. 8-8, which shows the form of the etched surface of pearlite. The width of these layers can be controlled by altering the cooling rate; faster cooling, such as by cooling a small bar in air or even oil quenching, will produce finer spacing of the layers of ferrite and iron carbide in pearlite. Narrowing of the layers with faster cooling occurs because the transformation in this *quasi-equilibrium* transformation takes place at a temperature lower than 727°C. Thus it is not a true equilibrium process, but because the products of the transformation are the same it is called quasi-equilibrium. When the transformation occurs at a lower temperature, diffusion is limited to shorter distances, the transformation begins at more locations throughout the austenite, and so the pearlite is finer.

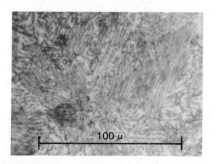

Figure 8-7 Optical micrograph of pearlite.

Exercise 8-7

Calculate the mass fraction of iron carbide in pearlite at room temperature.

Solution. From Fig. 8-6 the percentage of carbon in ferrite at room temperature is nil, so, moving across the $\alpha + \theta$ region from left to right, we find that the mass fraction of θ is

$$\frac{0.77 - 0}{6.69 - 0} = 0.115 = 11.5\%$$

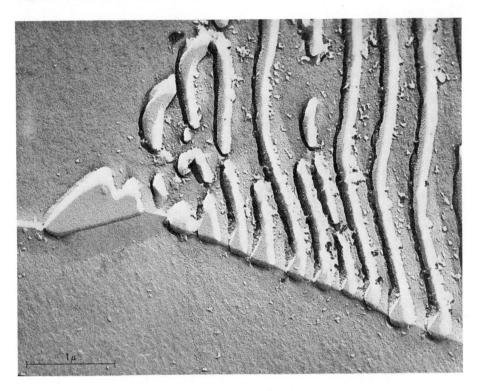

Figure 8-8 Electron micrograph of replica of polished and etched surface of annealed carbon steel with less than 0.77% carbon. Upper right portion is pearlite; note the irregularities in the formation of the layers of carbide and ferrite. The bottom and upper left regions are ferrite.

For *noneutectoid* compositions, the procedure used to calculate amounts of different microstructures is exactly the same as for the noneutectic systems. Figure 8-9 gives more detail in the low-carbon region of the Fe–C diagram than can be seen in Fig. 8-6. From Fig. 8-9, slow cooling of 15 kg of, for example, 0.40% C gives the following structures:

At 820°C: 100% austenite, or 15 kg of austenite.

At 728°C: Fraction $\gamma = \dfrac{0.40 - 0.02}{0.77 - 0.02} = 0.507 = 50.7\%$

$0.507 \times 15 = 7.60$ kg γ and 7.40 kg α.

At 726°C: Separate into two isolated systems:
    (a) $\gamma \rightarrow$ pearlite (abbreviated P), giving 7.60 kg P($\alpha + \theta$).
    (b) $\alpha \rightarrow \alpha$ (no change), 7.40 kg α. This is called *proeutectoid* or *free ferrite* to distinguish it from the ferrite in the pearlite.

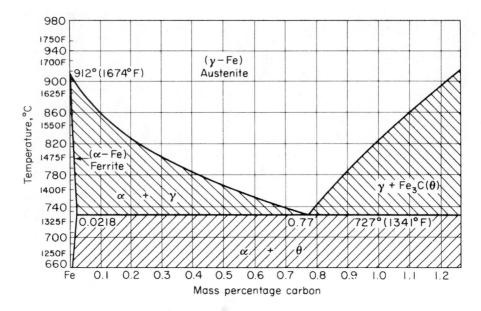

Figure 8-9 Iron–carbon equilibrium phase diagram, to 1.2% C. [From "Phase Diagrams," *Metals Handbook: Metallography, Structures and Phase Diagrams*, 8th ed., vol. 8, T. Lyman, ed. (American Society for Metals, 1973), p. 275.]

At 20°C: (a) Mass of pearlite is unchanged, 7.60 kg, but composition of ferrite changes from 0.02% C to virtually zero C, while the fraction of Fe$_3$C increases very slightly.

(b) $\alpha \rightarrow \alpha + \theta$. The final ferrite has virtually zero C. The very small amount of Fe$_3$C forms at the grain boundaries and in most cases can be neglected. Mass of grain-boundary Fe$_3$C is

$$\frac{0.02 - 0}{6.69 - 0} \times 7.40 \text{ kg} = 0.022 \text{ kg}$$

In calculating rough mass fractions for the 0.40% C–Fe, little error is introduced by neglecting the 0.02% C composition of ferrite at 726°C and approximating the eutectoid composition as 0.8% C. Thus the fractions of pearlite and free ferrite for this problem would be exactly 0.5 and would remain thus throughout the range from 726°C to 20°C. This simplification can be used for approximate calculation of mass fractions for carbon contents above about 0.15% without significant error. However, for calculating compositions, the values shown in the equilibrium diagram must be used without alteration.

Exercise 8-8

Calculate the fraction of pearlite at 20°C in 0.60% C–Fe that has been slowly cooled (*annealed*) in a furnace from austenite to room temperature.

Solution. From Fig. 8-9 and the foregoing simplification, the fraction of austenite at 728°C is $(0.60 - 0)/(0.80 - 0) = 0.75$. All austenite at 728°C goes to pearlite, so the fraction of pearlite at 726°C is 0.75. This fraction is unchanged by further cooling, so the fraction of pearlite at 20°C is 0.75.

Thus equilibrium-cooled steels will have pearlite fractions approximately as follows: 0.20% C steel, 25% pearlite; 0.40% C steel, 50% pearlite; 0.60% steel, 75% pearlite; eutectoid steel, 100% pearlite.

Compositions with carbon content greater than 0.77% produce pearlite plus iron carbide. For example, consider 15 kg of 1.00% C cooled from austenite:

At 860°C: 100% austenite, 15 kg.

At 728°C: Fraction $\theta = \dfrac{1.00 - 0.77}{6.69 - 0.77} = 0.0389 = 3.9\%$.

The other 96.1% is γ.
Mass $\theta = 0.0389 \times 15 = 0.58$ kg. This is found at the austenite grain boundaries.

At 726°C: Separate into two isolated systems.
(a) $\gamma \rightarrow$ pearlite $(\alpha + \theta)$, $(15 - 0.58) = 14.42$ kg.
(b) $\theta \rightarrow \theta$ (no change), which now is found at the boundaries of the prior austenite and thus comprises a three-dimensional network within the pearlite.

At 20°C: No further significant change occurs below 726°C under equilibrium cooling conditions.

SUMMARY

Equilibrium-phase diagrams reveal the composition and amount of each phase present under equilibrium conditions. These are the phases toward which a nonequilibrium system will tend to move if given high enough temperature and long enough time for the necessary diffusion of atoms. The form and distribution of the phases (the *microstructure*—see Chap. 9) are not given by the equilibrium-phase diagram and therefore must be found by experiment and observation. In many binary metal systems the equilibrium phases can also be produced by different nonequilibrium processes so as to have very different forms and distributions. The consequent differences in strength and hardness are discussed in Chap. 9.

REFERENCES

8-1. American Society for Metals, *Metals Handbook*, 9th ed. vol. 8, Metallography, Structures, and Phase Diagrams. Metals Park, Ohio: American Society for Metals, 1973. This volume contains an extensive and readily available collection of binary and ternary phase diagrams.

8-2. Birchenall, C. Ernest, *Physical Metallurgy*. New York: McGraw-Hill Book Co., 1959. Pages 87–92 give an introduction to ternary systems.

8-3. Brophy, Jere H., Robert M. Rose, and John Wulff, *Thermodynamics of Structure*. New York: John Wiley & Sons, 1964. Chapter 3 discusses the conditions for the formation of surfaces between phases, that is, phase boundaries. Chapter 6 describes nucleation and growth of second phases.

8-4. Reed-Hill, Robert E., *Physical Metallurgy Principles*. Princeton, N.J.: D. Van Nostrand Co., 1964. Chapter 13 thoroughly discusses all important types of phase diagrams.

8-5. Rhines, F. N., *Phase Diagrams in Metallurgy*. New York: McGraw-Hill Book Co., 1956. A test completely devoted to the subject of phase diagrams.

PROBLEMS

8-1. For the binary system shown in Fig. 8-10, calculate the mass of β in the eutectic at 50°C, if 3 kg of element A is mixed with 7 kg of element B and slowly cooled from 400°C.

8-2. Three 100-kg bars of element A and one 100-kg bar of element B are put in a crucible and heated to 400°C, then slowly cooled to 100°C. Using the equilibrium diagram in Fig. 8-10, find, at 100°C.:

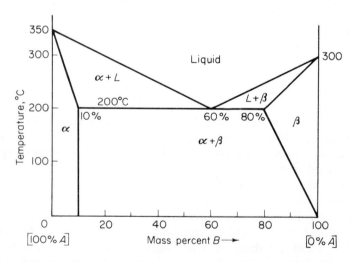

Figure 8-10 Equilibrium phase diagram for fictitious elements A and B (for use in problems).

(a) The mass of β solid solution in the eutectic.

(b) The mass of element B not in the eutectic.

8-3. A mixture of 3 kg of silver and 7 kg of copper is heated until completely melted and then cooled slowly to 900°C. Find:

(a) The mass of liquid present at 900°C.

(b) The mass of eutectic β (silver-rich) after cooling to 700°C.

(c) The total mass of β formed from the noneutectic α after this is cooled rapidly to 400°C and held until phase equilibrium again occurs.

8-4. Into a crucible are placed 20 kg of tin and 60 kg of lead (see Fig. 8-11). This mixture is heated until completely melted, stirred well, and cooled very slowly to 100°C. Find the following, for 100°C:

(a) The total mass of tin in the eutectic present in this crucible.

(b) The composition of the noneutectic α.

(c) The mass of noneutectic a.

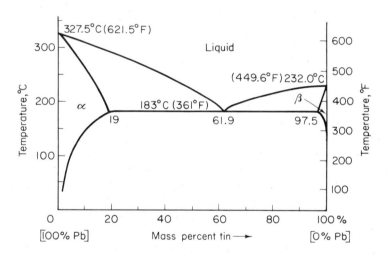

Figure 8-11 Lead–tin equilibrium phase diagram. [From "Phase Diagrams," *Metals Handbook: Metallography, Structures and Phase Diagrams*, 8th ed., vol. 8, T. Lyman, ed. (American Society for Metals, 1973), p. 330.]

8-5. In the lead–tin diagram (Fig. 8-11) what is the percentage β in the eutectic?

8-6. (a) If 40 kg of tin and 60 kg of lead (see Fig. 8-11) are placed in a crucible, heated to 350°C and slowly cooled to 180°C, what are the amounts of the microstructure(s) present?

(b) What is the composition of each phase present?

(c) If the mixture is then cooled to 50°C, what are the amounts of each phase present?

8-7. Compute the amount of each microstructure and the fraction of each phase present in a 0.3% carbon steel slowly cooled from the melt to each of the following temperatures:

(a) 930°C.

    (b) 760°C.

    (c) 700°C.

8-8. A 1.00% C steel that has been cooled slowly from 1100°C and then held at 740°C for 0.5 h is air-cooled to room temperature.

    (a) Name the phases present.

    (b) Describe the microstructure.

    (c) Calculate the fraction of noneutectoid carbide present.

8-9. A mixture of 20 kg of element A and 20 kg of element B (see Fig. 8-10) is melted in a crucible and cooled slowly to 0°C.

    (a) Calculate the change in mass of eutectic α between 199°C and 0°C. Is this an increase or a decrease?

    (b) Describe the physical process by which such a change in mass can occur.

8-10. A mixture of 6.49 kg of silver and 11.63 kg of copper is cooled slowly from 1100°C to 400°C. Calculate the mass of noneutectic α present at 400°C.

8-11. A mixture of 2.45 kg of lead and 1.43 kg of tin is heated to 350°C and slowly cooled to 100°C.

    (a) Calculate the total mass of α at 100°C.

    (b) Calculate the mass of eutectic β at 100°C.

    (c) Calculate the composition of noneutectic β at 100°C.

Multiple Phases

The effectiveness of second phases in interfering with dislocation motion and in forcing the generation of new dislocations has a marked influence on the strength of a solid. The size, shape, and distribution of a second phase, as well as the form in which the first phase occurs, will affect the ease of dislocation motion. This chapter examines the behavior of several metal alloys to provide examples of the control of strength through control of second phases.

PHASE BOUNDARIES

In Chap. 8 it was shown how two or more phases can be formed and why they coexist in equilibrium at moderate temperatures. A *microstructure* is a phase or combination of phases of specified geometry, unique enough to warrant its being assigned a special name. Any single-phase material may thus be called a microstructure, in which case the microstructure is also a phase. Mixtures of two or more phases that are, say, layered, or with one phase spheroidal, or with one phase disk-shaped or rod-shaped, may properly be designated as microstructures and thus may contain two or more phases. If two phases are intimately mixed, for example, the eutectic microstructure of Ag–Cu in Fig. 8-3(a), then the *phase-boundary area* within the *microconstituent* (i.e., the eutectic mixture) is extensive. In contrast, a microstructure consisting of two phases, each of normal grain size, will have much less phase-boundary area and the phase boundaries will occur between the microconstituents.

Phase boundaries are very effective barriers to dislocation motion for at least two reasons: (1) The atoms are displaced from their equilibrium positions by the forces acting on them by unlike atoms in a different crystal at some angle to the first, in much the same way as was described earlier for grain boundaries; and (2) if a dislocation does pass through the phase boundary, it is forced to move in a crystal of different orientation and Burgers vector, and possibly with a different crystal structure. Because only the orientation difference occurs at a grain boundary in a single-phase solid, with no change in magnitude of Burgers vector or crystal structure, grain boundaries are in general less effective barriers to dislocation motion than are phase boundaries. In addition to this argument, the second phases that form in many alloy systems are chemical compounds that have structures in which dislocation motion is much more difficult than in either pure constituent element. This further enhances the effectiveness of many phase boundaries as barriers to dislocation motion. This chapter discusses the marked effect of phase boundaries on strength.

MEAN FREE PATH

Chapter 5 described the effect of dislocation glide distance on strength. As the mean glide distance increases, the number of moving dislocations required for a given strain decreases. Thus the last dislocations have moved at a lower stress than would be required for the same strain but with shorter glide distance, since more dislocations would then be needed to produce the same strain. This is generally associated with a reduction in strength. The glide distance for dislocation motion in the softer phase of a multiple-phase alloy will vary considerably, depending on the location and orientation of each dislocation in the microstructure. If a series of random measurements of glide distance between harder phases is taken, the mean value of these measurements is termed the *mean free path*. Figure 9-1 shows the dependence of the yield strength of pearlite on the mean free path in the ferrite layers. If manganese is introduced into an iron–carbon alloy, it acts as if it were in some ways equivalent to carbon, so that the fraction carbon for the eutectoid composition is then lower than 0.77%. The data for Fig. 9-1 are for pearlite from 0.77% down to 0.56% C, and for manganese contents up to 3.5% (which is much higher than is normally present in steels). The range of yield strengths shown by the shaded area in Fig. 9-1 suggests that factors in addition to mean free path are influencing strength. Other strengthening mechanisms must be operating in addition to phase-boundary effects. This conclusion is generally valid for the more complex materials, since most of the strengthening mechanisms discussed in previous chapters can act simultaneously with phase boundaries to increase strength. In order to determine the strengthening effect of any one variable, the other variables should be closely controlled. Since this frequently is not possible, the individual contribution of each strengthening mechanism cannot be established in many materials.

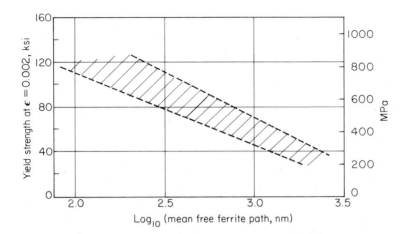

Figure 9-1 Variation of yield strength with mean free ferrite path in pearlite for a range of carbon and manganese contents. [From R. W. Guard, "Mechanisms of Fine-Particle Strengthening," *Strengthening Mechanisms in Solids* (Metals Park, Ohio: American Society for Metals, 1962), p. 261.]

The data shown in Fig. 9-1 are not in convenient form for estimating the strength of pearlite that will result from a given heat treatment. A rule-of-thumb procedure, based on measurements, can be used for plain-carbon steels in the form of bars or plates that are of the order 10 mm in diameter or thickness, as follows:

1. If furnace-cooled from austenite (*annealed*), the pearlite has relatively wide spacing and is termed *coarse pearlite* (P_c), with a Brinell hardness number (Bhn) of about 240. It has been observed that for ductile steels

$$S_u[\text{MPa}] \approx 3.45 \text{ Bhn} \tag{9-1}$$

   Thus S_u for coarse pearlite is about 830 MPa. Furnace-cooled eutectoid steel often exhibits some *spheroidization*, the formation of spheroidal carbide particles, as can be seen in Figs. 8-7 and 8-8. Therefore in such cases its hardness represents not the hardness of coarse pearlite but that of a mixture of coarse pearlite and spheroidite. Spheroidite has a lower hardness (about 150 Bhn) than coarse pearlite for reasons given in the next section.

2. If air-cooled from austenite (*normalized*), the pearlite has a narrower, intermediate spacing and is termed *medium pearlite* (P_m), with hardness of about 280 Bhn and $S_u \approx 970$ MPa. If a specimen of air-cooled pearlite is reheated to just below the eutectoid temperature, say 700°C, and held for a long time, say 50 ks, the layers of iron carbide will fully agglomerate into individual spheroid-shaped particles, called *spheroidite*. This mi-

crostructure is similar to that just discussed under furnace-cooling conditions, except that in the present case no layered pearlite remains.

3. If oil-quenched from austenite, the pearlite has very narrow spacing and is termed *fine pearlite* (P_f), with hardness of about 380 Bhn and $S_u \approx 1.3$ GPa.

4. For contrast, the hardness of free ferrite alone is about 80 Bhn.

These approximate values of Bhn are listed in Table 9-1.

For faster cooling rates, and in fact even for oil quenching of some high-carbon plain-carbon steels, the formation of martensite (see Chap. 10) instead of pearlite markedly alters the final hardness. The calculations of (1) through (4) above are only approximate and should be used with caution, not for design purposes, since variations in austenite grain size, cooling rate, specimen size, and composition can alter these results. For a design application the specific properties of the steel to be used must be known.

TABLE 9-1   Hardness of Ferrite, Spheroidite, and Pearlite*

Constituent	Abbreviation	Bhn (3000 kgf)	Rockwell Equivalent
Ferrite	α	80 (500 kgf)	47 R_B
Spheroidite	P_s	150	80 R_B
Coarse pearlite	P_c	240	22 R_C, 100 R_B
Medium pearlite	P_m	280	30 R_C
Fine pearlite	P_f	380	41 R_C

*Approximate hardness of constituents of plain-carbon steel, to be used only for rough estimate, not for design.

The data in Fig. 9-1 can be roughly approximated by the Petch equation, Eq. (6-4), to give the yield strength of pearlite as a function of mean free ferrite path $\bar{p}$ (μm):

$$S_y \approx 200 \text{ MPa} + (240 \text{ MPa})(\bar{p}[\mu\text{m}])^{-1/2} \tag{9-2}$$

The following exercise suggests that the effect of mean free ferrite path on yield strength can be approximated by Eq. (9-2).

Exercise 9-1

If the yield strength of pearlite is approximately 70% of its ultimate strength, estimate the spacing of coarse pearlite.

Solution. From (1) above, $S_u \approx 830$ MPa, $S_y = 581$ MPa, and from Eq. (9-2), $\bar{p} = 1/[(581 - 200)/240]^2 = 0.397$ μm. To check this, measure the pearlite spacing of the electron micrograph shown in Fig. 8-8.

The inverse relationship between strength and dislocation glide distance

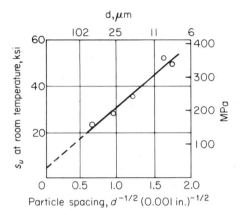

Figure 9-2 Tensile strength vs. (particle spacing)$^{-1/2}$ for a sintered aluminum powder. [From R. W. Guard, "Mechanisms of Fine-Particle Strengthening," *Strengthening Mechanisms in Solids* (Metals Park, Ohio: American Society for Metals, 1962), p. 268.]

holds for many other materials. For example, Fig. 9-2 shows the variation in strength with the reciprocal square root of the spacing of hard particles in aluminum powder that has been sintered (pressed together at high temperature) to make a single piece of metal. Mechanisms for strengthening by fine particles are discussed later in this chapter.

Exercise 9-2

After hot rolling at 900°C, plain-carbon structural-steel plates are stacked and allowed to cool. Describe two changes in this procedure that would increase their room-temperature strength.

Solution. Alterations in grain size or spacing of pearlite will change the strength. The grain size could be reduced by lowering the hot-rolling temperature. Hot rolling can be regarded in a simplified way as cold rolling and instantaneous recrystallization (although it actually is not), so that lowering the temperature at which recrystallization occurs will reduce the diffusion rate, decrease the grain size, and increase strength.

Strength increases caused by finer grain size are small compared with strength increases that result from changes such as in the pearlite spacing. Thus the cooling rate could be increased to produce finer pearlite, through cooling in still air without stacking, or cooling with a cold-air blast.

SHAPE OF HARDER PHASE

For a given amount—by mass or volume fraction—of second phase, the shape of the second phase can exert a substantial influence on the strength of the alloy. The two sketches in Fig. 9-3 show why this can be true. In Fig. 9-3(a) shading indicates those regions where dislocation motion in a particular direction will be impeded by the second-phase particles. Figure 9-3(b) shows how flattening these same particles would increase the interference to dislocation motion. In this example it is true that dislocation motion in the vertical direction would be made easier. However, in every other direction the particles

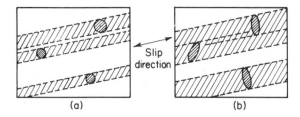

Figure 9-3 Schematic sketches to show the effect of shape of second phase: (a) spheroidal, and (b) flattened.

would interfere more; thus the mean free path would be smaller for Fig. 9-3(b) than for (a), and hence the strength would be higher.

The limiting case of the change shown in Fig. 9-3 occurs when each particle is flattened to a layer extending across the grain. Pearlite is such a material, although it is not made by flattening spheroidal particles. If a layered structure is changed to a spheroidal structure, the strength should decrease. This can be accomplished by heating pearlite to produce spheroidite, as discussed earlier. This microstructure of spheroidal carbide in ferrite is also called *spheroidized cementite*, as discussed earlier, and it has a hardness of about 150 Bhn. The microstructure of spheroidite, shown in Fig. 9-4, should be compared with the microstructure of pearlite shown in Fig. 8-8.

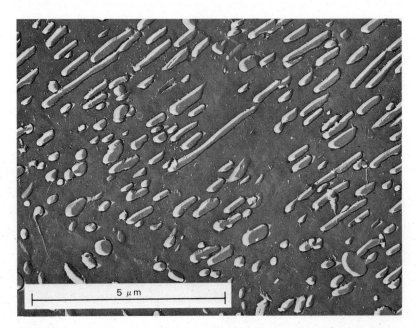

Figure 9-4 Electron micrograph of replica of polished and etched surface of spheroidite.

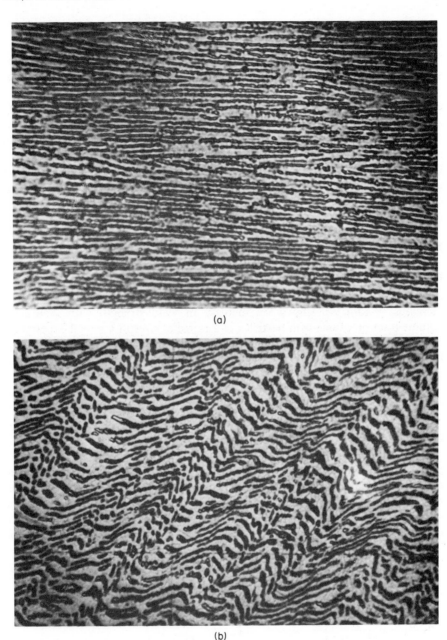

(a)

(b)

Figure 9-5 (a) Microstructure of undeformed, directionally solidified silver–copper eutectic alloy. (b) Microstructure of cut chip from machining test showing discontinuous shearing. The "built-in grid" of the eutectic is deformed into the chevron pattern associated with a double shearing process. [From W. B. H. Cooke and W. B. Rice, ASME, *J. Engr. Industry*, **95** (1973), p. 844. Photographs courtesy W. B. Rice.]

Control of the shape of the second phase holds substantial promise for higher strengths in some common alloys. By cooling a eutectic alloy so as to force solidification to progress along a nearly plane front, the characteristic shape of the eutectic microstructure can be changed. Figure 8-3(b) shows, for example, the fine lamellar shape of directionally solidified eutectic silver–copper. Microstructures such as these provide the possibility of achieving higher strengths without the difficulties of artificially dispersing long, high-strength whiskers or hard, fine particles in a metal to increase its strength (see the discussion on insoluble phases later in this chapter). The directional eutectic has the further advantage that good surface bonding exists between the two different phases. Figure 9-5 shows a directionally solidified eutectic used to exhibit flow patterns in machining.

AMOUNT OF HARDER PHASE

If more of the harder phase is present, whether as layers or spheroids or some other shape, the mean free path in the softer phase decreases and the strength increases. Figure 9-6 shows the increase in yield strength with increasing volume fraction of fine iron particles of a range of sizes in a model experiment using mercury at very low temperature. Such an increase is to be expected, as is the appearance of an upper limit to the strength. The sudden change in slope of the graph in Fig. 9-6 probably occurs because the stress has become high enough for the dislocations to move through the particles by shearing them, or to move around the particles by other mechanisms. (Some of the possible mechanisms that achieve this are described in later sections of this chapter.)

The strength of some two-phase alloys is found to be approximately proportional to the fraction of the second (harder) phase present. Pearlite, the eutectoid mixture of alternate layers of ferrite and iron carbide in steel, behaves as if it were a separate, harder constituent when it is mixed with free ferrite. Experiments demonstrate that a reasonable estimate of hardness and strength of a *hypoeutectoid* (less than eutectoid carbon) pearlitic steel can be

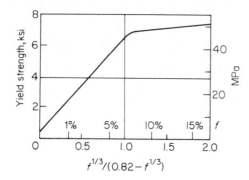

Figure 9-6 Yield strength of frozen mercury vs. volume fraction f of iron particles. [From W. H. Meiklejohn and R. E. Skoda, "Dispersion Hardening," *Acta Met.* (New York: Pergamon Press, 1960), **8**, p. 773. Reprinted by permission of the publisher.]

made by adding together the products of the mass fraction and hardness of each constituent. (The word *constituent* here refers to a component of the microstructure that is treated as a separate entity. Pearlite is such a component, but since it contains two phases, it cannot be called a phase. Sometimes the term *microconstituent* is used synonymously with constituent.) The general form of the method for calculating hardness is

$$\text{Bhn} \approx (\text{mass fraction pearlite})(\text{Bhn}_P) + (\text{mass fraction ferrite})(80) \qquad (9\text{-}3)$$

From Table 9-1, Bhn_P is 380 for fine pearlite, 280 for medium pearlite, and 240 for coarse pearlite. If spheroidite of eutectoid composition is present instead of pearlite, its Bhn is about 150.

Exercise 9-3

Calculate the hardness and strength of a furnace-cooled 0.20% C steel.

Solution. The mass fraction of austenite at 728°C is approximately $0.20/0.80 = 0.25$, which is equal to the mass fraction of pearlite at 20°C. The fraction of ferrite is $1 - 0.25 = 0.75$. Furnace cooling produces coarse pearlite (Bhn = 240), with no effect on ferrite properties (Bhn = 80). Thus Bhn = $(0.25)(240) + (0.75)(80) = 120$ Bhn, and $S_u = 3.45$ (Bhn) = 414 MPa. These values are very close to those measured for this steel.

FINE PARTICLES

For a given fraction of a harder second phase, the greatest strengthening effect might be presumed to occur when the particles are as small as possible and thus have the greatest possible phase-boundary surface area (see Fig. 9-3). The limiting case for the finest possible particles would be isolated atoms. This would then be solid-solution strengthening, which was shown in Chap. 7 to have only a moderate effect on strength. Observation and experiment have demonstrated instead that the greatest strengthening effect occurs for very small second-phase particles of maximum dimension two or three orders of magnitude larger than atomic dimensions. When a dislocation intersects an array of such fine particles, it does not behave in the same way as a dislocation that encounters the large surface of a phase boundary, where the dislocation is blocked along a large part of its length. Because *fine-particle strengthening* is one of the most important and effective strengthening mechanisms, an introduction to the theory, with examples of commercial alloys, is given in the paragraphs that follow.

Exercise 9-4

Which would be the most effective shape for fine particles, each having the same mass, for interference with dislocation motion: rods, thin plates, or spheres?

Solution. The answer depends on which dimensions are restricted. Rods are more effective if their diameter is the same as the thickness of the plates, because the rods would be much longer and thus have the higher surface area per unit volume. If the length of the rods is held the same as the diameter of the (circular) plates, then the diameter of the cylindrical rods would be relatively large and the plates would be more effective. In either case, for a given particular volume, spheres provide the minimum surface area, so they would be least effective in interfering with dislocation motion.

The portion of a dislocation that approaches a fine particle will be held up by the presence of this imperfection in the crystal structure, in the same manner as described earlier in this chapter for phase boundaries in general. Portions of the dislocation remote from the fine particle will not be held up, so the dislocation will start to bend around each particle. This situation is shown in Fig. 9-7(a), where the shaded circles repesent fine particles and shear stress τ is such as to cause the dislocation to move to the right. If further strain is to occur, the stress must be raised. (Recall the argument in Chap. 5 that the number of dislocations is very large, so the behavior described here occurs simultaneously in a small fraction, which will still be a fairly large number, of this very large number of dislocations.) Then one or more of the following can occur: (1) The dislocation will form loops around the particles and continue; (2) the dislocation will move over or under the particles and continue; (3) the particles will be sheared by the dislocation; or (4) the dislocation will not move very much, while other dislocations elsewhere in the crystal will move.

The first possibility was described by Orowan, in the sequence shown in Fig. 9-7. The shear stress τ required to move a dislocation through the space between particles of average spacing f is estimated from dislocation elasticity calculations to be

$$\tau \approx \frac{Gb}{f} \qquad\qquad (9\text{-}4)$$

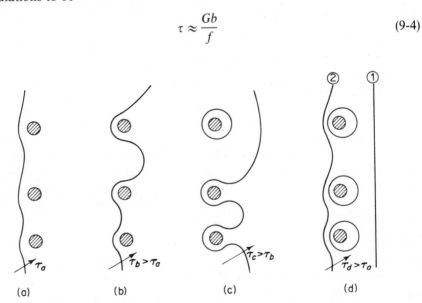

Figure 9-7 Orowan's model for fine-particle strengthening.

where G is the shear modulus of elasticity and b is the Burgers vector. As the stress increases, the dislocation can bow out between the particles [Fig. 9-7(b)]. Figure 9-8 shows bowed-out dislocations pinned by fine particles in a magnesium oxide matrix. When the shear stress is sufficient to produce a radius of curvature small enough to allow the dislocation segment to pass between two particles [Eq. (9-4)], the dislocation wraps completely around the particle until it touches on itself. At this instant the dislocation line snaps out into a more nearly straight line and leaves a dislocation loop around the particle. This process is shown in Fig. 9-7(c) and (d). With further increase in stress, the dislocation can either form secondary loops around each of the remaining particles or a single loop encircling both particles; the first case is shown here. [By a similar mechanism of bowing of a dislocation that is pinned at two points but otherwise lies out of the plane, a *Frank–Read source* can generate a large number of dislocations when the stress reaches the value of Eq. (9-4). The discussion of cross slip that follows, and Fig. 9-10, describe the essential geometry for the Frank–Read source.]

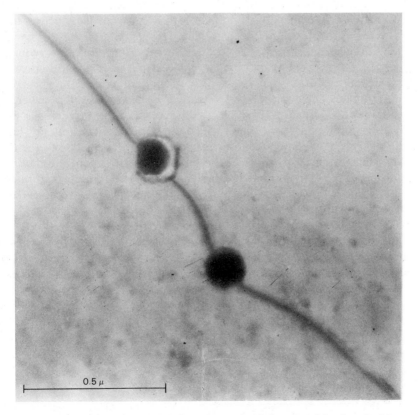

Figure 9-8 Dislocations in MgO bowed out after being pinned by fine particles (probably of magnesium nitride). (Photo courtesy of Gareth Thomas.)

Figure 9-7(d) shows the original dislocation "1" moving off to the right; the stress required to move it, once past the row of particles shown in the sketch, would be lower than the stress when particles are present, but in the real case any dislocation would be continuously encountering fine particles at many points along its length and would not experience a drop in stress. This model thus shows how fine particles require an increase in stress for dislocation motion.

Exercise 9-5

Estimate the particle spacing in an aluminum alloy having a yield shear strength of 400 MPa.

Solution. From Eq. (9-4), since aluminum is the matrix material, $G = 24$ GPa, $b = 286$ pm, and so $f = Gb/\tau = (24 \text{ GPa})(286 \text{ pm})/400 \text{ MPa} = 17$ nm. This agrees quite well with the particle spacing observed in aluminum alloys.

The model in Fig. 9-7 also exhibits another property observed in real alloys having fine particles: rapid strain strengthening. The loops that are left around the particles repel subsequent dislocations of like sign, and the effect is an apparent enlargement of each particle with a consequent reduction in the effective spacing between the particles. The geometry faced by the new approaching dislocation, shown as dislocation "2" in Fig. 9-7(d), thus requires a higher stress for dislocation motion around the particles already circled by loops. The rate of increase of stress with strain should therefore be greater than in a similar single-phase solid. Figure 9-9 shows this effect, where the stress–strain slope of the alloy with even relatively large particles is much greater than for the supersaturated single-phase alloy. [The initially higher strength of the single-phase alloy, represented by curve (a), is due to the effect of strain fields produced by supersaturation that impede dislocation motion; a single-phase alloy of Al–2 atomic percent Cu will always be supersaturated at or below room temperature. The alloy containing particles, represented by curve (b), is not supersaturated.]

The second alternative mechanism listed earlier in this section occurs when the dislocation moves on a different slip plane and thereby bypasses the fine particles by going over or under them. This possibility is limited to a dislocation line that is of pure screw form, since at moderate temperatures an

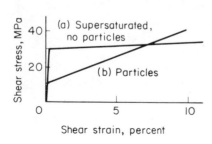

Figure 9-9 Strain hardening of single crystals of Al–2 atomic percent Cu at $-196°C$: (a) supersaturated solid solution produced by quenching; (b) with fine particles of CuAl$_2$ about 2.5 μm apart. [From G. Greetham and R. W. K. Honeycombe, "The Deformation of Single Crystals of Aluminum–4.5 Per Cent Copper Alloy," *J. Inst. Metals*, **89** (1960–61), p. 13. Reprinted by permission of the Metals Society.]

edge dislocation cannot move except on its slip plane. Motion of a screw dislocation onto a different plane, called *cross slip* (see Fig. 5-43), requires an increase in stress, since the dislocation must be extended and will hence have greater strain energy. Thus cross slip can also contribute to fine-particle strengthening and, to a certain extent, to an increased strain-hardening rate, since the longer dislocations will be more effective barriers to other dislocations.

Exercise 9-6

Why cannot a small portion of a screw dislocation cross-slip over a particle and then continue on?

Solution. If only a small portion of a screw dislocation cross-slips, the connecting dislocation lines are at least part edge (see Fig. 9-10). Dislocation climb (see Chap. 6) would then be required for motion of these edge components in the direction of continued slip. Climb is so slow as to be virtually impossible at room temperature. Therefore, further slip can result only from the bowed-out portion in Fig. 9-10(c), and the original dislocation is stopped. This bowed-out segment can form a Frank–Read source, as discussed earlier, but unless the stress is high enough to force the dislocation to bow out to a half circle, the requirement of Eq. (9-4), the source will not operate.

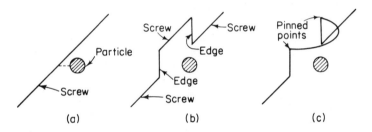

Figure 9-10 Solution to Exercise 9-5, showing restricted motion of the cross-slipped portion of a dislocation.

The third possibility, shearing of the fine particles, apparently occurs if the stress becomes great enough. Shearing of the particles also results in an increase in phase-boundary surface area, so a kind of strain hardening should occur when this mechanism operates. In Fig. 9-6 the curve becomes nearly flat when the volume fraction of iron particles is large enough to raise the stress to a level where some other mechanism such as this apparently comes into play. If this mechanism is the shearing of the particles, it explains the sudden change in slope and the very gradual increase in yield stress with further increase in the volume fraction of iron.

The last possibility, that other dislocations will begin to move, is trivial, since the other dislocations will sooner or later be blocked in the same way as the dislocation in Fig. 9-7(a). At that time a further increase in stress will be required to move the dislocations by (1) looping around the particles, (2) cross-slipping over the particles, or (3) shearing the particles.

Frequently new dislocations are observed to form at fine particles. These dislocations cause much cross slip and entanglement of other dislocations in the vicinity of the particles. This produces substantial strengthening of the same magnitude as for the processes already described but in a more complex manner. Although dislocation configurations in real materials are almost always more complicated than in the simple models described here, the reasons for strengthening remain the same.

ALUMINUM–COPPER ALLOYS

The classic example of fine-particle strengthening occurs at the aluminum-rich end of the aluminum–copper system. Figure 9-11 gives about half of the aluminum–copper equilibrium diagram. Note the similarity in general appearance to the full silver–copper system discussed in Chap. 8. This form is quite common, and in the Al–Cu case an intermetallic compound divides off a portion of the equilibrium diagram as if it were a separate diagram. Consider a mixture that is 4.6% Cu by mass (shown dashed), cooled slowly from the melt. At about 570°C, the microstructure becomes 100% α, which is an aluminum-rich substitutional solid solution. With further cooling, at about 510°C some θ phase begins to form at the grain boundaries. This is an *intermetallic compound* (a chemical combination of two metals) of the approximate atomic ratio of $CuAl_2$. During slow cooling through the two-phase region, θ continues to form at the grain boundaries. The final product after slow cooling is relatively weak and only moderately ductile, because the brittle θ exists at the grain

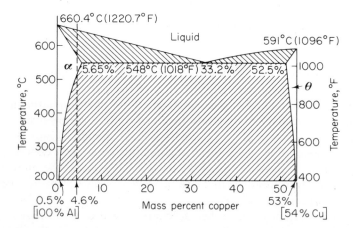

Figure 9-11 Aluminum-rich half of the aluminum–copper equilibrium phase diagram. [From Lowell A. Willey, "Phase Diagrams for Binary Alloy Systems," *Metals Handbook: Metallography, Structures and Phase Diagrams*, 8th ed., vol. 8, ed. T. Lyman (American Society for Metals, 1973), p. 259.]

boundaries. Cracking can occur in the grain boundary θ after plastic deformation and will thus reduce the ductility of the matrix.

If the single-phase α, for example at 540°C, is cooled very rapidly by water quenching to room temperature (the room-temperature equilibrium conditions are similar to the conditions at 200°C in Fig. 9-11), the θ phase will not have time to form. It has already been shown that formation of a new phase requires time and elevated temperature for diffusion of atoms. In this case, diffusion of Cu atoms into a region and Al atoms out of that region is required, until there is about one Cu atom for every two Al atoms. If time and elevated temperature are not provided, as in water quenching, the atoms of Cu and Al must remain about where they were at 540°C. Thus immediately after quenching, this alloy will still be a solid solution even though the phase diagram indicates that two phases would be present under equilibrium conditions. The stress–strain curve for this as-quenched alloy, aided by supersaturation effects, was shown in Fig. 9-9 and is repeated in Fig. 9-12 as curve (a). (Note that 2 atomic percent Cu in Al is 4.6 mass percent Cu, because the atomic mass of Cu is 2.36 times that of Al.)

If the supersaturated solid solution is held at room temperature or above

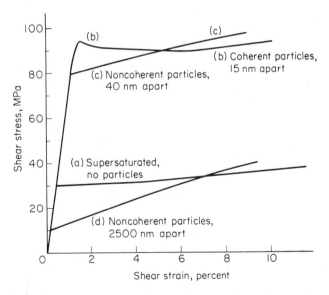

Figure 9-12 Shear stress–strain behavior of single crystals of Al–2 atomic percent Cu (4.6% by mass) at −196°C: (a) supersaturated solid solution produced by quenching; (b) coherent platelets about 15 nm apart, by aging 170 ks at 130°C; (c) noncoherent particles about 40 nm apart, by aging 99 ks at 190°C; (d) noncoherent particles about 2500 nm (2.5 μm) apart, by aging 170 ks at 250°C. [From G. Greetham and R. W. K. Honeycombe, "The Deformation of Single Crystals of Aluminum—4.5 Per Cent Copper Alloy," *J. Inst. Metals,* **89** (1960–61), p. 13. Reprinted by permission of the Metals Society.]

Figure 9-13 Thin-foil electron micrograph of Al–4% Cu solution heat-treated and aged 36 ks at 160°C. The very fine particles in the lower part of the photograph are GP zones, while the larger particles in the upper portion are coherent particles of $CuAl_2$, an intermediate state between the GP zone and the noncoherent $CuAl_2$ particle. This remarkable photograph was taken by *darkfield* electron microscopy, in which only the diffracted portion of the electron beam is imaged. [From G. Thomas, "Kikuchi Electron-Diffraction and Dark-Field Techniques in Electron-Microscopy Studies of Phase Transformations," *Trans. AIME,* **233** (1965), p. 1608. © 1965 by AIME and reprinted with its permission. Photo courtesy of Gareth Thomas.]

for a period of time (a process called *aging*), its microstructure tends to change toward the microstructure of the equilibrium phases. Several distinct stages have been identified, from which three are selected for discussion here. Curve (b) in Fig. 9-12 exhibits the very high strength that results from formation of extremely small particles, or platelets, of pure copper one atom thick, spaced about 15 nm apart. These particles are examples of *Guinier–Preston zones* (GP zones), which form according to the previously described mechanisms. Figure 9-13 shows these platelets (the smaller ones) as detected in an electron micrograph of a thin foil of alloy of similar composition and treatment. Note how very small the spacing between particles is.

One of the sources for the large strengthening effect of these particles is that they are *coherent* with the lattice, that is, the atoms of each platelet match up on a one-to-one basis with the surrounding atoms of α. This is shown schematically in Fig. 9-14(a). Because the lattice constant for pure copper (361 pm) is slightly smaller than for aluminum (404 pm) and the surrounding

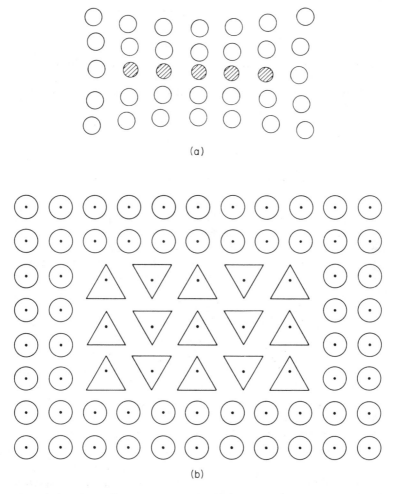

(a)

(b)

Figure 9-14 Schematic representation of coherent and noncoherent particles. (a) Coherent atoms of slightly smaller atomic size than the matrix. (b) Noncoherent particle of an intermetallic compound (triangles) of much larger molecular spacing than the spacing of atoms in the matrix.

matrix is nearly pure aluminum, the matrix is elastically strained by the particle. This strain field will impede dislocation motion, so that the effective size of the particles is increased; that is, a dislocation need not intersect the particle in order to have its motion restricted. The rate of strain hardening of curve (b) in Fig. 9-12 is very low, probably because the very high initial yield stress is high enough to shear the platelets (which are only one atom thick) in a plane not parallel to the platelets.

Aging at higher temperature produces larger particles that give curve (c) in Fig. 9-12. These particles are *noncoherent*, of approximate composition

CuAl$_2$, and spaced closely enough together to retain yield strength nearly as great as for the coherent platelets, with a much greater rate of strain hardening. Figure 9-14(b) shows how the strain in the surrounding matrix can be relatively small in a noncoherent particle, since the mismatch between the particle molecules and the matrix atoms is so great. While the effective size of these particles is not as much greater than their actual size as for the case of coherent particles, these particles are apparently stronger in shear than the coherent platelets, probably because the noncoherent particles are more than one atom thick.

Exercise 9-7

For what reason are larger particles less likely to be coherent?

Solution. Larger particle size causes the atomic mismatch at the phase boundary to be so large that the forces on the atoms at the surface of the particle become higher than the atomic bonding forces between the particle and matrix atoms. Thus the atoms at the phase boundary move nearer to their respective equilibrium positions, with the resulting gross mismatch of unlike atoms across the phase boundary shown in Fig. 9-14(b). In small particles the cumulative effect of the size mismatch is not so great [see Fig. 9-14(a)], so coherency is possible.

Curve (d) in Fig. 9-12 shows the effect of aging at a much higher temperature, which causes the fine particles to agglomerate to form fewer and larger noncoherent particles of consequently wider spacing. The marked reduction in yield strength and substantial strain hardening follow consistently the arguments already presented. This reduction of strength with excessive aging is called *overaging.* In commercial practice, overaging can be deliberately induced in order to stabilize an alloy to resist strength reduction at elevated temperatures. The designer must then accommodate to a lower strength at the outset but obtains, in exchange, a somewhat higher allowable operating temperature.

Another structure involving fine particles that is of great economic importance is tempered martensite. Because of its special nature, this structure will be discussed in detail in Chap. 10.

INSOLUBLE PHASES

The chief disadvantages of the microstructures described so far are (1) that they are limited to phases that precipitate from a supersaturated solid solution, and (2) that the fine particles and thin layers agglomerate at intermediate temperatures and are redissolved at higher temperatures. A much wider choice of properties ought to result from intimate mixtures of dissimilar solids, called *composites*, to produce a solid of desirable overall properties. Nature has many examples of structural composites, such as wood and bone, which combine rigidity and lightness of weight. Composites are covered more thoroughly in

Chap. 17, so the discussion here will serve only as an introduction. Two of the many developments of composites will be mentioned here for illustration; both depend on phase-boundary strengthening and the very small sizes of their constituents for their unique properties.

Very fine (less than 1 μm) thoria powder has been successfully dispersed throughout pure nickel, apparently by mixing the two in powder form and then sintering. The strength of this composite is not particularly high at intermediate temperatures, but because thoria retains its hardness to temperatures above the melting point of nickel (1453°C), the composite retains substantial strength to above 90% of the absolute melting temperature of nickel. This is called *dispersion strengthening* or *dispersion hardening*, terms reserved for the mechanical mixing of fine particles in a matrix, in contrast to natural processes. Figure 9-15 shows the strength of this composite at high temperatures in contrast to two other corrosion-resistant metal alloys. Thoria–nickel has not come into general use, chiefly because it is too brittle for normal engineering applications.

Continuous graphite filaments can be manufactured with a diameter of 7

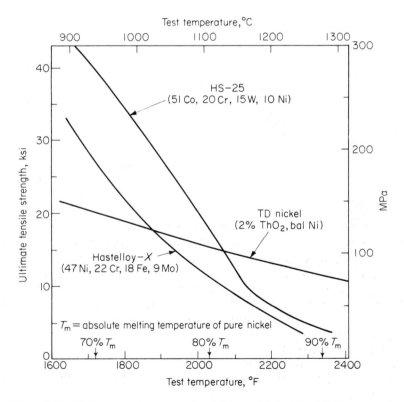

Figure 9-15 High-temperature strength of thoria–nickel (TD nickel) compared with two other high-temperature materials.

μm and a fracture strength of the order of 2 GPa. With a density of about 1.8 Mg/m^3, these filaments have strength/density of about 1.1 (km/s)^2, compared with that for high-strength low-alloy steel, 0.25 (km/s)^2, and for wrought aluminum alloys, about 0.2 (km/s)^2. However, fine filaments alone are of little use for engineering purposes unless they can be fabricated into a solid of useful shape. Thus manufactured *filamentary composites* consist of high-strength filaments surrounded by some castable matrix material. One useful matrix material for graphite is high-strength epoxy (a complex polymer—see Chaps. 12 and 13). Thus a part of useful shape, such as a thin sheet, can be made with aligned and closely spaced graphite filaments surrounded by epoxy. Figure 9-16 shows the fracture cross section of such a composite. The tensile strength of this composite parallel to the direction of the aligned filaments is about 1 GPa, less than that of the filament alone, but its overall density is less, about 1.5 Mg/m^3, so the strength/density is about 0.65 (km/s)^2.

Graphite–epoxy composites owe their high strength in part to the small distance across the diameter of the filament and through the epoxy between adjacent filaments. The magnitude of these dimensions is about 100 times the spacing of fine-particle-strengthened aluminum alloys, but on the other hand, the restriction of slip processes to the dimensions of the filament is a powerful deterrent to slip. We will see in Chap. 14 how the limitation of maximum

Figure 9-16 Fracture cross section of a graphite–epoxy composite, as seen in the scanning electron microscope. In this specimen the plane of fracture varies from filament to filament. Some of the filaments have pulled out of the surrounding epoxy matrix in the process of fracture, and cracking along the graphite–epoxy interface can also be seen. Both processes contribute to increased *fracture toughness* (see Chap. 14). The marker is 20 μm long. (Photograph courtesy of W. H. Durrant.)

possible crack size to no larger than the filament diameter is a powerful strengthening mechanism.

High cost, brittleness, and limited supply will continue to restrict the use of high-performance composites. The most likely uses will probably continue to be for those special applications requiring high strength/density, different properties in different directions in a part, or where unusual combinations of physical properties (such as low thermal conductivity) and strength are required. It may be expected that application of the fundamental rules of strengthening mechanisms will continue to produce better, more practical composites. (Chapter 17 covers composite materials in detail.)

REFERENCES

9-1. American Society for Metals, *Metals Handbook*, 9th ed., vol. 8, Metallography, Structures, and Phase Diagrams. Metals Park, Ohio: American Society for Metals, 1973.

9-2. American Society for Metals, *Strengthening Mechanisms in Solids*. Metals Park, Ohio: American Society for Metals, 1962. Chapter 9, "Mechanisms of Fine-Particle Strengthening," by R. W. Guard, evaluates existing data on the subject for several materials.

9-3. Caddell, R. M., *Deformation and Fracture of Solids*. Englewood Cliffs, N.J.: Prentice-Hall, 1980. Chapter 7 provides a thorough basic description of dislocation interaction with fine particles.

9-4. LeMay, Iain, *Principles of Mechanical Metallurgy*. New York: Elsevier North Holland, 1981. Chapter 6 discusses strengthening mechanisms by the action of multiple phases.

9-5. McLean, D., *Mechanical Properties of Metals*. New York: John Wiley & Sons, 1962. Chapter 6 includes a thorough development of the theory of fine-particle strengthening.

9-6. Smith, Charles O., *The Science of Engineering Materials*. Englewood Cliffs, N.J.: Prentice-Hall, 1977. Chapter 9 discusses solid-state reactions that lead to strengthening, and Chap. 10 discusses the formation and spacing of pearlite in steel.

PROBLEMS

9-1. Which of the strengthening mechanisms can occur in:
   (a) Pure elements?
   (b) Single-phase solids?
   (c) Multiple-phase solids?

9-2. One method of increasing the strength of copper–aluminum alloys is by precipitation hardening. Certain alloy compositions of copper and zinc (from 0 to 37% zinc) are called *alpha brasses*. Are these alloys susceptible to precipitation hardening? Explain your answer.

9-3. The *Bauschinger effect* consists of a reduction of yield strength in compression following initial yielding in tension (or in the reverse order, tension after compression). Give two explanations for the Bauschinger effect, one based on simple dislocation motion and the other based on the Orowan mechanism for fine-particle strengthening.

9-4. Assume that we can produce mixtures of a pure fcc metal (element Z) and a hard, brittle intermetallic compound (θ) in a variety of shapes and sizes. In Fig. 9-17 are five plausible true-stress–true-strain curves (A through E) and five microstructures (1 through 5, all at 500 ×). Match each microstructure with the most likely stress–strain curve, without duplication.

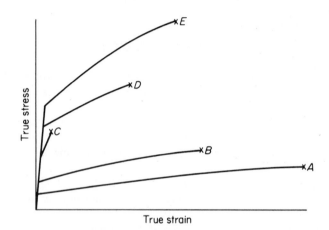

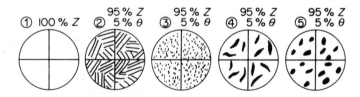

Figure 9-17

9-5. A metal alloy is in the form of a bar 5 mm in diameter. When furnace-cooled from the melt, it looks like microstructure 1 (see Fig. 9-18) at room temperature. (All microstructures are shown at 400 × ; the dark phase is the harder phase.) The bar is then moderately cold-worked.

(a) The metal is heated to the lowest temperature at which it is a solid single phase, then water-quenched. At the instant immediately following the quench, which microstructure in Fig. 9-18 most closely represents the probable appearance?

(b) If the specimen is moderately cold-worked and heated again to the highest temperature at which it is still a solid single phase, then water-quenched, which microstructure most closely represents the probable appearance?

(c) Which will have the greater yield strength, the specimen after treatment (a) or after treatment (b)?

(d) Following treatment (a), the specimen is held at room temperature for one month. Assuming that its microstructure has changed during this period, state which microstructure in Fig. 9-18 best represents it.

(e) What is this process called?

(f) If process (d) is carried out at a moderately elevated temperature for the same time, say 80°C for one month, describe any differences in the appearance of the microstructure that would result.

(g) Which specimen would have the highest strength: the one shown as 1; after treatment (a); after treatment (b); after treatment (d); or after treatment (f)?

(h) Describe one dislocation mechanism that could explain why the strongest microstructure of part (g) is stronger than the next-strongest microstructure.

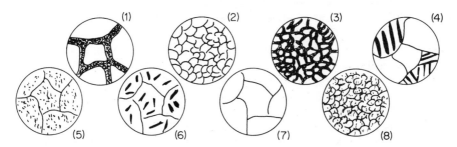

Figure 9-18

9-6. Calculate the room-temperature Brinell hardness of austenitized and furnace-cooled (that is, fully annealed) plain-carbon steel of:
(a) 0.20% C.
(b) 0.40% C.
(c) 0.70% C.

9-7. Calculate the Rockwell B or C hardness of plain-carbon steel of 0.30% C that is austenitized and then given each of the following treatments, each one taken separately:
(a) Air-cooled to 20°C.
(b) Furnace-cooled to 20°C.
(c) Oil-quenched to 20°C.
(d) Oil-quenched, heated to 700°C for 10 h, then air-cooled to 20°C. (Use the hardness conversions given in Appendix 4.)

9-8. You have available the following steels as bars 10 mm in diameter: 0.20, 0.40, and 0.77% C, all plain-carbon steels. These can be heated to any desired temperature and either furnace-cooled, air-cooled, or oil-quenched. Considering each requirement below as a separate problem, select a steel and describe a heat treatment (within the above limitations), including temperatures, that would produce (within ±10%) the following:
(a) Tensile strength = 655 MPa.
(b) Bhn = 100.
(c) Bhn = 160.

9-9. A piece of 0.54% C steel 10 mm in diameter is held at 885°C for 3.6 ks, then furnace-cooled to 20°C. Its final hardness is 92 R_B. A piece of 0.30% C steel 10 mm in diameter is held at 885°C for 3.6 ks, then oil-quenched to 20°C. Its final hardness is 92 R_B. Explain, from a fundamental point of view, why these two tests produce the same hardness from different steels.

9-10. A bar of 0.72% C steel is held at 885°C for 3.6 ks, then air-cooled to 20°C, at which temperature its hardness is found to be 27 R_C. It is then reheated to 700°C for 150 ks, then air-cooled to 20°C, at which temperature its hardness is found to be 81 R_B. Explain, from a fundamental point of view, why its hardness changed.

9-11. A mixture of 0.26% carbon, balance iron, of total mass 947 g is furnace-cooled from the melt. For each of the temperatures 1149°C, 780°C, and 726°C:
   (a) Calculate the mass of each phase present.
   (b) If the phases are mixed in the form of a characteristic microstructure, name the microstructure and calculate its total mass.

9-12. Calculate the percentage of pearlite present in a 1.10% C steel after it is slowly cooled from 900°C to 100°C.

9-13. Which has the higher hardness, 0.40% C steel that has been air-cooled from 885°C or 0.40% C steel that has been oil-quenched from 885°C? Briefly explain your answer from a dislocation-theory point of view.

9-14. Calculate the Brinell hardness number of a 0.50% C steel that has been oil-quenched from 900°C.

9-15. A set of schematic microstructures is given in Fig. 9-19. For each material and process listed below, write the letter of the microstructure from Fig. 9-19 that best describes the alloy after the process described. (An answer may be used more than once.)
   (a) 2024 aluminum (UNS A92024), 510°C for 17 ks, water-quenched to 20°C, time = 0.
   (b) 1100 aluminum (UNS A91100), 510°C for 12 ks, furnace-cooled to 20°C, time = 40 Ms.
   (c) 0.95% C steel (UNS G10950), air-cooled from 900°C to 20°C.
   (d) 2024 aluminum (UNS A92024), 510°C for 12 ks, water-quenched to 20°C, then 100°C for 3 ks, water-quenched to 20°C.

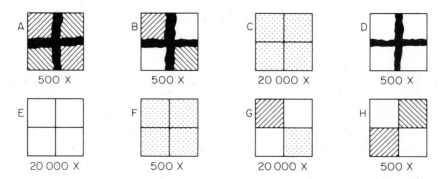

Figure 9-19

9-16. An aluminum alloy bar 6.49 mm in diameter is heat-treated so that the average spacing of $CuAl_2$ particles is 37.3 nm. How many 200-kg filing cabinets will it support simultaneously in tension before it breaks?

9-17. For each of the following two processes, taken separately, after the time(s) specified and at the final temperature stated:

(a) Name the phases present; if supersaturated, say so.

(b) Determine the composition of each phase.

(c) Describe the shape (spheroids, fine particles, flakes, etc.) or microstructure (as either eutectic, eutectoid, grain boundary, or free) of all phases present except the most prevalent phase.

Both specimens are bars 12 mm in diameter:

1. Aluminum, 4.8% copper, furnace-cooled from 680°C to 300°C.

2. Aluminum, 5.5% copper, furnace-cooled from 680°C to 540°C, water-quenched to 20°C, heated to 250°C for 600 ks, water-quenched to 20°C.

9-18. Draw schematic microstructures and label phases for specimen of 2011 aluminum (UNS A92011) 5 mm in diameter, after each of the following treatments and times, taken sequentially:

(a) 500°C for 3 h, furnace-cooled to 20°C.

(b) 500°C for 3 h, water-quenched to 20°C, time $= 0$.

(c) 200°C for 1 ks.

(d) 200°C for an additional 7 ks.

9-19. In an aluminum–copper alloy of shear strength 275 MPa, how far apart are the fine particles?

9-20. Two tensile specimens of 2017 aluminum (UNS A92017: composition 4% Cu, 0.5% Mg, 0.5% Mn, balance Al) give the following data:

Heat Treatment	Init. Dia.	Max. Load	Fract. Dia.
550°C, furnace-cooled	12.8 mm	31.1 kN	9.93 mm
550°C, water-quenched, 20°C for 100 Ms	12.8 mm	56.2 kN	10.2 mm

(a) Calculate the strength and true strain at fracture for both specimens.

(b) Describe precisely but briefly how the different heat treatments led to different strengths.

9-21. What limits the maximum temperature at which precipitation-hardened aluminum alloys can be used? Describe the mechanism involved.

9-22. A bar of 2024 aluminum (UNS A92024) is soaked at 510°C for 10.8 ks, then:

(a) water-quenched in "zero time" to 20°C; then

(b) held at 371°C for 0.3 ks and water-quenched; then

(c) held at 371°C for 3 ks more and water-quenched.

Draw a schematic microstructure at 20 000 × magnification of the microstructure immediately after each of the three processes, and label each phase with its name (not a Greek letter) or chemical symbol.

9-23. Compute the amount of each microstructure and the fraction of each phase present in a 0.30% C steel slowly cooled from the melt to each of the following

temperatures:
(a) 925°C.
(b) 760°C.
(c) 700°C.

9-24. A 1.00% C steel that has been cooled slowly from 1100°C and held at 730°C for 0.5 h is air-cooled to room temperature.
(a) Name the phases present.
(b) Describe the microstructure.
(c) Calculate the fraction of noneutectoid carbide present.

9-25. A plain-carbon steel of 0.60% C is heated to 885°C for 0.5 h and then air-cooled to 20°C, yielding a hardness of 220 Bhn. The same specimen is reheated to 885°C for 0.5 h, then oil-quenched to 20°C, and has a hardness of 260 Bhn. Explain, from a fundamental point of view, why the hardness changes.

9-26. Describe in detail a heat treatment (without cold work) that will give a hardness of approximately 21 R_C for a bar of 1040 steel 10 mm in diameter.

10 | Martensites and Tempered Martensites

The principal strengthening mechanisms were described in earlier chapters, but *martensite*, which in the Fe–C system is a microstructure of great hardness, is considered separately because it is both unique and important. The microstructures that can be produced from martensite are likewise important, for although they usually are not as hard as martensite itself, so-called *tempered martensite* structures can be produced with a wide range of strength and ductility that makes them very useful engineering materials. Furthermore, martensite and its products are often involved in the multiple-strengthening effects that are used in producing some of the ultrahigh-strength steels. The essential aspects of martensite strengthening are therefore presented in this chapter, accompanied whenever possible by a description of the basic strengthening mechanisms that contribute to the observed behavior. In many instances the relative magnitudes of the effects of several mechanisms acting together are not known because each mechanism cannot be tested separately.

Martensitic microstructures can also be produced in the iron–nickel system and in nonferrous alloy systems, such as copper–manganese, copper–zinc, nickel–titanium, indium–thallium. They are becoming of great interest, not because of any high hardness property, but because of their unusual thermoelastic mechanical behavior: Their elastic stiffness can change markedly over a relatively small temperature range, and this behavior, together with an associated phenomenon called the "shape memory effect" (discussed later in this chapter), can be exploited in the design of thermostatic control devices.

MARTENSITIC TRANSFORMATION IN Fe–C

If the rate at which a small specimen of plain-carbon eutectoid steel (0.77% C) is cooled from just above the eutectoid transformation temperature (727°C) is successively increased, the resulting room-temperature microstructures will be different. In Chap. 9 the pearlite spacing was shown to vary from coarse for furnace cooling to fine for oil quenching. If a still faster cooling process, such as water quenching, is used, the temperature of the specimen will drop so fast that when it reaches about 250°C, pearlite will not yet have begun to form. At about this temperature a small portion of the austenite transforms in a very short time, of the order of 100 ns, to a new phase that is body-centered tetragonal, called *martensite*. As the specimen continues to cool, martensite continues to form from the remaining austenite in very rapid local bursts, until at room temperature the structure is mostly martensite with a small amount (about 8%) of *retained austenite* still present. All the retained austenite at room temperature or below will be in a metastable state, unable to transform to martensite because the temperature is too high and unable to transform to the equilibrium products, ferrite and iron carbide, because the temperature is too low for the necessary atomic diffusion. If the specimen is then cooled to the

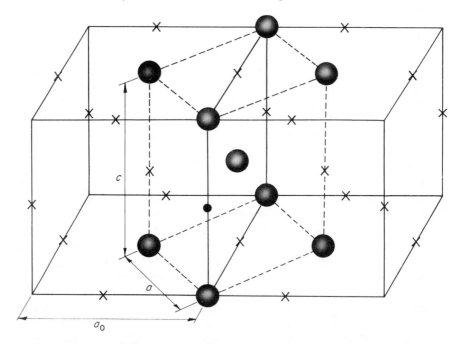

Figure 10-1 Face-centered cubic crystal of iron, shown as spheres, with lattice constant a_0. Dashed lines show how atoms from two adjacent fcc cells can be viewed as body-centered tetragonal, with lattice constants c and a. Possible interstitial lattice sites are indicated by $\times$, with a carbon atom (small black circle) shown in one of these sites.

temperature of liquid nitrogen ($-196°C$), more retained austenite will transform to martensite until only about 3% of the structure remains as retained austenite. Thus upon return to room temperature the 3% retained austenite will remain and will not change even if cooled again to $-196°C$.

The formation of martensite occurs as a *diffusionless* transformation and is therefore unique among the transformations considered in this book. It should be clear that virtually no diffusion can occur in iron in 100 ns at 250°C, so the atoms in martensite must be in almost the same positions as the atoms of the austenite from which the martensite was formed. Figure 10-1 shows that any fcc structure, in this case austenite, can also be considered body-centered tetragonal (bct). Only the atoms of one bct unit cell are drawn in this sketch; the atoms at the remaining corners and faces of the fcc cell are omitted. The fcc structure is ordinarily used to describe the structure of austenite because it is geometrically simpler than the bct structure.

Exercise 10-1

Find the ratio c/a of the bct cell in Fig. 10-1.

Solution. $c = a_0$ and $a = a_0/\sqrt{2}$, so $c/a = \sqrt{2}$. This is a tetragonal structure because all angles are 90°, two of the sides of the unit cell are equal in length, and the third (c) is different.

If no carbon is present in austenite, it will transform to a perfect bcc structure (ferrite) at temperatures below 912°C, irrespective of cooling rate. This transformation may be considered a squashing of the bct unit cell in Fig. 10-1, until $c = a = 286$ pm. This is accompanied by a slight change in the atomic radius of iron from the fcc structure to bcc structure.

If carbon is present in the austenite, it is found at the middle of a few of the edges of the fcc unit cubes and at the centers of the unit cubes (see Fig. 10-1). These two positions are crystallographically equivalent, since any iron atom can be considered to be either at a corner or on a face of a unit cell. Carbon atoms cannot occupy a very large proportion of the sites in equilibrium austenite, because the resulting structure would be severely strained and thus unstable. The sketch in Fig. 10-1 shows the possible sites for interstitial carbon in face-centered cubic iron (austenite), with a carbon atom situated at one of these sites.

When austenite that contains some carbon attempts to transform (at high cooling rates) to the bcc structure of pure ferritic iron, the carbon acts to prevent full contraction of the c-direction. Figure 10-2 shows a unit cell of martensite, in which $c > a$ because a carbon atom is situated along one of the edges parallel to the c-direction. (If there is no carbon present, then the austenite transforms to ferrite; ferrite might thus be thought of as martensite of zero carbon content.) The carbon atom is in the same position in the lattice that it occupied in the austenite lattice in Fig. 10-1; this emphasizes the diffusionless character of the martensite transformation. The presence of interstitial carbon along c-axis edges causes the c-direction to be slightly larger than the a-

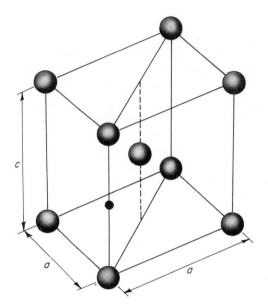

Figure 10-2 Martensite unit cell, where $c > a$. Spheres are iron atoms; the small, black circle is an interstitial carbon atom.

direction. The effect of the amount of carbon on the lattice parameters is shown in Fig. 10-3. As would be expected, an increase in carbon content increases c/a. All unit cells of a given martensite crystal have parallel c-axes.

Exercise 10-2

Calculate the number of empty sites that occur for every full site in 0.77% C martensite.

Solution. Let N be the number of atoms of an element, and W the weight of the element. Then

$$\frac{N_{Fe} \times (\text{atomic wt.})_{Fe}}{N_C \times (\text{atomic wt.})_C} = \frac{W_{Fe}}{W_C}$$

$$\frac{N_{Fe}}{N_C} = \frac{W_{Fe} \times (\text{atomic wt.})_C}{W_C \times (\text{atomic wt.})_{Fe}}$$

$$= \frac{99.23\% \times 12.0}{0.77\% \times 55.85} = 27.7$$

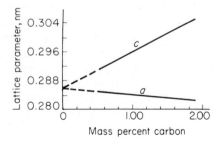

Figure 10-3 Variation in lattice parameters with carbon content of martensite. [From C. S. Roberts, "Effect of Carbon on the Volume Fractions and Lattice Parameters of Retained Austenite and Martensite," *Trans. AIME*, **197** (1953), p. 203. © 1953 by AIME and reprinted with its permission.]

In each unit cell of martensite (see Fig. 10-2) there are two atoms of iron and $(2 \times \frac{1}{2}) + (4 \times \frac{1}{4}) = 2$ possible whole interstitial sites for carbon. (Note that all of the sites are shared with one or three other unit cells.) Then for 0.77% C in Fe, $N_{Fe} = $ no. interstitial sites, and

$$\frac{\text{no. interstitial sites}}{\text{no. atoms of carbon}} = 27.7$$

$$\frac{\text{no. empty sites}}{\text{no. atoms of carbon}} = 27.7 - 1 = 26.7$$

It is not obvious from the exercise that such a small fraction of interstitial sites actually containing carbon atoms should cause all unit cells of iron to be tetragonal. Most unit cells contain no carbon at all. Yet X-ray diffraction studies of martensite crystals have given the data represented in Fig. 10-3. Thus we may conclude that the local straining by one carbon atom is so severe that its influence extends to all unit cells in the vicinity. The presence of one carbon atom along an edge parallel to the c-direction not only will elongate the unit cell but also will probably cause some local warping of the unit cell out of a perfect tetragonal shape. The presence of such large, highly localized strains in martensite must contribute substantially to its high hardness.

The manner in which martensite plates form from austenite is a complex process of local shearing, compression, and extension. Martensite in ferrous alloys normally occurs as very small, nonparallel, lens-shaped crystals. Figure 10-4(a) shows a photograph of the characteristic appearance of martensite (in this case Fe–Ni martensite), with some austenite that has not transformed. In the course of formation of a martensite crystal, the matrix surrounding it will inevitably be strained because the martensite has dimensions that differ from the austenite from which it is formed. If the surrounding matrix is already martensite, it will probably be subjected to an increase in local stresses, owing to the nonparallel occurrence of the martensite plates. If the surrounding matrix is austenite, it will probably deform plastically to accommodate the formation of the martensite plate. Thus the retained austenite may be locally strain-hardened as a result of the martensite transformation.

In those nonferrous alloy systems in which martensitic microstructures can be produced, the crystallographic changes are not necessarily the same as those occurring in the Fe–C system (i.e., not always fcc → bct). In the case of Cu–Mn and In–Th the changes are fcc → fct; in Ni–Ti the change is bcc → orthorhombic. The shear deformations associated with these nonferrous martensitic transformations are much smaller than in steels, and the cell dimensions of the martensite phases are not greatly different from the parent matrix phases. Consequently the differences in size between transformed and untransformed regions can be accommodated elastically within the body, so that such martensitic structures can respond reversibly to changes in applied stress and/or temperature [Fig. 10-4(b)]. This allows these alloys sometimes to display the so-called *shape-memory* (or *marmem*) effect, whereby, for example, a

(a)

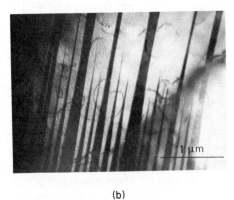

(b)

Figure 10-4 (a) Optical micrograph of martensite and retained austenite (white background) in Fe–32% Ni, cooled to −195°C. [From V. F. Zackay, ed., *High-Strength Materials* (New York: John Wiley & Sons, 1965), p. 404. Reprinted by permission of the publisher. Photograph by D. W. Stevens, courtesy of W. C. Leslie.] (b) Martensite plates in 60 Mn–40 Cu alloy heat-treated at 400°C for eight hours and water-quenched. In contrast to Fe–Ni martensite (where the needles form in a random fashion), here the plates are roughly parallel over large distances. This permits easy, reversible movement along the length of the needles in shape-memory alloys. Related nonferrous alloys display very high acoustic damping for similar reasons. Curly dislocation lines may be seen in the background to the picture. (Courtesy J. A. Hedley.)

bent strip of the alloy in its martensitic state will curl and uncurl as the temperature is altered.

Exercise 10-3

A trick metal spoon curls up when used to stir hot tea. Give a possible explanation for the observation.

Solution. It is likely that the spoon is made from a shape-memory metal such as one of the nonferrous martensitic memory alloys. The behavior described could be caused to occur in the following way: The spoon, in a curled-up shape, would be heat-treated in some appropriate manner in order to get it into the state where it has a latent memory. Such heat treatment might be a rapid quench to room temperature of the sort used in Fe–C martensites, but some nonferrous martensites can be produced in other ways; details do not matter here. The quenched structure will consist of the two phases of some transformed martensite and some untransformed matrix (analogous to retained austenite in the Fe–C system).

If the curled-up spoon is then flattened out at room temperature, it is a feature of shape-memory alloys that it will stay flat. At first sight, it might be thought that the spoon remains flat because the unbending strains have exceeded the yield strain. However, this is not the case, since in the martensitic state these alloys have relatively large elastic limits ($e \approx 0.02$ and greater). But if the elastic limit has not been exceeded, why does the spoon not resist permanent flattening and spring back? The reason is bound up in the peculiar behavior of the martensitic structures.

In the act of flattening, the acicular phase boundaries between the martensite induced by heat treatment and the untransformed matrix move so as to consume matrix material and convert it into more martensite. Such boundaries can move easily because of the comparatively small differences in sizes of crystal cells in the transformed and untransformed phases in these particular memory alloys. In the flattened condition there is thus a greater volume of martensite than in the curled state. Internal elastic stresses (akin to residual stresses) are induced when the martensite boundaries are caused to move by some external loading. Such internal stresses attempt to restore the original shape of the body on unloading, when the martensite interfaces would move back and the stress-induced martensite revert to matrix phase. Under a certain range of temperature (exploited in this spoon example) the elastic restoring forces are insufficient on their own to reverse the deformation, so that the spoon remains flat under the action of the internal stresses. However an appropriate increase in temperature will cause the spoon to "remember" its original shape and so curl back up, as a result of the changes in resistance to movement with temperature to the martensite/matrix phase boundaries. (Such resistance to movement is analogous to Peierls–Nabarro forces in dislocation mechanics.) The martensite is thereby reversibly restored, and hence the spoon bends up when used to stir hot tea.

Since the resistance to movement of the phase boundaries decreases with increasing temperature, attempts to flatten the spoon in a somewhat higher range of temper-

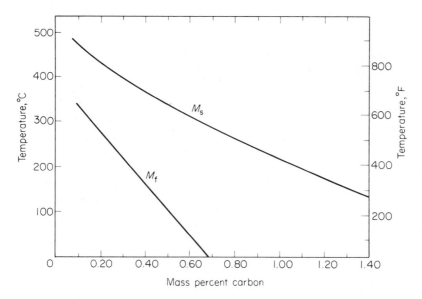

Figure 10-5   Effect of carbon content on temperatures for start (M_s) and finish (M_f) of martensite transformation. [From A. R. Troiano and A. B. Greninger, "The Martensite Transformation," *Metals Handbook*, ed. T. Lyman (Metals Park, Ohio: American Society for Metals, 1948), p. 265.]

atures would not succeed. However, the martensitic material would reversibly deform over quite large strains, as mentioned earlier, and this type of response in these solids is termed *superelastic* behavior.

Polymer spoons can also display similar large recoveries over certain ranges of temperature as a result of relief of residual stresses produced in forming processes. Vacuum-formed objects made from inappropriate sheet material have been known to flatten out if exposed to sunlight over long periods. Heat-shrinkable plastics, which are irradiated and quenched polymers, work for reasons similar to martensitic memory metals: They are deformed (expanded) when hot and then quenched while still in the expanded state. On reheating they remember their original, smaller shape and thus shrink tightly around whatever is being covered.

The amount of carbon present in Fe–C austenite influences the temperatures for the *start* of the martensitic transformation, called M_s, and for the *finish* of the transformation, called M_f. These are the temperatures for 1% and 99% martensite, respectively. Figure 10-5 shows the drop in these temperatures with increasing carbon content. This drop might be expected from energy considerations: With higher carbon content there will be more strain energy in the martensite that is to be formed, so to provide the necessary driving force to form martensite the transformation must take place at a lower temperature where the free energy difference between fcc and bcc is greater. Thus instead of transformation to the desired low-energy bcc ferrite, the absence of diffusion causes the fcc austenite to transform to a lower-energy, but still highly strained, martensite. Note that for carbon contents above about 0.65%, there will be some retained austenite present following quenching to room temperature.

Exercise 10-4

One step in the heat treatment of some high-strength steels is to cool them to very low temperature, well below room temperature. Why is this done?

Solution. This step usually follows a quenching process, which is intended to produce as much martensite as possible. Higher-carbon steels will have substantial amounts of retained austenite at room temperature (Fig. 10-5), so further cooling is necessary to transform this austenite to martensite. If the part is made to close tolerances and may subsequently be exposed to temperatures much lower than the original quench temperature, the increase in specific volume resulting from the additional martensite formation during subcooling can lead to dimensional misfit and failure. Because maintenance of very low temperatures entails an additional processing cost, such treatment is restricted to steels having very high hardness or strength and requiring close tolerances, where the extra expense is justified.

MECHANICAL PROPERTIES OF MARTENSITE

The hardness of iron–carbon martensite depends on the amount of carbon present. Figure 10-6(a) gives the Rockwell C hardness of thin wafers (about 2 mm thick) of plain-carbon steel that have been heated into the austenite range

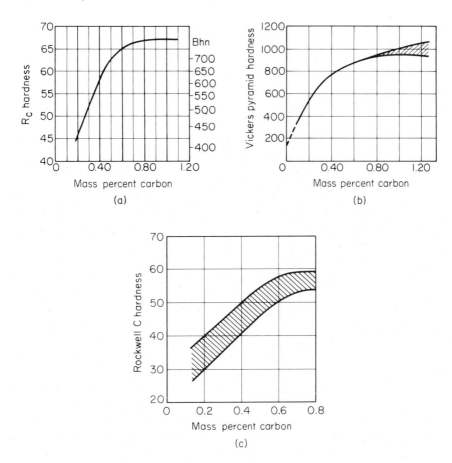

Figure 10-6 Effect of carbon content on hardness of iron–carbon martensite. (a) and (b), 99.9% martensite. [From Edgar C. Bain, *Functions of the Alloying Elements in Steel* (American Society for Metals, 1939), p. 36.] (c) 80% martensite. [From ASM Committee on Hardenability, "The Selection of Steel for Hardenability," *Metals Handbook: Properties and Selection of Metals*, 8th ed., vol. 1, ed. T. Lyman (Metals Park, Ohio: American Society for Metals, 1961), p. 189.]

and then water-quenched to room temperature. If the quenching is fast enough to form all martensite, the hardness can be reliably predicted from the carbon content of the steel. With ordinary quenching techniques it is not possible to obtain martensite in plain-carbon steels of less than about 0.20% C. (Reasons for this limitation are discussed later in this chapter.) With carbon content above about 0.80%, a saturation in the R_C hardness occurs. This does not give a true picture of the resistance to penetration, since the Vickers pyramid hardness (obtained with a diamond-shaped indenter, with hardness related inversely to the areas of indentation), shown in Fig. 10-6(b), continues to rise with increasing carbon. The R_C measurement of depth of indentation is apparently not sensitive to changes in the martensite in this range. The continuing increase in hardness shown in Fig. 10-6(b) is consistent with the argument that more carbon strains the lattice more and thus inhibits dislocation motion more effectively.

Quenching steel more than 2 mm thick usually results in less than 99% martensite (for reasons that are discussed later in this chapter), with consequent lower hardness. For example, Fig. 10-6(c) shows the range of hardness with carbon content when 80% martensite is produced. This hardness range can be considered typical of usual quenching procedures.

The strength of martensite is often much lower than would be predicted from the equation S_u [MPa] = 3.45 × Bhn. For example, Fig. 2-25 shows two instances where martensite produced by water quenching, with about 600 Bhn, has a strength of only about 700 MPa instead of the predicted 2 GPa. On the same graph is another water-quenched martensite specimen, lower in carbon, with about 310 Bhn and 830 MPa, which is much closer to the predicted behavior. This fickle behavior is related to the kind of fracture that occurs in martensite. Whereas pearlitic steel usually deforms plastically, reaches a maximum load, deforms further, and then fractures by ductile tearing, martensite fractures before plastic flow sets in, while the load is still increasing along a nearly elastic curve.

Figure 10-7 contrasts true stress–strain curves for a 0.40% C* low-alloy steel after several heat treatments. Curve (a) has the substantial deformation characteristic of a specimen composed of ferrite and pearlite. Curve (b), for which only the fracture point is shown, exhibits the usually small deformation that precedes fracture in as-quenched martensite. The energy required to fracture each specimen is related to the area under each curve in Fig. 10-7. Because the area under curve (b) is very small, as-quenched martensite is considered brittle. The area under curve (a) is large, indicating that a microstructure of ferrite and pearlite is usually ductile at room temperature.

The low tensile strength of as-quenched martensite is partly due to residual stresses that occur during quenching. These large residual stresses result from the nonuniform local strains during the transformation from austenite to martensite. When a specimen of martensite containing residual stress is loaded in tension, the residual tensile stresses can add to the applied tensile stresses to cause fracture at a lower applied tensile stress than would be required to cause fracture in the absence of residual stresses. If much plastic deformation oc-

*The American Iron and Steel Institute (AISI) and the Society of Automotive Engineers (SAE) designate plain and low-alloy steels by use of a simple code, in which the first two digits indicate the extent of alloy content (where 10 refers to plain-carbon steels) and the remaining digits indicate the amount of carbon present, in hundredths of one percent. Hence 1040 is a plain-carbon steel of about 0.40% C, and 4140 is a low-alloy steel (AISI-SAE tables give the alloy composition range) of about 0.40% C. The Unified Numbering System (UNS) for metals and alloys has been adopted by the American Society for Testing and Materials (ASTM), SAE, and AISI. This system is expected to come into general use with time; thus the UNS designation is listed frequently in this book.

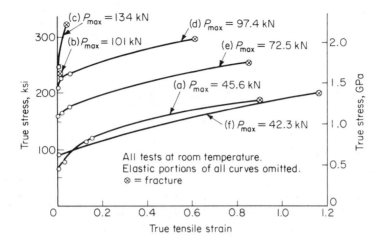

Figure 10-7 True-stress–true-strain curves at room temperature for tensile specimens (9.07 mm initial diameter) of a low-alloy steel of 0.40% C (AISI 4140, UNS G41400), initially heated into the austenite range: (a) furnace-cooled (mostly coarse pearlite); (b) oil-quenched (martensite); (c) oil-quenched, then heated to 230°C for 1.8 ks (stress-relieved martensite); (d) oil-quenched, 430°C for 1.8 ks (tempered martensite); (e) oil-quenched, 590°C for 1.8 ks (tempered martensite); (f) oil-quenched, 700°C for 36 ks (mostly spheroidite). (Data by William G. Ovens.)

curred in the martensite before fracture, the residual stresses would be greatly reduced and the load for fracture would be increased. Thus it is the brittle nature of the fracture of martensite that is indirectly responsible for the frequent lowering of its strength. Oil quenching, with the lower residual stresses that result from a slower cooling rate, is therefore preferable to water quenching if the steel is such that martensite can be produced at the slower cooling rate.

The hardness test thus provides a truer measure than does the tension test of the ability of martensite to resist slip. A conical or spherical hardness indenter will cause compressive stresses in all three directions (triaxial compression) in the specimen. Slip occurs because the compressive stresses are not of the same magnitude in all three directions. Tensile fracture cannot occur without tensile stresses, so in materials without flaws or cracks there is no immediate fracture problem to interfere with plastic deformation in the hardness test. Furthermore, the hardness test is not substantially influenced by residual stresses because the large plastic deformation that occurs in the presence of the triaxial compressive stresses wipes out most of the residual stresses.

Exercise 10-5

Construct Mohr's circle for stress for (a) a uniaxial tensile test, and (b) a hardness test where $\sigma_1 > \sigma_2 > \sigma_3$ and all are compressive. Select magnitudes of shear stress for slip (τ_{slip}) and tensile stress for fracture (σ_f) so that fracture occurs before slip in (a) and slip occurs before fracture in (b).

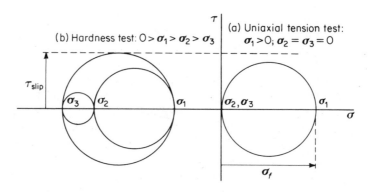

Figure 10-8 Solution to Exercise 10-5 showing in a simple way why a brittle material does not fracture under moderate triaxial compression.

Solution. The construction is shown in Fig. 10-8. (See Chap. 2 for the Mohr's circle construction, if review is required.) The radius of the largest circle for each case gives the maximum shear stress. Thus, for the tensile test, if $\sigma_f < 2\tau_{slip}$, tensile fracture will occur. In the hardness test, slip will always occur first. It should therefore be clear from this construction how plastic deformation can occur in the hardness test of very brittle metals.

Heating martensite to the range 150–260°C for 2 ks apparently allows the residual stresses from quenching to be relieved by local slip. Some changes on a very fine scale also occur in the microstructure of martensite, but these do not substantially reduce the hardness and may aid in the relief of residual stresses. Curve (c) in Fig. 10-7 shows the marked increase in strength that occurs in oil-quenched AISI-SAE 4140 steel after *stress relieving* at 230°C. In this temperature range the residual stresses resulting from rapid nonequilibrium cooling and the austenite-to-martensite transformation are substantially reduced with negligible change in the microstructure. This results in an increase of strength over as-quenched martensite, which fails prematurely with virtually no plastic deformation. In order to discriminate here between these two similar-appearing microstructures, the stress-relieved microstructure is called *stress-relieved martensite*. The strength of stress-relieved martensite is now much closer to the value predicted from the hardness test, although the fracture strain remains small. In Fig. 2-25 four of the five highest strengths are for steels that have been quenched and then tempered in the range 200–260°C.

TEMPERED MARTENSITE

If quenched specimens of steel are heated to temperatures above about 250°C, the excess carbon in the martensite precipitates out in a series of steps that terminates with formation of fine particles of Fe_3C in a matrix of bcc ferrite.

The ferrite can be thought of as martensite with almost all of the carbon removed, as was pointed out earlier in this chapter. Because this process takes place at relatively low temperatures, diffusion is quite limited. Carbon can thus diffuse only very short distances, so the Fe_3C forms as very fine particles that are uniformly distributed throughout the ferrite. The resulting microstructure gives an ideal combination of high strength and ductility, which is here called *tempered martensite* (although no martensite is present when the precipitation of Fe_3C is complete). Note that "to temper" means to moderate; in this case the terminology refers to the reduction of brittleness.

Curves (d) and (e) in Fig. 10-7 show the range of mechanical behavior that can be expected of tempered martensite of UNS G41400. The hardness of tempered martensite is more closely related to strength than was observed for as-quenched martensite. Most of the data in Fig. 2-25 in the 900–1700 MPa strength range are for tempered martensite. Proper selection of tempering temperature can provide (within limits) a wide range in strength, hardness, and ductility because the higher tempering temperature allows a higher rate of diffusion with a consequent larger size and wider spacing of the Fe_3C particles. Consequently the strength and hardness of tempered martensite depend on both the percentage of carbon and the tempering temperature.

Figure 10-9 shows the variation of hardness with tempering temperature for steels of a range of carbon content. If lengthy tempering occurs at just below the eutectoid transformation temperature, as for curve (f) in Fig. 10-7, the Fe_3C forms as very large spheroidal particles of extremely wide spacing. The mixture of ferrite and this form of Fe_3C is virtually the same as spheroidite (see Chap. 9), so this microstructure can be considered a mixture of free ferrite (α) and spheroidite (P_s, of 0.77% C). This structure has lower strength and higher ductility than any tempered martensite or pearlitic microstructure. Curve (f) in Fig. 10-7 exhibits this structure's characteristic properties. Microstructures of ferrite + spheroidite can also be produced by first forming ferrite + pearlite and allowing the Fe_3C to agglomerate at about 700°C into large spheroids, as described in Chap. 9. The test results shown in Fig. 10-9 for tempering at 700°C do not represent the hardness of spheroidite (e.g., for 0.74% C steel) because the samples were tempered for only 1.8 ks. Complete spheroidization, forming larger Fe_3C particles, requires conditions of the order of 10 h (36 ks) at 700°C.

The ductility of tempered martensite increases and the strength decreases with increasing tempering temperature, except in the range of about 250–400°C. Tempering in this range decreases notch ductility (as measured in a special notched-impact test; see Chaps. 2 and 14) to values lower than can be achieved through tempering at either 250 or 400°C. This loss of ductility, known in the past as *500°F embrittlement* or *blue brittleness*, is accompanied by a reduction of strength below that from tempering at 500°F (260°C), so there are no compensating gains in mechanical properties to warrant tempering in the range of about 250–400°C. (For consistency with the use of degrees

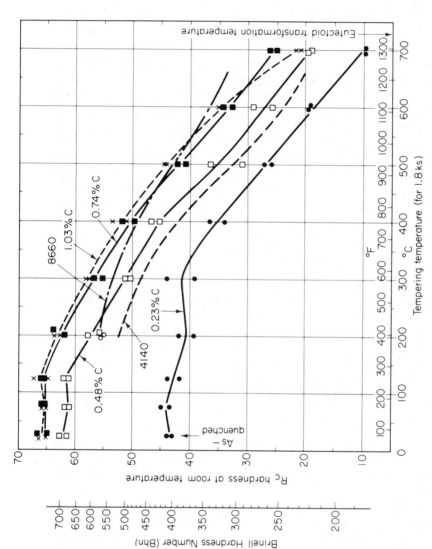

Figure 10-9 Hardness at 20°C as a function of tempering temperature (all tempered for 1.8 ks) for steels of a range of carbon and alloy content. Water quenching (plain-carbon steels) was in ice water; oil quenching (low-alloy steels) was at 20°C. [Data on plain-carbon steels by Roger A. Heimbuch. Data on low-alloy steels from American Society for Metals, *Source Book on Industrial Alloy and Engineering Data* (Metals Park, Ohio: American Society for Metals, 1978).]

celsius, 500°F embrittlement may also be termed *250°C embrittlement*, since 250°C is a rough approximation to 500°F.) The cause of this embrittlement is associated with the diffusion of interstitials (probably nitrogen and carbon) to dislocations, which are then prevented from moving freely in such a way as to promote high local stresses and brittleness.

In some low-alloy steels that have been quenched to produce martensite, tempering in a temperature range above 550°C promotes precipitation of very fine particles, which are usually carbides that are formed with some of the alloying elements. Important carbide-forming elements are titanium, vanadium, molybdenum, and tungsten. These fine particles act to inhibit dislocation motion and therefore can reverse the trend toward lower strength with higher tempering temperature. This is called *secondary hardening*. This behavior can be seen in a mild way in Fig. 10-9, where the hardness of 8660 (UNS G86600) at the higher tempering temperatures, resulting from carbide precipitation, not only exceeds the hardness of the 0.74% C steel but also exceeds the hardness of the 1.03% C steel. Secondary hardening is much more marked than this in certain other low-alloy steels.

Exercise 10-6

How well do the tempered martensites hold their strength at elevated temperatures?

Solution. Not very well, in general, because the fine particles of Fe_3C agglomerate in a manner analogous to $CuAl_2$ (see Chap. 9), and the strength drops. Secondary hardening provides an alternative mechanism for maintaining strength at higher temperatures, and high-temperature materials make extensive use of stable fine particles.

HARDENABILITY

Tempered martensite has a good balance between high strength and moderate ductility, and it is used where the requirement for high strength justifies the greater processing cost than for pearlitic steels. Tempered martensite can be made only from martensite. Likewise, stress-relieved martensite of high carbon content is the only practical structure having hardness in the 55–65 R_C range. The conditions that promote the formation of martensite are therefore of great engineering interest, and a substantial amount of knowledge on the subject is available in the form of articles, books, and handbooks. *Hardenability* is a measure of the ease with which martensite can be formed in a given steel; this property can be measured quantitatively in several satisfactory ways.

The time required for transformation of austenite to pearlite becomes shorter when transformation occurs at temperatures progressively further below the eutectoid temperature; the time reaches a minimum at about 550°C, and at lower temperatures diffusion is so slow that pearlite does not form at all. If a piece of eutectoid steel is cooled from austenite so fast that when it has reached 550°C no pearlite has formed, the microstructure will remain austenite during further rapid cooling until it reaches the M_s temperature, which is

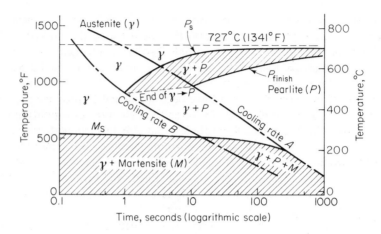

Figure 10-10 Continuous cooling transformations of a eutectoid plain-carbon steel.

about 270°C for plain-carbon steels with 0.77% C (see Fig. 10-5). Martensite then begins to form from the austenite, and at room temperature this steel will be mostly martensite with a small fraction of retained austenite. The ease of forming martensite is thus seen to be related solely to the ease of avoiding the formation of pearlite, and thus it bears no relation to the speed of the austenite-to-martensite transformation, which is always very rapid.

The effects of cooling rate and temperature on microstructure are shown in Fig. 10-10 for small pieces of eutectoid plain-carbon steel cooled as described in the preceding paragraph. When time is plotted on a logarithmic scale, continuous cooling curves are nearly straight lines. The dashed line labeled "cooling rate A" represents the fastest cooling rate that will still result in all pearlite at room temperature. Specimens cooled at "cooling rate B" and at faster rates result in all martensite (plus retained austenite) at room temperature, in the manner described in the preceding paragraph. Cooling rates between curves A and B produce a mixture of microstructures, since the transformation of austenite to pearlite does not have enough time at high temperature to reach completion. The lower edge of the shaded $\gamma + P$ region, labeled P_{finish} in Fig. 10-10, represents the finish of pearlite formation. Nothing further happens for cooling rates between curves A and B until the temperature reaches the M_s line and martensite starts to form. At room temperature, after cooling at a rate between A and B, four phases will be present: austenite, ferrite and carbide (together as pearlite), and martensite.

Exercise 10-7

How can four phases be present in Fig. 10-10 when the equilibrium phase diagram gives only two, ferrite and carbide?

Solution. The equilibrium diagram applies only to equilibrium and near-equilibrium cooling, certainly not to fast quenching where diffusion is limited. Thus the phases present after fast cooling are neither limited to nor described by the equilibrium diagram. In this case, a diagram such as Fig. 10-10 tells what phases are present.

The final microstructure after quenching therefore depends on two factors: (1) the rate at which austenite transforms to pearlite, a factor that is a property of the steel and can be specified by the limiting cooling rates A and B that span the cooling-rate range from all pearlite to no pearlite, respectively; and (2) the actual rate at which a given grain of steel is cooled. The first of these factors is just another way of describing hardenability, for if the transformation of austenite to pearlite can be suppressed (that is, the whole shaded region $\gamma + P$ is moved to the right, so cooling rate B is also moved to the right in Fig. 10-10), the minimum cooling rate that will produce martensite is slower. A slower cooling rate is easier to achieve, so such a steel has a higher hardenability. Parts that are thick, say 200 mm, where the cooling rate in the middle of the cross section must be slower than at the outside, can be transformed to martensite only if a high-hardenability steel is used. Figures such as Fig. 10-10 are variously called TTT curves, S curves, or Bain curves.

Many alloying elements have been found that increase hardenability; carbon (higher carbon measurably increases hardenability in plain-carbon steels), manganese, molybdenum, chromium, silicon, and nickel are common examples. The alloying elements other than carbon have negligible influence on the mechanical properties of martensite. The primary purpose of these alloying elements is to slow the formation of pearlite by interfering with diffusion and thus to enable martensite to be produced at slower cooling rates, the nose of the diagram being shifted to the right. Steels possessing this feature are said to have better hardenability. Another purpose of alloy additions, particularly of strong carbide-forming elements such as titanium, vanadium, and molybdenum, is to promote secondary hardening during tempering, as described in the preceding section. When the alloying elements that are added total no more than 5 to 10% of the mass of iron present, the steel is called a *low-alloy steel*. Such a steel should not be confused with an *alloy steel*, which usually has over 10% alloying elements and often has a markedly different microstructure—such as all stable austenite at room temperature—as well as unique properties such as high corrosion resistance. The line of demarcation between plain-carbon and low-alloy steels remains fuzzy, since most commercial plain-carbon steels have greater mass of other alloying elements than carbon present. For example, the standard composition of a nominally 0.40% plain-carbon steel has 0.60–0.90% Mn, an element that is very effective in increasing hardenability but has little effect in these amounts. (The small amount of manganese in plain-carbon steels acts preferentially to combine with sulfur as MnS and thus prevents the formation of brittle FeS.)

The continuous cooling-transformation diagrams of two common low-

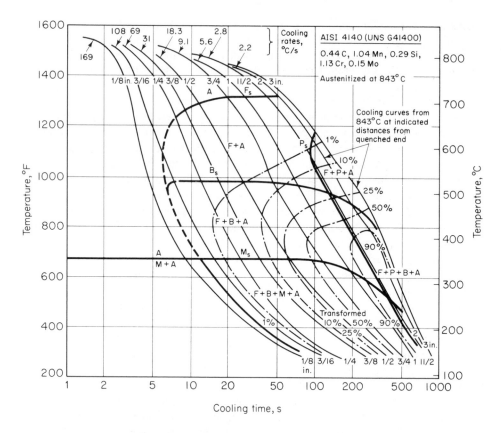

Figure 10-11 Continuous cooling transformations for AISI 4140 steel (UNS G41400). Cooling rates are at 700°C. A = austenite, F = ferrite, P = pearlite, B = bainite, M = martensite. [From Homer Research Laboratories, Bethlehem Steel Co., reprinted in *Source Book on Industrial Alloy and Engineering Data* (Metals Park, Ohio: American Society for Metals, 1978), pp. 125, 129.]

alloy steels are shown in Figs. 10-11 and 10-12 for a wide range of cooling rates. (The cooling rates are also given here as distances from the quenched end of an end-quenched bar, a procedure that is described in the following paragraphs.) Figures 10-11 and 10-12 contain two more structures than the plain-carbon eutectoid steel in Fig. 10-10: *ferrite* and *bainite*. Ferrite is absent from Fig. 10-10 because there is no free ferrite produced during quasi-equilibrium cooling of eutectoid steel. Bainite is a microstructure of very fine particles of carbide in a matrix of ferrite that forms directly from austenite. Bainite and tempered martensite are thus of similar microstructure, but tempered martensite is formed only from the tempering of martensite and bainite is formed directly from the cooling of austenite.

Figure 10-12 shows how bainite can be formed in a low-alloy steel by

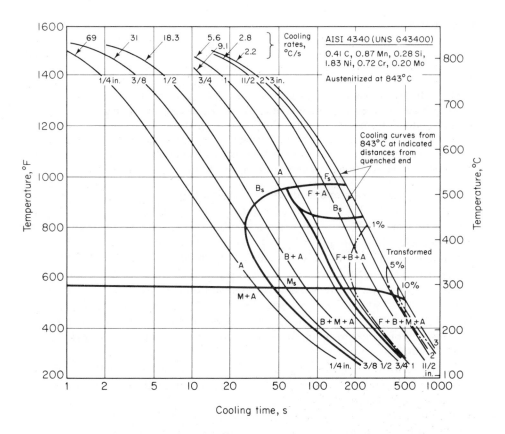

Figure 10-12 Continuous cooling transformation for AISI 4340 steel (UNS G43400). Cooling rates are at 700°C. A = austenite, F = ferrite, B = bainite, M = martensite. [From Homer Research Laboratories, Bethlehem Steel Co., reprinted in *Source Book on Industrial Alloy and Engineering Data* (Metals Park, Ohio: American Society for Metals, 1978), pp. 125, 129.]

continuous cooling. It is possible to produce bainite in plain-carbon steel by *isothermal transformation*, which simply means transformation at constant temperature. (Contrast this with the continuous cooling transformation in Figs. 10-10 through 10-12.) For example, if a plain-carbon eutectoid steel is quenched very rapidly from 800°C to 400°C, it remains austenite. (We can use Fig. 10-10 to establish this, as the quenching process up to this point is continuous cooling; but from this point on we cannot use Fig. 10-10, because we will no longer be cooling.) If the specimen is held at 400°C for, say, 1000 s, the austenite will transform to ferrite and carbide in the form of bainite. This occurs because (1) the temperature is too low to allow sufficient diffusion to produce pearlite, and (2) the temperature is above the M_s, so that the austenite will not transform to martensite.

Exercise 10-8

How would you determine the time required to produce all bainite at 400°C?

Solution. A series of tests on very small specimens would be run: (1) Quench from 800°C to 400°C; (2) hold at 400°C for time t; (3) water-quench to 20°C; (4) measure hardness. If any austenite remains at time t, it will be transformed to martensite when water-quenched (except for a fraction of retained austenite), giving higher hardness than for all bainite. For times greater than the time t required just to produce all bainite, no martensite will be produced and the hardness should remain substantially constant. The bainite can be established by metallographic examination. This procedure is used to establish *isothermal transformation diagrams* for steels.

The amount of nonmartensite transformation products is indicated in Fig. 10-11 for AISI 4140 steel at several levels from 1% to 90%. Thus a cooling rate of 108°C/s at 700°C leads to about 99% martensite when cooled to 150°C, a cooling rate of 18.3°C/s leads to slightly more than 50% martensite, and a cooling rate of 2.8°C/s yields less than 10% martensite. By contrast, in Fig. 10-12 for AISI 4340 steel, a cooling rate of 2.8°C/s at 700°C produces more than 95% martensite. Thus a cooling rate that produces more than 95% martensite in 4340 steel will produce less than 10% martensite in 4140 steel. Thus AISI 4340 has a much higher *hardenability* than AISI 4140 (a result of the presence of nickel in 4340, in addition to the chromium and molybdenum of AISI 4140).

The grain size of the austenite will also influence hardenability. If the grain size is small, the large surface area of grain boundaries means large surface energy, and there are also more grain-boundary sites for the nucleation of pearlite. For both reasons, fine-grained austenite forms pearlite faster than coarse-grained austenite and hence has a lower hardenability.

The second factor cited above that strongly influences the final microstructure is the actual rate of cooling of a given grain of steel. The specimens used in Fig. 10-10 were very small, so that the cooling rate was nearly the same throughout. As the thickness increases, so does the difference in cooling rate between the center of a piece of steel and the surface. It is therefore possible to have a quenched piece of steel where the cooling rate at the surface is faster than rate B in Fig. 10-10, producing martensite (and retained austenite), while the cooling rate at the center is slower than rate A, producing all pearlite (and ferrite, if the steel is below 0.77% C). Thus the dimensions of the piece and the quenching medium (such as water, oil, or air) will control the cooling rate at a given point in the steel.

The *Jominy* end-quench hardenability test was developed to measure with a single test the hardness, and hence the microstructure, that result from a wide range of cooling rates. This test consists of water quenching one end of a specimen [1 in. (25.4 mm) in diameter and 4 in. (102 mm) long] that has been heated about 40°C above the lowest temperature at which the steel is all austenite. The physical setup for this test is shown in Fig. 10-13. It can be seen that the cooling rate at the quenched end is very rapid and progressively

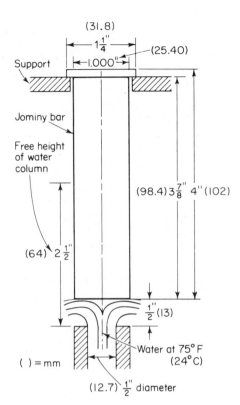

Figure 10-13 Jominy end-quench hardenability test.

decreases with distance from the quenched end. (Details can be obtained from Ref. 10-1.) Two parallel flats, 0.015 in. (380 μm) deep, are ground on the cylindrical surface for hardness measurements. The Rockwell C hardness along the flats is measured and plotted vs. distance at $\frac{1}{16}$-in. (1.6-mm) intervals from the quenched end of the bar. Such *Jominy hardenability curves* are shown in Fig. 10-14 for a 0.40% C plain-carbon steel (1040) and the 0.40% C low-alloy steel (4140) used for the tensile results given in Fig. 10-7 and included in the hardness results of Fig. 10-9. The higher hardenability 4340 steel is also included in Fig. 10-4. The shaded areas represent the usual ranges of hardness that result from the allowable variations of carbon content and alloy content for these steels. Note that the three steels have about the same hardness at the quenched end because the hardness of the 100% martensite is controlled only by the 0.40% carbon content. The cooling rate at the quenched end is fast enough to form martensite in most plain-carbon and all low-alloy steels.

Comparison of Jominy curves for different steels is based on little or no variation in thermal properties, such as thermal conductivity and specific heat. For plain-carbon and low-alloy steels there appears to be no substantial change in these properties from that of iron. Thus at any particular distance from the quenched end, the cooling rate in the Jominy test is the same for all

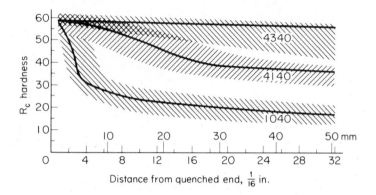

Figure 10-14 Jominy hardenability curves for three 0.40% C steels: AISI 1040 (UNS G10400), AISI 4140 (UNS G41400), and AISI 4340 (UNS G43400). Shaded regions show ranges; solid curves show typical values.

of these steels. Each point therefore experiences a unique cooling rate, which can be plotted as the dashed lines in Figs. 10-11 and 10-12. The drop in hardness for greater distances from the quenched end corresponds to the decreasing amount of martensite that results as the cooling rate is decreased from curve B to curve A and slower in Fig. 10-10, and from *Jominy distances* ranging from $\frac{1}{8}$ in. to 3 in. in Figs. 10-11 and 10-12.

Exercise 10-9

Why would a designer specify a particular low-alloy steel, for example AISI 4140, in place of a plain-carbon steel such as AISI 1040 when both can be quenched to about 58 R_C (see Fig. 10-14)?

Solution. If a part is small enough to be quenched in water to the hardness desired, then the plain-carbon steel may be adequate. However, in practice a low-alloy steel is usually specified even in this case, to be oil-quenched, because high residual stresses and the consequent possibility of cracking can result from the high cooling rate in water quenching. For larger parts it is impossible to achieve very much martensite in regions away from the surface of AISI 1040 steel, so to obtain mostly martensite a higher-hardenability steel must be used. Compare the typical hardness values for the three steels in Fig. 10-14 at a Jominy distance of 16 ($\frac{16}{16} = 1$ in., a cooling rate of 5.6°C/s at 700°C): 21 R_C for 1040, 42 R_C for 4140, and 57 R_C for 4340. Note that high-hardenability steel has one important disadvantage besides higher cost: Welding requires special precautions to avoid cracking because so much martensite forms during cooling. Thus the designer should not specify a steel of higher hardenability than is needed.

MECHANICAL PROPERTIES OF TEMPERED AISI 4140

Most mechanical properties are changed by tempering, as discussed above and shown in Figs. 10-7 and 10-9. Figure 10-15 shows several property changes

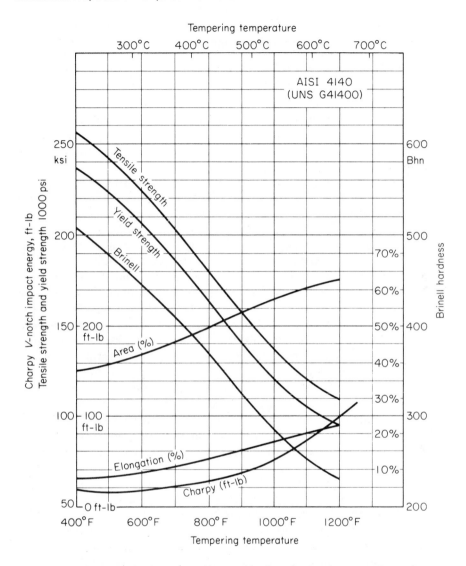

Figure 10-15 Mechanical properties as a function of tempering temperatures for oil-quenched AISI 4140 (UNS G41400) bars of 12.7 mm diameter. "Area (%)" is cold work at fracture, r_f. "Elongation (%)" is nominal strain to fracture, e_f. [From Republic Steel Corp., "Alloy Steels," *Source Book on Industrial Alloy and Engineering Data* (Metals Park, Ohio: American Society for Metals, 1978), p. 36.]

from tempering AISI-SAE 4140 steel over a range of temperatures following oil quenching from 840°C. The trade-offs from tempering can be seen in Fig. 10-15, where one gains ductility in exchange for reductions in strength and hardness. The dip in Charpy impact energy in the temperature range from 200 to 400°C, already described as 250°C embrittlement, usually leads to a choice of tempering either above or below this temperature range.

ULTRAHIGH-STRENGTH STEELS

Many of the strengthening mechanisms that have been described can be combined to produce metals of unusually high strength. Two examples are the nickel–martensite *maraging* steels and the *ausforming* of certain high-hardenability steels.

Martensites of iron and nickel can be produced in a manner similar to martensites of iron and carbon, although the cooling rates required are much slower and the final structure is less hard. Figure 10-4(a) shows the appearance of iron–nickel martensite. A high-nickel steel that has been vacuum-melted (for control of impurities) and that will produce iron–nickel martensite upon air cooling may have, for example, the following composition: 17.5% Ni, 12.5% Co, 3.75% Mo, 1.8% Ti, 0.15% Al, balance Fe plus very small amounts of C, Si, and Mn. After air cooling to form martensite, the ultimate tensile strength is approximately 1 GPa. When heated to 480°C for 11 ks, a very fine precipitate forms from the alloying elements to increase the room-temperature yield strength to 2.4 GPa, while retaining the original martensite. This is called *maraging*. Note the similarity of this process to secondary hardening, where fine particles are formed on heating after quenching. The chief difference between maraging and secondary hardening is that in maraging the martensite is stable and does not transform to tempered martensite. The practical advantages of a maraging steel are that the martensite can be formed with slow cooling, and this relatively soft structure can then be cold-worked or machined. The subsequent aging process takes place with virtually no dimensional changes and at a temperature that is so low that the quality of the finished surface is not destroyed. The final structure represents a combination of retained austenite, iron–nickel martensite, and finely dispersed alloy particles. This steel has excellent fracture toughness.

Ausforming combines cold work and tempered martensite strengthening in a unique way. It is possible to deform some high-hardenability steels while they are still austenite in a temperature range below the eutectoid temperature, so that recrystallization does not occur and neither pearlite nor bainite is formed. After being cooled to form martensite and then tempered, these steels can have extremely high strengths. For example, a 5% Cr steel, Vascojet 1000, air-cooled from 1000°C to 20°C, deformed 90% at 590°C, tempered at 565°C, has the following room-temperature properties: $S_u = 2.6$ GPa, $S_y = 2.3$ GPa, $r_f = 40\%$, $e_f = 8\%$, and 63 R_C hardness. This ausforming steel retains acceptable properties to 550°C, where, for example, its strength is 1.8 GPa.

Although the reasons for this exceptional behavior are not fully understood, the "cold-worked" tempered martensite microstructure may have the fine particles of carbide preferentially located near the dislocations formed during ausforming so as to be unusually effective in preventing dislocation motion. It is also possible that aging effects are important here, since fine alloy-carbide particles have been found to influence the hardness in some instances for ordinary martensites.

REFERENCES

10-1. American Society for Metals, *Metals Handbook*, 9th ed., vol. 1, Properties and Selection: Irons and Steels. Metals Park, Ohio: American Society for Metals, 1978.

10-2. American Society for Metals, *Metals Handbook*, 9th ed., vol. 2, Properties and Selection: Nonferrous Alloys and Pure Metals. Metals Park, Ohio: American Society for Metals, 1979.

10-3. American Society for Metals, *Source Book on Industrial Alloy and Engineering Data*. Metals Park, Ohio: American Society for Metals, 1978. A book of useful engineering data, mostly as tables and graphs, of metals and alloys of industrial importance such as iron and steel, stainless, refractory metals, aluminum, copper, magnesium, and titanium.

10-4. Birchenall, C. Ernest, *Physical Metallurgy*. New York: McGraw-Hill Book Co., 1959. Pages 290–309 give details on the formation of martensite, tempered martensite, and bainite.

10-5. Reed-Hill, Robert E., *Physical Metallurgy Principles*. Princeton, N.J.: D. Van Nostrand Co., 1964. Chapter 15 is particularly valuable for its detailed description of the martensite transformation. See also Chap. 16, "The Iron–Carbon System," and Chap. 17, "The Hardening of Steel," which provide much more detail than is possible here.

PROBLEMS

10-1. What microstructures result from each of the following treatments, taken independently, of a previously annealed 0.77% C steel?
   (a) Heat to 740°C, water-quench.
   (b) Heat to 820°C, air-cool.
   (c) Heat to 150°C, water-quench.
   (d) Heat to 760°C, furnace-cool.
   (e) Heat to 800°C, water-quench; reheat to 480°C for 2 ks, air-cool.

10-2. For each of the following treatments, taken consecutively, calculate (or estimate, if necessary) the Brinell hardness number. Start with a bar of 0.60% C steel 10 mm in diameter of unknown history:
   (a) Heat to 850°C for 2 ks, air-cool to 20°C.
   (b) Heat to 815°C for 2 ks, water-quench to 20°C.
   (c) Heat to 120°C for 4 ks, oil-quench to 20°C.
   (d) Heat to 480°C for 2 ks, air-cool to 20°C.
   (e) Heat to 700°C for 40 ks, water-quench to 20°C.
   (f) Heat to 815°C for 2 ks, furnace-cool to 20°C.

10-3. Find the tensile strength of 0.40% C steel after each of the following heat treatments, taken independently:

(a) Heat to 850°C, water-quench to 20°C, and stress-relieve (no residual stresses present).

(b) Heat to 730°C, air-cool to 20°C.

(c) Heat to 730°C, water-quench to 20°C, and stress-relieve (no residual stresses present).

(d) Heat to 870°C, quench in molten salt to 740°C, oil-quench to 20°C.

10-4. Estimate the maximum tensile load that initially annealed bars of 0.30% C steel 6 mm in diameter would support after the following heat treatments, each taken separately:

(a) Furnace-cool from 850°C.

(b) Oil-quench from 850°C.

(c) Water-quench from 850°C and stress-relieve.

10-5. If you heat a piece of cold-worked 0.40% C steel to 850°C and water-quench it, it becomes harder. If you heat a piece of cold-worked alpha brass to 500°C and water-quench it, it becomes softer. Explain why similar heat treatment produces opposite effects in the two alloys.

10-6. Small pieces of plain-carbon steels are given the treatments listed below. What are the resulting microstructures and hardnesses?

No.	Percentage of Carbon	Austenitizing Temp., °C	Quenching Medium
1	0.77	760	Oil
2	0.77	760	Water
3	0.20	760	Water
4	0.40	840	Water

10-7. The surface microstructure of a bar of 0.40% C steel 6 mm in diameter, water-quenched from 730°C, contains one half ferrite and one half martensite. (This is an example of a *duplexed* microstructure.)

(a) What is the carbon content of the martensite immediately after quenching?

(b) What is the carbon content of the ferrite immediately after quenching?

(c) Estimate the surface hardness of this bar. Show your method.

10-8. A ball-bearing company makes high-precision bearing races (the housing for ball bearings) of AISI-SAE 52100 steel (UNS G52986), a low-alloy steel of 1.00% C. These are first rough-machined, then austenitized and oil-quenched, then cooled to −75°C and warmed to room temperature before the final precision-grinding operation. What might occur if the −75°C cooling is omitted? Explain why.

10-9. Tempering of martensite at 430°C produces a precipitate, and the final hardness is lower than the hardness before tempering. In contrast, precipitation in 5% Cu–Al alloys at 150°C (age hardening) *increases* the hardness. Explain this difference in behavior of the two alloys.

10-10. A production part 75 mm in diameter made of AISI-SAE 1040 steel (UNS G10400), water-quenched from 850°C, is hard enough at the outside surface but too soft at the center.

    (a) What changes would you recommend?

    (b) In practice, parts of this size are virtually never water-quenched from austenite. Why not?

10-11. The surface of a bar of AISI-SAE 1040 steel (UNS G10400) 50 mm in diameter, water-quenched from 870°C, is near 43 R_C.

    (a) If this bar were 100% martensite, what would be its hardness?

    (b) Why is this bar not 100% martensite?

10-12. A part made of AISI 4140 steel (UNS G41400) fractured in service under impact loading. Several other identical structures were to be constructed, and the owner wanted to avoid future fractures. These parts had been oil-quenched from 840°C, tempered at 480°C, and air-cooled, resulting in 1.1 GPa strength. The designer had specified this treatment so these parts would be extra strong—to be on the "safe" side—although the design stress, including all safety factors, was only 0.96 GPa. What would you recommend that the owner do to reduce the chances for failure of the structures that were still to be built?

10-13. Is it possible for two steels to have the same hardenability but different hardnesses? Explain very briefly.

10-14. Two complete Jominy curves, for steels *A* and *B*, are given in Fig. 10-16.

    (a) Which steel has the higher hardenability, *A* or *B*?

    (b) Which would probably be better for use in knife blades?

    (c) Which would give more uniform strength after quenching?

    (d) Which has the higher percentage of alloying elements other than carbon?

    (e) Which will have more pearlite present at the center of a bar 40 mm in diameter that has been water-quenched from 900°C?

    (f) Which contains harder martensite?

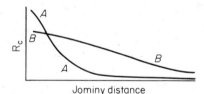

Figure 10-16   Jominy distance

10-15. Name or describe each of the phases present and the microstructure in which these phases are found after the treatment given:

    (a) 0.20% C, 99.8% Fe, slowly cooled from the melt.

    (b) 1.20% C, 98.8% Fe, slowly cooled from the melt to 750°C, then water-quenched.

    (c) Alloy (b), given the same treatment, then reheated to 430°C for 2 ks.

    (d) AISI 1040 steel (UNS G10400), quenched from 840°C to 480°C, held for 600 s, then oil-quenched.

    (e) AIAI 1095 steel (UNS G10950), quenched very rapidly from 870°C to 150°C.

    (f) Alloy (e), given the same treatment, then reheated to 315°C for 2 ks.

10-16. Isothermal (constant-temperature) transformation data for very small pieces of 0.77% C plain-carbon steel initially at 800°C are given below for two bath temperatures, each followed by water quenching to 20°C.

Bath Temp.: 675°C		Bath Temp.: 370°C	
Time, s	R_C	Time, s	R_C
5	63	5	63
10	63	10	61
15	53	15	52
20	43	60	50
30	27	100	45
60	24	150	43

Name the microstructure(s) and phase(s) present at the end of:
(a) 5 s at 675°C.
(b) 5 s at 370°C.
(c) 15 s at 675°C.
(d) 15 s at 370°C.
(e) Explain, using the fundamental ideas of strengthening mechanisms, why the hardness after 60 s at 675°C and water quenching is lower than the hardness after 150 s at 370°C and water quenching. Start by describing the final microstructures of these two specimens.

10-17. For each treatment listed below (assuming bars 6 mm in diameter), write the number of the schematic microstructure shown in Fig. 10-17 that most accurately depicts the final microstructure. A microstructure may be used in more than one answer. In all cases α and β are used to indicate any possible solid solutions, and θ is used to indicate an intermetallic compound or harder phase of some other type. Note magnifications.
(a) 70% Cu–30% Zn, cooled slowly from the melt, heated to 315°C for 2 ks, oil-quenched.
(b) 5% Cu–95% Al, water-quenched from 510°C, heated to 260°C for 4 ks, oil-quenched.
(c) 5% Cu–95% Al, cooled slowly from the melt to 20°C.
(d) 99.6% Fe–0.4% C, cooled slowly from the melt to 20°C.

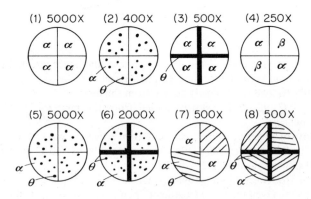

Figure 10-17

(e) 99.6% Fe–0.4% C, water-quenched from 840°C.

(f) 99.6% Fe–0.4% C, oil-quenched from 840°C, then heated to 700°C for 20 h.

(g) 99.6% Fe–0.4% C, water-quenched from 840°C, then heated to 425°C for 2 ks.

(h) AISI 4140 steel (UNS G41400), oil-quenched from 840°C.

(i) 99% Fe–1% C, cooled slowly from 870°C to 740°C, then water-quenched.

(j) AISI 1040 steel (UNS G10400), from the outer surface of an end-quenched Jominy bar at position $J = 24$.

10-18. Taking into account the functional requirements and general material characteristics and properties, match each of the following materials with the most appropriate example of use.

Materials	**Examples of Use**
(1) Aluminum alloy	**(a)** Die casting, cooking utensils, hydraulic tubing
(2) Brass	**(b)** Common nails
(3) 1010 steel	**(c)** Roller bearings
(4) 1040 steel	**(d)** Gears, crankshafts, low-quality hand tools
(5) 1080 steel	**(e)** Radiator cores, springs, tubes
(6) 10120 steel	**(f)** High-quality wrenches, hammers, pliers, axes
(7) 4140 steel	**(g)** Moderate, uniform-strength shafts, high-quality bolts
(8) 52100 steel	**(h)** Shear blades, knife edges, files

10-19. An AISI 1040 steel (UNS G10400), oil-quenched from 850°C, has a hardness of 28 R_C; the same heat treatment for an AISI 4140 steel (UNS G41400) gives a hardness of 56 R_C.

(a) Describe the differences in microstructure.

(b) Show how these differences lead to the difference in hardness.

(c) Explain why the 4140 gave a different structure for the same heat treatment.

10-20. An AISI 4140 steel (UNS G41400) bar 12.8 mm in diameter is oil-quenched from 870°C, then reheated to 540°C for 2 ks, and air-cooled to 20°C. Calculate the maximum load that this bar will support in tension.

10-21. You are given a piece of AISI 1050 steel (UNS 10500) in the shape of a short bar 8 mm in diameter.

(a) Describe a heat treatment, giving exact temperatures, times, and cooling rates (e.g., quenching in oil, water; air-cool, furnace-cool) to achieve the maximum possible hardness in this steel, without changing its chemical composition. What value of hardness can you achieve?

(b) Describe a heat treatment in the same way as in (a) to achieve the *lowest* possible hardness for this steel. What hardness would you expect?

10-22. Tensile tests on three specimens of AISI 4140 steel (UNS 41400) gave the following data (initial diameter of each is 9.07 mm):

No.	Heat Treatment	R_C	Max. Load, kN	Fract. Dia., mm
1	Oil-quench (OQ)	53	73.4	9.04
2	OQ, 233°C for 2 ks	48	113	6.50
3	OQ, 566°C for 2 ks	27	66.7	5.87

(a) Describe the final microstructure of all three specimens, and give the fractions of each phase present.

(b) Explain why the maximum load for specimen 2 is greater than for specimen 1. Be specific as to cause.

(c) Explain from a fundamental point of view why the true strain at fracture of specimen 3 is greater than for specimen 2.

10-23. Draw the approximate curve of Charpy V-notch energy measured at 20°C vs. tempering temperature for AISI 4140 steel (UNS G41400) that has been first oil-quenched from 850°C. Explain all parts of this curve.

10-24. An AISI 4140 steel (UNS 41400) is continuously cooled from 820°C to 20°C at a rate so as to give a hardness of 41 R_C. Name all phases present at 20°C.

10-25. A plain-carbon steel of 0.38% C is heated to 820°C for 2 ks, oil-quenched to 20°C, heated to 517°C for 2.5 ks, water-quenched to 20°C, subcooled to −60°C for 10 ks, and air-heated to 20°C. Estimate its final Bhn.

10-26. A plain-carbon steel of 0.77% C is heated to 760°C for 2 ks, cooled to 350°C in 1 s, held at 350°C for 1 ks, and oil-quenched to 20°C. Name the microstructure(s), sketch the microstructure(s), and label the phase(s) that result.

10-27. An unknown steel of unknown heat treatment gives the following hardness data: as received, 27 R_C; heated to 885°C and water-quenched, 56 R_C; heated to 885°C and oil-quenched, 49 R_C.

(a) What is the approximate carbon content?
    (i) 0.20% C.
    (ii) 0.40% C.
    (iii) 0.80% C.

(b) What is the alloy content?
    (i) Plain-carbon steel.
    (ii) Low-alloy steel.

(c) What was the as-received condition?
    (i) Annealed.
    (ii) Normalized.
    (iii) Quenched to martensite.

11 | Stainless and Heat-Resistant Steels

Stainless steels were developed chiefly in the twentieth century, to provide a low-cost alloy of iron that is resistant to many types of corrosion. In addition, some alloys of this class of steels maintain their strength and corrosion resistance at elevated temperatures. This chapter covers the general classes of stainless and heat-resistant steels, describes their strengthening mechanisms, and examines the most common causes of their embrittlement or loss of corrosion resistance. Because it is impossible to cover the large number of different alloys thoroughly in a single chapter, only a few of the most common stainless steels are described as examples.

For reasons described below, stainless steels are essentially iron–chromium alloys that may or may not contain other elements such as carbon, nickel, and molybdenum. These elements can cause the stable crystal structure of the iron to be either fcc or bcc, so we use the terms "austenite" and "ferrite," respectively; but the austenite and ferrite are not the same as in Fe–C. We are here not dealing with a simple Fe–C alloy as in previous chapters, so we should not be surprised that the behavior of Fe–Cr alloys is markedly different.

STAINLESSNESS

Monnartz and Borchers first discovered in 1911 that the corrosion rate of iron–chromium steels drops to virtually zero when the chromium content reaches 12%. This phenomenon can be seen in Fig. 11-1, taken from later

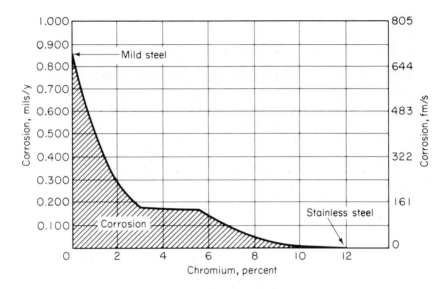

Figure 11-1 Corosion rate for Fe–Cr alloys in an outdoor industrial atmosphere, measured for 10 y (315 Ms). [From R. Franks, "Chromium Steels of Low-Carbon Content," *Trans. ASM*, **35** (1945), p. 616, disc. p. 636. Copyright American Society for Metals. Reprinted with permission.]

work by Franks. In a 10-year (315 Ms) outdoor test, mild steel corroded away at the rate of about 680 fm/s, Fe–4% Cr at about 140 fm/s, and Fe–12% Cr at virtually zero rate. Above 12% chromium, *passivity* occurs. This is the formation of a surface condition that inhibits corrosion completely. For this reason, virtually all stainless steels have a chromium content of at least 12%.

IRON–CHROMIUM–CARBON EQUILIBRIUM

Figure 11-2 is a pseudobinary equilibrium diagram of iron and chromium, developed by Zapffe (Ref. 11-7). It is "pseudo" because the carbon content varies to match the different alloys in this figure. Nevertheless, Fig. 11-2 (combined later with Fig. 11-3) shows clearly why the different classes of stainless steels behave in different ways.

The small region in the upper left of Fig. 11-2, which represents austenite (Fe–Cr), shows the limits of austenite in a pure Fe–Cr alloy, of composition up to 12.7% Cr. However, with small amounts of carbon present, this *austenite loop* expands to about 18% Cr (for about 1% C content) to form the larger loop labeled "austenite (Fe–Cr–C)." Alloys of this composition range (12–18% Cr) can usually be air-cooled to form martensite and are thus the *martensitic stainless steels*. In general, the alloys of this class having higher Cr also have higher C, since increased Cr leads to a reduction of the range of the austenite loop, which must be compensated for by increased C, which increases the size

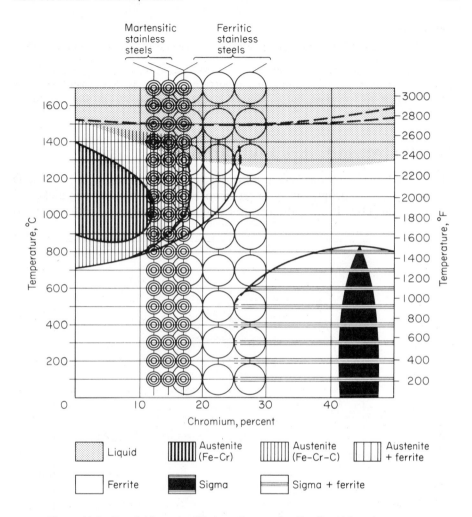

Figure 11-2 Pseudobinary equilibrium diagram for Fe–Cr, with carbon content varying to balance chromium. [From Carl A. Zapffe, "Constitution of Stainless Steels," *Stainless Steels*, American Society for Metals, 1949, p. 114.]

of the austenite loop. There is thus a balance required between chromium and carbon contents.

Fe–Cr steels above 12.7% Cr and with zero carbon are ferritic at all temperatures, as are Fe–Cr–C steels ranging from 12.7 to 18% Cr if the carbon content is low enough. Above 18% Cr, the alloys of higher carbon may still be partially austenitic, but usually they are ferritic at 20°C. Thus low-carbon alloys ranging from 12 to 30% Cr are the *ferritic stainless steels*. The ferritic stainless steels will always have lower carbon contents than martensitic stainless steels of the same chromium content.

Exercise 11-1

On what basis would a designer decide between martensitic and ferritic stainless steel?

Solution. Ferritic stainless steel can be cold-worked at 20°C, so it is appropriate for applications requiring cold forming and moderate stress levels. Martensitic stainless steel cannot be cold-worked, but it is much stronger and harder and thus useful for structural applications.

IRON–CHROMIUM–NICKEL EQUILIBRIUM

Addition of nickel to an Fe–Cr alloy leads to an extension of the stable austenite region to progressively lower temperatures with increasing nickel, as shown in Fig. 11-3. In fact, the transformation from austenite to ferrite is so

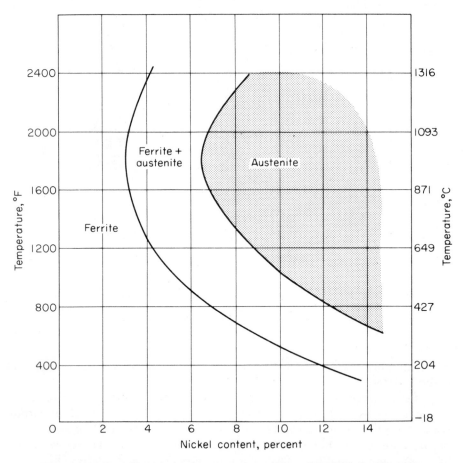

Figure 11-3 The effect of nickel content on an alloy of Fe–18% Cr. [From J. Gordon Parr and Albert Hanson, *An Introduction to Stainless Steel*, American Society for Metals, 1965, p. 27.]

sluggish that it normally does not occur for nickel contents above about 8% in the Fe–18% Cr alloy, for example. Thus we have another method, the addition of nickel, for extending the austenite range of Fe–Cr alloys, but in this case the austenite is stable at room temperature. These alloys are the *austenitic stainless steels*. Because of the stability of the austenite, these alloys cannot normally be heat-treated to increase strength.

The only other austenite-promoting element of consequence is manganese, which can be used in place of some of the nickel to produce a stable austenitic alloy. We will consider these as part of the austenitic class.

MARTENSITIC STAINLESS STEELS (Fe–Cr–C)

This class has chromium content from about 12 to 18% and carbon content from about 0.15 to 1.20%. Recall that in general the carbon content must increase with the chromium content in order to permit the expansion of the austenite loop.

From this point onward there will be extensive discussion of the effects of chemical composition on behavior. For simplicity we shall often omit the "%" sign when giving compositions; it should always be clear from the context.

One of the most used martensitic stainless steels is alloy Type 410 (UNS S41000). (Throughout this chapter we will designate this alloy by its traditional number, "410," but the new UNS designation is provided for ready reference. Compositions of the common stainless steels are given in Appendix 7.) The composition of Type 410 is 11.5–13.5 Cr and 0.15 max C. (Most of the common stainless steels have composition limits ranging as follows: 1–2 max Mn, 0.5–1.5 max Si, 0.040–0.060 max P, and 0.030 max S. These limits are omitted in this discussion; see Appendix 7 for exact limits.) It is used as a general-purpose, heat-treatable stainless steel, for such applications as machine parts, pump shafts, cutlery, jet engine parts, screws, and valves. Most martensitic stainless steels can be air-cooled from austenite to produce martensite, but the austenitizing temperature for the 12–14% Cr grades must be closely controlled to avoid the two-phase region *above* the austenite loop (see Fig. 11-2). The recommended austenitizing temperature range for Type 410 is 930–1010°C, followed by air or oil quenching. The part can then either be stress-relieved at 200–370°C or tempered.

Typical properties of Type 410 following quenching and stress relieving are 1.3 GPa strength, 1.0 GPa yield strength, 375 Bhn, 15% elongation at fracture, and 54 J Izod impact energy.

The mechanical properties of alloy Type 410 after tempering over a range of temperatures are given in Figs. 11-4 and 11-5. Note the severe temper embrittlement that occurs in the 370–600°C range for this alloy. As this temperature range also leads to decreased corrosion resistance, recommended tempering temperatures are either in the 200–370°C range or the 560–610°C range. As is usual for martensitic steels (see Chap. 10), tempering in the lower

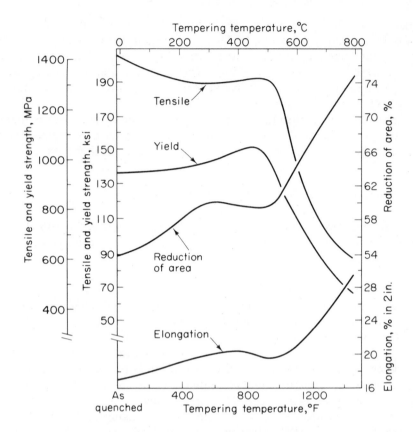

Figure 11-4 Mechanical properties of Type 410 after austenitizing, quenching, and tempering. [From *The Making, Shaping and Treating of Steel*, 8th ed., ed. H. E. McGannon. (Pittsburgh, PA: U.S. Steel Corp., 1964.) Copyright 1964 by United States Steel Corporation.]

range gives higher strength, hardness, and lower toughness, while tempering in the upper range gives the opposite.

Although retained austenite is not a serious problem in this alloy, the martensitic stainless steels with higher chromium and carbon must be subcooled immediately after quenching. Cooling to about −75°C forces the retained austenite to transform to martensite and avoids later service embrittlement or dimensional changes that might occur if much retained austenite remains.

The alloys of this class with higher chromium and carbon are particularly susceptible to *hydrogen embrittlement* (see Chap. 18). Hydrogen can enter the steel during melting or heat treating or during such processes as electroplating. The hydrogen can cause premature cracking and unexpected failure of a part. Hydrogen-generating processes should be avoided, or subsequent

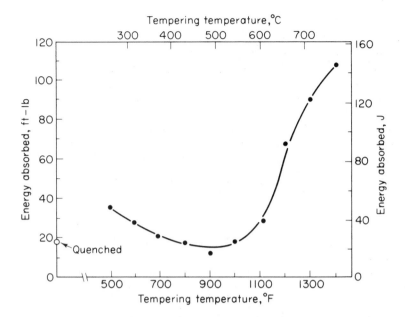

Figure 11-5 Izod impact test results for Type 410 quenched from 980°C and tempered 3 h at indicated temperature. [From W. C. Rion, Jr., *Stainless Steel Information Manual for the Savannah River Plant*, vol. I, Properties, USAEC Report DP-860 (July 1964) NASA N64-33060. Reprinted with permission.]

heating in the range 200–370°C can be used to drive the hydrogen from the metal.

Martensitic stainless steels of higher chromium (approaching 18% Cr) and carbon contents have tensile strengths up to 1.9 GPa, yield strengths to 1.8 GPa, and hardness to 600 Bhn. They are also more corrosion-resistant and more expensive than the 12% Cr martensitic steels.

FERRITIC STAINLESS STEELS (Fe–Cr)

When the carbon content is reduced for any given chromium content, the austenite loop contracts and the alloy becomes partially or completely ferritic when in the 1000°C range. The steels with carbon content lower than in martensitic stainless steels, but with chromium content in the range 12–30%, are usually ferritic.

The most commonly used ferritic stainless steel is alloy Type 430 (UNS S43000). Its typical composition is 16–18 Cr and 0.12 max C. (Note that this has both higher Cr and lower C than the martensitic Type 410 discussed in the previous section.) This alloy is very commonly used in corrosion-resistant nonstructural applications requiring cold forming, such as kitchen sinks, decorative trim, annealing baskets, and nitric acid tanks. While more expensive (because of higher Cr content) than the martensitic stainless steels, the ferritic stainless steels are more corrosion-resistant. Heat treatment consists of annealing at 700–790°C followed by air cooling or water quenching. This avoids the formation of austenite by the lower annealing temperature.

The rapid cooling after annealing is essential to avoid what was traditionally called *885°F embrittlement*, which we shall call *475°C embrittlement* following the suggestion of Parr and Hanson (Ref. 11-5). When held in the temperature range of 450–550°C, ferritic stainless steels experience a loss in toughness and an increase in hardness. The loss in toughness is so severe that the steels are useless, unless the embrittlement is removed by holding at 600°C and quenching. The mechanism is not well understood, being attributed either to formation of a coherent precipitate or to atomic-level ordering, but the effect is more marked with the higher-chromium grades. Slow cooling through the 450–550°C range is thus to be avoided. For this reason the ferritic stainless steels are not generally welded, unless it is possible to remove the effects of 475°C embrittlement by subsequent heating and quenching. As other satisfactory weldable grades of stainless steel are available, the ferritic grades are usually used for unwelded parts.

Another problem that precludes welding of the ferritic grades is grain growth. As temperature increases, so does grain size. In the martensitic grades, reheating will lead to a phase change to austenite, with consequent fine grain size. But since the ferritic grades cannot be heated to force a phase transformation, they can only achieve grain-size reduction by cold work and recrystallization. This is usually impractical, so the ferritic grades are not usually welded. Grain growth can be controlled by addition of nitrogen.

Typical mechanical properties of Type 430 are tensile strength 520 MPa, yield strength 300 MPa, and 155 Bhn in the annealed condition, and tensile strength 620–900 MPa and yield strength 450–850 MPa in the cold-worked condition. Nominal strain to fracture of 30% for gage length 51 mm in the annealed condition drops to 20% to 2% in the cold-worked condition. It is clear that the ferritic stainless steels would not usually be the choice for structural applications because of their low strength. The higher-Cr members of this class have cold-worked strengths lower than Type 430, but they are more corrosion-resistant.

Two other new problems occur in the ferritic grades: *sigma-phase embrittlement*, and *sensitization*. In Fig. 11-2, at 46% Cr and temperatures below 821°C, the *sigma phase* occurs. This phase is an intermetallic Fe–Cr compound that is hard and brittle and extends to include Cr contents in the 20–30% range. Fortunately, sigma forms slowly, so it is not generally present unless the alloy is held between 500 and 800°C for many hours. The sigma forms first at grain boundaries and leads to an embrittled part. The sigma can be dissolved by holding at 900°C and quenching.

Sensitization, while more significant in the austenitic stainless steels, does occur in the ferritic grades. In alloys with chromium contents above 16%, rapid cooling after holding at temperatures above 925°C leads to accelerated grain boundary attack when in a corrosive medium, and eventual failure. As sensitization is more of a problem with austenitic stainless steels, it is discussed further in the next section. Sensitization in ferritic stainless steels can be avoided by not heating to 925°C (note the low annealing temperature range of 700–790°C), which means that welding is undesirable. We now have three reasons for not welding the ferritic grades: 475°C embrittlement, grain growth, and sensitization.

AUSTENITIC STAINLESS STEELS (Fe–Cr–Ni)

Figure 11-3 suggests that a very high nickel content would be required for austenite to be stable at 20°C, but this is misleading. The austenite-to-ferrite transformation in Fe–Cr–Ni alloys is very slow, and in fact an 18% Cr alloy with 8% Ni is stable austenite at 20°C. Most of the austenitic grades are based on variations of this basic "18–8" austenitic stainless steel. It cannot be hardened by heat treatment, but some alloys show remarkable strengthening by cold work, having strain-hardening exponents n (in $\sigma = K\epsilon^n$) of 0.4 $\sim$ 0.5. This leads to difficulty in mechanical working, which is due to the rapid increase in flow stress for small deformations (see Chap. 2). The austenitic grades, of 16 to 30% Cr and 6 to 20% Ni, are the most corrosion-resistant of all the stainless steels and do not suffer embrittlement at very low temperatures. They are generally nonmagnetic and are for this reason used in some electrical applications.

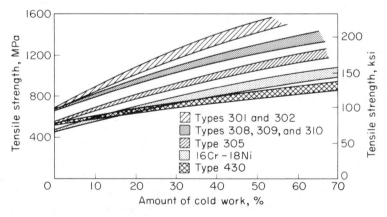

Figure 11-6 Effect of cold rolling on tensile strength of selected stainless steels. [From ASM Committee on Wrought Stainless Steels, "Wrought Stainless Steels," *Metals Handbook*, vol. 3, 9th ed., David Benjamin, Senior Editor; American Society for Metals, 1980, p. 33.]

The basic 18–8 alloy is either Type 301 (UNS S30100) or Type 302 (UNS S30200). Type 301 has typically 16–18 Cr, 6–8 Ni, and 0.15 max C. Type 302 is the same except that the Cr limits are 1% higher and the Ni is 2% higher. We shall here consider these a single alloy type. This stainless steel is annealed in the range 1000–1120°C, then air-cooled or water-quenched. In this condition Type 301 has tensile strength 760 MPa, yield strength 275 MPa, and elongation at fracture 60% over 51-mm gage length. The influence of cold work on Type 301 (and some other stainless steels) is shown in Fig. 11-6. This remarkable increase in strength with cold work is largely attributable to the transformation of unstable austenite to martensite as a result of the cold work. The balance of the strength increase results directly from plastic deformation. (Note that stainless steel introduces no new strengthening mechanisms but combines them in new ways.) Types 301 and 302 are widely used where parts are cold-formed and require higher corrosion resistance, strength, and ductility than for ferritic Type 430.

Exercise 11-2

Compare the costs of the common grades of martensitic, austenitic, and ferritic stainless steels.

Solution. Remember that iron is by far the cheapest metal in any steel. Chromium and nickel are much more expensive. Thus 18 Cr–8 Ni austenitic stainless steel will be the most expensive, ferritic grades such as the 17 Cr Type 430 will be less expensive, and martensitic 12.5 Cr Type 410 the cheapest.

The chief problem with the austenitic grades is *sensitization*, already mentioned as a minor problem with the ferritic grades. When heated in the range 425 to 800°C, chromium carbide precipitates at the grain boundaries. This is called sensitization because it leads to a local depletion of the chromium content in the vicinity of the grain boundaries to below 12% Cr, with a resulting loss in passivity. The alloy then experiences rapid corrosion at the grain boundaries and can literally fall apart. Sensitization is reversible if, prior to service, the part is heated to about 1100°C, soaked long enough to redissolve the chromium carbides, and then water-quenched. If possible, the temperature range of 425–800°C should be avoided altogether, but this is not possible during processes such as welding. In these cases an austenitic stainless steel must be used that is less susceptible to sensitization.

Many of the austenitic alloys were developed in order to control the problem of sensitization. The two most effective means for controlling sensitization are (1) to keep the carbon content very low so that little chromium is lost to chromium carbide, and (2) to introduce other elements that will preferentially combine with the carbon before the chromium does, and thus effectively tie up the carbon.

Alloys Type 304 (UNS S30400) and Type 304L (UNS S30403) reduce but do not eliminate the sensitization problem through limiting carbon to low levels. The composition of Type 304 is 18–20 Cr, 8–10.5 Ni, 0.08 max C. Type

304L differs significantly only in having an extra-low carbon content, a maximum of 0.03%. These grades are used where welding precludes use of Types 301 and 302, as in welded cryogenic (very low temperature) vessels. Comparison of the compositions of Types 302 and 304 shows that the only differences are that Type 304 has 1% more Cr and half the C. The establishment of two entirely separate alloys having progressively lower carbon demonstrates the importance of the sensitization problem. The mechanical properties of Types 304 and 304L are close to each other and only slightly lower than those of Type 302; for Type 304, tensile strength is 580 MPa, yield strength is 290 MPa, and elongation at fracture is 55%. The substantial differences in strain hardening between Type 301 and 304 can be seen in Fig. 11-7. Clearly, the higher Cr and Ni and lower C of Type 304 permit less strain-induced transformation of austenite to martensite, with consequently less strain hardening.

Type 347 (UNS S34700) controls sensitization by small additions of niobium. Its composition is 17–19 Cr, 9–13 Ni, 0.08 max C, and Nb = 10 × C min. The intent of the composition is to be reasonably sure that enough niobium (which is a strong carbide former) is present to tie up the carbon before it can combine with any chromium. Notice that, except for a small decrease in chromium and an increase in nickel, the Nb requirement is the only change from the composition of Type 304. (Type 321, UNS S32100, is virtually identical to Type 347 except that titanium is used as a stabilizer instead of niobium.) Annealed mechanical properties of Types 347 and 321 are very similar to those of Type 304. Since titanium reacts rapidly with oxygen at high temperature, Type 321 cannot be used for electric arc welding. Types 347

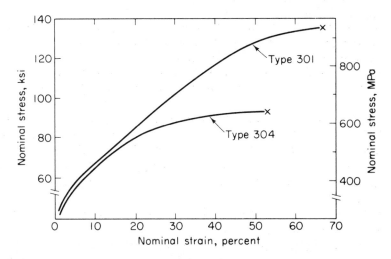

Figure 11-7 Stress–strain curves for austenitic Types 301 and 304. [From K. G. Brickner, "Stainless Steels for Room and Cryogenic Temperatures," *Metals Engineering Quarterly* (May 1968). Copyright American Society for Metals. Reprinted with permission.]

and 321 are used for high-temperature welded parts such as aircraft exhaust manifolds, boiler shells, pressure vessels, and welded tank cars for chemicals (347).

Type 316 (UNS S31600) has molybdenum added to expand the passivity range, which thus reduces pitting and increases resistance to corrosion from neutral chloride solutions (such as sea water). Thus Type 316 is one of the more popular of the stabilized austenitic strainless steels. Type 316L (UNS S31603) is extensively used for prosthetic implants in the human body. The composition of Type 316 is 16–18 Cr, 10–14 Ni, 0.08 max C, and 2–3 Mo. (316L has 0.03 max C.) The room-temperature mechanical properties of Type 316 are virtually identical to those of Type 304.

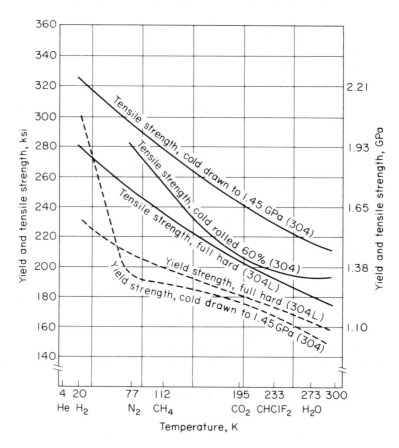

Figure 11-8 Change in mechanical properties with decreasing temperatures for cold-worked Types 304 and 304L. [From *Chromium–Nickel Stainless Steel Data*, Section I: Wrought Stainless Steels, Bulletin C. "Mechanical and Physical Properties of Austenitic Chromium–Nickel Stainless Steels at Subzero Temperatures," The International Nickel Co., Inc. (1963). Reprinted courtesy of The International Nickel Company, Inc.]

The low-temperature mechanical properties of the austenitic grades are unique among the stainless steels. For example, Fig. 11-8 shows the influence of temperatures down to 20 K on some properties of Types 304 and 304L.

For some of the austenitic alloys that contain some ferrite when welded, sigma formation can occur through extended heating in the range 500 to 800°C. The sigma can be redissolved in the same way as for the ferritic stainless steels, through soaking at 900°C and quenching.

Exercise 11-3

For what reasons are the austenitic stainless steels widely used?

Solution. Although they are the most expensive of the stainless steels, the austenitic grades can be (moderately) cold-worked, many can be welded, they have excellent corrosion resistance, and their strength, ductility, and elevated temperature properties are good.

PRECIPITATION-HARDENED STAINLESS STEELS

Developed during the 1940s, the precipitation-hardened stainless steels make possible the precision fabrication of a corrosion-resistant part of high strength and hardness. None of the other grades allows this: The martensitic stainless steels suffer dimensional changes when undergoing the martensite transformation; the ferritic grades are never very strong; and the austenitic grades achieve their highest strength as a result of cold forming.

The principles of precipitation hardening, as already described in Chap. 9 for the Al–Cu system, apply equally well to Ni–Al and Ni–Ti: The nickel-rich ends of both of these systems exhibit decreasing solubility with decreasing temperature in the same fashion as the aluminum-rich end of the Al–Cu system (Fig. 9-11). Thus, if either Al or Ti is present in a martensitic stainless steel that contains Ni, fine particles of intermetallic compounds of Ni–Al or Ni–Ti will be formed upon aging at intermediate temperatures. This aging takes place at temperatures well below the austenitizing temperatures required to produce martensite, so the hardening and strengthening that result from the formation of fine particles during aging can take place after fabrication, forming, and machining.

A common precipitation-hardened stainless steel is 17-7 PH (UNS S17700), of the following composition: 16–18 Cr, 6.5–7.75 Ni, 0.09 max C, and 0.75–1.50 Al. This alloy is austenitized at $954 \pm 14°C$ and air-cooled to 20°C. In order to force the transformation of as much austenite as possible to martensite, it is then cooled to below $-68°C$ for 29 ks. The aging (hardening) temperature required for fine-particle precipitation of Ni–Al will depend on the strength level desired. Figure 11-9 shows the variation with temperature of strength and elongation for aging 17-7 PH for 3.6 ks. For example, actual recommended treatments are as follows: For a minimum tensile strength of 1.45 GPa, heat to $510 \pm 6°C$; for strength of 1.03 GPa, $593 \pm 6°C$. Thus 17-7

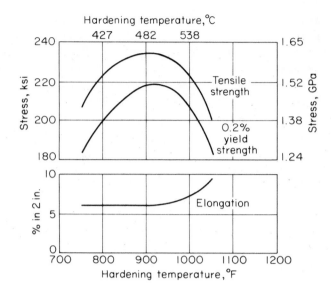

Figure 11-9 Tensile and yield strengths, and nominal strain at fracture as a function of aging temperature for 17-7 PH (UNS S17700). The specimens were first heated at 954°C for 600 s; air-cooled, liquid-cooled to −73°C for 29 ks; hardened for 3.6 ks at indicated temperatures. [From ASM Committee on Heat Treating of Stainless Steel, "Heat Treating of Stainless Steel," *Metals Handbook*, 8th ed., vol. 2, ed. T. Lyman (Metals Park, Ohio: American Society for Metals, 1964), p. 251.]

PH, for 510°C aging, can have a yield strength of 1.5 GPa and a hardness of 47 Rockwell C. Since it can be cold-drawn prior to this final heat treatment, it is useful for such applications as springs, knives, and pressure vessels.

The precipitation-hardening stainless steels have slightly better corrosion resistance than the martensitic stainless steels because of generally higher chromium content. At high strength levels, the precipitation-hardening steels are susceptible to stress-corrosion cracking in some environments (see Chap. 18), and they are notch-sensitive. Thus in practice these steels are often deliberately overaged to lower strengths in order to avoid stress-corrosion problems and to increase their notch toughness.

ELEVATED TEMPERATURE PROPERTIES OF STAINLESS STEELS

Many of the stainless steels of the four classes described in the foregoing sections retain their corrosion-resistant properties to elevated temperatures. Some of them also retain excellent strength at elevated temperatures. For these reasons stainless steels are frequently used for service above room temperature where plain-carbon steels are inadequate. Figure 11-10 shows the short-term

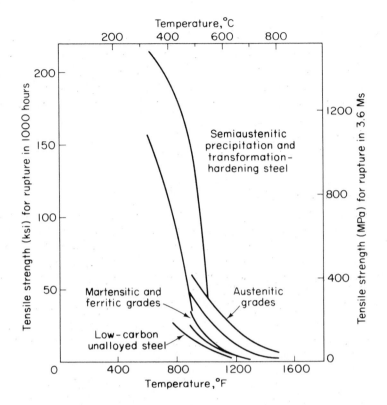

Figure 11-10 Short-term strength of stainless steels vs. temperature. [From T. D. Parker, "Strength of Stainless Steels at Elevated Temperature," *Metals Engineering Quarterly*, May 1968. Copyright American Society for Metals. Reprinted with permission.]

strength of stainless steels, taken from an ordinary tension test, over a range of temperatures from 300 to 800°C. The precipitation-hardening grades are clear choices at temperatures up to about 500°C, above which the austenitic grades are superior.

One measure of high-temperature performance of materials is the resistance to *creep*. (This is discussed in detail in Chaps. 12 and 13.) Creep is the slow deformation of a material under stress at elevated temperature that occurs with virtually no strain hardening. The *creep rate* in this case is defined as the total nominal strain divided by the time at which a specimen is exposed to a fixed tensile stress and temperature. (In general, creep rate is the instantaneous slope of the ϵ vs. t curve; see Chap. 12.) Figure 11-11 shows the stress that will produce a fixed creep rate in a number of different stainless steels. Compositions and other properties of the numbered alloys can be found in Appendix 7 and in Refs. 11-4, 11-5, and 11-6. Type 316 has become one of the

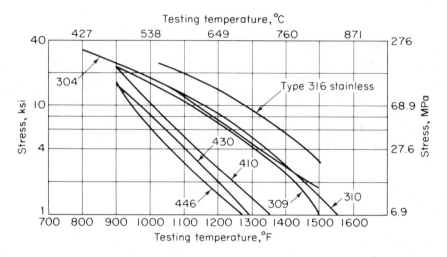

Figure 11-11 Stress to produce a creep rate of 0.1%/1000 h (2.78×10^{-10}/s) vs. testing temperature. [From J. Gordon Parr and Albert Hanson, *An Introduction to Stainless Steel*, American Society for Metals, 1965, p. 38.]

most commonly used stainless steels for elevated-temperature service, at least in part because of the 2–3% molybdenum in its composition.

HEAT-RESISTANT STEELS

In addition to conventional grades of stainless steels for use at elevated temperatures, there is a special group of oxidation- and corrosion-resistant steels called *heat-resistant steels*. This means that these steels are capable of resisting operating temperatures above 650°C. In general, these steels are an extension of the ordinary grades of stainless steels into higher alloy contents and less iron. Thus there are alloys of Fe–Cr (10–30% Cr), Fe–Cr–Ni (>18% Cr and >7% Ni), Fe–Ni–Cr (>25% Ni and >10% Cr), as well as nickel-base and cobalt-base alloys. Carbon content is usually higher than in the ordinary stainless steels because of the need for some strength at high temperatures, but the heat-resistant steels are not selected for high strength at high temperatures. (Chapter 16 discusses such materials.)

The most widely used heat-resistant stainless steel is an austenitic grade designated Type HH when cast, and Type 309 when wrought (UNS S30900). Its composition is 22–24 Cr, 12–15 Ni, and 0.20 max C. It is used for many high-temperature applications, particularly because of its excellent oxidation resistance to 1100°C. Its strength at elevated temperatures is exceeded by several other stainless steels (see Fig. 11-11), so its chief applications are in low-stress parts: annealing trays, exhaust manifolds, and radiant tubes.

SUMMARY

Iron will become corrosion-resistant if it contains more than 12% Cr. The wide range of stainless steel alloys represents solutions to the problems of embrittlement, sensitization, high-temperature failure, low strength, low formability, and high cost. No single alloy provides the best solution to all the problems.

REFERENCES

11-1. American Society for Metals, *Metals Handbook*, 8th ed., vol. 1. Metals Park, Ohio: American Society for Metals, 1961, pp. 437–636.

11-2. American Society for Metals, *Metals Handbook*, 8th ed., vol. 2. Metals Park, Ohio: American Society for Metals, 1964, pp. 243–270.

11-3. American Society for Metals, *Metals Handbook*, 9th ed., vol. 3. Metals Park, Ohio: American Society for Metals, 1980.

11-4. American Society for Metals, *Source Book on Stainless Steels*. Metals Park, Ohio: American Society for Metals, 1976.

11-5. Parr, J. Gordon, and Albert Hanson, *An Introduction to Stainless Steel*. Metals Park, Ohio: American Society for Metals, 1965.

11-6. Peckner, D., and I. M. Bernstein, *Handbook of Stainless Steels*. New York: McGraw-Hill, 1977.

11-7. Zapffe, Carl A., *Stainless Steels*. Metals Park, Ohio: American Society for Metals, 1949.

PROBLEMS

11-1. Explain why it is often desirable for stainless steels with higher chromium to have higher carbon content.

11-2. What is the *essential* requirement for "stainlessness" in an iron-base alloy?

11-3. Describe the chief advantages of martensitic stainless steel over austenitic stainless steel.

11-4. How would you specify change(s) in the composition of a ferritic stainless steel so that it would become a martensitic stainless steel?

11-5. Figure 11-3 shows that an iron alloy of, for example, 18% Cr and 8% Ni is ferritic at temperatures below about 300°C. Then why is an 18-8 steel called an austenitic steel?

11-6. Which stainless steel has the best corrosion resistance? Why is it not therefore used for all applications?

11-7. Which stainless steel has the highest hardness? What are its concomitant disadvantages?

11-8. Why are some martensitic stainless steels subcooled below room temperature after heat treatment? Which martensitic stainless steels are so treated?

11-9. What limitations does 475°C embrittlement in ferritic stainless steels impose on its use?

11-10. What is the most important advantage of precipitation-hardened stainless steel over martensitic stainless steel?

11-11. Explain the reason(s) for the expansion of the basic 18 Cr–8 Ni austenitic stainless (e.g., Types 301, 302) composition into 304, 304L, 316, and 347 (which are the most common of this set).

11-12. For what reason(s) do austenitic stainless steels have large values of n in $\sigma = K\epsilon^n$?

11-13. What is *sigma-phase embrittlement*? In which stainless steel(s) can it occur?

11-14. Describe two advantages of ferritic over austenitic stainless steels.

11-15. Speculate as to why an austenitic stainless steel, Type 309, is probably a reasonable choice for heat-resistant applications. Describe disadvantages of (a) ferritic, (b) martensitic, (c) precipitation-hardening stainless steels for heat-resistant applications.

11-16. Describe the most important problems and disadvantages associated with use of stainless steel for automobile fenders, and include in your discussion each of the reasonable possible choices for this application.

11-17. Steels of Type 410 and 440C are martensitic, whereas Type 430 is ferritic.
   (a) Compare the compositions of the three steels, and describe why each has the microstructure it has.
   (b) Which has the highest hardness? Why?

12

Rate- and Temperature-Dependent Mechanical Properties: Creep and Viscoelasticity

No mention was made in Chap. 2 of the possibility of basic strength properties being rate-dependent. There was no reference to time in $\sigma = K\epsilon^n$, and by implication, sudden application of a stress that is then kept constant produces the corresponding strain "in no time." On testing machines we do not expect a different answer for extension under a given load if we leave the load on and go back the next day. For most of the materials so far considered, such a description of mechanical behavior is broadly correct, and room temperature strengths and hardnesses are "fixed." There are exceptions, such as in steel prestressing wires used under high stresses for the reinforcement of concrete; we also saw in Chap. 9 that the strength of precipitation-hardened alloys during aging does vary with time. Even then, those alloys chosen for use at room temperature do not appreciably change their strengths under ambient conditions. Reasons for such deliberate choices are obvious: It would be foolish in an engineering design to allow the strength of parts to change as time went by.

Nevertheless, it was implied in Chap. 4, when discussing the structure of solids, that in the cooling down from liquid to crystals or liquid through molasses to glassy solids, the mechanical properties did vary with temperature and rate of deformation—particularly at elevated temperatures approaching the melting/freezing range of crystals or at temperatures above the glass-transition temperature for amorphous solids. In broadest terms, materials are stiffer and stronger at lower than at higher temperatures. If experiments, using conventional testing machines, demonstrate that the common engineering solids are comparatively unaffected by changes of temperature and by changes in speed of deformation under ambient conditions, room temperature for them must be far removed from their melting points or glass-transition temperatures. Thus steels melt at about 1450°C, which is much greater than room temperature (about 23°C); brasses melt at about 900°C; many aluminum alloys at about 600°C; glass fuses at about 1100°C; and many ceramics and minerals melt at over 2000°C. On the other hand, the melting points of other metals and alloys are much closer to room temperature; for example, lead melts at 327°C, tin at 232°C, plumber's solder (76 Pb/33 Sn) at 275°C, Wood's metal (50 Bi/25 Pb/12.5 Sn/12.5 Cd) at 66°C, and, of course, mercury (which is a metal) has a melting point of -39°C, so that it is liquid at room temperature. Most common polymers have glass-transition temperatures T_g comparatively close to room temperature (e.g., for nylon 60°C, for polystyrene 100°C); indeed, for some, room temperature is above T_g (e.g., for low-density polyethylene, -20°C, for polypropylene, 0°C); elastomers by definition have T_g below room temperature, that of polyisobutylene (butyl rubber) being -70°C. Butter melts at 31°C.

We should not be surprised if the mechanical properties of all these solids are markedly affected by temperature and speed of deformation. Some toffee behaves in a brittle fashion when hit with a hammer but is pliable in the hands when deformed slowly. Silly putty (silicone) responds quite differently when deformed quickly rather than slowly, as does chewing gum. Even some of the solids that are far removed from their melting points at room temperature can have different strengths under "impact" or dynamic conditions (i.e., under very high rates of strain). At higher temperatures the strengths of all solids become more and more rate-dependent.

An appreciation of this behavior is necessary either because we are looking for solids with which to make items that have to operate at high temperatures (e.g., gas turbines) or because we want to deform things more easily (as in the "hot working" of metals). The traditional room-temperature description of solids, as given in Chap. 2, is inadequate for such purposes, and we must broaden our coverage to include stress–strain curves over wide spectra of temperature and rates of deformation. Furthermore, we may have to introduce new methods of testing and new types of description of mechanical behavior. In this way, so-called *creep testing* and the phenomenon of *relaxation* play their part.

It is important to realize from the outset that "high temperature" in this context is defined relative to the melting temperature or glass-transition temperature of the solid concerned, so that room temperature may be "high" for some solids. It is useful for this field of study to introduce the concept of *homologous temperature*, defined as the ratio of the temperature of interest to the melting temperature, where the temperatures are in degrees kelvin (K). Thus zinc at room temperature is at $(23 + 273)/(419 + 273) = 0.43$ of the melting point; this may be written $0.43T_m$ and called "0.43 homologous." Low-carbon steel is at $(23 + 273)/(1450 + 273) = 0.17T_m$. Since homologous temperatures serve merely to bracket temperature ranges, the choice of liquidus or solidus for alloys is not critical.

STRAIN RATE

Whenever strengths depend on time, the concept of strain rate is a convenient way of bringing time into the description of mechanical behavior. The average strain rate in a tensile test is the strain divided by the time taken to reach that strain level. This time depends on the velocity of the testing machine crosshead, or on how rapidly the load is applied by the testing machine, or in an engineering design by how quickly a bridge, for example, becomes strained by a train running over it. It is just as appropriate to talk about load rates or stress rates, but it is common practice to use strain rates, and of course it is possible to convert from one to another.

Both engineering (nominal) strains and true (logarithmic) strains were employed in Chap. 2, so we have

$$\text{nominal strain rate} = \frac{e}{t}$$

$$\text{true strain rate} = \frac{\epsilon}{t}$$

The units of both are "per second," or s^{-1}, as strain is dimensionless. In the elastic range both formulas will give the same value for strain rate, because in that range of deformation $\epsilon = \log(1 + e) \approx e$. In the plastic range, however, there can be wide differences between ϵ and e, and since ϵ was considered to be the more meaningful measure of strain, we shall limit our discussion to rates of true strain (and, if necessary, to rates of true stress).

To account for possible variations in strain rate during a test, we can rewrite the foregoing as

$$\text{Instantaneous strain rate} = \frac{d\epsilon}{dt} = \dot\epsilon$$

where the fluxional (dot) notation is used for derivatives with respect to time. Since

$$d\epsilon = \frac{dL}{L}, \qquad \dot\epsilon = \left(\frac{1}{L}\right)\frac{dL}{dt} = \frac{V}{L} \tag{12-1}$$

where V is the crosshead velocity of the testing machine. Furthermore, since $L = L_0 \exp \epsilon$, where L_0 is the initial gage length now stretched to length L,

$$\dot\epsilon = \frac{V}{L_0 \exp \epsilon} \tag{12-2}$$

For small strains (including the elastic region) exp $\epsilon = 1$, so

$$\dot{\epsilon} = \frac{V}{L_0} \tag{12-3}$$

Note that Eq. (12-2) suggests a constant strain rate for small strains but a diminishing strain rate in tension as plastic deformation continues. In compression, increasing strain rates through the plastic range are suggested, since L decreases as ϵ increases.

These expressions assume *uniform deformation*, so that they cannot be applied when barreling occurs in compression, and particularly when necking occurs in tension (where the deformation becomes concentrated in a small zone). The differences in strain rate within, and outside, necked regions play important roles in Lüders band formation, "polymer drawing," and super-plasticity; these are discussed later.

A test piece of gage length 75 mm, strained by a crosshead moving at 200 μm/s, has $\dot{\epsilon} = (200 \times 10^{-6})/(75 \times 10^{-3}) \approx 3 \times 10^{-3} \text{ s}^{-1}$. Traditional laboratory testing machines (whatever their load capacity) produce strain rates no greater than about 10^{-1} s^{-1} on test pieces of traditional size. Special machines are required to produce greater strain rates.

Exercise 12-1

Determine the average elastic strain rate for a material with a 0.2% offset yield strength of 150 MPa, if loaded at the constant load rate of 3 kN/s. The gage section is 11.3 mm in diameter.

Solution. The cross-sectional area is $\pi(11.3)^2/4 = 100 \text{ mm}^2$. Therefore the yield load is $150 \times 10^6 \times 100 \times 10^{-6} = 15$ kN. The time taken to reach yield load is $(15 \times 10^3)/(3 \times 10^3) = 5$ s. Thus

$$\dot{\epsilon} = \frac{0.002}{5} = 4 \times 10^{-4} \text{ s}^{-1}$$

Exercise 12-2

If the material in Exercise 12-1 follows $\sigma = 230\epsilon^{0.3}$ MPa, determine the strain rate in the plastic range up to ultimate.

Solution. The load at ultimate is

$$230(0.3)^{0.3} \times 10^6 \times 100 \times 10^{-6} \exp(-0.3) = 117 \text{ kN}$$

because the area at ultimate is $A_0 \exp(-n)$. The time to reach maximum load, at a loading rate of 3 kN/s, is 39 s. Thus, if we neglect the yield strain, the average strain rate between yield and ultimate is $0.3/(39) = 7.7 \times 10^{-3} \text{ s}^{-1}$. Notice that the strain rate has increased by an order of magnitude.

Exercise 12-2 well illustrates how difficult it is to conduct a constant strain rate test through both the elastic and the plastic regions. During the work-hardening stage after the onset of gross yielding, the test piece is less stiff and the load varies much more slowly than in the elastic region. Much greater strain increments are obtained for the same stress increments. In order to complete tests in reasonable times, information rates have to be increased after yield. Remember that one way of picking up the yield point on old-fashioned hydraulic testing machines is to note when the load needle fails to keep up with the "pacing disk" that is set for some chosen load rate, and that this effect occurs because of changes in stiffness of the testpiece at yield.

Despite these changes in rate of deformation during testing, it turns out that the room-temperature σ vs. ϵ curves of most traditional engineering metals are *not* much affected by the value of $\dot\epsilon$, provided that the strain rate is *less* than about 10^{-1} s^{-1}. The values of σ_y and σ_u do not depend very much on how fast the test is performed or on how much it is speeded up after yield. At higher rates of strain, changes in the level of the room-temperature σ vs. ϵ curve can occur for some alloys. Since such higher strain rates can be produced only by means of special testing machines, rates of deformation are not always mentioned in connection with the strength and hardness of solids at low homologous temperatures.

BEHAVIOR OF METALS

If σ vs. ϵ tests are performed at different strain rates and temperatures, it is found that both the shape and the level of the curves are affected but in opposite senses; that is, increase of strain rate increases stress levels and sometimes work-hardening rate, but when deformation is carried out at elevated homologous temperatures, the whole level of the σ vs. ϵ curve drops.

If we may still describe the true stress–strain behavior by an equation of the form $\sigma = K\epsilon^n$, it means that K must vary with both rate and temperature. The values quoted in Chap. 2 and Appendix 3 merely reflect temperatures around 23°C and strain rates around $10^{-4} \sim 10^{-2}$ s^{-1}. Figure 12-1 shows how K depends on $\dot\epsilon$ at various temperatures for annealed 2024 aluminum. Clearly, temperature always seems to affect K, but strain rate only becomes significant at temperatures greater than 200°C (which is about $0.58T_m$); for temperatures less than that, K is essentially independent of $\dot\epsilon$, being, for example, some 330 MPa at 28°C.

The logarithmic scales in Fig. 12-1 suggest that

$$K = C\dot\epsilon^m \qquad (12\text{-}4)$$

where C (the intercept at $\dot\epsilon = 1$) decreases as the temperature T increases. The slope of the plot, m, is called the *strain-rate sensitivity* and itself changes with temperature, as shown by the increasing slopes with increasing temperature.

To compare the level of stress at two strain rates, we can write

$$\left(\frac{\sigma_2}{\sigma_1}\right) = \left(\frac{\dot\epsilon_2}{\dot\epsilon_1}\right)^m$$

where the subscripts 1 and 2 refer to the different rate conditions.

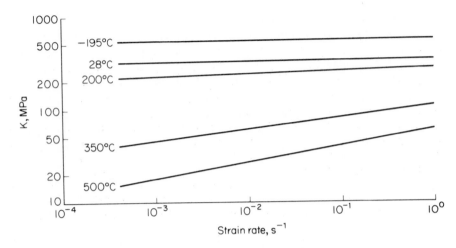

Figure 12-1 Variation of K in $\sigma = K\epsilon^n$ with strain rate and temperature for 2024 aluminum. [After D. S. Fields and W. A. Backofen, *Trans. ASM*, **55** (1959), p. 950.]

In Fig. 12-1, $m = 0.005$ (a very shallow slope) at room temperature. Increasing the strain rate by a factor of 10^4 produces an increase in flow stress of less than 5%. This is why precipitation-hardened aluminum alloys, copper alloys, and magnesium alloys are considered rate-insensitive at room temper-

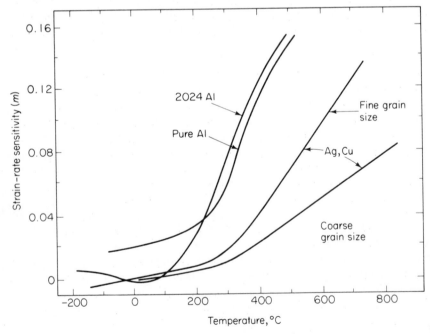

Figure 12-2 Effect of temperature on strain-rate sensitivity index (m).

ature. Other materials, particularly annealed bcc metals, aluminum, and titanium, show greater rate sensitivity. For example, annealed 1010 steel having a yield strength of 210 MPa at $\dot{\epsilon} = 10^{-3}$ s^{-1} has a flow stress of 450 MPa at $\dot{\epsilon} = 10^2$ s^{-1}. This corresponds with $m = 0.06$, from $(450/210) = (10^2/10^{-3})^m$. Such greater strain-rate sensitivity disappears the more work-hardened the material. When cold-rolled to a strain of $\epsilon_{cw} = 1$, 1010 steel changes its yield strength only from 630 MPa at 10^{-3} s^{-1} to 735 MPa at 10^2 s^{-1}; again, high-purity aluminum cold-rolled to 50% reduction of area has yield strengths of 98 MPa at $\dot{\epsilon} = 10^{-3}$ s^{-1} and 133 MPa at 10^3 s^{-1}. For both m is then roughly 0.01 to 0.02.

At higher homologous temperatures, m increases for all metals, some results being shown in Fig. 12-2. At 600°C copper may have an m of 0.1, and the flow stress is increased by 10% when the strain rate is merely doubled.

Exercise 12-3

The range of crosshead speeds on some testing machines can be altered by three orders of magnitude. What is the minimum temperature at which pure aluminum test pieces would show yield strengths differing by a factor of 2 between the slowest and fastest available crosshead speeds?

Solution. At some chosen fixed yield strain, yield strengths differing by a factor of 2 means that the K-values must differ by the same factor. Thus in Eq. (12-4) if $K_2 = 2K_1$, $\log 2 = m \log (\dot{\epsilon}_2/\dot{\epsilon}_1)$. Assuming that $\dot{\epsilon}$ is directly proportional to crosshead velocity, and using the fact that the fastest speed is 1000 times the slowest, we have $\log 2 = m \log 10^3$, and so $m = 0.1$. Inspection of Fig. 12-2 shows that such a sensitivity is displayed at about 400°C.

The value of C in Eq. (12-4) decreases as the temperature increases, and semilogarithmic plotting of C and T shows that C varies according to $\exp (1/T)$. As discussed later, the form of this temperature dependence is a manifestation of *thermal activation* on an atomic level. Thus, in general,

$$\sigma = K_0 \exp \left(\frac{1}{T}\right)\dot{\epsilon}^m\epsilon^n \tag{12-5}$$

This is the full version of the simple $\sigma = K\epsilon^n$ empirical relation; both m and n vary with temperature, m increasing and n decreasing as the temperature is raised.

In σ vs. ϵ experiments at low homologous temperatures (say, less than $0.5T_m$), rapid application of fixed stresses produces rapid strains, which are achieved very quickly along the lines in Fig. 12-3(a). At higher temperatures these strains do not stop suddenly but increase slowly in the time following the application of stress and go over into slow plastic flow, as in Fig. 12-3(b). This flow is called *creep*. Sometimes the creep stops, and further loads have to be added to cause further deformation. Even so, the supplementary creep components of strain can be appreciable. On other occasions, particularly at high homologous temperatures, creep strains never settle down, and rapid plastic

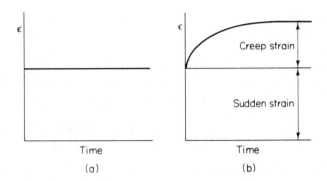

Figure 12-3 Strains resulting from the rapid application of a fixed stress (a) for $T < 0.5T_m$ and (b) for $T > 0.5T_m$, showing occurrence of creep strains in the higher temperature range.

flow continues for a short or long time until the test piece breaks, *without the need for additional loads.* For example, Fig. 12-4 shows the effect of a constant force of 89 N (produced by a hanging mass of 9.08 kg) on a Pb–Sn solder wire 3.2 mm in diameter at room temperature, which is $0.65T_m$ for this alloy. Note the rapid extension in the early stages, followed by a deceleration to a steady rate of creep extension, followed by a final gradual acceleration to fracture.

Under different loads the creepy Pb–Sn alloy behaves differently. For larger loads the creep rate increases, and fracture occurs in a shorter time (Fig. 12-4). At very high loads the transition between the deceleration and acceleration phases is sometimes only a point of inflection; at lower loads than shown

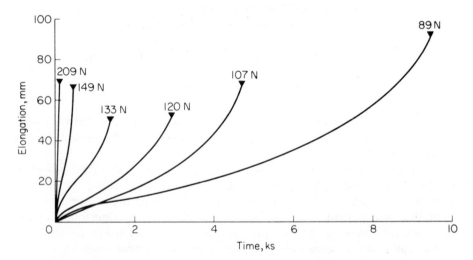

Figure 12-4 Creep of commercial Pb–Sn solder wire of initial diameter 3.2 mm and length approximately 90 mm, at 20°C. The axial force applied to each specimen is given. (Data courtesy of W. H. Durrant.)

in Fig. 12-4 the extension tends asymptotically to a fixed value, with no final fracture. In general, however, there is a more-or-less straight extended portion of the creep curve between the concave deceleration and convex acceleration parts. Accordingly, creep curves are conventionally subdivided into four parts:

1. the sudden extension upon initial application of load,
2. the period of deceleration (called *transient* or *primary* creep),
3. the period of constant creep rate (*steady-state* or *secondary* creep), and
4. the period of acceleration (*tertiary* creep), which ends with fracture.

In most cases tertiary creep follows after necking, with its decrease of cross-sectional area, internal hole formation, and consequent increase of stress, akin to conventional σ vs. ϵ curves. For reasons of practical convenience tensile creep tests are performed with the load and not the stress being kept constant, although tertiary creep can usually be eliminated under the correct conditions.

The effect of increased temperature on creep curves is similar to that of increased load (or stress): Extensions occur more quickly at higher homologous temperatures. On the other hand, this same solder wire would have a conventional σ vs. ϵ curve, which would be essentially time-independent, at sufficiently low temperatures (say, at $-100°C$).

Whenever the time effects of creep become appreciable, conventional stress–strain tests are not usually performed, and dead-load tests as just described are used to characterize mechanical behavior (i.e., strain as a function of time at constant stress). There is an overlap temperature range between the two regions, when σ vs. ϵ curve *can* describe the behavior, provided that adequate attention is paid to rate and temperature effects on K, as described earlier in this section. Then creep is beginning to be felt, but creep strains settle down after a time; this is before creep entirely dictates mechanical behavior.

Orowan has linked the σ vs. ϵ curve in this region with transient creep, as shown in Fig. 12-5. He suggested in the construction of the σ vs. ϵ curve that the stress is applied in small increments $\Delta\sigma$. During each corresponding time interval Δt, creep occurs as in a conventional creep test under the same current average stress level. The σ vs. ϵ curve is then the plot of the accumulated sum of the stress increments vs. the accumulated sum of the creep strain increments. In Fig. 12-5, $\Delta\sigma/\Delta t$ is constant for purposes of illustration, that is, it is a constant-stress-rate test.

The decrease in creep rate in the transient part is caused by work hardening, but competition can arise from thermal softening. Steady-state creep occurs when the two effects are balanced.

Creep occurs in metals because mechanisms come into play that allow easier flow but are absent at low homologous temperatures. Dislocation movement by climb (described in Chap. 6) is one such mechanism. Dislocations piled up against a barrier on some slip plane can move on if somehow they can jump over the barrier by moving to the next slip plane above. This can

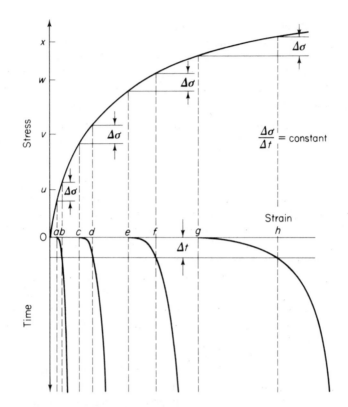

Figure 12-5 Orowan's description of the connection between transient creep at various stresses and the monotonic stress–strain curve. At every time increment Δt, transient creep strains ab, cd, ef, gh, ... are produced by average stress levels Ou, Ov, Ow, Ox, ... The monotonic σ/ϵ curve is the accumulated sum of all the stress increments $\Delta\sigma$ vs. the accumulated sum of all the creep-strain increments $\Delta\epsilon$.

come about by migration of vacancies (holes or missing atom sites) to the "bottom" of the extra half-plane of the dislocation, taking out one row of atoms, and effectively moving the base of the dislocation up onto the slip plane above. There are many vacancies naturally occurring in crystals, and diffusion of vacancies is a process helped by increasing temperature (i.e., it is thermally activated), so that it contributes more to creep as the temperature rises.

Another event that is thermally activated is atomic mobility. At low homologous temperatures, atoms may be thought of as fixed in the crystal lattice, but as the temperature rises atoms can change places with one another. This process, called *self-diffusion*, comes into play significantly at temperatures around $0.5T_m$ and higher. Atomic mobility is particularly marked in grain boundaries (where the crystal lattice positions are mismatched) more than in the grains themselves. The enhanced atomic mobility of grain boundaries at high homologous temperatures makes them behave in the fashion of a viscous fluid, and the associated grain boundary sliding contributes significantly to creep strains. The inverse exponential temperature terms in Eq. (12-5) represent the collective effects of thermal activation.

Empirical relations between creep strain and time, as may be determined from experiments in which masses are hung on wires, take on the following form (among others):

$$\dot{\epsilon}_{transient} = A_1 t^{-2/3} \tag{12-6}$$

where A_1 is some constant depending on the stress; the rate decelerates as time goes by. The constant is a function of stress and temperature as first studied by Andrade in 1910. Mott has explained this in terms of the climb of dislocations. It will be shown later how transient creep can be linked to viscoelasticity and how steady-state creep can be linked to non-Newtonian flow.

Exercise 12-4

Describe the influence of grain size on creep resistance.

Solution. Clearly, if grain boundary sliding is a significant contributor to creep strains, the latter are reduced if the regions in which sliding can take place are kept as small as possible, that is, if the grain boundary area is small. Thus, other things being equal, larger grain sizes have smaller creep rates.

BEHAVIOR OF AMORPHOUS SOLIDS

Amorphous materials stiffen and harden progressively as the temperature drops, unlike crystalline materials, whose mechanical properties change markedly at the melting/freezing point. There is a gradual transition for noncrystalline solids, as the temperature drops, from liquid behavior through molasses behavior to that of elastic solids. Liquids often display simple Newtonian viscosity, in which (for the purpose of this book) shear stress τ is proportional to the *rate* of shear strain $d\gamma/dt$ or $\dot{\gamma}$, that is,

$$\tau = \eta\dot{\gamma} \tag{12-7}$$

The constant of proportionality η is called the *coefficient of viscosity* or *viscosity* for the liquid. Note that here the response is related to the *rate* of displacement, whereas for materials displaying simple elastic or plastic behavior the response is related to the displacement (strain) itself.

As the temperature drops and the liquid thickens toward an amorphous solid, the shear stress ceases to be directly proportional to the strain rate, and instead we have nonproportional (non-Newtonian) behavior given by

$$\tau = \eta(\dot{\gamma})^q \tag{12-8}$$

This expression, like the linear Newtonian relationship, may be integrated to give the magnitude of the strain at any time:

$$dy = \left(\frac{\tau}{\eta}\right)^{1/q} dt$$

or

$$\gamma = \int \left(\frac{\tau}{\eta}\right)^{1/q} dt \tag{12-9}$$

The applied τ may vary with time (as in a simple tension test where the load, and hence stress, is increased as the test proceeds). Integration of Eq. (12-9) would require an expression for τ as a function of t in order to evaluate the change of strain in some period of time. Note, however, that even with τ constant, γ *increases with time* according to

$$\gamma = \left(\frac{\tau}{\eta}\right)^{1/q} t \tag{12-10}$$

(We assume that $\gamma = 0$ at the reference observation starting time when $t = 0$.)

This is similar to the creep behavior of metals, such as given by the simple Andrade relation for $\dot{\epsilon}(t)$, Eq. (12-6), when integrated. The magnitude of induced strain depends on τ, of course (greater stresses producing greater strains), but particularly it depends on $1/q$. When this index is very small, γ is small even if τ is large; that is, the amorphous material is then not very rate-dependent. On the other hand, if $1/q$ is large, great strains are produced in short times even by very small stresses.

Exercise 12-5

An amorphous solid follows $\tau = 800\dot{\gamma}^2$ Pa at some temperature above T_g. What strain is produced 20 s after applying a uniform shear stress of 80 Pa to a block of the solid?

Solution. From Eq. (12-10),

$$\gamma = \left(\frac{80}{800}\right)^{1/2} \cdot 20 = 6.32$$

Note that were the index 0.2 instead of 2, γ would be 2×10^{-4}, which is considerably smaller.

In all these expressions the strains are irreversible, and even after some period following the removal of τ, the amorphous body retains its deformed shape. However, at lower temperatures (particularly below T_g for polymers) elasticity comes into play. Then the deformation at any instant has two components: an irreversible flow component (as already discussed) and a reversible elastic component. The relative magnitude of the component deformations or strains depends on the character of the solid and the temperature. At lower and lower temperatures the flow component diminishes and the amorphous material behaves more and more like a time-independent elastic body with all deformation reversible. The range of mechanical properties shown by different polymers at room temperature—some, such as polymethylmethacrylate (PMMA), being virtually elastic in response, others, such as low-density polyethylene, being very creepy—reflects the relative importance of the flow and elastic components of strain. We have

$$dy_{total} = dy_{flow} + dy_{elastic} \tag{12-11}$$

Here dy_{flow} comes from $\tau = \eta(\dot{\gamma})^q$ and $dy_{elastic}$ comes from $\tau = G\gamma$, where G is the shear modulus for the amorphous solid. Methods of measuring these mechanical properties are discussed later.

In many cases of interest the magnitude of the recoverable elastic strains is small in comparison with the irreversible strains. This applies to metal deformation problems in which the plastic strains swamp the elastic strains, and it applies to the behavior of amorphous solids above T_g. Sometimes, however, $dy_{elastic}$ and dy_{flow} are comparable in magnitude. This is particularly true for polymers below T_g, since, as discussed in Chap. 13, their elastic moduli (G in $\tau = G\gamma$ or E in $\sigma = E\epsilon$) are much smaller than those of metals, and the associated elastic strains at a given stress level are greater, becoming comparable with the flow components of strains. Thus around T_g the subject of *viscoelasticity* (i.e., mechanical behavior with combined flow and elastic components) is important.

In the transient creep of metals, where the total strains are small, reversible and irreversible components are comparable in magnitude, so the mechanics of primary creep and viscoelasticity are similar. In steady-state creep, however, the elastic component of total strain can usually be ignored, so that secondary creep relations may be thought of as rather complicated non-Newtonian flow.

Dividing Eq. (12-11) by time, we obtain

$$\begin{aligned} \dot{\gamma}_{total} &= \dot{\gamma}_{flow} + \dot{\gamma}_{elastic} \\ &= \left(\frac{\tau}{\eta}\right)^{1/q} + \left(\frac{1}{G}\right)\dot{\tau} \end{aligned} \tag{12-12}$$

If, at $t = 0$, a stress τ^* is applied to a body following Eq. (12-12) and then kept constant for some time t^*, there would be an instantaneous strain τ^*/G, which would remain unchanged throughout t^* (since elastic deformations occur without reference to time), followed by a steadily increasing flow component given by $\gamma_{flow} = (\tau^*/\eta)^{1/q}t$ during the period $0 < t < t^*$. This is shown in Fig. 12-6. The manner in which γ increases with t depends on q, concave-up slopes being produced for $q > 1$ and vice versa; a linear change in γ with t is produced for Newtonian flow with $q = 1$. If the stress τ^* is removed at time t^*, $\gamma_{elastic} = \tau^*/G$ is recovered instantaneously. However, the accumulated flow component remains. Clerk Maxwell first discussed this type of behavior with combined elastic and Newtonian components of deformations, so that a material following Eq. (12-12) is called a *Maxwell solid*.

There is a mechanical analogue for a Maxwell solid, which consists of an

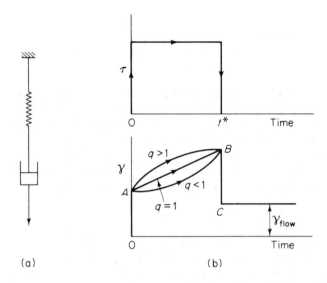

(a)   (b)

Figure 12-6 (a) Mechanical analogue, consisting of a spring and dashpot in series, which describes a Maxwell solid. (b) Strain (or displacement) vs. time response to a step application and removal of stress (or load).

elastic spring in series with a dashpot (i.e., viscous damper or type of shock absorber); see Fig. 12-6(a). The same load (stress) is experienced by both, and the total deflection (strain) is the sum of the two components. A dashpot can be made from a piston containing holes, sliding in a bath of oil; it is viscous and "takes time to get going," so that in Fig. 12-6(b) OA reflects the spring elongation under stress, and the (straight) path AB the subsequent extension of the dashpot piston rod. The spring returns on removal of the load, moving the dashpot bodily with it (BC); however, the dashpot piston rod end remains in its extended position and can only return to the starting position if a load is applied in the opposite direction from before.

Some polymers do not behave as if they were Maxwell solids. Rather, they exhibit *delayed elasticity*; that is, instead of attaining an elastic strain instantaneously there is some sort of retardation, which delays the process. Long polymer chains, tangled up like spaghetti, must uncurl as they are deformed, and this can produce delayed elastic effects.

A mechanical analogue, first described by Voigt, which displays such an effect is shown in Fig. 12-7(a). This time the spring and dashpot are in parallel, rather than in series. Thus they share a common displacement (strain) rather than a common load or stress. We have, at some strain γ,

$$\tau_{\text{total}} = \tau_{\text{elastic}} + \tau_{\text{flow}}$$
$$= G\gamma + \eta(\dot{\gamma})^q \tag{12-13}$$

For a constant stress τ^* applied for a time t^*, Eq. (12-13) becomes a

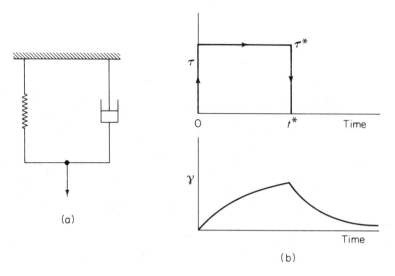

Figure 12-7 (a) Mechanical analogue, consisting of a spring and dashpot in parallel, which describes a Voigt solid. (b) Strain (or displacement) vs. time response to a step application and removal of stress (or load).

linear differential equation for the simplified case of $q = 1$ (i.e., for a Newtonian dashpot). The solution is

$$\gamma = \left[\frac{\tau^*}{G}\right]\left[1 - \exp\left(\frac{-Gt}{\eta}\right)\right] \tag{12-14}$$

which describes how the strain γ grows with time t for the duration of τ^*. Figure 12-7(b) displays the behavior, which is an exponential increase toward the elastic strain (τ^*/G) that would have been attained instantaneously in the absence of the retarding dashpot. Removal of τ^* produces an exponential return to zero strain.

A combination of the Voigt and Maxwell solid behaviors into a four-element model [Fig. 12-8(a)] embodies many of the essential characteristics of the mechanical response of polymers: instantaneous deformation, deformation decaying to a constant rate, followed by some permanent deformation after stress removal. The total strain in the combination during the application of a constant τ^* is

$$\gamma_{\text{total}} = \frac{\tau^*}{G_1} + \frac{\tau^* t}{\eta_1} + \left(\frac{\tau^*}{G_2}\right)\left[1 - \exp\left(\frac{-G_2 t}{\eta_2}\right)\right] \tag{12-15}$$

where subscripts 1 and 2 are used to differentiate between the two spring and dashpot systems. Figure 12-8(b) shows the complete stressed–unstressed strain–time history for the four-element model. Precise simulation of real polymers usually requires much more complex models, but all of the essential features are present in the *Voigt–Maxwell model*.

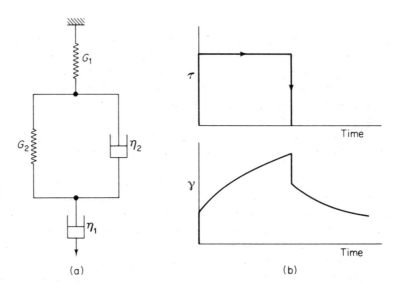

Figure 12-8 (a) Maxwell and Voigt analogues combined into a four-elemen. model. (b) Strain or displacement vs. time response to a step application and removal of stress (or load).

It should be emphasized that the springs and dashpots have *no* physical meaning*; rather, microstructural characteristics of polymers lead to the same sort of response as these conceptual analogue models. Values for the G's and η's of the analogue models would be obtained such as by curve-fitting experimental γ vs. t data to Eq. (12-15). The microstructural behavior of creepy metals leads to a similar state of affairs, except that there the elastic and recoverable components of strain are usually negligible because of the relatively large elastic moduli of metals, even at high homologous temperatures.

Exercise 12-6

A tensile force of 500 N was applied to a polymer rod, 5 mm × 5 mm × 500 mm long. The rod stretched to 508 mm very quickly, then continued to lengthen so that by the same time on the next day it was 550 mm long. Another day later, it was 570 mm long. The load was then taken off, and three days after that the rod had shortened to 525 mm. Determine the parameters of a Voigt–Maxwell tensile analogue that describes this behavior.

Solution.   In terms of tensile stresses and strains

$$\epsilon = \frac{\sigma}{E_1} + \frac{\sigma t}{\eta_1} + \left(\frac{\sigma}{E_2}\right)\left[1 - \exp\left(\frac{-E_2 t}{\eta_2}\right)\right]$$

*Fracture in polymers is not, as a student once asserted, when the piston falls out of the dashpot.

At time zero

$$\epsilon = \frac{8}{500} = \frac{500}{25 \times 10^{-6}E_1}; \qquad E_1 = 1.25 \text{ GPa}$$

Twenty-four hours later

$$\epsilon = \frac{50}{500}$$

$$= \frac{8}{500} + \frac{500 \times 24 \times 60 \times 60}{25 \times 10^{-6}\eta_1} + \left(\frac{500}{25 \times 10^{-6}E_2}\right)\left[1 - \exp\left(\frac{-24 \times 60 \times 60E_2}{\eta_2}\right)\right]$$

Forty-eight hours later

$$\epsilon = \frac{70}{500}$$

$$= \frac{8}{500} + \frac{500 \times 48 \times 60 \times 60}{25 \times 10^{-6}\eta_1} + \left(\frac{500}{25 \times 10^{-6}E_2}\right)\left[1 - \exp\left(\frac{-48 \times 60 \times 60E_2}{\eta_2}\right)\right]$$

One hundred-twenty hours later (i.e., after 72 hours recovery)

$$\epsilon = \frac{70 - 25}{500} = \frac{70}{500} \exp\left(\frac{-72 \times 60 \times 60E_2}{\eta_2}\right)$$

Hence

$$E_2 = 80 \text{ MPa}, \qquad \eta_1 = 35 \text{ TPa-s}, \qquad \eta_2 = 47 \text{ TPa-s}$$

STRESS RELAXATION

Creep is the increase of γ with t at constant τ, such as illustrated by a mass hanging on a wire or a hot turbine blade stretching under centrifugal force. The only way in which such increase in γ can be eliminated (for a material in a given state) is to reduce τ as t increases. In a Maxwell solid, for example, $\dot{\gamma}_{\text{total}} = 0$ if γ_{total} remains constant, so that

$$\left(\frac{\tau}{\eta}\right)^{1/q} + \left(\frac{1}{G}\right)\dot{\tau} = 0 \qquad (12\text{-}16)$$

For $q = 1$ the solution of this differential equation is

$$\tau = \tau^* \exp\left(\frac{-Gt}{\eta}\right) \qquad (12\text{-}17)$$

where τ^* is the initial stress which produced a strain γ^* that is now remaining constant. Equation (12-17) gives the manner in which τ^* must be decreased to keep γ^* fixed, as shown in Fig. 12-9. When a time-dependent body is strained

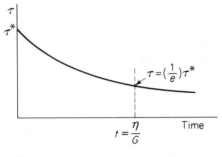

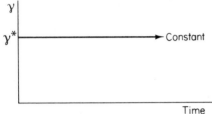

Figure 12-9 Stress relaxation where stress falls with time at constant strain.

and the strain then held (as by driving the crosshead of a screw-driven testing machine to some new position), the stress itself naturally decays according to some relationship like Eq. (12-17). Internal flow takes place so as to decrease the stress, the process being called *stress relaxation*. It is characteristic of all creepy materials.

Note that when $t = \eta/G$ in Eq. (12-17), $\tau = \tau^* \exp(-1) = 0.37\tau^*$; η/G is called the *relaxation time* (i.e., the time required for the stress to relax to $1/e$ of its initial value). It is similar to time constants of alternating-current circuits or vibrating springs (hence, again, the spring/dashpot analogues of viscoelastic behavior).

Exercise 12-7

Show how the value of η/G in a Maxwell solid can predict, for a given loading duration, whether the behavior will be predominantly elastic, predominantly flow, or a mixture of both.

Solution. $\gamma_{elastic} = \tau/G$; $\gamma_{flow} = (\tau/\eta)t$. Thus $\gamma_{flow} = [(G/\eta)t]\gamma_{elastic}$. When η/G is large, $\gamma_{flow} \ll \gamma_{elastic}$; when η/G is small, $\gamma_{flow} \gg \gamma_{elastic}$ for the same t. When η/G is large, a long time elapses before much relaxation takes place, that is, the solid is hardly rate-dependent at all; the opposite effect occurs for small η/G. When η/G and t are comparable, the mixed behavior of the Maxwell model comes into play.

The ratio of relaxation time to observation time (η/Gt) is sometimes called the *Deborah number*. It is zero for ideal fluids, infinite for ideal solids, and unity at T_g.

The presence of creep and/or relaxation in solids causes difficulties in the experimental measurement of elastic moduli. For example, if a mass is hung on a wire in order to determine a pair of σ/ϵ values that would give E by division, it is clear that in the presence of creep the value of ϵ depends on how soon it is measured after the application of the stress. That is, in Fig. 12-4 any value up the rising strain curve might be taken. Similar considerations would apply to a soft (e.g., hydraulic) testing machine (see Chaps. 2 and 14). Or, in a hard machine where a fixed strain might be put on, the associated stress will relax away from the value first seen after the attainment of the chosen strain level, along the lines of Fig. 12-9.

It is customary therefore in polymers to talk of *relaxed* and *unrelaxed* *moduli.* Unrelaxed moduli are those obtained from instantaneous readings of stress and strain and are also the same as "dynamic" or high-strain-rate moduli such as relate to the passage of sound waves through a polymer where the molecules do not have time to creep about. The moduli of Voigt–Maxwell models and the like are strictly unrelaxed moduli. Relaxed moduli, on the other hand, come about from dividing the stress by that strain attained some considerable time after first stressing. They are thus much smaller than unrelaxed or dynamic moduli. As described in Chap. 13, the elastic properties of some polymers are given in terms of a so-called 10-second modulus, which is stress divided by the strain reached 10 seconds after first loading.

Exercise 12-8

A bar of low-density polyethylene is suspended from two fine wires 25 mm apart and forced to vibrate by sending pulses down one of the wires. The time taken for a sound wave to travel along the bar was measured by picking up the rod vibration in the second wire. If that time was 54 μs and the density of the polymer is 920 kg/m^3, calculate the dynamic modulus of the polymer.

Solution. The velocity of sound in a solid is given by $\sqrt{E/\rho}$, where E is the modulus and ρ the density. Thus, if the time taken to travel 25 mm is 54 μs, the velocity must be $25 \times 10^{-3}/54 \times 10^{-6} = 463$ m/s. Therefore $463 = \sqrt{E/920}$, from which $E = 200$ MPa.

Exercise 12-9

Contrast the relative effects of strain rate and temperature on the moduli of metals and polymers.

Solution. At a given temperature, particularly in the range below half homologous, the elastic moduli of most metals are not much affected by rate. In contrast, polymers can be quite rate-dependent, the extremes being given by the relaxed and unrelaxed values. For example, PMMA at room temperature ($< T_g$) displays approximately 6 GPa in tests the durations of which are microseconds, but only about 3 GPa in slower tests taking 10 seconds or so. Other glassy polymers (and semicrystalline polymers) behave similarly, their moduli changing by up to a factor of 2. Some rubbery polymers of exceptionally high molecular weight flow over long periods of time, which means that their moduli tend to zero. Occasionally the application of a high strain rate to a rubbery polymer can transform the material through the T_g into the glassy region, in which low-rate rubbery moduli of less than 2 MPa can become high-rate glassy moduli of perhaps 2 GPa in magnitude.

CREEP AND HARDNESS

If a hard indenter is pressed into the surface of a solid to form a plastic indentation, the mean pressure p over the indentation may be written

$$p = c\sigma_y$$

where, as explained in Chaps. 2 and 19, c is a proportionality factor that depends on the shape of the indenter ($2.5 < c < 3.0$ most often) and σ_y is the uniaxial compressive yield stress as locally augmented by the indentation process itself. That is, using $\sigma = K\epsilon^n$, we have

$$p = (cK)\epsilon_{av}^n \tag{12-18}$$

where ϵ_{av} is the average plastic strain produced in the vicinity of the indenter. For ball indenters, $\epsilon_{av} = 0.2(d/D)$, where d is the diameter of the projected impression and D the diameter of the indenter. In such cases the representative strain increases as the ball is pushed further in, which is physically sensible. With pointed indenters, however, where the indentations are geometrically similar whatever their depth, the representative strain is a fixed value; for example, a Vickers pyramid produces a strain of 0.08. That strain corresponds to an equivalent (d/D) for a ball indentation of 0.4, or an impression of diameter rather less than the radius of the ball; this is typical of recommended hardness testing practice.

When the yield stress is rate- and temperature-dependent, the apparent hardness will change, as the longer the load is left on the bigger will be the size of the indentation. Using Eq. (12-5), we may rewrite Eq. (12-18) as

$$p = cK_0 \exp\left(\frac{1}{T}\right) t^{-m_\epsilon} \epsilon_{av}^{n+n} \tag{12-19}$$

Thus the hardness diminishes with t, giving linear logarithmic plots of p vs. t, as shown in Fig. 12-10. It is remarkable in Fig. 12-10 that the creep process during indentation is not greatly dependent on indenter shape, except for the 60° cone. It turns out that, apart from the 60° cone, the deformation zones below all the other indenters resemble a radial compression (Fig. 12-11), and the reason that the behavior appears similar is that the change of hardness with time is a measure of the rate of progress of the elastic/plastic boundary "hemisphere" into the solid. Since this plastic zone is continuously increasing as time progresses, it is not surprising that a relationship exists between transient creep and hardness, via Eq. (12-5), which is itself related to Eq. (12-6) and Fig. 12-5.

The time dependence of indentation hardness at high homologous temperatures can be used as a simple means of selecting creep-resistant alloys.

In the case of solids with relatively low moduli, such as polymers, inden-

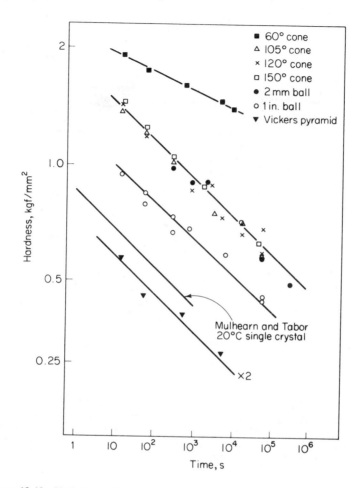

Figure 12-10 Variation of hardness with loading time for indium polycrystalline specimens at room temperature. Except for the 60° cone, the slope is the same for conical, pyramidal, and spherical indenters. (The behavior is also similar to that observed by Mulhearn and Tabor on a single crystal using a spherical indenter.) Evidently, the creep process during indentation is not greatly dependent on indenter shape. [From A. G. Atkins, A. A. dos S. Silverio, and D. Tabor, "Indentation Hardness and the Creep of Solids," *J. Inst. Metals*, **94** (1966), p. 369. Reprinted by permission of the Metals Society.]

tation pressures are lower than would be expected from either $p = c\sigma_y$ or Eq. (12-19), and elastic recovery of the indentations (principally in the depth of indentation) affects the results. This arises because the elastic strains in the region of the indentation are no longer negligible compared with the plastic strains; only when the strain imposed by the indenter is 50–100 times greater

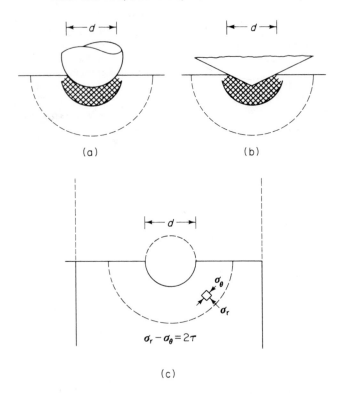

Figure 12-11 Plastic deformation mode for: (a) a spherical indenter; (b) a conical indenter; (c) represents the equivalent plastic expansion of a hemispherical cavity into a semiinfinite solid. [From A. G. Atkins, A. A. dos S. Silverio, and D. Tabor, "Indentation Hardness and the Creep of Solids," *J. Inst. Metals*, **94** (1966), p. 369. Reprinted by permission of the Metals Society.]

than the yield strain does the simple $p = c\sigma_y$ relation hold. Even so, as shown by K. L. Johnson, proper c factors can be developed for smaller ratios of applied strain to yield strain.

POLYMER DRAWING, LÜDERS BANDS, AND METAL SUPERPLASTICITY

In simple tension tests of common engineering metals, elongations are limited by necking. The true strain at ultimate when necking commences is numerically equal to the work-hardening index n, as shown in Chap. 2. For mild steel, $n = 0.1 \sim 0.25$, so the uniform elongation at the maximum load is no more than 25%; even with brass, for which $n = 0.4 \sim 0.5$, the elongation is limited to 50%. Thereafter deformation is concentrated in the necked regions leading to fracture. There are circumstances with some metals and some polymers under which a neck forms but does not give immediate fracture. Rather the neck goes along the test bar in a stable fashion, the thinned-down neck being "fed" material from the adjacent shoulders (Fig. 12-12). The drawn material in the neck is usually very strong and often anisotropic; it is also highly fibrous in nature (compare internal necks in metals). Some high-strength nylon fibers are produced in this fashion, as are glass filaments by "hot-drawing" glass rods. Chewing gum behaves similarly. In so-called *dieless drawing* of hot metals, a sharp temperature gradient is maintained at one section of the rod under tension, and significant reductions in area may be obtained at that region.

An explanation for this type of behavior is available when strain-rate sensitivity is taken into account. For material inside and outside a neck Eq. (12-4) may be rewritten, for constant K,

$$\frac{\dot{\epsilon}_w}{\dot{\epsilon}_s} = \left(\frac{\sigma_w}{\sigma_s}\right)^{1/m} = \left(\frac{\text{cross section shoulder}}{\text{cross section neck}}\right)^{1/m}$$

where the subscripts w and s refer to neck and shoulder, respectively.

Suppose, for example, that at an early stage of necking the cross section in the necked region is 1% less than elsewhere. Then $\sigma_w = (1/0.99)\sigma_s$ and $\dot{\epsilon}_w = (1.01)^{1/m}\dot{\epsilon}_s$.

For many common engineering materials at room temperature $m = 0.005$ (Fig. 12-2), so

$$\dot{\epsilon}_w = (1.01)^{200}\dot{\epsilon}_s = 7.3\dot{\epsilon}_s$$

This means that the necked regions will stretch much more rapidly and little deformation occurs in the shoulders; in fact, these elastically unload after ultimate in the usual type of tension test. By the time a 2% difference in cross-sectional area develops, $\dot{\epsilon}_w = 52\dot{\epsilon}_s$, so that the process is self-accelerating. However, were m to have a much larger value, such as 0.5, we have for a 1% area difference

$$\dot{\epsilon}_w = (1.01)^2\dot{\epsilon}_s = 1.02\dot{\epsilon}_s$$

That is, the rate of deformation in the necked region is only 2% greater than elsewhere. This means that a great deal of deformation will continue in the

Figure 12-12   Polymer drawing, that is, the stable growth of a neck in high-density polyethylene.

shoulders despite the attempted localization of the deformation at the first place the change in areas took place. Even when the neck has developed to the extent that its cross section is down to 90% of the shoulders, its rate of deformation is only $(1/0.9)^2 = 23\%$ greater. With such slow localization of flow, a very large amount of deformation continues throughout the piece.

At neck initiation, $dP = 0$, as discussed in Chap. 2. The subsequent load drop bottoms out, however, and $dP = 0$ again when the neck stabilizes and draws out. Both instances correspond with $d\sigma/d\epsilon = \sigma$, but the types of tangency are different (Fig. 12-13). The first has the σ vs. ϵ curve "inside" the tangent line, and, as with traditional engineering metals under tension at room temperature, this corresponds with the strain ϵ_A at neck initiation. The second has the σ vs. ϵ curve "outside" the tangent line and corresponds with stable neck propagation at that fixed value of true strain, ϵ_B. It should be noted that extension of the σ vs. ϵ curve temporarily stops at ϵ_B, because the growth of the stable neck does *not* further strain the material in the neck. Only when the neck would have run to both gripped ends of the testpiece would the rest of the curve be picked up. In contrast, the curve of steady load vs. extension (on the chart of the testing machine) goes on for as long as the neck grows. Since a specimen 100 mm long will have a shorter ΔL region than a specimen 1 m long, conversions of ΔL into e values or into "extension ratios" (L_f/L_0) are quite meaningless because of the same objections that were raised in Chap. 2

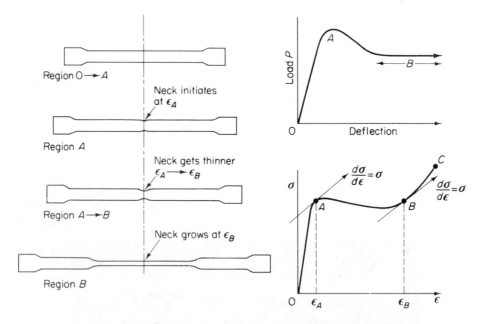

Figure 12-13 Mechanics of (isothermal) stable and unstable necks; both occur at $d\sigma/d\epsilon = \sigma$, but the types of tangency are different. Unstable necks at A have the σ vs. ϵ curve "inside" the tangent line. Stable necks at B have the curve "outside" the tangent line.

to the use of percentage elongation without reference to gage length in tension tests. Even so, many polymer data are still published in terms of extension ratios or as "many hundreds of percent" extension.

There is a similarity between regions of *discontinuous yielding* in ductile metals and the growth of stable necks in polymers. In a highly polished steel sample, discontinuous yielding will be seen to be accompanied by a series of striations that ripple along the gage section in the period ab in Fig. 2-20. Such Lüders bands, as they are called, may be thought of as a series of stable necks. Careful measurement will show many pairs of tangency points satisfying $d\sigma/d\epsilon = \sigma$ in the discontinuous yielding region at the $dP = 0$ local maxima and minima. The strain-hardening part of the σ vs. ϵ curve beyond initial yielding up to ultimate and then to fracture, follows upon exhaustion of the supply of Lüders bands. Note that, strictly speaking, the same problems, as in polymers, arise in regard to the meaning of strain in terms of ΔL during Lüders band propagation, except that whereas stable necks in polymers occur at large strains (see Fig. 13-9), discontinuous yielding in ductile metals takes place at very low strain levels, so the resulting errors in converting from true to nominal strains are small. That is, the strain range ab in Fig. 2-20(c) is, say, $0.001 \sim 0.02$ but the strain range $\epsilon_A \sim \epsilon_B$ in Fig. 12-13 may be $0.1 \sim 0.3$. Note that not all polymers display stable necks, and many have P vs. ΔL and σ vs. ϵ curves like those for brass in Fig. 2-2; details are given in Chap. 13.

Exercise 12-10

Lüders bands were observed in the 1 mm $\times$ 10 mm $\times$ 100 mm gage length of a polished low-carbon steel test piece. The extension of the test piece was then 100 μm. Determine the extent of Lüders band propagation in the gage length, given that Lüders band strain is approximately 0.02 within the band.

Solution. The apparent strain of the test piece is $(100 \times 10^{-6})/(100 \times 10^{-3}) = 0.001$. The local strain in a Lüders band is given as 0.02. If those parts of the gage section outside the bands have essentially zero strain, the extension of 100 μm strain is produced by some regions with zero strain and the band regions with 0.02 strain. To simplify matters, let us assume that the Lüders bands are perpendicular to the applied load (they usually are not), in which case the accumulated length of gage section taken up by the Lüders bands must be $100 \times 10^{-6}/0.02 = 5$ mm. That is, of the 100-mm gage length, all the Lüders ripples add up to 5 mm.

In principle, the values of ϵ_A and ϵ_B for polymers and Lüders strains may be deduced by differentiating a relation such as $\sigma = K_0 \exp(1/T)\epsilon^m\epsilon^n$, which additionally takes account of the possibility of the bar heating up from the work of deformation. They are not predicted from differentiation of $\sigma = K\epsilon^n$, because that is a single-valued function describing the work hardening after initial yield (i.e., the region buf in Fig. 2-20).

Great strain-rate sensitivity is also displayed by some metals under certain conditions, when it is known as *superplasticity*. The circumstances that promote the behavior are those often associated with large rates of creep in metals, namely, high homologous temperatures, low strain rates, and a very

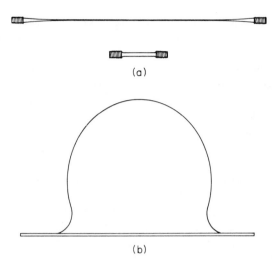

Figure 12-14 (a) Sketch of a tensile bar of 80 Zn–20 Al extended 10 times its original length at 350°C and 200 μm/s. (b) Bulged sheet of 80 Zn–20 Al alloy. Sheet thickness 0.75 mm; diameter of hole approximately 100 mm. [After W. A. Backofen, I. R. Turner, and D. H. Avery, *Trans. ASM* **60** (1964), p. 980.]

fine grain size. Indeed, a good superplastic metal is the antithesis of a good creep-resistant alloy. It appears that quasi-viscous mechanisms such as grain boundary sliding may contribute significantly to the strain-rate sensitivity and account for an appreciable fraction of the total deformation.

Reduction in area of over 90% (and associated tremendous elongations) have been reported for superplastic alloys in tension tests performed under the correct conditions. For example, at 300°C (approximately $0.8T_m$), an 80 Zn–20 Al alloy displays $m > 0.6$ when $10^{-3} < \dot{\epsilon} < 10^{-1}$ s^{-1}. This means that formability limits of superplastic sheet metals can be increased to such an extent that it is possible to use forming techniques that are normally employed only with glass and plastics (Chap. 13). Figure 12-14 is a sketch (to scale) showing the tensile elongation and forming capability of the Zn–Al superplastic alloy. Care must be taken to ensure that the material remains superplastic during forming. Grain growth (which would reduce the grain boundary area) must be avoided. Two-phase (duplex) microstructures are especially useful in this regard; thus eutectic Pb–Sn is another superplastic solid under the right conditions.

REFERENCES

12-1. Alexander, J. M., and R. C. Brewer, *Manufacturing Properties of Materials.* London: Van Nostrand, 1963.

12-2. Cottrell, A. H., *The Mechanical Properties of Matter.* New York: John Wiley & Sons, 1964.

12-3. Johnson, K. L., "The Correlation of Indentation Experiments," *Journal of the Mechanics and Physics of Solids*, **18** (1970), p. 115.

12-4. McClintock, F. A., and A. S. Argon, *Mechanical Behavior of Materials*. Reading, Mass.: Addison-Wesley, 1966.

12-5. McCrum, N. G., B. E. Read, and G. Williams, *Anelastic and Dielectric Effects in Polymeric Solids*. New York: John Wiley & Sons, 1967.

PROBLEMS

12-1. The data for σ vs. ϵ over a range of $\dot\epsilon$ in Fig. 12-15 were obtained on low-carbon steel. Determine the parameters in a relation $\sigma = \text{constant} \cdot \epsilon^n \dot\epsilon^m$.

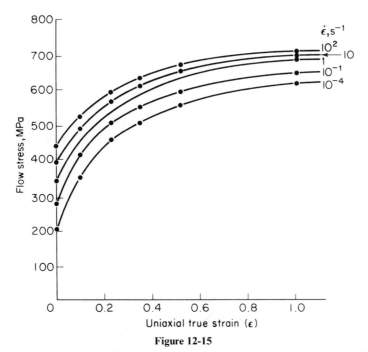

Figure 12-15

12-2. For the two-element model shown in Fig. 12-16 calculate the required minimum value of η in order to limit the total strain of the system in 10 Ms to 0.003.

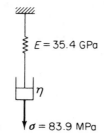

Figure 12-16 $\sigma = 83.9$ MPa

12-3. The load displacement behavior of a certain polymer is described by a spring and dashpot in series as shown in Fig. 12-17. For the force application P as shown from time t_1 to time t_4, plot the corresponding displacement x within that time interval.

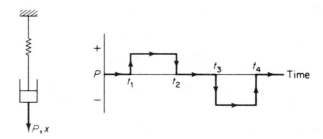

Figure 12-17

12-4. A steel beam subjected to bending loads in the laboratory deforms almost instantaneously when a load is applied, and the deformation then remains constant with time. A lead–tin solder wire in tension will deform by a small amount when a load is applied and will continue to deform with time until fracture occurs. Why does the lead–tin solder behave differently from the steel?

12-5. In a constant-load tensile test on 50 Pb–50 Sn solder at room temperature the specimen exhibits initial behavior of the form $\epsilon = C[1 - \exp(-Et/\eta)]$, where ϵ is the strain, C is a constant, E and η are constants, and t is time. Then, for most of the test, the strain approximates the form $d\epsilon/dt = $ constant. Explain the reasons for the *change* in behavior, in terms of basic changes that occur in the structure of the metal.

12-6. The deflection–time curve of the lead–tin alloy shown in Fig. 12-4 can show a slightly upturned shape throughout the "steady-state creep" region. Why is this?

12-7. A lead–tin wire, initially 100.0 mm long and 3.18 mm in diameter, is loaded in tension with a constant force of 118 N at 20°C. After 1.74 ks its length is 119.0 mm, and 0.90 ks later its length is 129.5 mm.
   (a) Estimate the magnitude of the steady-state viscosity η_1 (in Pa-s) if the specimen is considered a four-element Voigt–Maxwell model.
   (b) Is this a liquid or a solid?

12-8. Results for a creep test of Pb–Sn solder are given in Fig. 12-18. Calculate η_1 and E_1 if we assume this specimen approximates the Voigt–Maxwell four-element model shown.

12-9. The strain–time curve for a viscoelastic polymer under a tensile stress of 100 MPa is shown in Fig. 12-19. Regarding the polymer as the Voigt–Maxwell model shown:
   (a) Calculate the value of E_1.
   (b) Calculate the value of η_1.

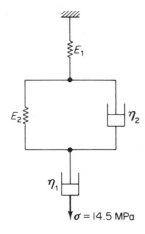

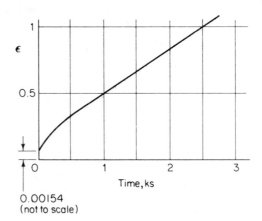

Figure 12-18

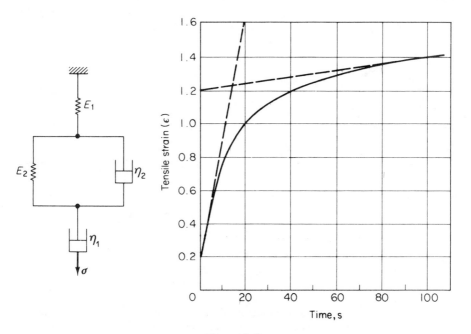

Figure 12-19

12-10. The strain–time curve for a four-element Voigt–Maxwell model is given in Fig. 12-20. Calculate the strain after 1 ks for a model identical in every respect except that E_1 is one fourth the E_1 of the model shown.

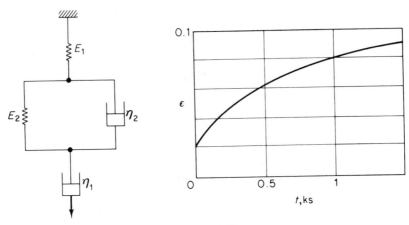

Figure 12-20

12-11. The relative ductility or brittleness of silly putty depends on the rate at which it is stressed. Which of the three models shown in Fig. 12-21 best describes the behavior of silly putty? Explain why you reject the other models.

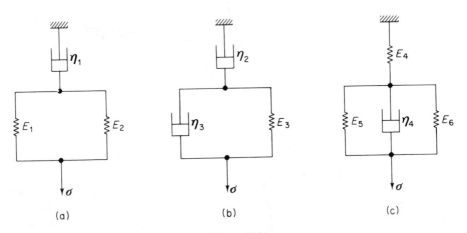

Figure 12-21

12-12. A four-element viscoelastic model produces the strain–time curve shown in Fig. 12-22. Draw to *reasonably accurate scale* the strain–time curve, on the same coordinates, for tripling E_1, halving η_1, and lowering σ_A by 20%.

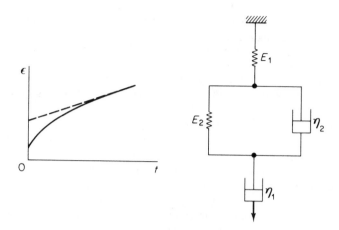

Figure 12-22

12-13. A round tensile testpiece of some polymer necked (as a metal often necks in tension), but instead of necking on down and fracturing, the polymer neck stabilized and "drew out" at constant diameter. Explain why this happens with reference to the load–extension or stress–strain curve for the polymer. The following loads and diameters applied to such a stable neck test:

	Diameter	Load
Initial condition	10 mm	0
Neck initiates	8 mm	1 kN
Neck grows steadily	6 mm	560 N

Sketch this behavior on a true-stress–true-strain (σ–ϵ) plot, marking off important values of σ and ϵ, and indicating any significant aspects of the diagram.

12-14. The relaxation time of polyisobutylene is some 200 s at 50°C. Determine the relative magnitudes of the elastic and flow components of strain under loads the duration of which are 1 s, 10 s, 100 s, 1000 s. The same polymer at 50°C is held at a constant strain for 10 ks. Sketch the percentage fall in the applied stress from its initial value over the 10-ks time interval.

12-15. A series of ball indentation experiments under a load of 5 kgf was carried out at room temperature on a block of tin. The load was left on for various lengths of time, and the following were the diameters of the hardness impressions produced: 1.13 mm after 10 s; 1.26 mm after 100 s; 1.39 mm after 1000 s; and 1.51 mm after 3200 s. Obtain from these data an estimate for the strain-rate sensitivity of tin at room temperature.

12-16. Strip is reduced in thickness with no sideways spread from t_{in} to t_{out} by pulling through tapered dies (sheet drawing). If the speed of the strip at the exit of the die is V_{out}, show that the mean strain rate for the process is $2\,V_{out}\,(1 - r)$ tan

α/t_{out}, where α is the semiangle of the die and r is the percentage reduction in area for the drawing process.

Determine the mean strain rate when $V = 1$ m/s, $r = 20\%$, $\alpha = 8^0$, and $t_{out} = 2$ mm. Comment on the magnitude of this strain rate as compared with quasi-static rates of testing, and using the data given in Fig. 12-15, determine by how much the working loads in the process are increased when rate effects are taken into account.

12-17. A bar of polyethylene 10 mm in diameter reduces uniformly to 9.5 mm diameter and then forms a stable neck 8.2 mm in diameter that propagates down the bar. Show that percentage elongation is an unsuitable parameter for characterizing the neck formation, and provide a sensible parameter.

12-18. A four-element Voigt–Maxwell model for deformation of a polymer has elements E_1 and η_1 in series with E_2 and η_2 in parallel.

   (a) Draw the strain–time curve for this system, with the correct shape but without numbers; label this "curve (a)."

   (b) Increase η_2; on the same coordinates as (a) draw the new curve with a dotted line; label this "curve (b)."

   (c) Decrease E_1, using original values of E_2, η_2, and η_1, and draw a new curve with long dashes on the same coordinates; label this "curve (c)."

13 | Polymers and Design with Time-Dependent Solids

The mechanical properties of polymeric solids (commonly referred to as plastics) have been touched on in earlier chapters. It was shown that a whole range of stress–strain–time responses was possible, depending on conditions, from very creepy behavior to stiff elastic behavior, with the possibility of a rubbery range between these. The structure of the polymer and its makeup, together with temperature and rate of deformation, determine to a great extent the mechanical behavior.

Many polymers are now accepted as regular engineering materials and given the loose description of "engineering plastics." The term is clearly artificial because applicability depends on circumstances, but it probably originated to distinguish those polymers that could be substituted satisfactorily for metals in certain designs from those plastics with inadequate properties. When polymers are used in design, they fulfill a need that either cannot be met with traditional materials or cannot be met so cheaply otherwise. The mass production of intricate, accurate parts by injection molding is one feature available with thermoplastics of which advantage is often taken.

Examples of plastics used successfully in engineering contexts are polyvinyl chloride (PVC) in large-diameter pipes, polyolefins in small-diameter pressure pipes, nylons and acetals in small gear wheels, polypropylene (PP) in hinges, and high-density polyethylene (HDPE) in artificial hip joints. In most of these applications the choice of material is determined by some particular virtue sufficiently valuable to offset the relatively low moduli and strengths. Other load-bearing products designed in polymers include bottle crates, vegetable trays, and other stacking containers injection-molded from polypropylene or high-density polyethylene; pallets thermoformed from acrylonitrile butadiene styrene (ABS) sheet or molded from HDPE or PP; and chairs made by a variety of techniques from materials that include PP, PVC, ABS, polystyrene, acrylic, and HDPE.

The limitations imposed by creep on the wide application of polymers led to intensive research programs in the 1950s and 1960s aimed at increasing the temperatures at which polymers could be used without sacrifice of mechanical properties. Research was spurred by aerospace and defense needs. Even so, few of the newly developed materials have really reached significant production levels (in tonnage) in comparison with the polyethylenes, polyvinyl chlorides, polypropylenes, or ABS. Of those that have, nylons and polyacetals account for a significant part of the consumption of "new" engineering plastics, followed by polycarbonate. Of the rest of the newer materials, only modified polyphenylene oxide, PPO (e.g., GE Noryl), and thermoplastic polyesters [polybutylene terephthalate (PBT), polyethylene terephthalate (PETP)] are used in reasonable amounts; others, such as the fluorocarbons, polysylphones, and polyamides, are expensive and thus find only specialized applications.

In engineering design with polymers it would be ideal if the region where the behavior is primarily elastic, with small rate effects, could always be used. Conventional strength-of-materials design calculations could then be employed to a reasonable approximation. When other factors (corrosion resistance, low mass, cost, and so on) suggest the use of polymers whose behavior would be time-dependent in use, such polymer parts must be stressed merely to levels that produce only tolerable creep. Examples of such calculations (which also apply, of course, to creepy metals) are given later.

Although polymer parts are often cast (molded) directly to shape, some polymers may be subjected to the same sort of mechanical working processes (e.g., rolling, drawing, and extrusion) as metals, with interesting changes in mechanical properties. Before considering mechanical behavior and design with time-dependent solids, however, some discussion must be given to the molecular structure of typical polymers and how structure influences mechanical properties.

POLYMER STRUCTURE

Polymerization produces giant long molecules, often joining together thousands of *monomer units*, like a length of beads on a string. The molecular forces that hold individual links together in the chain are primary valence bonds (Chap. 4), and it is secondary bonding that attracts separate polymer chains together. Although these secondary cohesive bonds are relatively weak, there are so many of them between the many backbone strands that a tightly cohering mass results. Differences in the strength of secondary bonding produce different types of mechanical behavior in polymers. For example, weak secondary bonding gives a rubbery and stretchy polymer; stronger interchain cohesive forces give stiffer and stronger plastics. The substitution of different groups in the polymer chain produces stronger secondary bonding and thus polymers with altered mechanical properties. For example, acrylonitrile is essentially polyethylene $(CH_2)_n$, but with alternate H–C–H units replaced by H–C–N units.

In addition to the foregoing simple *linear polymers*, it is possible for chains to grow out sideways from the main chain backbone, as branches on a tree do. Furthermore, the branches may link across to other main chains, giving *ladder* or *network* polymers (Fig. 4-3). In those circumstances, main chains are held together not only by secondary valence forces but also by the stronger primary forces along the "rungs" of the (three-dimensional) ladder. This, in effect, produces a single, strong, giant molecule.

As explained in Chap. 4, the more molecules interfere with one another in the liquid state, the more viscous the liquid and the less chance there is for a crystalline solid to be formed with a clear-cut freezing/melting point on cooling. Thus branched and network polymers are more "amorphous" than simple linear polymers. Polymers of the class called *thermosetting plastics* start out in the liquid state but solidify under the influence of heat and/or a catalyst hardener, by forming cross links between linear main chains, which reduce the mobility of these strands. One example of a thermoset is epoxy glue. Once formed, thermosets tend to char and degrade on heating. In contrast, polymers such as simple linear polymers and glasses, which soften to a toffee consistency on heating, and which moreover return to their earlier state on cooling, are called *thermoplastics*.

The properties of polymers depend strongly on their detailed structural shape, where the chemistry of the monomer and repeat patterns are paramount. Particularly important is chain geometry. Chains of mers (a *mer* is one basic molecular unit) are usually sketched in textbooks as being planar, but strictly they are three-dimensional. The bonding directions of a carbon atom may be thought of as acting through the corners of a tetrahedron, so even a molecularly simple chain zigzags in space. For example, the chemical structure of polyethylene, $(CH_2)_n$, is usually drawn as in Fig. 13-1(a), whereas it really looks more as in Fig. 13-1(b). Polymethylmethacrylate is similar to polyethylene, but the side groups, instead of being simple hydrogen atoms, are more bulky methyl (CH_3) and $(COOCH_3)$ groups (Fig. 13-2). In truth, PMMA is not very complicated compared with many other polymers.

Chains in general wander in many directions and become raveled and matted, and the usual analogies for polymers are tangled string or masses of spaghetti. It follows that the mechanical properties of polymeric materials are markedly dependent on the flexibility of the joints in chains and the ease with which the chains can slide over one another during deformation; time-dependent behavior results, in part, from the uncurling of long, tangled chains, which gives delayed elastic effects. The behavior is further modified when the chains are cross-linked to one another as in network polymers.

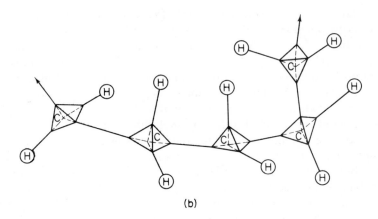

(a)

(b)

Figure 13-1 (a) Planar and (b) three-dimensional representation of polyethylene; bonding directions of carbon atoms act as if through the corners of a tetrahedron.

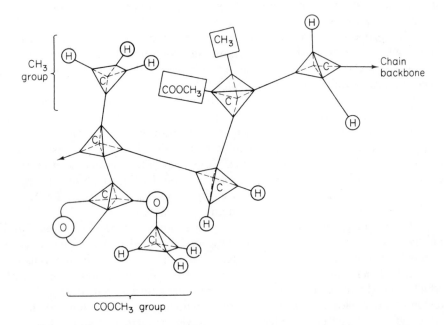

Figure 13-2 Polymethylmethacrylate (PMMA) molecule, consisting of methyl (CH_3) and ($COOCH_3$) side groups on the carbon backbone chain.

Again, the behavior is modified if there are crystalline regions in the polymers. Although most bulk polymers are amorphous, it is possible for regular arrays of osculating parallel chains to order themselves and form a lattice in certain regions of a polymer, especially when the polymer consists of short, simple, linear chains or has symmetrical mers, with no lumpy side groups and with no cross-linking to other chains.

Exercise 13-1

The volume vs. temperature plot for a certain polymer cooled at two widely different rates is shown in Fig. 13-3. At room temperature which condition, (A) or (B), is likely to be more brittle?

Solution.   Condition (B) is amorphous, and (A) is crystalline, because of the distinct change in volume at a single (melting) temperature. Condition (A) is less likely to show viscoelastic behavior and thus is likely to be less ductile.

Of common polymers, polyethylene with its simple structure of chains less than about 100 mers long, has a strong tendency to crystallize, achieving perhaps 80% overall crystallinity. Others, such as nylon, propylene (of the so-called isotactic type), and PVC, are partially crystalline (say 50–70%), whereas polymethylmethacrylate, with lumpy side groups, is virtually impossible to crystallize. It will be appreciated that the more complex the molecular chain system, the less chance that crystalline regions (having a regular and repeated three-dimensional pattern) will easily be found. For similar reasons branched and network polymers do not readily crystallize. We note that, as a rule, partially crystalline polymers are opaque or at most translucent. Amorphous polymers, on the other hand, are usually transparent. This provides a rough but convenient means of identification: Remember that window glass is transparent and glasses are amorphous. In a bulk polymer containing microcrystalline regions, a single molecular chain may extend through both the crystalline and the noncrystalline zones.

The arrayed regions are called *crystallites*, and it is found that the parallel chains are not simply linear but fold back and forth in pleats (called a

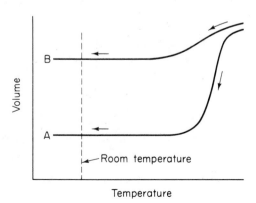

Figure 13-3 Volume vs. temperature plots of two polymers (Exercise 13-1).

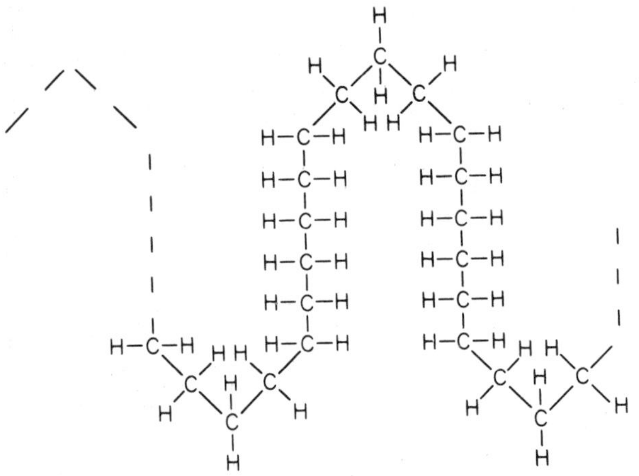

Figure 13-4 "Chinese cracker" arrangement of crystallites in crystalline regions of polyethylene.

Chinese cracker structure by Frank by analogy with the firework), as shown in Fig. 13-4. Here again, the simple planar drawing is in reality a closely packed, three-dimensional figure. The platelike folded areas are called lamellae and individually extend for some 10 nm. Structured aggregates of crystallites occur on a macroscopic scale, and lamellae appear in plan view to flow on edge radially outwards from small regions, like spokes in a wheel. The resulting rosette structure is called a *spherulite* (Fig. 13-5).

Figure 13-5 Spherulite in polyethylene, which is a structured aggregate of crystallites on macroscopic scale. (Photograph courtesy of G. S. Yeh.)

Polymers that have some crystallinity *whiten* on straining; that is, the highly stressed areas change color. The drawn necks in high-density PE are very white; if a hardness ball is pressed into a lump of PVC, the distribution of nonuniform strains is seen. The probable cause of whitening is the mechanical breakup of spherulites leading to light scattering. If a whitened region is warmed, the color disappears and the process is reversed.

BEHAVIOR RELATIVE TO GLASS-TRANSITION TEMPERATURE

As described in Chap. 4, the mechanical properties of amorphous polymers may be conveniently split into two principal regimes: above T_g and below T_g. Above T_g time-dependent creepy behavior is evident, whereas as the temperature drops below T_g time and rate effects gradually become less marked and creep is low (Fig. 4-4 and Chap. 12). Near T_g the behavior is often called *leathery*. From what has been said previously the glass transition is caused by changes in the molecular mobility of the main polymer backbone chains; diminished creep below T_g comes about from a stiffening of these main chains in comparison with their flexibility above T_g. Any small creep and rate effects that are present below T_g arise from molecular mobility of polymer branch side chains. At low enough temperatures, a "secondary transition" may occur, after which the mobility of even these side chains is restricted. At temperatures below the secondary transition, the polymer is then essentially elastic and brittle, akin to some time-independent metals or ceramics.

Figure 13-6(a) shows the changes in modulus that occur with changes in molecular mobility at the secondary and primary transition points as the temperature changes. The modulus falls by a factor of 3 to 5 as side-chain mobility comes into play, and by a further factor of 10 to 500 as main-chain mobility occurs [the ordinate in Fig. 13-6(a) is a logarithmic scale]. The existence and position of the transitions and their magnitudes depend on the detailed molecular structure, the conditions of measurement, and indeed on the history of the sample. In a way, this statement is no different from the observation that recrystallization temperatures in metals depend on the amount of prior cold work. In the case of partially crystalline polymers, it is only the amorphous part that may show such transitions. Consequently the diminutions in modulus for them are less marked, and modulus–temperature plots for crystalline polymers often show only minor drops before melting. Exercise 12-9 concerns the relative effects of strain rate and temperature on the moduli of metals and polymers.

Other properties, such as toughness and damping capacity, also change at transition temperatures. It is a feature of these materials that their ability to damp out vibrations increases markedly near the transition temperatures when compared with their damping capacity at temperatures much below and

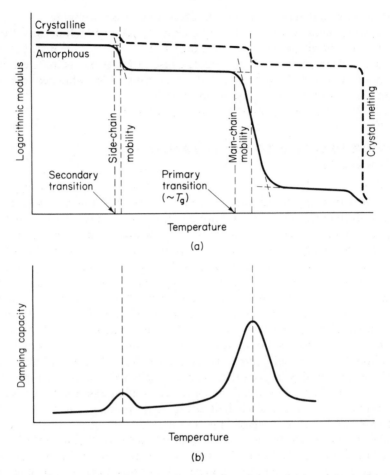

Figure 13-6 Schematic changes with temperature of (a) modulus and (b) damping capacity for polymers. Temperatures at which the damping peaks occur usually coincide with the inflection in the modulus–temperature curve; the glass-transition temperature T_g is usually defined as the intersection of tangents, so that the primary and secondary transition temperatures differ somewhat from the temperatures of maximum damping.

above. For example, the oscillations of a torsional pendulum, the suspension of which is made from a polymer displaying transitions, die away much more quickly at those transition temperatures, and plots of damping capability have associated peaks, as shown schematically also in Fig. 13-6(b). The bounce of rubber balls, measured at different temperatures, will be the least at temperatures near their primary and secondary transition temperatures, because of the high damping and low restitution. The temperatures at which the damping peaks occur usually coincide with the points of inflection of the modulus–temperature curve; as T_g is conventionally defined in terms of the intersection of the tangents at the "bends" in the modulus–temperature curve, T_g and the temperature at peak damping are not the same. For many polymers the difference is of the order 25°C.

Phenomenologically, the mechanical behavior of a creeping polymer above the secondary transition, and certainly above the primary glass transition, is similar to a creeping metal. However, because the elastic moduli of polymers are in general much smaller than those of metals, the recoverable elastic strains of polymers may be comparable in magnitude with the irreversible creep strains, which is not the case with most creeping metals of engineering importance. The combination of viscous sliding and normal elastic behavior leads to *viscoelasticity*: Large amounts of deformation can occur for quite low stresses, as described in Chap. 12.

Generally speaking, the more complicated the molecule and the more cross linking present, the higher the glass-transition temperature. The following are the approximate T_g values for some common polymers: polyisobutylene (butyl rubber), $-70°C$; polyethylene, $-20°C$; polypropylene, 0°C; nylon, 60°C; PVC, 80°C; polymethylmethacrylate, 100°C; polystyrene, 100°C; PTFE, 125°C; polycarbonate, 150°C. Thus, at room temperature, nylon, PVC, PMMA (plexiglas), polystyrene, PTFE, and polycarbonate display quasi-elastic behavior, with comparatively small time effects, whereas all the rest would display marked viscoelasticity. Obviously, if nylon were warmed above its respective T_g, it would become creepy. Conversely, the behavior of polypropylene much below 0°C is time-independent, and so on.

The glass-transition temperature of polymers may be reduced by adding a plasticizer, which is a liquid in which the polymer is soluble but that does not combine with the polymer at the molecular level. For example, the glassy polymer cellulose nitrate is softened (T_g is lowered) by the addition of camphor, to give celluloid; toughness has been produced at the expense of rigidity. Plastic raincoats are made from plasticized PVC. Another technique is to mix two polymers together, to give a *copolymer*, whose T_g will lie between those of its components. The corresponding mechanical behavior is affected by changes in T_g, so that polymers can be tailored to suit particular applications.

Exercise 13-2

Select T_g relative to 23°C for the following applications: (a) a child's rubber ball for maximum bounce, (b) a polymer for electrical insulation in the Arctic, (c) rubber fenders for a boat dock in Miami, (d) a polymer with the highest possible room-temperature strength, (e) rubber for automobile tires.

Solution.   (a) $\ll 23°C$ (b) $\ll 23°C$ (c) $\approx 23°C$ (d) $\gg 23°C$ (e) $\ll 23°C$

ELASTICITY

Since, along most polymer chains, the backbone is formed of carbon bonds, one might think that polymers ought to have the same stiffness as other substances with carbon–carbon bonds, such as diamond. The Young's modulus of diamond is some 760 GPa, but diamond has a specific gravity of about 3.5, whereas that of many polymers is about unity. So, making allowances for the differences in density (and hence fewer bonds per unit volume), we might say that polymers should have moduli of perhaps $(1/3.5)\,(760) = 207$ GPa.

In practice the moduli of typical polymers are about 700 MPa. Only with certain natural polymers in fiber form, such as the cellulose molecule in wood, is the modulus at all close to the anticipated value. Why?

The reasons for the discrepancies should be clear from the clues just given:

1. Diamond has ordered crystal structure.
2. Fibers, although amorphous, are usually oriented with the chain backbones along the fibers.
3. Regular, amorphous tangles of string or spaghetti have no orientation whatever.

Polymer chains zigzag in space and, like coil springs, must be more flexible than the same length of chain bonds stretched straight. The exceptional deformability of the rubber molecule is due to the helical structure of the isoprene carbon chain. In gutta-percha, another form of the rubber polyisoprene, the carbon chain is zigzag rather than helical, and the substance is much less deformable. A coil spring made from metal wire is much more compliant than the same length of straight wire. Even in a highly crystalline polymer, where a fairer comparison could be made between theory and practice, the "Chinese cracker" structure of lamellae can never be as stiff as a straight chain.

Wide-angle X-ray examinations show that the rolling process orients molecular chains nearly parallel to the rolling direction, and improvements in polymer stiffness along the rolling direction can be marked. For example, the Young's modulus of undeformed polypropylene is perhaps 710 MPa; a rolling reduction of area of 87% at about 120°C can increase the modulus more than sixfold, precise values depending on strain rate and temperature (cold work also being possible). The Dupont company sells material of this type for strapping. Other polymers, such as polyoxymethylene and nylon-66, also improve in modulus under similar processing conditions. In the case of polyethylene, cold rolling up to 25% reduction of area produces a decrease in modulus at first, followed by a subsequent increase (Fig. 13-7). (Note also the effect of the speed of the testing crosshead on the results.) The reduction is probably associated with a breakup of the crystalline spherulite structures without observable chain orientation, the latter only coming into play at higher amounts of cold work.

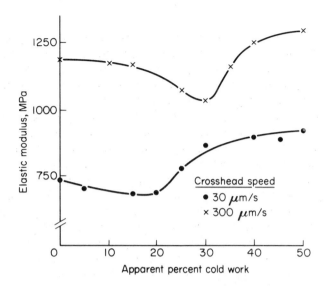

Figure 13-7   The effect of cold work on the elastic modulus of high-density poly-ethylene at two crosshead speeds. [From R. M. Caddell, T. Bates, and G. S. Yeh, *Mat. Sci. and Engr.*, **9** (1972), p. 233, by permission.]

Other mechanical properties (particularly strength) are affected over these different ranges of prestrain, as discussed later (Fig. 13-13), and other polymers may be affected differently.

Exercise 13-3

The following values of elastic modulus E were obtained by measuring the deflection of a cantilever beam ($\delta = PL^3/3EI$) of polyethylene with a constant load on its end, after the times indicated.

Time, s	30	60	90	120
E, MPa	24	22	20	19

Explain why the calculated values of E change with time.

Solution.   With more time the viscoelastic deformation increases (Chap. 12). Thus the tip deflection of the cantilever is greater for the same load, so in $\delta = PL^3/3EI$ the apparent E must decrease.

The example indicates a need to standardize rates of loading, time under load, and so on; hence it is customary with polymers to quote a *ten-second* modulus, which relates deflection and load 10 s after reasonably rapid application of the load. The data in Fig. 13-6 would typically be 10-s moduli.

BRITTLE BEHAVIOR AND CRAZING

Polymers much below their T_g behave in a brittle fashion, with virtually linear stress–strain curves up to fracture. Rate effects are not very marked. At room temperature polystyrene and PMMA are typical examples. Both these solids display a characteristic called crazing (Fig. 13-8).

Crazes, which were first observed in 1949 by Sauer and coworkers in PS, appear to be tiny, widely distributed, cracklike defects from which bright reflections are produced. They are *not* true cracks, however, but rather regions of highly plastically deformed material interspersed with voids. The microstructure of a craze resembles that of a sponge and when stressed appears as a set of oriented microfibrils of diameters 20–40 nm elongated in the direction of maximum tensile stress. The formation of crazes in glassy polymers is the localized plastic response to stress concentrations at stresses lower than the overall yield stress (see next section); unless suppressed by compressive hydro-

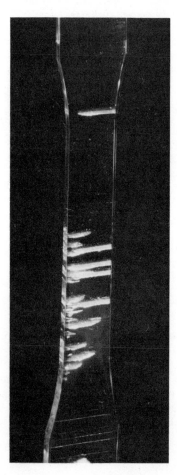

Figure 13-8 Crazes in tensile test piece of polystyrene. [From R. P. Kambour, *Macrom. Rev. I*, **1** (1973), by permission. Photograph courtesy R. P. Kambour.]

static stress fields, crazes often form at a tensile stress level of about $\sigma_y/2$. The formation of crazes is linked to the tensile hydrostatic component of stress and cavitation processes.

That crazes are not simple cracks is confirmed by the fact that test pieces cut from highly crazed material show respectable strengths and do not, as might be expected, have little strength. In real cracks in glassy polymers, crazed structures are often found in the highly deformed region around the crack tip (see Chap. 14).

YIELD AND FLOW STRESS

There is some confusion in terminology associated with the mechanical behavior of polymers, a result of the lower moduli of polymers compared with, say, metals and of the phenomenon of the drawing of stable necks that sometimes occurs for polymers (Chap. 12). It is necessary to spell out these differences fully, to avoid misunderstanding the polymer strength data given in handbooks and the scientific literature.

Engineers have traditionally differentiated between yield stress and ultimate tensile strength in metals, but in polymers these terms are sometimes confused.

The common definitions of yield strength (S_y) and tensile strength (S_u) of ductile metals are illustrated in Fig. 13-9 with reference to a load–extension (or nominal-stress–nominal-strain) curve for an annealed low-carbon steel. For all practical purposes, yield strength, elastic limit and limit of proportionality are given by the same stress S_y, where the associated strain is most often less than about 0.2%. There is essentially no change in cross-sectional area between 0 and Y, so that nominal stress and true stress are the same at yield.

The maximum load point U, at which an unstable neck initiates, gives the ultimate tensile strength (or tensile strength) S_u. The associated strain is numerically equal to the work-hardening index n (Chap. 2) and might be quite considerable, say $n = 0.2$ or about 20% reduction in area. There is a marked but uniform reduction in cross-sectional area between Y and U; the true stress at ultimate σ_u is thus greater than S_u. The unstable neck at U *always* leads to fracture in the region where this neck first develops.

The flat portion of the curve at Y is really a series of ripples in the load trace, with the associated propagation of Lüders bands (see Chap. 12). With certain materials and testing arrangements, a stress spike can sometimes occur at Y, giving an upper and lower yield point. For design purposes, S_y as shown in Fig. 13-9 is usually the important strength parameter. As discussed in Chap. 2, S_y may be increased by cold working; when the amount of prestrain exceeds n, the yield and maximum load points coincide, with immediate necking.

Most other metals display nominal stress–strain curves like the one shown in Fig. 13-10, where there is a gradual transition from elastic to plastic

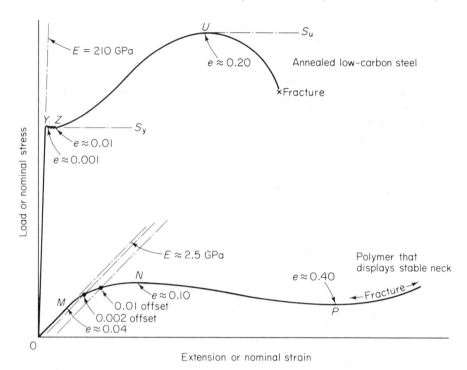

Figure 13-9   Nominal stress–strain curves for a polymer that displays stable neck propagation ("cold drawing") as compared with an annealed plain low-carbon steel that exhibits a pronounced yield point. [From C. S. Lee, R. M. Caddell, and A. G. Atkins, *Mat. Sci. and Engr.*, **18** (1975), p. 213.]

behavior. The flat region at Y in Fig. 13-9 is not in evidence, and the yield point is defined by an offset (proof) method. The associated yield strain is still of the order 0.2%. The reduction in area to ultimate can be as high as 40% for those metals, such as some brasses and stainless steels, which have $0.35 \leq n \leq 0.5$. Furthermore, S_y can again be increased by cold working.

A typical load–extension (or nominal stress–strain) plot for polymers such as polyethylene, nylon, or polyvinyl chloride, which "draw" at room temperature, is shown in the lower half of Fig. 13-9; this behavior was discussed in Chap. 12, where the reasons for the phenomenon in terms of strain-rate sensitivity were investigated. There is a departure from linearity at M (where the strain may be some 1%), and the load curve rises to a local maximum at N (where the strain may be 10%), at which point the stable neck initiates. The load then falls as the neck is reduced in cross-sectional area, until stability is reached and the neck propagates along the test piece at the essentially constant load P. Subsequently, after the neck has propagated the length of the test bar, the load increases again and fracture eventually ensues under rising load, a situation akin to failure in brittle materials. The particular

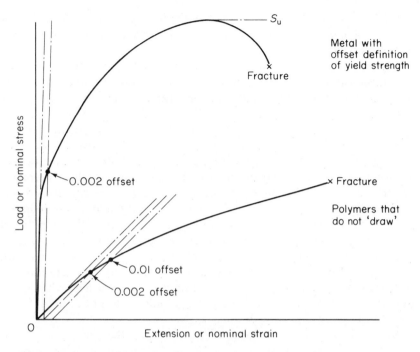

Figure 13-10 Nominal stress–strain curve for a polymer that does not display localized necking (i.e., does not "cold draw") as compared with a ductile metal that does not exhibit a pronounced yield point. [From C. S. Lee, R. M. Caddell, and A. G. Atkins, *Mat. Sci. and Engr.*, **18** (1975), p. 213.]

polymer fracture load may be less than or greater than the load at N, depending on the interaction of the geometry of the propagated neck and the original shoulders of the test piece. If the drawn section is cut out and a new test piece made from it, the strength of this highly oriented material becomes an extrapolation of the *true-stress–true-strain* curve beyond P.

After the local load maximum at N, measurements and definitions of strain in terms of *length*, such as the nominal strain given by $(L - L_0)/L_0$ or the "draw ratio" L/L_0, are quite meaningless because the reduction is nonuniform. Clearly, *any* value is possible depending on the reference gage length (the percentage elongation in tensile tests of metals is likewise meaningless without reference to the starting gage length). It is most sensible beyond N to measure the strain in terms of percentage reduction of area $(A_0 - A)/A_0$, or true strain $\epsilon = \ln(A_0/A) = 2 \ln(D_0/D)$. Then during most of the stable neck propagation, changes in true stress and true strain are minor until the neck has propagated to the shoulders. It is worth noting that, if the initial neck formation is very sudden, it may be very difficult experimentally to obtain points between N and P. Thus, in Fig. 13-14 (later), it was impossible with micrometers to pick up the neck-formation strains over the range 0.05 to 0.50 for the as-received polycarbonate. The region NP in Fig. 13-9 and its growth of *one* stable neck corresponds with the region YZ on the low-carbon steel curve, where there are many local maxima and minima in the load, corresponding to the propagation of many Lüders bands.

Polymers that do not draw (such as polymers containing cold work) have continuously rising load–extension (nominal stress–strain) curves, similar to the one in Fig. 13-10 for metals, but the strains are larger and the moduli smaller.

The deformation and strains at N and P (Fig. 13-9) in ductile polymers that draw are much greater than those at Y and Z. Moreover, the range of strain between M and N is extremely large compared with the corresponding regions for ductile metals that occur near Y. The point at which permanent deformation sets in is somewhere between M and N. The curvature over a large range of strains between M and N and the local maximum in load at N have led some workers to call N the ultimate tensile strength for polymers. Reference to Fig. 13-9 shows that this is erroneous, if we mean that ultimate tensile strength is followed by an unstable neck and fracture, as is the case with ductile metals. However, a point such as N *may* indicate the maximum *load*, and in that context the use of "tensile strength" is consistent with a long-standing definition. However, it is clear that N and P do not represent upper and lower yield points in the traditional sense.

Others identify N as the local yield point, and the similarity of events between N and P and those between Y and Z makes such an approach justified. However, the strains at N are much larger than at Y, and from a design point of view M may be more meaningful as a limiting strain; even then, deformations are greater than traditional metal yield strains. However, the principal objection to using N as a yield value relates to those polymer load–extension curves, such as in Fig. 13-10, that show *no* load maximum. The offset definition of yield allows a consistent method of determination whatever the shape of the load–extension diagram.

Exercise 13-4

Tangled string and spaghetti are analogies for long-chain polymers. Explain the mechanical behavior for which each model is appropriate.

Solution. The tangled or knotted string analogy describes rubbery behavior, where the material may be stretched appreciably but then becomes stiff and cannot be moved further without breaking some of the string (compare a rubber band). The spaghetti analogy describes the situation where continuous appreciable sliding is possible between the chains, so that viscous flow comes into play as well as elasticity, which it eventually swamps at higher and higher temperatures.

Viscous sliding is more inhibited the longer the chain length of simple linear polymers, and it is particularly restricted when there is cross linking between the chains. Thus the spaghetti model reverts to tangled string. The vulcanization of rubber, where sulfur is added to produce cross linking, allows a range of properties to be produced ranging from soft, compliant natural rubber to the harder, stiffer tire rubbers.

Appendix 8 gives some representative mechanical properties for various polymers. The two groupings in the table relate to thermoplastics (first part) and thermosets. These limited data must be used with caution and are merely indicators of mechanical properties; most of the thermoset data happen to relate to reinforced ("high-impact") grades. Processing conditions, strain rates, temperature, number and type of side branches, and so on affect the results. The glass-transition temperatures are only guides; in the case of partially crystalline polymers, T_g relates to the amorphous phase. Unlike the case with metals, it is not really possible to compare, except in broadest terms, a given polymer (of which there may be numerous grades) with another polymer from these few data. The trade names given are in no way exhaustive, different suppliers having other names. Furthermore, manufacturing processes have changed over the years so that, for example, commercial PMMA is slightly different from what it formerly was. Similarly, the designations of high- and low-density versions of PE have lost much of their meaning.

Elastomers are not covered in Appendix 8. Their properties vary widely, with elastic moduli between 1 and 4 MPa and tensile strengths between about 5 and 30 MPa. The various names and types of synthetic rubber include SBR (styrene butadiene copolymer), butyl rubber (isobutene/isoprene), neoprene (chloropene), silicone rubbers (substituted silicones), urethane rubbers (polyester urethanes), and thiokol (polysulphides).

Both polymeric and nonpolymeric additives, such as asbestos, cellulose, fabric, flock, glass, graphite, and wood-flour, are used in commercial polymers in amounts up to 50%. These are over and above the small amounts of additives employed to protect against environmental attack or as pigments and dyes. Polymeric additives are often used in the form of fine dispersions of rubbery particles in a matrix of stiffer polymer; this composite structure promotes resistance to crack propagation and can give high impact strength, ABS being the classic example. In a different manner up to 15% of PTFE can be added to nylon to improve wear resistance. Nonpolymeric additives increase the moduli of polymers but can sometimes reduce the impact strength. The coefficient of thermal expansion is often reduced, and this is of benefit not only in engineering design but also in reduced shrinkage during the solidification and cooling stages in molding. The most successful type of additive for engineering plastics is the nonpolymeric fiber (carbon, glass, and so on). Chapter 17 concentrates on such filament-reinforced composites. Figure 13-23 (later) shows how the incorporation of glass fibers affects creep data for polymers (in this case polypropylene).

THE EFFECT OF MECHANICAL PROCESSING ON PROPERTIES

If polymers are rolled, drawn, or extruded, their properties can be changed. This behavior is similar to that of ductile metals, except that the trends are not always the same. The response of high-density polyethylene sheet subjected to rolling at room temperature may be used as an illustration. Rolling here is used in the same sense as rolling of metals and should not be confused with the calendaring process used to produce the sheet from the raw polymer in the first place. Polyethylene is a material which, in the as-received condition, displays stable neck growth with stress-whitening in the neck. The tensile strength at neck initiation is about 21 MPa, when the associated true strain is about 0.1. The first difference between the rolling of a ductile metal and of this polymer is the amount of elastic recovery on exit from the mill; the thickness of the rolled strip is greater than the roll gap setting. The recovery takes place

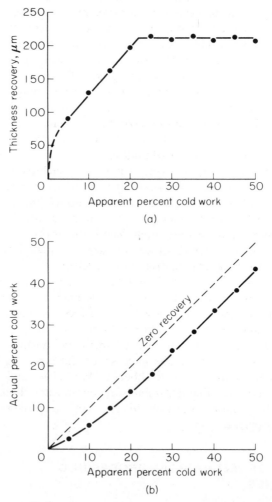

Figure 13-11 (a) Thickness recovery after one week as a function of apparent cold work. (b) Comparison of actual vs. apparent percent cold work of high-density polyethylene. [From R. M. Caddell, T. Bates, and G. S. Yeh, *Mat. Sci. and Engr.*, **9** (1972), p. 223, by permission.]

over a period of time but eventually stabilizes out. Figure 13-11(a) shows the thickness recovery after one week for PE sheet of starting thickness 3.3 mm rolled with rolls 75 mm in diameter to different nominal percentage reductions in area (as defined by the roll gap setting). As the absolute recovery does not increase beyond some 25% cold work, this means that the actual cold work is directly correlated with the apparent cold work in the manner of Fig. 13-11(b). With other polymers the recovery shows a general correlation with T_g, being smaller the higher the glass-transition temperature. This is sensible in terms of recovery and relaxation behavior.

Tensile tests on PE specimens containing up to 50% cold work show that the onset of necking and stress-whitening is retarded until larger strains (Fig. 13-12) for greater cold work; note that higher strain rates reduce these strains at ultimate load. However, the tensile strengths (maximum load divided by starting gage cross-sectional area) in such tests on PE stay comparatively constant for a given testing strain rate; the true stress at ultimate is, of course, much greater for larger cold work, since the uniform area reduction to ultimate is greater, as shown in Fig. 13-13. Data for two crosshead speeds varying by two orders of magnitude are shown, out to ultimate load only in this figure; that is, the endpoints correspond with $d\sigma/d\epsilon = \sigma$. This behavior should be contrasted with the behavior of ductile metals (Chap. 2), for which prior cold work increases the tensile strength and reduces the true strain at ultimate. The effect of cold work on the elastic modulus of rolled PE has been mentioned

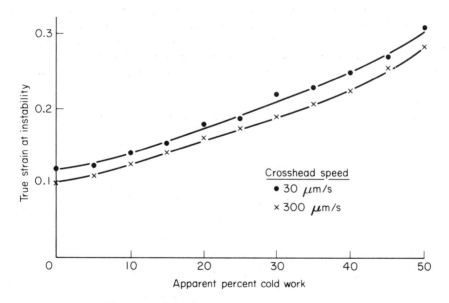

Figure 13-12 Effect of cold work on the true strain at tensile instability for two strain rates. [From R. M. Caddell, T. Bates, and G. S. Yeh, *Mat. Sci. and Engr.*, **9** (1972), p. 223, by permission.]

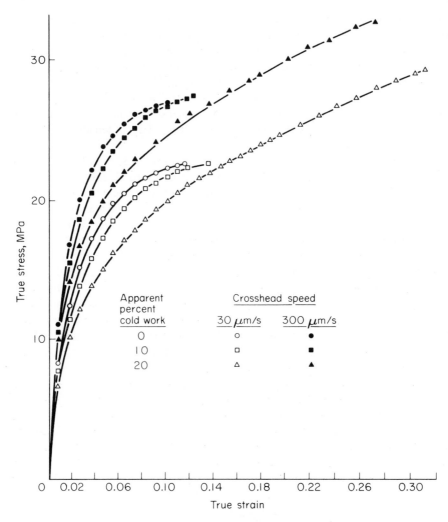

Figure 13-13 Typical true-stress–true-strain behavior of polyethylene cold-rolled to different amounts and tested at two strain rates. [From R. M. Caddell, T. Bates, and G. S. Yeh, *Mat. Sci. and Engr.*, **9** (1972), p. 223, by permission.]

earlier, and the change in behavior at 25% cold work seems to relate to breakup of crystallites at heavy reductions in area.

What does happen for increasing prestrains is that the load drop (Fig. 13-9) associated with neck formation, which bottoms out when the neck stabilizes and begins to draw, gets progressively smaller. The drawing load in as-received PE might be about 0.6 of the maximum load, whereas at 50% cold work it might be 0.8. With some other polymers, necking and the load drop are eliminated at smaller amounts of prestrain, for example, at some 30% cold work for polycarbonate and polysulfone, and about 40% cold work for ABS and PPO when unidirectionally rolled. Delay or elimination of necking is extremely useful for deep-drawing operations in sheet, which may thereby be taken to much larger strains.

Extrusion and drawing have similar effects. In the case of polycarbonate, cold working by extrusion eliminates necking, and the tensile strength thereafter increases from 66 MPa in the as-received condition through 81 MPa at 18% reduction to about 120 MPa at heavy reductions. The fracture strains continuously decrease with increasing cold work, but the true stress at fracture increases at first and subsequently drops (Fig. 13-14). In this figure, points satisfying the necking criterion $d\sigma/d\epsilon = \sigma$ may only be located on the as-received σ vs. ϵ curve, which is therefore the only condition that displays stable neck drawing; note, as remarked earlier, that no data were picked up over the strain range $0.05 \sim 0.50$ in this case, because of the rapidity with which the specimen necked.

After extrusion polycarbonate, nylon 6-6, and PE are all weaker in compression than before, and pronounced directional effects are introduced by the

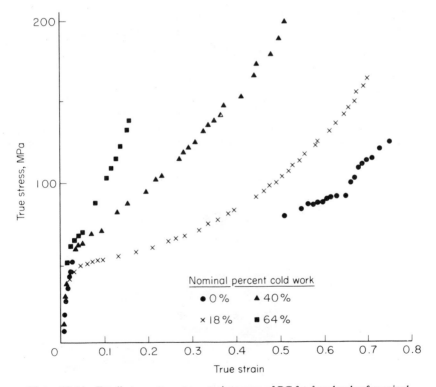

Figure 13-14 Tensile true-stress–true-strain curves of PC for four levels of nominal percent cold work by extrusion. [From C. S. Lee, R. M. Caddell, and A. G. Atkins, *Mat. Sci. and Engr.*, **18** (1975), p. 213.]

forced flow through the throat of the die with axial material being weaker than material transverse to the die. This may be associated with buckling of oriented chains.

Extrusion increases the elastic modulus from 2.4 GPa for as-received rod through 2.7 GPa for 18% reduction up to some 3.6 GPa for 64% reduction in area. This is of interest to designers, but, unfortunately, small amounts of cold working (18%) produce substantial decreases in 1% offset yield stress, from 63 MPa to 46 MPa. Although more cold work increases the yield stress above this value, it never goes above the original strength; for example, at 64% reduction it is about 62 MPa. However, if the cold-worked material is heat-treated at 100°C for a couple of hours, the yield strength may be increased to about 76 MPa with only a slight reduction in modulus. Heat treating of polymers is a relatively unexplored area, but it promises to provide means of obtaining specially desired combinations of mechanical properties, namely, in the foregoing case, higher strengths combined with higher stiffnesses.

FLOW UNDER COMBINED STRESSES

As described in Chap. 2, many stress situations in practice do not conform to the simple tests performed in the laboratory. Hence there is need to study the behavior of materials when stressed in two or more directions and to relate the results to the simple measurements for application to other cases. Sheets of material may be subjected to biaxial tension by pressurizing one side with a fluid, a procedure akin to blowing up a balloon. In the laboratory there are hydraulic machines that bulge disks into dome shapes, producing the results for various polymers shown in Fig. 13-15. The four sectioned PVC specimens at the top illustrate the sequence of doming that accompanies increasing pressure. Note the peculiar "Asian hat" shape of the PE specimen at the right; this is typical of the latter stages of bulging of this polymer.

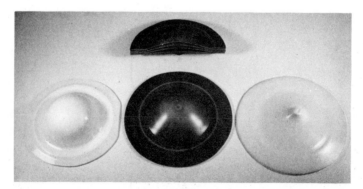

Figure 13-15 Bulged specimens of ABS/PVC alloy, PVC, and PE (left to right). Sectional specimens at center top show bulging sequence under increasing pressure. (Photograph courtesy of W. H. Durrant.)

At the pole of the dome there is balanced tension, where the stress, as in surface tension bubbles, is given by

$$\sigma = \frac{pR}{2t}$$

where p is the pressure, R the radius of the dome at the pole, and t the thickness of the sheet [Fig. 13-16(a)]. Using ideas of superposition, we may view the balanced tensile stress as a hydrostatic tension coupled with simple compression of magnitude $pR/2t$ through the sheet thickness [Fig. 13-16(b)]; the algebraic sum of the stresses in each direction gives the original equal biaxial tension. Thus a bulge test gives the same results as a *simple compression* test, providing that hydrostatic stresses do not affect flow. Devices exist for measuring the strain through the thickness, so that equivalent compressive stress–strain curves may be generated from bulging experiments, as shown for

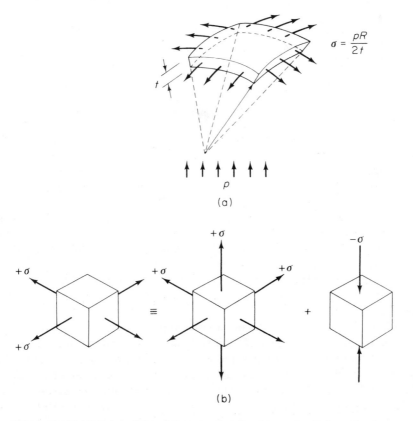

Figure 13-16 (a) Balanced biaxial tension produced at pole of dome in sheet bulged over circular hole. (b) Equivalent stress state showing that bulging is equivalent to frictionless simple compression.

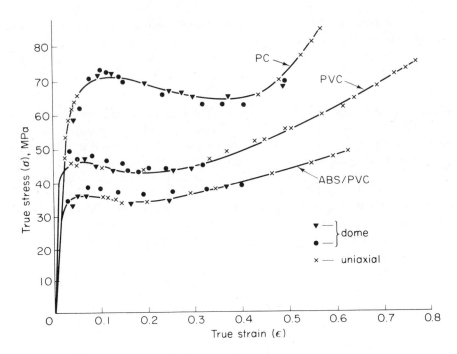

Figure 13-17 True-stress–true-strain behavior of PC, PVC, and an ABS/PVC lam-
inate subjected to uniaxial and balanced biaxial tension. [From R. Raghava, R. M.
Caddell, L. Buege, and A. G. Atkins, *J. Macromol. Sci. Phys.*, **B6**(4) (1972), p. 655.
Reprinted by permission of Marcel Dekker.]

various polymers in Fig. 13-17. Superimposed on the figure are simple uniaxial
tensile stress–strain data (see Fig. 13-14 for other PC data). The agreement is
very good, suggesting that for these polymers uniaxial tensile and compressive
behavior are similar. Note that bulging experiments produce equivalent compressive stresses unaffected by friction; in this they differ from conventional
compression tests, which have platen friction.

The rise and fall in the curves display the characteristics discussed earlier,
particularly the growth of stable necks, which develop into thin disks at the
pole (Fig. 13-15).

Figure 13-18 shows the true stress–strain behavior of high-density PE
subjected to uniaxial and balanced biaxial tension (equivalent to uniaxial compression). Here there is *not* the same agreement between bulge data and tension data (note that the strains in this figure extend to much larger values than
in Fig. 13-13, which went merely as far as neck initiation). The reason concerns
the fact that flow in many polymers is really pressure-dependent, that is,
compaction makes it more difficult for them to flow in tension, or hydrostatic
tension makes it more difficult for them to flow in compression. The assumption about equivalent stress states is not valid in such cases; hence the discrep-

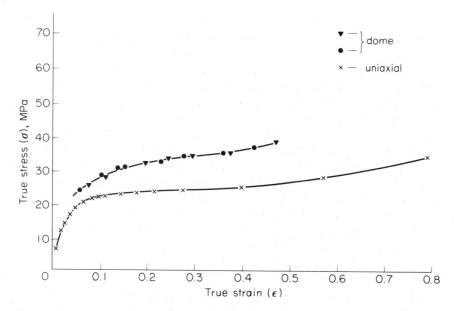

Figure 13-18 True-stress–true-strain behavior of high-density PE subjected to uni-axial and balanced biaxial tension. [From R. Raghava, R. M. Caddell, L. Buege, and A. G. Atkins, *J. Macromol. Sci. Phys.*, **B6**(4) (1972), p. 655. Reprinted by permission of Marcel Dekker.]

ancy in Fig. 13-18 for PE, in which the effect is more marked than for the polymers in Fig. 13-17.

Experiments to investigate the pressure dependency of flow in polymers may be performed under combined stress states by means of thin-walled tubes subjected to combinations of axial loading, internal or external pressure, and torsion. *Yield criteria* (i.e., which combinations of σ and τ on various planes and in various directions will cause yielding) may be established from the results of such experiments.

DESIGN WITH TIME-DEPENDENT SOLIDS (POLYMERS AND METALS)

When deformations due to creep are only minor, and when an article has to operate under load for less than 1 ks, say, the modulus measured in a tensile test at constant strain rate may be used in standard strength-of-materials formulas in order to arrive at suitable proportions. Many room-temperature plastics design calculations can be performed in this way, and the usefulness of the data may be modestly extended in scope to cover a range of test temper-atures, but always when the time scale is short-term. For longer durations of loading, where time-dependent effects are more marked, consideration must be given to the mechanical behavior of metals and polymers expressed in terms of *creep*, *relaxation*, and *stress rupture*. As explained in Chap. 12, creep relates to the increase of strain with increasing time at constant stress; relaxation relates to the decrease in stress at constant strain with increasing time; and stress rupture data give the times for which a body may sustain a given stress before breaking. These three types of information are appropriate to different design criteria, as follows:

1. Creep strain data are appropriate for situations where the maintenance of accurate sizes is important (e.g., turbine blades, casings, fittings).

2. Stress relaxation data are appropriate for situations where stress levels must be maintained (e.g., bolts in tension, washers, flanges in compression).

3. Stress rupture data are appropriate for situations where breakage must be avoided but where shape stability is not vital (e.g., bottles, pipes, pressure vessels, and tubes).

Figure 13-19(a) shows room-temperature tensile creep curves for a PVC compound, and Fig. 13-19(b) gives the stress-relaxation cross plot derived from the creep data. It is also possible to perform the other cross plot of stress vs. strain at constant times, but as it is not so useful in design, it is not shown here. Similar sets of curves apply at other temperatures. The fact that the creep strain–time curves at low stress levels are nearly horizontal for short loading times in Fig. 13-19(a) is the justification for using constant rate moduli in strength-of-materials design stress calculations when the time scale is short.

From creep curves such as in Fig. 13-19(a) may be derived *creep modulus* data. The creep modulus is defined as stress ÷ creep strain at some chosen strain level and may be plotted against time in order to display changes in stiffness. Figure 13-20 shows creep moduli for a different plastic (an acetal copolymer) at two temperatures. The 10-s modulus mentioned earlier is but one point on such curves.

If the right-hand end failure points are taken from creep curves such as those in Fig. 13-19(a) and plotted as breaking stress vs. time to failure at that stress, stress-rupture curves are produced. Figure 13-21 gives such curves for a high-density polyethylene compound. When using such data for polymers, care must be taken to note that precise values depend on processing conditions and other factors, so that no single failure curve is characteristic of a particular material. A variation on the stress-rupture approach is to postulate different "failure" criteria (such as some maximum allowable strain) connected with the evidence of optical failure by crazing, or avoidance of excessive deflections, and so on.

Similar types of creep, relaxation, and stress-rupture curves are available for creepy metals, as shown, for example, in Fig. 13-22, which relates to a chrome–molybdenum steel at 600°C.

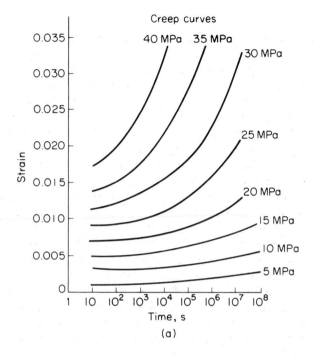

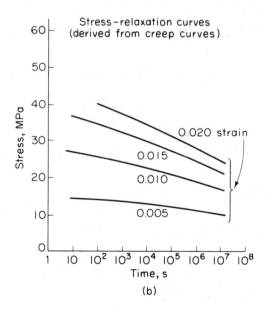

Figure 13-19 Tensile creep at 20°C of a PVC compound. [From S. Turner, *Plast. Rub. Mater. and Applications*, **2**(3) (1977), p. 90, by permission.]

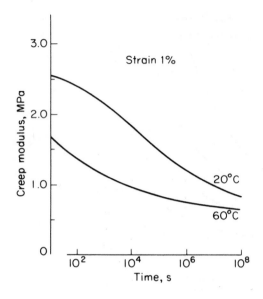

Figure 13-20 Creep modulus vs. logarithmic time for an acetal copolymer. [From P. C. Powell, *Engineers Digest*, **38**(6) (1977), p. 13, by permission.]

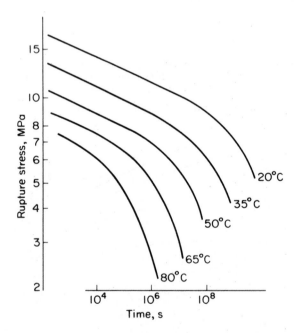

Figure 13-21 Creep rupture data for an HDPE pipe compound. [From P. C. Powell, *Engineers Digest*, **38**(6) (1977), p. 13, by permission.]

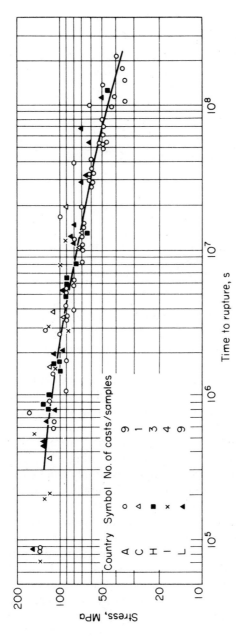

Figure 13-22 Creep-rupture data for 2.25% CrMo steel at 600°C. Test results from five countries embracing a total of 26 casts. Scatter of data falls within 20% of mean line. [From M. P. E. Desvaux, *Engr. Mater. and Design* (1976), p. 38, by permission.]

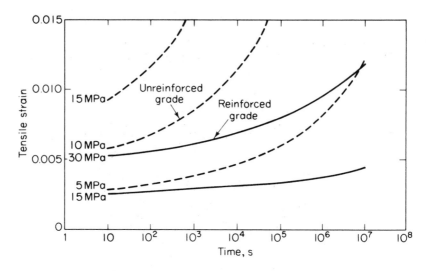

Figure 13-23 The effect of reinforcing fibers on the tensile creep of polypropylene at 23°C. [From P. L. Clegg and S. Turner, *Metallurgist and Materials Technologist,* 7(12) (1975), p. 618. Courtesy *The Metallurgist and Materials Technologist.*]

As mentioned earlier in this chapter, the incorporation of fibrous additives affects the creep behavior of polymers. Figure 13-23 gives data for a plain and a fiber-reinforced polypropylene. It is seen that the presence of fibers changes the magnitude of the strain response to a stress but does *not* change the basic time dependence to any significant degree. In contrast, fibers reduce the creep sensitivity to temperature quite strongly, mainly because the modulus of the fibers remains constant while the modulus of the matrix falls, slowly in some regions but by perhaps two orders of magnitude as the temperature rises through the glass-rubber transition regions (the region associated with main-chain mobility). Thus the effective reinforcing efficiency of the fibers increases as the temperature rises, though this is only true where the coupling between fibers and matrix is good; where the coupling is poor, the mechanical properties of a fiber-reinforced grade are worse than those of the unreinforced grade at high temperatures because the matrix expands away from the fibers and the overall performance becomes that of a matrix with holes.

The time dependence of mechanical properties introduces several complications into the design process for thermoplastics, such as calculation difficulties when the applied forces fluctuate. Even in simple cases, the fact that elastic moduli and yield stresses have been replaced by viscoelastic functions means that the designer must choose the appropriate datum from an array of values rather than use a simple established constant. An associated difficulty is that of comparing the cost-effectiveness of different materials. For instance, the relative cost per unit stiffness of two materials with rather different viscoelastic behavior will depend on whether service stresses will act for short or long periods and on the service temperature. Much of the market conflict between polypropylene and high-density polyethylene turns on their similar yet distinctly different viscoelastic characteristics. The short-term modulus values of the two classes of material are similar at room temperature (the actual magnitudes depending on the make and grade), but the modulus ratio will change and favor polypropylene for stresses of long duration and for elevated temperatures. The superior load-bearing capability of the polypropylenes under stresses of long duration is part of the reason why beer crates and soft-drink crates are molded predominantly out of polypropylene in Great Britain and predominantly out of high-density polyethylene in most of the rest of Europe; in British breweries, crates are stacked very high to meet seasonal peak demand, whereas in Continental breweries the stacks are generally lower, because the demand is steady.

Moment-by-moment design calculations for creepy solids under varying stresses are really very complicated, as the compliances vary with time and with stress, so that the strain response depends on the history of loading. For related reasons (which are beyond the scope of this book) proper stress-relaxation data are *not* strictly the same as cross-plotted creep data. For most practical purposes, however, the differences are not marked unless a polymer is being used near its transition points (e.g., the glass transition), when moduli and Poisson's ratios may change by an order of magnitude within a short period of time.

In order to simplify safe design procedures, a "worst-case" principle is often employed, which amounts to using the least stiffness shown by the body in strength-of-materials calculations. The steps in the procedure are:

1. to identify the maximum service temperature and the maximum (continuous) duration for which the maximum load is applied;
2. to calculate the maximum stress(es) in the proposed design;
3. to read, interpolating as necessary, the strain from the appropriate creep curve relevant to the worst case;
4. to calculate the creep modulus;
5. to use this creep modulus in the formulas for elasticity to predict deflections, deformation, stability, and load relaxation.

For example, a polyacetal cantilever is 100 mm long by 6 mm wide and carries a constant force of 10 N at the free end. What beam depth for a rectangular cross section is required so that the tip deflection should not exceed 2 mm over 2 years at 20°C? We require a modulus E, which depends on strain, time, and temperature, to use in the strength-of-materials expression for cantilever deflection, namely, $\delta = PL^3/3EI$, where δ is the deflection, P the load, L the length of the beam, and $I = th^3/12$, the second moment of area (moment of inertia) of the section of thickness t and depth h. Assuming that strains in bending are small, we may use a 1% strain creep modulus taken from Fig. 13-20. At 20°C the modulus after 2 years (63 Ms) is some 850 MPa. Thus $2 \times 10^{-3} = 10 \times (100 \times 10^{-3})^3/3 \times 850 \times 10^6 \times I$, from which $h = \sqrt[3]{12I/t} = 15.8$ mm. This gives a beam L/h ratio of 6, which is arguably on the borderline of strength-of-materials applicability. Nevertheless, it is clear the beam is overstiff in the short term in order to be adequately stiff after long periods.

Exercise 13-5

A polyacetal beam spring consists of a quarter-circle of radius 30 mm, width 6 mm, and depth 3 mm. It is deflected by a fixed amount, 3 mm. In what manner does the thrust exerted by the deformed spring vary with time and temperature? (The tip thrust of a spring with such a geometry is given by $P = 4\delta EI/\pi R^3$, where R is the radius of the arc.)

Solution. From Fig. 13-20 the 1% creep moduli after 100 s, 100 ks, and 100 Ms are 2.4 GPa, 1.5 GPa, and 830 MPa, respectively, at 20°C, and 1.3 GPa, 830 MPa, and 630 MPa, respectively, at 60°C. Substitution of the fixed deflection and geometry of the beam in the expression for thrust gives $P = 1.91 \times 10^{-9}E$. Thus at 20°C, the respective tip forces are 4.5 N, 2.86 N, and 1.59 N; and at 60°C, they are 2.5 N, 1.59 N, and 1.2 N. Clearly, if a thrust such as 4.5 N is required over long periods, design based on short-term data is inadequate.

    Creep rupture design calculations proceed along the following lines. Suppose that an extruded high-density polyethylene pipe of mean diameter 100 mm is required to withstand water at 65°C for 500 hours at a maximum pressure of 600 kPa. What would be the minimum thickness of the pipe wall? Using Fig. 13-21 as an indicator of behavior, we find that the expected failure stress after 500 h (1.8 Ms) at 65°C is about 4.5 MPa. If the design factor of safety is 1.3 (typical of HDPE water pipe applications) the design working stress becomes $4.5/1.3 = 3.45$ MPa. Using the "thin-walled cylinder" hoop stress formula for the pipe, [i.e., $\sigma = pD/(2t)$, where σ is the hoop stress, p the fluid pressure, D the pipe diameter, and t the wall thickness], we have $3.45 \times 10^6 = (600 \times 10^3 \times 10 \times 10^{-3})/(2t)$, from which $t = 870$ μm.

    A design life of 50 years is required for underground gas piping made from polymers. With metals, stationary gas turbines used under steady conditions (for pumping oil and gas, for example) work for at least 20,000 hours before major overhauls; and civil aeroengines, with regular thermal cycling, go some 4000 h between repairs. Very often the cause of taking out of service is not creep but other effects (such as erosion of the blades), so many components are reused and therefore are designed for a cumulatively longer creep life.

    Because of the length of time required and costs involved in acquiring creep data, there is a temptation to extrapolate test results taken over a few hundred hours or a thousand hours out to 10^5 hours (which is about 11.4 years). Extrapolation can be a dangerous procedure, creep rupture data, for example, perhaps not falling on a curve extrapolated from short times, thereby giving a false indication of component life. The consequence of poor design in

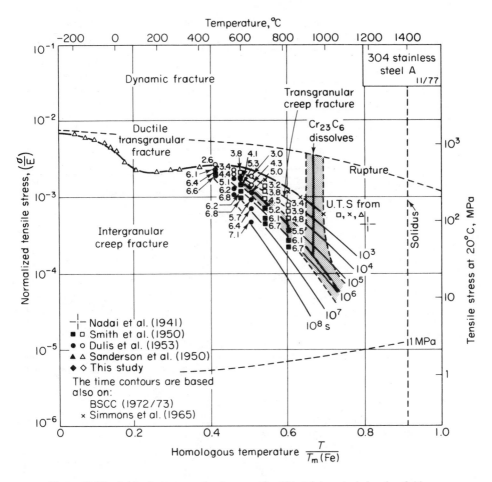

Figure 13-24 Ashby fracture mechanism map for 304 stainless steel showing fields of dominant micromechanisms of fracture at different temperatures and stresses. (M. F. Ashby and coworkers, by permission.)

boiler or turbine components based on unproved data can be expensive failures. In power plant steel, "safe" extrapolation never uses more than a factor of three between the duration of actual data and the required design stress.

Ashby and coworkers have constructed *fracture-mechanism maps* which are relevant to this matter. In their usual presentation the maps are diagrams with tensile stress as one axis and temperature as the other, showing the fields of dominance of a given micromechanism of fracture. Figure 13-24 shows a typical diagram, in this case for 304 stainless steel; further details on the mechanisms of fracture will be given in Chap. 14. Superimposed on the fields are contours of constant time to fracture. The maps may be cross-plotted as tensile stress vs. failure time with lines of constant temperature superimposed

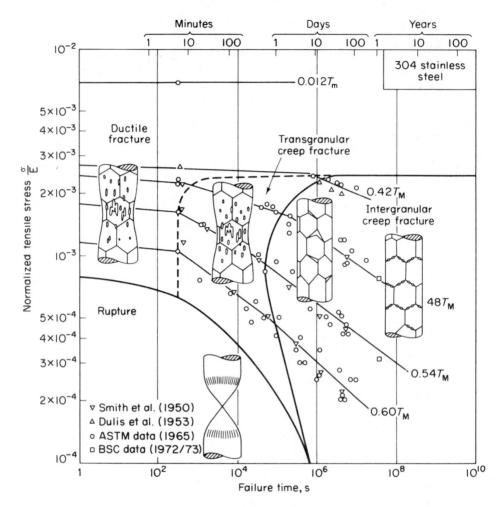

Figure 13-25 Cross plot of Fig. 13-24 as tensile stress vs. time to failure with lines of constant homologous temperature superimposed. (M. F. Ashby and coworkers, by permission.)

(Fig. 13-25). The validity of extrapolation of short-term results at constant temperature to long times concerns whether the lines of constant temperature in the likes of Fig. 13-25 have "unknown" downward kinks at long times. Such dangerous occurrences have been observed in some materials such as Type 316 stainless steel and some superalloys.

REFERENCES

13-1. Alfrey, Turner, and Edward F. Gurnee, *Organic Polymers*. Englewood Cliffs, N.J.: Prentice-Hall, 1967.

13-2. Harris, B., and A. R. Bunsell, *Structure and Properties of Engineering Materials.* London: Longmans, 1977.

13-3. Kambour, R. P., "A Review of Crazing and Fracture in Thermoplastics," *J. Polymer Sci.: Macromolecular Reviews,* **7** (1973), p. 1.

13-4. Lever, A. E., and J. Rhys, *The Properties and Testing of Plastics Materials.* London: Temple Press, 1962.

13-5. Powell, P. C., "Designing Load-Bearing Plastics Products," *Engineers' Digest,* **38** (1977), 6, p. 13.

13-6. Tebbatt, T. E., "Plastics in Engineering," *Engineering Materials and Design,* 1976, 9, p. 36.

13-7. Treloar, L. R. G., *The Physics of Rubber Elasticity,* 3rd ed. Oxford: Clarendon Press, 1975.

13-8. Ward, J. M., *Mechanical Properties of Solid Polymers.* New York: Wiley-Interscience, 1971.

13-9. Williams, J. G., *Stress Analysis of Polymers.* New York: John Wiley/Halstead, 1973.

PROBLEMS

13-1. Poly(α-methylstyrene) exhibits a higher glass-transition temperature than does polystyrene. Explain. (The structures of these two polymers are shown in Fig. 13-26.)

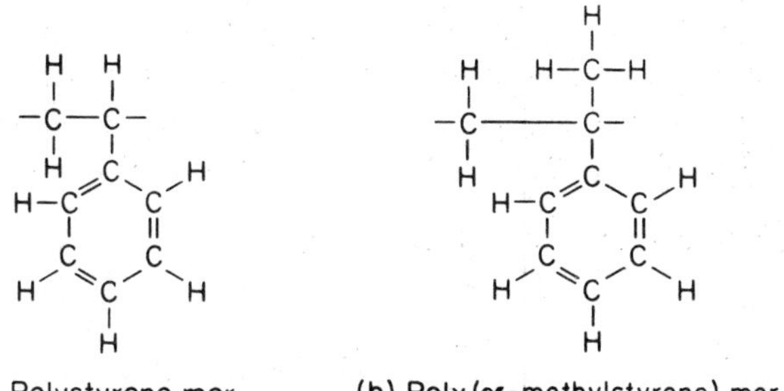

(a) Polystyrene mer   (b) Poly(α-methylstyrene) mer

Figure 13-26

13-2. Describe briefly how increasing each of the following will change the strength of a polymer:
   (a) Crystallinity.
   (b) Cross linking.
   (c) Branching.

13-3. Thin samples of translucent penton (chlorinated polyether) on copper strips are heated on a hotplate, then cooled either by water quenching or by air cooling. Qualitative indications of the ductility and hardness of the samples cooled in different ways may be obtained by bending the supporting copper strips. Describe any likely differences in mechanical properties and explain their origins.

13-4. Using the four-element model of a polymer, describe the effect of
    (a) Increasing strain rate on measured elastic modulus.
    (b) Increasing temperature on measured elastic modulus.
    (c) Long-term constant stress on deformation.

13-5. On the coordinates in Fig. 13-27 is drawn the curve for the four-element model of a polymer. Draw on these same coordinates the curve for the same polymer at a slightly lower temperature (say, 20°C lower) and at the same applied stress.

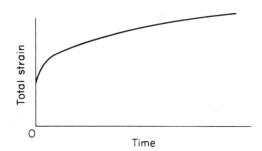

Time   **Figure 13-27**

13-6. A cantilevered polyethylene beam exhibited an elastic modulus under a 200-g load of 29.6 MPa after 20 s, whereas under a 500-g load for the same time the elastic modulus was 27.7 MPa. Why are the moduli different?

13-7. The rebound height of bounced rubber balls reached a minimum when tested over a wide temperature range. With what property, independently obtained, is this minimum best correlated?

13-8. What is the approximate relationship between the temperature (call it T_a) of maximum energy absorption in a polymer and its glass-transition temperature (T_g)?

13-9. Describe briefly two important changes in the mechanical properties of polymers that occur (a) when the temperature is changed from 20°C below the glass-transition temperature (T_g) to 20°C above T_g, and (b) when the temperature is changed from 150°C above the glass-transition temperature to 10°C below T_g.

13-10. The minimum rebound for a ball of butyl polymer occurred at about -18°C. The minimum rebound for a natural rubber ball was at about -40°C.
    (a) Which would you guess to be the better bouncing ball at 21°C? Why?
    (b) Which ball has the higher glass-transition temperature?

13-11. The rebound height of rubber balls dropped from 500 mm is measured as follows:

	Material	T_g, °C	Rebound Height, mm			
			-196°C	0°C	21°C	100°C
A	S-B rubber	-63	420	120	250	350
B	Polysar butyl	-40	350	5	10	220

  (a) Which rubber will have the higher rebound at 50°C, A or B?

  (b) Which rubber will have the higher rebound at −40°C, A or B?

  (c) Which rubber would you expect to have the higher strength at −50°C, and why?

13-12. Two bending tests were made 300 s apart on a polyethylene cantilevered beam. Results are plotted in Fig. 13-28; the same load (250 g) was used each time. Explain (a) the difference in deflection and (b) the difference in shape of the curve. Consider fundamental mechanisms.

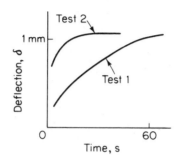

Figure 13-28

13-13. Cis-polybutadiene is used for superballs, styrene butadiene is used for regular car tires, natural rubber is used for truck and plane tires, and butyl rubber is used for inner tubes. Explain why the particular rubbers are chosen for those jobs.

13-14. Why can polypropylene trash can lids crack in winter in certain parts of the United States?

13-15. Explain why thermoplastics (particularly those reinforced with fibers or particles) can be used satisfactorily in the severe environment of engine compartments of vehicles, even though the conditions are beyond their traditional performance ranges.

13-16. A PVC component is bolted onto a metal frame at 20°C and thereby squashed to 2% compressive strain. If the torque applied to secure the part were T, what torque would be required just to begin to compress the PVC component (a) 1 day later? (b) 1 year later? (Use the data in Fig. 13-19.)

13-17. A pipe made of polystyrene has outer diameter 75 mm and inner diameter 70 mm. It is used to transport a fluid under different service loading conditions. When used in a flat, horizontal position, the service life of the pipe is very adequate; yet, in another application requiring a small bend, an identical pipe fails in too short a time. Explain the most likely reason for this difference, including the probable mechanisms that lead to failure.

14 Fracture

We have all broken things. Sometimes we want fracture to occur, as when we split wood, drill holes, plow fields, or tear paper. Often we do not want fracture to occur, as when stones hit car windshields, china vases fall to the floor, bolts snap off upon tightening, or ships break in two.

In this chapter we explore different types of fracture and the factors affecting the manner in which things break. Explanations are given for the observed discrepancies between ideal fracture strengths and those measured in practice. The concept of fracture toughness is introduced, along with the salient features of the mechanics of cracking. It is pointed out that fracture behavior can depend markedly on the method of testing or in-service conditions, so that simple classifications of "brittle" and "ductile" must be used with caution. Brittle fracture in large, fabricated structures made of materials that behave in a ductile fashion in the laboratory is discussed, and it is shown that stresses for cracking in such large structures can sometimes be much smaller than design stresses based on yielding.

FRACTURE CLASSIFICATIONS

Broadly speaking, fractures may be classified according to the amount of energy required: *brittle fracture* requires little energy, and *ductile fracture* requires much energy. In brittle fracture the broken parts usually may be refitted together to the original dimensions, with essentially no permanent deformation, the behavior being elastic. In ductile fracture, before rupture there is considerable plastic deformation that is not recovered, and often it is impossible to refit the severed parts.

Consider the simple tensile fractures shown in Fig. 14-1, which can be produced in the laboratory. Figure 14-1(a) shows the response of a rod of glass, cast iron, or blackboard chalk at room temperature. The test piece will break with great suddenness, and the fractured surface will be *normal* to the applied *tensile* stress. On the stress vs. strain diagram, OA is essentially straight; there may be small curvature in the region of A, but permanent deformation is extremely small. We can define OB as the brittle fracture stress, σ_f (represented by OT on the Mohr's circle diagram), and OC as the fracture strain, ϵ_f, which is mostly elastic. Steels at low temperatures fracture in a brittle manner, individual crystals failing by *cleavage*, that is, splitting of atom planes. In contrast, Fig. 14-1(b) shows what can happen if rods of a soft metal such as lead or indium are pulled at room temperature. Intense *shear* slides off planes of atoms (in the classical concepts of movement of dislocations) until only a vanishing cross section is left. There is permanent deformation preceding rupture throughout the test piece, as can be seen by the plastic deformation in path OD, and it is impossible to remake the original rod. The tensile stress at fracture (σ_f) is OF and the fracture strain (ϵ_f) is OG and is chiefly plastic. The driving stress in this instance is the shear fracture stress, given by the radius of the Mohr's circle, by $OS/2$.

In compression similar effects can be seen with concrete and rocks on the

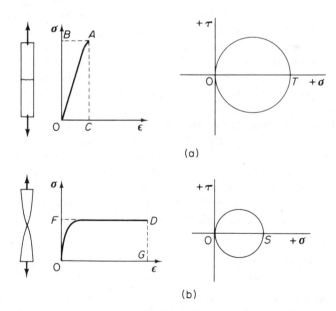

(a)

(b)

Figure 14-1 Simple tensile fractures and associated Mohr's circles: (a) perfectly brittle, produced by normal stresses; (b) perfectly ductile, produced by shear stresses.

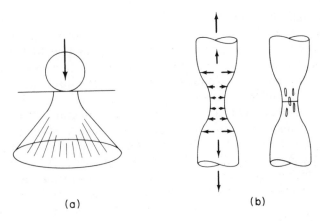

(a)   (b)

Figure 14-2 Fractures produced under three-dimensional stress states: (a) brittle cone cracks in glass under a ball indenter; (b) cup-and-cone fracture in the tensile test of a moderately ductile metal.

one hand, or ductile metals on the other, except that the overall behavior is affected by friction at the loading platens of the testing machine.

Most commonly experienced fractures involve two- or three-dimensional stress states. Then ideally brittle fractures *seem* to take place along the trajectories of maximum principal tensile stress. For example, if a hard ball is pressed into a block of glass, fracture occurs along a cone, the semiangle of which relates to the position of a peak in tensile stress [Fig. 14-2(a)]. This is analogous to the damage caused by projectiles hitting glass windows or protective shields. Ductile fracture under complex stresses can often be difficult to interpret. The commonest example known to students is the "cup-and-cone" fracture of a typical tensile test on a ductile metal [Fig. 14-2(b)]. Once the neck forms, the stress system ceases to be uniaxial tension and changes to triaxial tension. The material above and below the neck is under lower stress than the neck itself and acts to prevent radial contraction of the metal in the neck. Under these conditions three-dimensional tensile stresses *inside* the neck eventually open up internal holes around inclusions or hard second-phase particles, and when the inside area is reduced by the joining of these holes, the rest slides off around the edges. The cup is very rough from the internal necks and holes, and the cone is smooth from sliding off.

Exercise 14-1

Construct the Mohr's circle for torsion testing and demonstrate why materials that are brittle in tension sometimes yield rather than crack in torsion.

Solution. The Mohr's circle for torsion, which is pure shear, is a circle centered on the origin of stress. The Mohr's circle for simple tension is shown in Fig. 14-3, with maximum tensile stress given by OA. Brittle fracture is related to the maximum tensile stress in a system. For the same maximum tensile stress the size of the circle for torsion

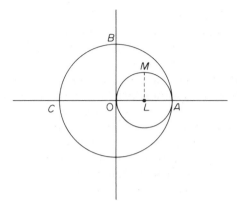

Figure 14-3 Solution for Exercise 14-1, showing Mohr's circles for tension and torsion.

will be twice as big as the simple tension circle (when the radius, OB, of the one is numerically the same as the diameter, OA, of the other). Then, however, the shear stresses in torsion are much bigger than in simple tension ($OB = 2LM$). Consequently, in torsion the material shear yield stress may be reached before its tensile fracture stress. Furthermore, the mean hydrostatic stress in torsion is "less tensile" than the mean stress in tension [in torsion the hydrostatic stress is $OA + (-OC) + 0 = 0$; in tension it is $(OA + 0 + 0)/3 = OA/3$].

FACTORS AFFECTING THE TYPE OF FRACTURE

The sort of fracture produced in a solid depends very much on conditions. The more significant controlling factors are as follows:

Decrease of *temperature* below half the absolute melting temperature tends to promote brittleness in most metals and ceramics, as does increase of *strain rate*; polymers do not always follow this trend (see Chap. 13). The state of stress is important, and although in-service situations may be complex, the mean hydrostatic stress component [i.e., $\sigma_m = (\sigma_1 + \sigma_2 + \sigma_3)/3$], sometimes called the *triaxiality*, is a good indicator to the type of fracture obtained. *Tensile* values for σ_m encourage brittle fracture. In contrast, very brittle substances such as stone can sometimes be made to flow plastically under large *compressive* σ_m. The effects of residual stresses (e.g., resulting from manufacturing processes or nonuniform heating and cooling) and prestrain on fracture can also be thought of in terms of their influence on promoting triaxiality.

Microstructure is an important factor in fracture. Broadly speaking, a coarse grain size encourages brittleness at temperatures below the recrystallization range. Likewise, aging and irradiation cause brittleness. However, some alloying elements tend to promote ductile behavior, in part because of their influence on the conditions for starting very small cracks. For those microstructural variables that are well understood, the one common factor in causing brittleness appears to be the encouragement of void formation and growth of microcracks. Hostile environments produce fractures that would not take place in clean conditions. Aspects of this, such as stress-corrosion cracking and hydrogen embrittlement, are discussed in Chaps. 18 and 19.

The interplay of strain rate and temperature influences fracture processes; that is, high temperatures and low strain rates generally promote ductile behavior in metals, but low temperatures and high strain rates generally encourage brittleness. The brittle stress–strain curves in Fig. 14-1(a) gradually change to the ductile stress–strain curves in Fig. 14-1(b) as the temperature is increased and/or the strain rate decreased. If a third axis of increasing temperature and decreasing strain rate is added to the normal σ–ϵ plot, we obtain the set of curves shown schematically in Fig. 14-4. Ashby fracture mechanism

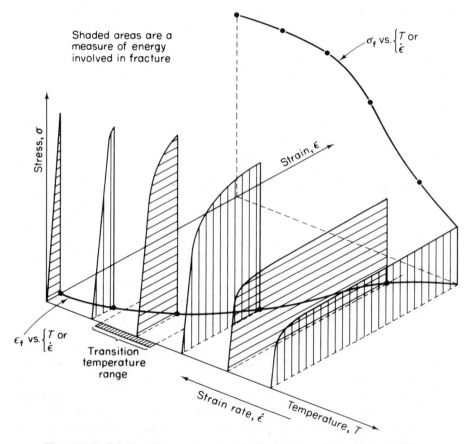

Figure 14-4 Relations between stress, strain, strain rate, and temperature in fracture. Brittle stress–strain curves transform to ductile curves at increasing temperature or decreasing strain rate (and vice versa) for most materials. Shaded areas under σ–ϵ curves are some measure of the energy involved in fracture. Change from brittle to ductile behavior takes place over a reasonably narrow span of temperature, called the transition temperature range; the position of this temperature range can vary with type of test piece and conditions of measurement.

maps were mentioned in Chap. 13 (Figs. 13-24 and 13-25). The fracture stress can be projected into the stress–temperature or stress–strain-rate plane, which shows the changes in σ_f. Similarly, the fracture strain can be projected into the strain–temperature or strain–strain-rate plane. A similarly shaped curve (with values increasing as the temperature is raised or the strain rate reduced) is obtained if the areas under the separate σ–ϵ curves are plotted against temperature/strain rate; as shown in Chap. 2 such areas are a measure of the energy per unit volume involved in flow and fracture. The change from brittle to ductile fracture in given types of test piece, that is, from low ϵ_f and relatively low fracture energies to high values, takes place over a fairly narrow span of temperature, called the *transition temperature range* (Fig. 14-5).

Notched-impact tests, rather than tensile σ–ϵ curves, are conventionally used to measure the transition temperature range. Notches produce large tensile σ_m and hence encourage brittle behavior and reduce the plastic deformation preceding fracture. The energy of a heavy pendulum, swinging under the standardized conditions, that is absorbed in fracturing the notched specimen is measured. When the specimen is broken in three-point bending, the test is called a Charpy measurement; if the specimen is mounted as a notched cantilever, it is referred to as an Izod test (see Chap. 2). The energies thus determined are a rough measure of the area under a tensile σ–ϵ curve, as distorted by the complex nature of the deformation at the notch.

The simple interpretation of the impact data is that even under severe triaxial conditions, ductile behavior will occur at temperatures greater than the

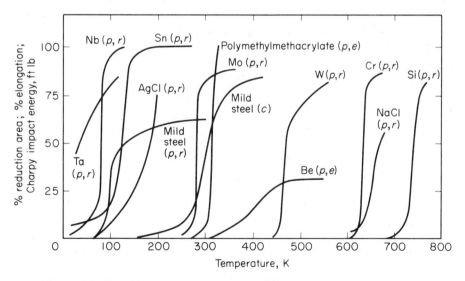

Figure 14-5 Transition temperature ranges for different materials (p, plain tension; r, reduction of area; e, elongation; c, Charpy V-notch impact energy). [From A. H. Cottrell, *Mechanical Properties of Matter* (New York: John Wiley & Sons, Inc., 1964), p. 358. Reprinted by permission of the publisher.]

transition temperature range. In general terms, therefore, the safe operating temperatures for structures must be those that are higher than the transition temperature. It must be emphasized, however, that notch-impact tests, such as the Charpy and Izod tests, are qualitative and produce no design values analogous to yield stress, for example. These tests are useful in giving a guide to the relative ranking of various materials, chiefly steels, in their resistance to failure at discontinuities and changes in section, but design against cracking ought really to relate to the material property called *fracture toughness*, dealt with in the following sections, which can give working design stresses in the presence of cracks.

Before doing so, however, let us compare fracture strengths measured in practice with those predicted by theory.

IDEAL FRACTURE STRENGTH
AND MEASURED VALUES

Can we give theoretical predictions for fracture strength? How do the experimental values for σ_f and ϵ_f live up to our expectations? If the actual behavior differs, why? In an unstressed situation the repulsive and attractive forces between atoms in a crystal balance out to zero at a particular equilibrium spacing of the atoms. If a crystal is strained, however, the net force resisting the external pull increases from zero as the atoms are moved away from their equilibrium locations. Eventually a maximum force is reached, and then the force between atoms falls off at greater spacing, as shown in Fig. 14-6. The ideal strength of the material in terms of cohesion is related to that maximum. Typically, the curve for applied force, or stress, vs. distance looks as shown in the figure. Although σ is *not* periodic, we can approximate that part of the curve of interest to us as $\sigma = \sigma_{max} \sin(2\pi x/\lambda)$ where λ is an artificial wavelength

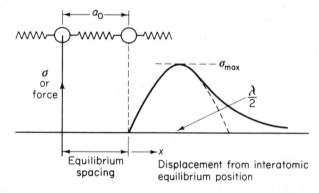

Figure 14-6 Forces or stresses generated in a crystal when the atoms are moved away from their equilibrium positions. σ is not periodic, but the region of interest can be approximated by $\sigma = \sigma_{max} \sin(2\pi x/\lambda)$.

and x the distance the atoms move from their equilibrium interatomic spacing a_0 during a tensile pull. For small x the crystal strain increments are dx/a_0, and in the elastic region around the origin $d\sigma \div dx/a_0 = E$, so that $d\sigma/dx = E/a_0 = (2\pi/\lambda)\sigma_{max}\cos(2\pi x/\lambda)$. Since $\cos(2\pi x/\lambda) \approx 1$ for small x/λ, $\sigma_{max} = \lambda E/2\pi a_0$. We do not really know what λ is, but a fair guess would be that $\lambda \approx a_0$. Hence

$$\sigma_{max} = \frac{E}{2\pi} \tag{14-1}$$

say, $E/10$. Note the similarity between this calculation for pulling atoms apart and the calculation in Chap. 5 for sliding atoms over one another; in that case, $\tau = \tau_{max}\sin(2\pi x/b)$ really is periodic.

Measured values of σ_f are usually about $E/10^4$ for most materials, unless they are in fiber or "whisker" form, in which case the measured values have been known to approach $E/15$ (Chaps. 16 and 17). In the same way that crystal dislocations were introduced to explain the discrepancy between theoretical and actual shear strengths, the presence of *microcracks* or *flaws* is postulated to explain the discrepancies between theoretical and actual brittle strengths. Statistically, a flaw of given size is less likely to occur in a small lump of material, which explains the *size effect*, according to which thin fibers are stronger than thick rods. The flaws are not figments of the imagination and have been observed in, for example, glass and low-carbon steel.

The flaws act as stress concentrators, locally raising the stresses to the theoretical level, even though away from the flaw the stresses may be low. The bonds adjacent to the crack tip can then be broken, and the crack opens. This is shown schematically in Fig. 14-7. All missing bonds go to the two atoms

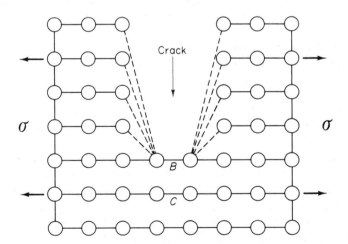

Figure 14-7   Stress concentration produced by flaw; all missing bonds are imposed on the atoms located around B and C.

labeled *B* and *C*, giving rise to a *stress concentration*. The crack eventually spreads, and the process repeats itself. Obviously, cracking is easiest when the applied stress is perpendicular to the crack, again leading to the concept of maximum tensile stresses for brittle fracture.

Exercise 14-2

When an elastic material is strained, strain energy is stored in the stretched chemical bonds; on fracture this energy is released. The chemical energy is no different from the "explosive energy." Why are tensile tests not therefore explosive?

Solution.   Real materials achieve only a fraction of their theoretical strength because of microflaws. The amount of energy released on fracture is thus only a fraction of the energy that would be produced by an equivalent mass of explosive. In the case of fibers and whiskers, which do achieve a high fraction of their theoretical strengths, there can be an explosion on testing and the fibers vanish into dust, not merely into a couple of pieces. Only the fact that their volume is small prevents their testing from being dangerous.

THE MECHANICS OF CRACKING

What happens when we load a piece of material containing a flaw? Can we predict when fracture will take place? Why do some materials yield before cracking and others crack before yielding? Can we prevent cracks from initiating and propagating?

Although many objects break with a bang and fly apart on fracture, cracking is not always necessarily catastrophic. Sometimes when an ice-covered pond is walked on, cracks propagate slowly; the growth of fatigue cracks is not fast; and there is one case where members of a ship's crew put a chalk mark each day at the tip of a growing crack in the deck. The precise response depends on the size of the piece of material, the method of loading, temperature, strain rate, environment, and so on.

Consider the problem of crack initiation and propagation in a system where the cracking loads produce general stresses in the system that are no-where greater than the yield stress. That is, the system behaves essentially as an elastic reversible body. Brittle materials behave in this fashion, although as we shall see, materials commonly thought of as ductile *can* follow this type of behavior under certain circumstances.

During loading, before the crack runs, the load deflects the piece of material containing the crack and puts *elastic strain energy* into the material—the same as energy in a spring. The magnitude of the strain energy is given by the area under the load/extension line. When a critical state is reached, the crack begins to extend; the process of cracking may be violently fast or slow. During the propagation of the crack the strain energy of the system changes, and kinetic energy can be generated if the crack runs very quickly. If all the interchanges of energy are balanced out, an amount will be found lost after fracture. In other words, *work must be done to open up the crack*. The specific work of cracking, the work done in developing the crack per unit area, is called the *fracture toughness* (R) of the solid. It is a material property, like E or yield strength, and likewise depends on rate, temperature, and environment.

Exercise 14-3

Calculate the fracture toughness R for the case where 40 mJ of energy is required to create a crack 14 mm long in a plate 4 mm thick.

Solution. $R = \dfrac{40 \text{ mJ}}{(14 \text{ mm})(4 \text{ mm})} = 714 \text{ J/m}^2$

Note that the conventional interpretation of "crack area" used in the calculation is the thickness times the growth in crack length. The actual total *new* crack area, on both halves of the cracked part, is twice this value.

Consider quasi-static cracking, cracking that is so slow that kinetic energy is negligible. Think of a split beam of material put into a testing machine, as shown in Fig. 14-8, with a crack (say, a fine saw cut) of initial length L_1 and area A_1. A typical plot of load P vs. crosshead displacement δ would be Oa prior to any propagation of the crack. The slope of Oa gives the stiffness of the cracked test piece (the reciprocal of the stiffness slope is called the compliance). Longer starter cracks make the body more compliant; that is, the slopes of lines such as Oa become smaller as the crack length or area becomes bigger. (Conversion of P into stress and δ into strain would *not* give Young's modulus for any of the lines. The loading/unloading behavior is nevertheless elastically reversible up and down lines such as Oa.) Imagine that the load increases to point b in this way [Fig. 14-8(b)], and then cracking starts.

Perhaps the crack propagates in such a way that δ remains constant, that is, the load drops down to c when the crack increases its area from A_1 to A_2 (which, since $A = L \times t$, where t is the thickness of the sheet, is the same as the crack lengthening from its starting length L_1 to L_2). This situation is shown in Fig. 14-8(c). The strain energy before the crack ran was the area Obe; after the crack has lengthened to L_2, it is area Oce, since the elastic stiffness line with longer crack L_2 is given by Oc. No extra work has been fed to the system, since the load has not moved its point of application, that is, δ_b is fixed. If kinetic energy is negligible, the area Obc has fed the crack with the work necessary to open it up. This toughness work is equal to $R(A_2 - A_1) = Rt(L_2 - L_1)$ if R is the fracture toughness. Thus $R = (\text{area } Obc)/(A_2 - A_1)$.

If, on the other hand, the crack propagates at constant load P_b, the displacement approaches δ_d as the crack lengthens from L_1 to L_2 [Fig. 14-8(d)]. Again the energy at the start is area Obe; at the finish it is now Odf, since Ocd is still the elastic stiffness line when crack length is L_2. This time, however, since P_b has moved through a distance $(\delta_d - \delta_b)$, external work has been

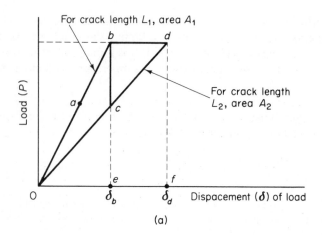

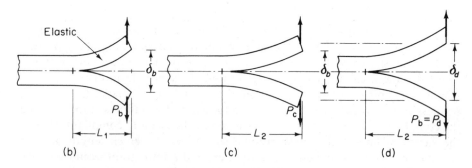

Figure 14-8 (a) Elastic load-displacement P–δ plot and energy interchanges during crack propagation. Ob is the stiffness of the body containing a crack of area A_1 (or length L_1 in a plate of uniform thickness); Od is the stiffness of the body containing a crack of area A_2 (or L_2), where $A_2 > A_1$. (b) Shape at the start of crack growth when $L = L_1$. (c) Shape at the end of crack growth (when $L = L_2$) with constant deflection δ. (d) Shape at the end of crack growth (when $L = L_2$) with constant load $P_b = P_d$. [The cracked body shown is called a *double cantilever beam* (DCB) test piece in fracture testing; see Fig. 14-11.]

done, represented by area $bdfe$, and energy has been fed into the system (i.e., to the crack *and* to the test piece).

Then cracking work = areas $(Obe + bdfe)$ − area Odf = area Obd. And $R(A_2 - A_1) = $ area Obd, so $R = $ (area $Obd)/(A_2 - A_1)$.

Notice that the constant-load case and the constant-deflection case approach the same value of work done to open the crack, as the increments of crack opening or deflection are made smaller. The work Obd for the constant-load case differs from the work Obc only by the amount bdc, which disappears as bd and bc approach zero.

In general for many geometries of cracked bodies the actual situation will usually lie between the two extremes shown in Fig. 14-8 and will follow a

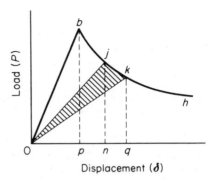

Figure 14-9 P–δ relations during cracking for many cracked bodies. $bjkh \ldots$ is the toughness locus, and the Gurney sector work area Ojk is the work done in propagating the crack from area corresponding with stiffness Oj to area giving stiffness Ok. The shape of $bjkh \ldots$ varies with cracked-body geometry and loading conditions (see Fig. 14-14).

curve such as bh in Fig. 14-9. Then, as shown by Gurney, the work done in opening up the crack from the crack area corresponding to point j to the crack area corresponding with k is the sector Ojk.

Sometimes (but not often) crack propagation takes place under *increasing* load, in which case the likes of $bjkh$ point upwards; the conical fracture ball indentation example shown in Fig. 14-2(a) happens to be one of these cases. All the foregoing arguments about sector work areas apply as before. Another example of cracking under increasing load is discussed later, in connection with Fig. 14-14.

Thus if we have load–extension–crack-length data we can find the toughness. Conversely, if we know something about a material's fracture toughness, it is possible to predict the likelihood of cracking in a given situation, or in other words to predict the *strength of cracked parts*.

Exercise 14-4

A cracked plate of a certain material was stressed until it began to crack at a constant load of 500 N. Just before the crack started to run, the deflection of the plate was 2.5 mm; when the crack stopped, it had become 3 mm.

(a) Calculate the work done in propagating the crack.

(b) Calculate the fracture toughness of the material if the crack area increased by 1200 mm^2 during propagation.

Solution. The toughness work area during propagation is a triangle, similar to Obd in Fig. 14-8, with base $(3.0 - 2.5) = 0.5$ mm and depth 500 N. (a) The work is thus $500(0.5 \times 10^{-3})/2 = 125$ mJ. (b) The fracture toughness is thus $125 \times 10^{-3}/1200 \times 10^{-6} = 104$ J/m^2.

The energy balance, which has just been described graphically, can be developed mathematically as follows:

Consider events during an increment of cracking where the external displacement of the cracked structure is $\Delta\delta$ and the associated increment of crack area is ΔA (Fig. 14-10). The external load does work $P\Delta\delta$. The crack absorbs energy $R\Delta A$. The strain energy Λ of the structure changes by $\Delta\Lambda$.

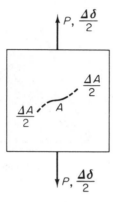

Figure 14-10 Events during an increment of cracking in which the displacement changes by $\Delta\delta$ and the crack area increases by ΔA.

The strain energy is given by a triangular area in Fig. 14-9 such as $Ojn = \frac{1}{2}P\delta$ at any instant, so that $\Delta\Lambda = \Delta\frac{1}{2}(P\delta)$.

Thus, equating the rate of doing external work to that of doing internal work for quasi-static cracking and neglecting kinetic effects, we have

$$P\Delta\delta = \Delta\Lambda + R\Delta A = \Delta\left(\frac{P\delta}{2}\right) + R\Delta A \qquad (14\text{-}2)$$

Hence
$$P\Delta\delta = \tfrac{1}{2}P\Delta\delta + \tfrac{1}{2}\delta\Delta P + R\Delta A$$

so
$$P\Delta\delta - \delta\Delta P = 2R\Delta A$$

Dividing both sides by P^2, we obtain

$$\frac{d}{dA}\left(\frac{\delta}{P}\right) = \frac{2R}{P^2}$$

that is,
$$P^2_{\text{crack}} = \frac{2R}{d/dA(\delta/P)} \qquad (14\text{-}3)$$

Equation (14-3) describes the relationship between P, δ, and A during cracking; that is, it is the algebraic expression for a line such as $bjkh$ in Fig. 14-8, which is the locus of all (P, δ) necessary to keep cracking just going. The quantity $d/dA(\delta/P)$ is the rate of change of compliance of the cracked structure with respect to increase of crack area and is given from the slope of an experimentally determined compliance vs. crack area curve; it generally changes with A for most cracked bodies.

Equation (14-3) is often written $P^2_{\text{crack}} = 2G_c/[d/dA(\delta/P)]$, where G_c is called the *critical strain energy release rate*. G_c is a measure of the work available from the body to feed the crack were the crack to propagate. In stable cracking $G_c = R$ at initiation, that is, for the first incremental sector area at b in Fig. 14-9. The idea of strain energy being released is readily understood from the constant displacement case in Fig. 14-8, as Oce after propagation is less than Obe before. However, in the constant-load case shown in the same figure, the strain energy after propagation (Odf) is *greater* than that before (Obe), so the use of the term "release rate" is misleading: when Λ increases at the same time that energy has been fed into the crack, it occurs as a result of the contribution of the external work term. For Λ to be constant along a path such as $bjkh$ in Fig. 14-9, the path would be hyperbolic ($P\delta =$ constant). Therefore, strictly speaking, strain energy is released during cracking only when $bjkh$ falls at a faster rate than an hyperbola.

The compliance of a structure can be measured for different starter crack lengths and $d/dA(\delta/P)$ developed graphically, or sometimes analytical expressions can be used. For example, consider the splitting in the middle of a thin, slender beam of thickness t and depth $2h$, as shown in Fig. 14-11(a). Let us assume that the load/deflection response can be represented by strength of materials beam theory. Then, for a cantilever beam,

$$\text{deflection} = \frac{PL^3}{3EI}$$

where L is the length, E is Young's modulus, and the second moment of area (moment of inertia) is $I = th^3/12$. This is the deflection of one arm of the test piece, so the observed deflection in splitting the beam δ will be twice this value. Thus

$$\delta = \frac{2PL^3}{3EI} \quad \text{or} \quad \frac{\delta}{P} = \frac{2L^3}{3EI} \tag{14-4}$$

The compliance is thus proportional to L^3, and a plot of experimental compliances vs. crack area would look something like Fig. 14-11(b). Since $dA = tdL$, $d/dA(\delta/P) = (1/t)d/dL(2L^3/3EI) = 2L^2/EIt$, which is an expression for the slopes in Fig. 14-11(b) at various A or L.

Therefore, in Eq. (14-1) for this situation

$$P^2_{\text{crack}} = \frac{2R}{d/dA(\delta/P)} = \frac{REIt}{L^2} \tag{14-5}$$

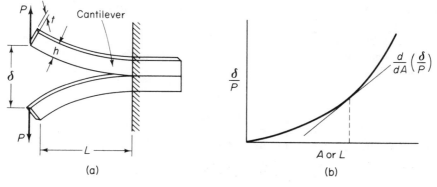

Figure 14-11 (a) Splitting a slender beam by opposing forces; (b) the associated compliances (δ/P) at various crack areas (lengths).

If the material cracks at constant R, P decreases as the crack becomes longer, and falls along a constant R locus such as *bjkh* in Fig. 14-9.

Exercise 14-5

Two strips of metal are glued together as shown in Fig. 14-12. A patch of glue was omitted over the central section. The joint is pulled apart by equal and opposite central forces as shown. At what load will the glued joint start to fail? You may assume beam theory for the metal strips, which have $E = 210$ GPa. The fracture toughness of the glue is 300 J/m².

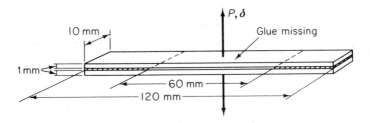

Figure 14-12   The glued joint discussed in Exercise 14-5.

Solution. Both strips above and below the unglued section will deform as *encastré* beams, the deflection of each being $PL^3/192EI$. Consequently the compliance of the glued two-beam structure is given by $\delta/P = L^3/96EI$. Using $dA = t\,dL$, we have $d/dA(\delta/P) = L^2/32EIt$, and in Eq. (14-3), $P_{crack}^2 = 64REIt/L^2$. Thus

$$P_{crack}^2 = \frac{64 \times 300 \times 210 \times 10^9 \times (10 \times 10^{-3})^2(1 \times 10^{-3})^3}{12 \times (60 \times 10^{-3})^2}$$

so $P_{crack} = 97$ N.

Another case, of great classical interest, is the so-called Griffith cracking problem, which relates to a small elliptical slit of length $2L$ in a very large plate loaded by a uniform tensile stress σ on an opposite pair of boundaries. Figure 14-13 shows the geometry and symbols used in the following analysis.

It may be shown that the stiffness P/δ of such a cracked plate (in terms of the equivalent remote boundary load $P = \sigma wt$ and the remote boundary displacement δ) is given by

$$\frac{P}{\delta} = \frac{Etw}{h(1 + 2\pi L^2/hw)} \tag{14-6}$$

where E is the Young's modulus of flawless material. Then, since the crack area in this case is given by $2Lt$, $dA = 2t\,dL$ and

$$\frac{d}{dA}\left(\frac{\delta}{P}\right) = \frac{4\pi hL}{2t^2w^2hE} = \frac{2\pi L}{t^2w^2E}$$

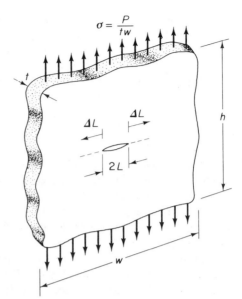

Figure 14-13 The Griffith cracking problem of a small elliptical slit in a very large plate loaded at the remote boundaries.

Hence, in Eq. (14-1), for the load to initiate cracking

$$P^2_{\text{crack}} = \frac{2Rt^2w^2E}{2\pi L}$$

or

$$\sigma^2_{\text{crack}} = \frac{ER}{\pi L} \quad \text{or} \quad \sigma_{\text{crack}} = \sqrt{ER/\pi L} \tag{14-7}$$

This is the Griffith equation for brittle fracture, written in terms of the fracture toughness R. In Griffith's experiments at the end of World War I the measured strengths of glass rods were correlated with surface flaws of different sizes (i.e., different L). For small L, σ was very high; and in almost perfect fine filaments of glass with very small L, σ approached the theoretical value for the strength of glass (approximately $E/10$).

Note the general form of the Griffith equation and the expression for splitting a beam. Both expressions suggest that a critical stress or load must be attained before cracking is initiated. The critical value squared varies directly with the modulus and toughness, and inversely with the crack length. The latter suggests a *critical crack length* below which the crack will not run, for a given applied load or stress. Equivalently the stress that that causes a known crack to propagate in an object is hence a measure of the strength of that cracked object.

Exercise 14-6

The strength of beer bottle glass is about 170 MPa. What is the Griffith flaw size associated with that strength? For glass, use $E = 70$ GPa and $R \approx 20$ J/m^2.

Solution. Using $\sigma^2_{\text{crack}} = ER/\pi L$ and noting that $E \approx 70$ GPa with $R \approx 20$ J/m^2 for glass, we have $170^2 \times 10^{12} = 70 \times 10^9 \times 20/\pi L$. Hence $L \approx 15 \times 10^{-6}$, or the Griffith flaw size is some 30 μm.

The use of the expression for a small crack in a large plate for a beer bottle geometry is justified in this case by the smallness of L.

CRACK STABILITY

The split beam case and the Griffith case require decreasing loads to keep the cracks propagating once initiated (i.e., $P \propto L^{-1/2}$ or $\sigma \propto L^{-1/2}$). In fact, many (but not all) cracking situations follow this general rule. It would seem therefore that if the loading agency remains at its initiation value (e.g., loading by hanging masses, a skater on an ice-covered pond, a train on a bridge), complete fracture is assured.

Attempts to decrease the load in some way as the crack propagates [and thus follow relationships such as Eq. (14-5) or (14-7)] by unloading the cracked structure may or may not be successful. It is common experience that some objects break in an uncontrollable fashion whereas cracks may be observed to move slowly in other situations. For example, a partially split block of wood still has to be pulled apart (with decreasing loads) to break it; one can tear a telephone directory in two only by tearing the slightly displaced pages sequentially. On the other hand, cracks produced by thermal shock (pouring hot liquids into cold glasses, for example) are disconcertingly catastrophic, as is the case of a skater who falls through the ice.

The case of the ice skater on a pond is useful in introducing the concepts of stability of cracking and the stiffness of testing systems. Consider the following three possibilities: (1) a skater who falls through ice that has cracked because of his mass (the vertical displacement δ of the skater is greater than 0), (2) a skater who happens to be under an overhanging tree and who can hang onto a bough to prevent himself dropping through the ice, even though the ice sheet may be breaking ($\delta = 0$), (3) a skater who not only can grab onto a tree as soon as he sees cracks appear but also can haul himself up off the ice sheet ($\delta < 0$).

In all cases the strain energy in the ice sheet that helps to feed the cracks is caused by the deflection of the ice under the skater. In the first case the cracking is catastrophic or *unstable* because the descent of the skater's body provides *more* energy to the ice sheet than is required for balanced *stable* cracking (i.e., for crack propagation to continue, the load should fall once the cracks start off, but in this case the load cannot diminish). In the second case the skater's body is prevented from dropping, and if he grabs onto the (rigid) tree as soon as he sees cracks initiating, he may prevent further cracking, since his body is locked at a fixed height and can supply no further energy to the system. In the third case, if he quickly lifts himself off the ice sheet upon sight of the first cracks, he may prevent catastrophic cracking altogether, in that he removes energy from the ice sheet by changing the way his mass is supported (by his feet and hence the ice, or by his arms and hence the tree).

We shall see later that, depending on the geometry of the cracked body, stable cracking under decreasing load may be possible even though energy is being fed into the system (δ increasing); this occurs in the split beam problem discussed in the previous section. In other cases, of which the deflection of a cracked sheet of ice is one, stable cracking is impossible when additional energy is fed into the system, and crack propagation can only be controlled when energy is removed from the system.

Testing machines in the laboratory are usually either *load-controlled* (the operator can adjust the load precisely but must take whatever displacement happens to be associated with that load) or *displacement-controlled* (the displacement is regulated and the load adjusts itself for that displacement). Machines that apply the load hydraulically are in the former category, and screw-driven machines are in the latter.

All testing machines themselves deflect under load, so that the displacement of the crosshead is usually an overestimate of the stretch in a tension test piece or the reduction in height of a compression test piece. In *hard* testing machines (displacement-controlled) the difference between crosshead displacement and specimen deflection may be quite small, but in *soft* (load-controlled) testing machines, which deflect much under load, the difference may be great. Displacement-controlled screw machines are stiffer than load-controlled hydraulic machines. During a test, the more a testing machine deflects, the more it stores strain energy in its frame. In cracking experiments this energy is available to feed a crack once started. If the specimen geometry is such that it is capable itself, from its own deflected strain energy, of adequately feeding the propagating crack, the extra energy released from the machine may tip the balance and cause the crack to become unstable and accelerate. Clearly soft machines, which have lower compliances than hard machines, are more likely to produce instabilities because they contain more energy at a given load. Cracking experiments are best performed in the hardest machines available.

Soft and hard situations are analogous to cases (1) and (2), respectively, of the ice skater. However, there are still situations where even in the stiffest testing system cracks will be unstable, as the crossheads of few conventional testing machines can reverse rapidly during a test, so that the machine is incapable of reducing the total energy in the cracked system as the skater does in case (3).

The foregoing considerations of crack stability can be expressed more formally as follows: If we use the P-axis and the radial lines of constant crack area emanating from the origin as a set of (P, L) coordinates, the relationship for the split-beam case in Eq. (14-5) can be plotted as shown in Fig. 14-14(a).

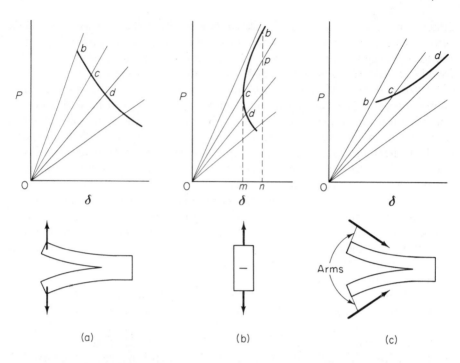

Figure 14-14 Stability of cracking. Constant R loci for different geometries of cracked body: (a) split beam; (b) Griffith; (c) split beam loaded at an angle. Cases (a) and (c) are stable. Case (b) is unstable, as there is so much work available for cracking that the system could afford to give up energy and still have enough to feed the crack, but it is usually impossible to remove the excess from the system.

Equivalently, L can be eliminated from Eq. (14-5) with the use of beam theory for the deflection [Eq. (14-4)] to give

$$P^2_{\text{crack}} = \frac{REIt}{L^2} = \frac{2R^{3/2}t^{3/2}\sqrt{EI}}{3\delta} \tag{14-8}$$

which is the previous relationship now in terms of the cartesian (P, δ) coordinates.

Similarly, the Griffith relation may be plotted on (P, δ) coordinates to give a constant-R locus of the shape shown in Fig. 14-14(b). In general, different cracked-structure geometries will give different shapes to the locus, including the possibility of cracking under increasing load as shown in Fig. 14-14(c), which occurs for a split beam loaded at an angle. In all the diagrams sector areas such as Obc and Ocd represent the work done in propagating the crack from its initial length to longer lengths, where Ob gives the compliance of the initially cracked structure, Oc, Od, etc. being the compliances of the solid with longer cracks.

Hence cracking may take place with both P and δ increasing [Fig.

14-14(c)]; with P decreasing and δ increasing [Fig. 14-14(a)]; and with both P and δ decreasing [Fig. 14-14(b)]. The latter case is intriguing and seems at first sight not sensible. What it means is that there is *so much* energy available in the structure at crack initiation that work can be *taken out* of the structure and still leave enough to feed the crack. That is, in Fig. 14-14(b) for stable cracking along *bc*, strain energy given by the area *mcbn* on the diagram must be removed from the specimen. The strain energy before cracking, put in by the external stress, is *Obn*; the recoverable strain energy, if it were possible to unload the specimen during crack propagation, is *Ocm*. As mentioned earlier, however, most testing machines cannot go rapidly backwards in a test, so that the excess energy cannot be released from the structure, which therefore results in an *unstable* accelerating crack, giving catastrophic rupture. A violent drop in load at fracture down *bn* would be observed on the testing machine chart, and the sector area *Obp* in Fig. 14-14(b), which is greater than the equilibrium value *Obc*, represents fracture work plus the kinetic energy of the accelerating crack.

The conclusion from the foregoing is not only that must there be enough energy available for cracking but also that the whole system must be capable of doing work on the crack at the exact *rate* that the crack wants to be fed. In simple terms, the "spring" (including the testing machine) must relax and give up its energy at just the appropriate rate—not too quickly, lest the crack receive too much energy and "go bang," nor too slowly, or the crack will not run at all. All this is the realm of *crack stability*. A *stable crack* is one that has the energy rate balance just right. The subsequent crack velocity, and questions of stability in general, depend among other things on the loading method, the toughness, and the geometry of the cracked piece. A consequence of the action of soft systems is that increments of load must always be positive during loading (i.e., $\Delta P > 0$). Thus stable cracks can only be expected for those cracked structures where cracking takes place under increasing load. Therefore the arrangement in Fig. 14-14(c) would be stable in a soft testing machine, but not those in Fig. 14-14(a) and (b). Again, a hard machine operates under the condition $\Delta\delta > 0$, so that stable cracks occur where δ is increasing during cracking [i.e., Fig. 14-14(a) and (c), but not Fig. 14-14(b)]. The Griffith case in Fig. 14-14(b) and all those where the R locus points backwards are inherently unstable.

Fluctuations in the load/deflection traces of cracking experiments can be caused by rate-dependent effects in the fracture toughness of the material, that is, when R varies with crack velocity. If R increases as the crack velocity increases, unstable situations may be made stable. The excess energy at crack initiation may be fully dissipated later in the cracking process by the ever-larger amounts of work required to generate new crack area. In this way, a crack that would otherwise shoot through a test piece to break it completely may stop after reaching a certain length and not propagate further, until the external displacement is again increased. Such an effect is called *crack arrest*,

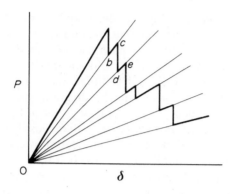

Figure 14-15 Sawtooth P–δ behavior during the cracking of materials similar to some epoxy resins and composites. Rate effects cause instability but are followed by crack arrest.

and use of it can be made in the safe design of structures. Again, stable situations can become unstable if R decreases as the velocity decreases. In the testing of epoxy resins and composites (Chaps. 12, 13, and 17) a sawtooth behavior pattern is often seen. This is a combination of both types of rate effect, in which a stable situation is at first made unstable but is followed by crack arrest (Fig. 14-15). Further propagation requires, in this case, both increasing load and increasing displacement (up to points c and e from points b and d).

CRACK TIP PLASTICITY

No molecular or atomistic interpretation has been developed thus far for R. What is happening at the crack tip? Where is the work of fracture going? Why do we have to supply energy at all? Conversely, why does a cracked part have any strength at all?

Clearly both brittle and ductile fracture involve separation of surfaces and *creation of new surfaces* (i.e., the atoms labeled B and C in Fig. 14-7 become exposed as the crack develops).

Griffith in 1920 first recognized that the exposure of interior atoms, when new surfaces are created by the act of crack spreading, requires work to be done in order to give those new surfaces their *surface free energy*. The surface free energy, analogous to surface tension in a liquid, is the work done in breaking the bonds between a unit area of atoms, thereby giving two mirror surfaces. Equivalently we can say that atoms in the surface have "unbalanced" bonds, which require more energy than interior atoms. (See the discussion in Chap. 5).

The work must be something to do with the area under the force or stress vs. atom displacement curve given in Fig. 14-6. The work to cause fracture will be

$$\int_0^{\lambda/2} \sigma_{\max} \sin\left(\frac{2\pi x}{\lambda}\right) dx = \frac{\lambda \sigma_{\max}}{\pi} \tag{14-9}$$

This is shared between the two surfaces, the surface energy of *each* of which is given the symbol γ_s. (Note that in this case this is a total work referenced to *both* surfaces; R is defined with reference to *one* side of a double surface, i.e., crack. See Exercise 14-3.) Therefore

$$2\gamma_s = \frac{\lambda \sigma_{max}}{\pi} \tag{14-10}$$

that is,

$$\sigma_{max} = \frac{2\pi\gamma_s}{\lambda}$$

or, using

$$\sigma_{max} = \frac{\lambda E}{2\pi a_0}$$

[of which Eq. (14-1) is an approximation], we get

$$\sigma_{max} = \sqrt{\frac{E\gamma_s}{a_0}} \tag{14-11}$$

where a_0 is the equilibrium spacing of atoms. Note that Eq. (14-11) could be derived from Eq. (14-7) if $R = 2\gamma_s$ and $L = 2a_0/\pi$. Thus Eq. (14-11) gives the theoretical maximum stress to *initiate* a crack when no crack exists other than the equilibrium spacing of atoms. All these expressions for σ_{max} amount to the same thing in the end and suggest that σ_{max} ought to be approximately $E/10$.

For example, for steel at very low temperatures the following values probably apply: $a_0 = 0.2$ nm, $\gamma_s = 1$ J/m^2, $E = 210$ GPa. So

$$\sigma_{max} = \sqrt{\frac{210 \times 10^9 \times 1}{0.2 \times 10^{-9}}} \approx 32 \text{ GPa} \approx \frac{E}{7}$$

At this juncture, having been introduced to γ_s, the reader may be wondering what the difference is between γ_s and R (apart from the "definition" of dividing by twice the crack area or by only the area). If, when a crack opens, we create new surfaces to each of which γ_s must be furnished, isn't that the lost energy of fracture?

To investigate this, think of fracture as a process of breaking a series of miniature tensile specimens ahead of the crack (Fig. 14-16). Each specimen must be taken to its fracture strain before the crack can move on. Obviously

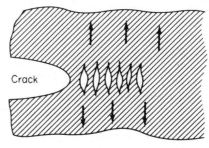

Figure 14-16 Oversimplified idea of fracture as a process of breaking a series of miniature tensile test pieces in the path of the crack.

materials differ in this imitation tensile behavior. Some involve large plastic strains, with correspondingly large amounts of irreversible work before they break; others merely snap elastically with little strain prior to fracture [Fig. 14-1(a)].

With the latter brittle materials such as glass (which Griffith used for his experiments) it is found that R and γ_s have similar orders of magnitude. Hence the Griffith equation (14-7) is more commonly written with $2\gamma_s$ replacing R, which gives

$$\sigma_{\text{crack}} = \sqrt{\frac{ER}{\pi L}} = \sqrt{\frac{2E\gamma_s}{\pi L}} = \sqrt{\frac{E\gamma_s}{L}}$$

With typical ductile or quasi-brittle materials, measured values for R are very much larger than the surface free energy γ_s; in fact, $R = 1000\ \gamma_s$ or greater. This means that work, much beyond that necessary for merely the

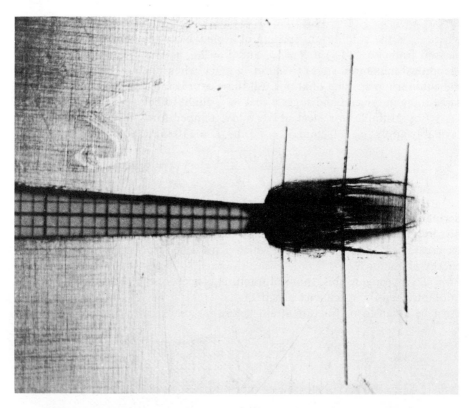

Figure 14-17 Orowan plastic zones at tips of cracks. In this illustration the zones of irreversible deformation, which form a contiguous "boundary" layer to the propagating crack, are shown for a crack in polycarbonate, but they were first explored in low-carbon steels. The vertical lines were drawn on the photograph to indicate the progress of the crack.)

creation of new surfaces, is required for crack propagation in these materials. Where does this work go? From the model just suggested it was postulated and then demonstrated by Orowan and Felbeck that there exists a *plastic zone* around most crack tips, in which the strains locally are brought up to ϵ_f or thereabouts as the crack propagates (remember that stress and strain are concentrated at crack tips; Fig. 14-17). The plastic zone is part and parcel of the cracking process, and cracking in these materials is impossible without it. The work involved in plastically deforming this volume of material is the cause of R being much greater than γ_s.[*]

The size of the crack tip plastic zone is quite small, and it must be emphasized that it is only a *local* region of plasticity; the rest of the material during cracking is often far below yielding. This is very important, because parts are often designed on the basis of the average stress at regions remote from the crack (the presence of which may not be known). We can think of the plastic zone as disappearing to zero in the most brittle materials and of there being a range of zone sizes up to very large in the most ductile materials.

Take care to note, however, that the idea of "little tensile test pieces" is oversimplified; real cracking is more complex, but the model serves to illustrate some salient points.

Exercise 14-7

Estimate the size of the plastic zone at the tip of a crack in low-carbon steel, for which $R = 100 \text{ kJ/m}^2$.

Solution. The plastic work done per unit volume up to some strain level is the area under the σ–ϵ curve up to that strain level (Chap. 2), which, if $\sigma = K\epsilon^n$, becomes $K\epsilon^{n+1}/(n+1)$. For a typical low-carbon steel $K \approx 1$ GPa, $n = 0.15$, and the strain to fracture in a tensile test is approximately 0.6. Hence the work per unit volume to fracture is $10^9 (0.6)^{1.15}/1.15 \approx 500 \text{ MJ/m}^3$.

If we assume that the shape taken by the plastic zone at the tip of a crack is something like a cylinder of radius r, the volume of material newly plastically deformed when the crack moves by unit distance will be given by a cylinder with crescent-shaped cross section. When the unit crack advance is less than r, the area of the crescent-shaped cross section is small compared with πr^2. The crescent area is thus approximately $(1/r)(\pi r^2) = \pi r$ when $1 < 2r$.

Consequently for unit depth of cylinder perpendicular to the plane of a plate (as in Fig. 14-17), the plastic work done if the newly deformed zone is strained to ϵ_f will be $500 \pi r$ MJ. But for unit crack advance across a crack front of unit depth, the work required is numerically equal to R or 100 kJ/m^2 in this case. If it is justifiable to equate

---

[*]The symbol γ (without subscript) is sometimes used to denote the *work of fracture*, which is obtained by taking the *total* area under a load/deflection curve during cracking (such as that in Fig. 14-9) and dividing by twice the crack growth area, that is, the load/deflection diagram is considered one big sector area. In these circumstances $R = 2\gamma$, but unlike the Gurney method, this procedure is not capable of picking up possible variations in R along the path of cracking and/or variations in R with velocity, or whether some of the work has been dissipated in plastic flow remote from the crack tip.

the two works, we have $r = 100 \times 10^3/500 \times 10^6\pi \approx 65$ μm. This dimension is, in fact, reasonably typical of the thickness of strained boundary layers found contiguous with cracks in some steels.

BRITTLE CRACKING IN LARGE STRUCTURES

At the time Griffith did his work, it was not recognized that brittle fracture was possible in materials, such as plain-carbon steel, that are normally considered ductile. Of course, it was known that at low temperatures (below the transition range measured with Charpy specimens) steel does become brittle, but during World War II welded ships started to break by cracking even though their strength-of-materials design stresses were considerably below yield and even though metal taken afterwards from near the crack had ductile σ–ϵ behavior with respectable values for ϵ_f. What was going on? Steel was not supposed to be brittle like glass or cast iron, the temperatures were not all that cold, and evidently most of the ship was really not overstressed. The ship had not yielded all over and necked before cracking!

It is now realized that fabricated items (ships, bridges, oil storage tanks, pipelines, airplanes, etc.) are never perfect. Somewhere a weld might not be as good as it ought to be; perhaps a metallurgical error has allowed a fissure to be rolled over in the hot mill; glued joints may not be up to specification. Consequently we now must identify any macroflaws, or even really large cracks in large fabricated structures (e.g., hatches in ships' decks have even been shown to be bad stress concentrators on occasion). Hence, if the conditions are favorable, cracking can initiate at these flaws and break the structure, even though stresses away from the crack may be small and *below the yield value*.

When are the conditions "favorable"? In traditional design, the stresses to which a structure is subjected are calculated from the working loads on the assumption that the structure is "perfect," with no flaws, no preexisting cracks, no locally brittle areas caused by wrong heat treating, and so on. These design stresses are usually arranged to be well below the yield stress. However, in the presence of a flaw, the applied stress necessary to open up the flaw and propagate it could be less than the design stress. The stress σ that must be applied to the structure to initiate cracking would be obtained from appropriate formulas such as Eqs. (14-5) or (14-7). Most often it would take the general form $\sigma^2 \propto ER/L$, where L is the crack length, the constant of proportionality depending on specimen shape and crack orientation.

Consider two versions of the structure, the large prototype itself and a model that might be tested in the laboratory. For simplicity, consider a geometrically similar model where all dimensions (height, width, thickness, crack length) are $(1/\lambda)$ of those of the prototype; λ is called the scaling factor. Using subscripts m and p, for model and prototype, we find that the stresses to cause cracking in the two sizes of structure will be

$$\sigma_m^2 \propto \frac{E_m R_m}{L_m} \quad \text{and} \quad \sigma_p^2 \propto \frac{E_p R_p}{L_p}$$

If the fracture toughness and Young's modulus of the material are assumed to be the same in the laboratory and in the fabricated structure, then

$$\sigma_p^2 = \left(\frac{L_m}{L_p}\right)\sigma_m^2 \quad \text{or} \quad \sigma_p = \left(\frac{1}{\lambda^{1/2}}\right)\sigma_m \qquad (14\text{-}12)$$

Since $\lambda = L_p/L_m$, we see that the stress necessary to cause cracking in the prototype is less than the stress required to do the same thing in the model. For example, if $\lambda = 25$, the prototype cracking stress is $\frac{1}{5}$ of that required in the model. No such scaling effect appears for yielding, flow stresses being independent of the size of the body according to theory. Consequently cracking may occur before yielding in a large structure, whereas in the laboratory a model of the same structure may yield before cracking (Fig. 14-18). Consequently what might appear to be a safe ductile material in the laboratory may perform quite differently and unsafely in a large fabricated structure. Hence the small tensile bars, taken from locations adjacent to the cracks in welded ships, showed ductile behavior. This is the origin of brittle fracture in large structures. In a way it is a logical extension of old ideas on stress concentration in design of machine parts; there, it was known that even in "perfect" steel, sharp corners in small pieces led to failure; now we look at sharp corners on a gross scale.

In the simple derivation of Eq. (14-12) it was assumed that the scaling factor λ was the same for length, height, width, and crack length. Clearly this is

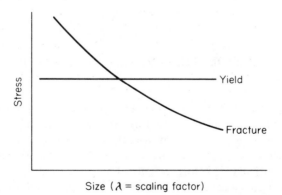

Size (λ = scaling factor)

Figure 14-18 Scale effects in brittle fracture. The larger the structure, the smaller the stress to cause cracking, but no size effect enters calculations for yielding. Thus a material that yields in the laboratory (small λ) may crack when fabricated into a large structure (large λ).

not always so: Ships' steel can be tested in the laboratory in its full thickness, but the areal extent of hull or decking cannot be duplicated; again, annular cross sections of pipelines can be tested in the laboratory but practical lengths of about a kilometer cannot, and so on. Even so, it is possible to perform the more complicated calculations for scaling laws with four different λ's, and it is usually found that $\sigma_p < \sigma_m$,

The possibility of brittle fracture in normally ductile materials is confusing if it is assumed that fracture in ductile materials must involve gross yielding (as in a tensile test of a ductile solid). Such ductile fractures can occur, of course, as in metal forming when we try to push the process to its limit (e.g., in rolling or deep-drawing car bodies, as discussed earlier), and in opening cans of beer, sardines, and so on, but these fractures are expected. It is the fact that we are *not* expecting fracture when the design stress is below yield in a ductile material that makes it dangerous.

Exercise 14-8

A cylindrical pressure vessel 6 m long, 3 m in diameter, and 20 mm thick wall exploded at 5 MPa. The steel from which it was made had the following properties: $E = 210$ GPa, $\sigma_y = 1.4$ GPa, $R = 130$ kJ/m^2. Show why failure was not expected on the basis of strength-of-materials design calculations, but that the failure was compatible with the existence of a longitudinal welding crack about 120 mm long.

Solution. The hoop stress in the cylinder is given by

$$\frac{pD}{2t} = \frac{5 \times 10^6 \times 3}{2 \times 20 \times 10^{-3}} = 375 \text{ MPa}$$

This is between a third and a quarter of the yield stress, so failure should not be expected.

Using the small-crack-in-a-large-sheet expression for cracking stress, we have

$$\sigma_{\text{crack}} = \sqrt{\frac{ER}{\pi L}} = \sqrt{\frac{210 \times 10^9 \times 130 \times 10^3}{\pi \times 60 \times 10^{-3}}} = 380 \text{ MPa}$$

As this is comparable with the design hoop stress, the failure of the pressure vessel is not unexpected.

The foregoing exercise indicates the importance of knowing whether flaws are present in a built-up structure and whether they are likely to be dangerous or not. It is customary nowadays to use stringent crack-inspection procedures (such as X-ray, ultrasonic, or other nondestructive methods) in an attempt to identify flaws before a critical part is used. If we want to use materials at a high strength level, yet clearly want cracking stresses to be greater than the design stress, we must look for very small cracks, since the crack initiation stress is inversely proportional to flaw size. Conversely, if the inspection equipment is only reliable for a certain lower L, this sets an upper value on the safe working stress in a component of given geometry.

Sometimes the cracking stresses associated with the flaws when found are very high, in which case it is occasionally permissible to leave the flaws; at other times the cracking stresses are unacceptably low and unsafe, so that the component or structure must be repaired and the flaws eliminated or reduced in size. Bad welds, for example, may be ground away and the cavity refilled with fresh metal.

FRACTURE MECHANICS

The loads or stresses that will cause a crack to propagate have so far been determined from the energy and compliance method [Eq. (14-3)]. Alternatively (and nowadays much more commonly) they may be arrived at from considerations of the detailed stress fields around the tip of a flaw and in the cracked body generally. This second line of attack is the discipline called *fracture mechanics*. Irwin realized that relations connecting the applied stress and the size and orientation of a flaw can be written in terms of the applied *stress-intensity factor K*, which is an alternative parameter to G and is related to it, as will be explained presently. For example, a fracture mechanics expression for the stress-intensity factor at the tips of the crack in a center-cracked panel of finite width (Fig. 14-19) is

$$K = \sigma\sqrt{\pi L}\ \sqrt{\frac{2b\ \tan^{-1}(\pi L/2b)}{\pi L}} \qquad (14\text{-}13)$$

It will be seen from the dimensions of this expression that the stress-intensity factor K has the peculiar units of $N/m^{3/2}$, which also may be written as $Pa\sqrt{m}$. K has the form it has from the mathematics of the nonuniform

Figure 14-19 A center-cracked panel of finite width $2b$ with crack length $2L$ loaded remotely in the axial direction perpendicular to the crack.

stress fields around the crack tip; it must not be confused with the stress-*concentration* factor, which relates to a multiplying factor (dimensionless) by which the local average stress at a change in cross section is increased above the average stress.

Experiments show that cracking occurs when K [determined from formulas such as Eq. (14-13) for the test piece being employed] attains a critical value, called K_c, which is characteristic of the material. In fact, it may be shown that $K_c^2 = ER = EG_c$, which is the product of the two separate physical properties E and R of a material (hence the peculiar units for K), so the attainment of a critical K_c for propagation is the same as satisfying energy-derived equations such as (14-5) or (14-7) at the material's R level. The link between the two methods is that the distribution of stresses and strains gives an integrated strain energy for the cracked body, which, in the same manner as described for G_c, feeds the crack at the correct rate at first propagation. The $d/dA(\delta/P)$ term in Eq. (14-3) has been already worked out in the fracture-mechanics expression for a particular geometry. Thus the fracture-mechanics formula for the DCB specimen is written as $P = (K/L)(It)^{1/2}$; compare Eq. (14-5).

Notice that when Eq. (14-13) is rewritten for small L/b [when tan $(\pi L/2b) \approx \pi L/2b$], with also $K_c = \sqrt{ER}$, we have

$$\sqrt{ER} = \sigma\sqrt{\pi L} \quad \text{or} \quad \sigma = \sqrt{\frac{ER}{\pi L}}$$

Not unexpectedly, this is the same as the Griffith expression (14-7) for a small center crack in a large plate. In general, $\sigma^2 = YER/L$ or $\sigma = K\sqrt{Y/L}$, where Y is constant for given cracked-body geometry and type of loading; values for Y for numerous bodies of different shape, crack orientation, and types of loading may be found in handbooks.

Fracture toughness (R or G_c) and critical stress-intensity factors K_c are now routinely measured for materials specifications and quoted in codes of practice for design for use with fracture-mechanics formulas. The designer utilizes fracture-toughness concepts by first calculating the value of the applied K as a function of stress, crack length, and geometry, using solutions already worked out and available in handbooks. He then finds from published data (or measures) the value of K_c for the material being used in the design and equates his calculated K to K_c. This tells him the safe value of stress for a specified value of stress.

MAXIMUM TENSILE STRESS FRACTURE CRITERIA

A criterion for fracture based on the attainment of some characteristic maximum tensile stress has been used with a certain degree of success, particularly in very brittle solids. It should be clear, however, that quantitative criteria for fracture based merely on some maximum tensile stress (or combination of stresses) address only part of the necessary and sufficient conditions for fracture. No consideration is given to the energetics of crack propagation. Although maximum stress theories may be intuitively appealing and *may* work reasonably well under the circumstances, they should not be expected to apply to large structures containing critical flaws.

Consider the tensile strength of brittle materials such as glass, ice, ceramics, rocks, and epoxies. It is given by the fracture load divided by the tensile bar cross-sectional area, or if determined by bending, from the simple beam theory outer fiber stress (the latter bending strength is called the *modulus of rupture* in the ceramics field). The cracking load in such tests is really the solution of Eq. (14-3) for the particular specimen geometry and the material's natural distribution of flaws (from which fracture starts). The cracking loads are likely to be quite different in different geometries. It is not surprising therefore that different answers for tensile strength of the same material are obtained from different tests, irrespective of the scatter in data resulting from the statistical distribution of natural flaws in these solids (Chap. 16). Consequently, in maximum-stress cracking theories, what is the "real" reference tensile strength to be used? It seems more satisfactory to work in terms of the fundamental fracture toughness parameter and take every case on its own merit.

GENERALIZED YIELDING, NOTCH BRITTLENESS, AND TOUGHNESS/STRENGTH RATIOS

In the split-beam example in Fig. 14-11, it was assumed that the stresses at regions remote from the crack tip were considerably below the yield strength of the material, so that all irreversible deformation was concentrated in zones contiguous to the crack tip. Such irreversible work is, of course, the origin of the material fracture toughness. If the arms of the split beam are not sufficiently deep, however, stresses greater than the yield stress can be produced, so that irreversible work (which has nothing to do with the process of cracking) must also be performed. The approximate maximum stress in the arms is given by engineers' bending theory as

$$\sigma = \frac{My}{I}$$

From Eq. (14-5) the moment to produce cracking is

$$P^2 L^2 = M^2 = REIt$$

Thus the depth of the one arm when $\sigma = \sigma_y$ (the yield stress) is

$$h = \frac{3ER}{\sigma_y^2}$$

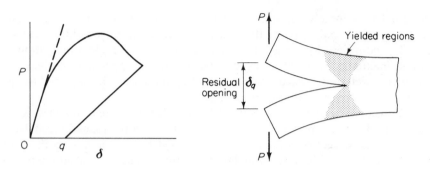

Figure 14-20 $P-\delta$ diagram for combined flow and fracture, where yielding remote from the crack faces occurs during (or at least before) cracking. Unloading lines do not return to the origin.

In other words, if $h < 3ER/\sigma_y^2$, the arms will yield before the body cracks. Depending on the complex interaction of work hardening and the loads to propagate the crack subsequently, a load–deflection plot under these circumstances might look as in Fig. 14-20. There is appreciable curvature, and the unloading line does not go back to the origin; the split beam remains yawed open, with permanent deflection Oq, because of the *generalized yielding*. Clearly it is difficult, from the P/δ diagram, to differentiate between the cracking work and the yielding work performed away from the crack tip. (Hence one criticism of the use of γ, "the work of fracture" from the total area under a $P-\delta$ plot.) Only test pieces having $h > 3ER/\sigma_y^2$ will crack reversibly. We see therefore that the behavior of the same material can be markedly changed simply by altering the geometry of the arms.

Again, changes of specimen thickness can profoundly affect the behavior. Thick sections tend to portray elastic reversible behavior, with no generalized yielding, because of triaxial constraint. That is, the central sections of thick test pieces tend toward *plane-strain* behavior, with no deformation across the thickness. This is accompanied by a tensile mean hydrostatic stress that promotes fracture and thus effectively a low R. Thinner *plane-stress* sections, on the other hand, neck at the crack tip (analogous to necking in a round tensile test bar) and have higher resistance to fracture, often caused in part by the presence of additional generalized yielding.

The differences in behavior are shown up in the types of fracture surface produced. Plane-strain fractures tend to be *flat* but are accompanied by *shear lips* (compare tensile cup-and-cone fractures), as in Fig. 14-21, when plane-stress edge effects come into play (note the sideways contraction of the specimen on the right-hand side of the figure). Most fractures in practice are "mixed mode" in character. These effects have long been known in the fracture surfaces of Charpy specimens. Since plane-strain fractures display lowest resistance to crack propagation, which is important for conservative design in the presence of cracks, the American Society for Testing and Materials

Figure 14-21   Fracture surfaces of Charpy V-notch test specimens of a very brittle steel (left) and a moderately ductile steel (right). The initial machined notch lies along the bottom edge of each specimen in the photograph. The restriction of plastic flow prior to fracture led to plane-strain conditions in the brittle steel, whereas the ductile specimen of the same dimensions allowed sufficient plastic flow to permit conditions of plane stress near the free edges of the cross section. The 45° shear fracture that results, called a *shear lip*, and the associated plastic contraction of the specimen, are evident along the three sides of the fracture. (Photograph courtesy of W. H. Durrant.)

(ASTM) has specified a minimum plate thickness for toughness test pieces so as to ensure plane-strain fractures. The thickness is given by the empirical expression $2.5(ER/\sigma_y^2)$ or $2.5(K_c/\sigma_y)^2$; when thicknesses are at least as great as this, it can be shown that the size of the plastic zone at the crack tip is being kept very small.

The fractography of brittle and ductile fractures in general is discussed in Chap. 19.

It is often stated that the area under a tensile curve out to fracture is a measure of fracture toughness. Although, with test pieces of identical size in the laboratory, a relative ranking between materials can be obtained (in the same way that impact data are used), the differences in behavior mentioned in the foregoing as well as the odd scaling effects are not always brought out by merely looking at areas under stress–strain curves. The classification of materials as brittle or ductile on the basis of some arbitrary laboratory test may therefore be misleading. In Charpy tests, for example, marked differences in the transition temperature range for steels are produced by varying the type and depth of notch. Many solids can be made to behave in a brittle manner in one type of test yet in a ductile manner in some other test. We have seen that fracture behavior may be altered by changing specimen thickness in tests that measure R: Thick test pieces have small crack tip plastic zones, and the applied cracking stresses are much lower than the material yield stress, whereas thin specimens have large necked zones and often yielded regions remote from the crack tip. Again, the fracture scaling laws that favor cracking instead of yielding in large pieces complicate our simple use of the words "brittle" and "ductile."

Perhaps we should classify solids in terms of a toughness/strength ratio rather than merely toughness or strength alone. A useful concept in this connection is *notch brittleness*. Consider pulling a notched piece of a material usually considered to be ductile, such as a flat tensile bar with a saw cut (Fig. 14-22). The behavior might be brittle with easy cracking, if the conditions are favorable. On the other hand, the stress on the so-called net section (i.e., the section of material left where the notch has been made and on which the greatest average stress in the system occurs) may get as high as the yield strength of the material before cracking commences. This suggests that yielding, rather than cracking, would occur at the net section in the same way that yielding took place in the arms of the split beam. However, we must remember the arguments about a triaxial state of stress in the neck of a tensile bar when the shoulder material constrains material from flowing into the neck and hence prevents elongation. Something similar happens in the notched sheet, and the result is that we have to apply *more* than the simple yield stress in order to get plastic flow. It can be shown that the average net stress necessary to promote plastic flow in the presence of deep notches (where the stress in the shoulders is much lower) can be as high as $3\sigma_y$; the situation is a sort of "tensile" hardness test (Chap. 2).

A *notch-brittle* material is one that is ductile in an ordinary laboratory test but brittle in notched tests with large plastic constraint factors (i.e., in tests with deep notches). Hence, if σ_{crack} is the boundary cracking stress and σ_y the yield stress, we have (1) a *simply brittle* material if $\sigma_{crack} < \sigma_y$; (2) a *notch-brittle* material if $\sigma_y < \sigma_{crack} < 3\sigma_y$; (3) a *simply ductile* material if $\sigma_{crack} > 3\sigma_y$. Since the cracking stress will be given by some formula such as $\sigma_{crack}^2 = Y(ER/L)$, where Y is a constant for the specimen geometry, the performance of a material in this test depends on the relative values of σ_y, $\sqrt{YER/L}$, and $3\sigma_y$. That is, a simply brittle solid has $ER/\sigma_y^2 < L/Y$; a simply ductile material has $ER/\sigma_y^2 > 9L/Y$; and a notch-brittle solid has ER/σ_y^2 between these values. Knowing the appropriate L/Y for a chosen type of test piece permits the value of ER/σ_y^2 for the solid to be bracketed, and different types of notched specimen can, in principle, refine the answer. However, it is customary nowadays actu-

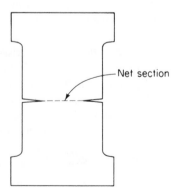

Net section

Figure 14-22 Tensile test of a notched specimen to assess notch brittleness.

ally to measure K_c or R independently from standard toughness tests, as described earlier in this chapter, rather than use notch-brittleness tests to characterize the behavior.

Nevertheless, the toughness/strength ratio given by ER/σ_y^2 [i.e., $(K_c/\sigma_y)^2$ in terms of the fracture-mechanics critical K_c parameter] is a most useful classification for materials. It has already appeared in this chapter in connection with transitions from cracking to yielding, and in the ASTM thickness requirement for plane strain fracture. Appendix 9 gives some representative values of K_c/σ_y. We see a gradual ranking, from materials traditionally considered brittle to those commonly considered ductile. However, note that although a material possessing a high ratio is more prone to display ductile behavior before fracturing, such materials are not immune from brittle fracture in very large fabricated structures. The Liberty ships and T-2 tankers of World War II were made from steel having K_c/σ_y probably about $0.5\sqrt{m}$ at 20°C, but of the order $0.1\sqrt{m}$ at $-10°C$.)

Exercise 14-9

Explain why notch-sensitivity data for low-carbon steels are often erratic.

Solution. One of the types of test piece that may be used to investigate notch brittleness is the round tensile bar containing deep peripheral V notches. The cracking expression for such a test piece is given approximately by $\sigma_n^2 \approx 2ER/\pi D$, where σ_n is the net section stress at cracking and D the gage section diameter. For bars 10 mm in diameter, with $E = 210$ GPa, if we take a constant R at about 130 kJ/m^2, $\sigma_n \to 1.3$ GPa; for 20-mm bars the cracking stress is 930 MPa and for 30-mm bars 750 MPa. The yield strength range of common steels is some $200 \sim 500$ MPa, and the constrained yield strength range, given by $3\sigma_y$, is then $600 \sim 1500$ MPa. Clearly there are steel test pieces for which σ_n and $3\sigma_y$ will be comparable (10-mm steel bars of yield point approximately 400 MPa; 20-mm bars of approximately 300 MPa; and 30-mm bars of approximately 250 MPa). Consequently some tensile bars may crack, while other nominally identical specimens may flow; similar behavior is found with Charpy and Izod steel specimens. Notch brittleness under these conditions is sensitive to slight variations in composition, previous heat treatment, and mechanical work, and to temperature. Change of specimen size evidently may move the solid from the boundary between notch sensitivity and simple ductility and may also change ranking comparisons made with other materials having the changed test piece size.

Note the "fracture scale effect" in the results; even the material of lowest yield strength will crack if the test piece is made big enough.

FRACTURE IN VERY THIN SHEETS

If a notched sheet of material is very thin, another type of fracture can occur. As before, the stress becomes concentrated at the tip of the notch. The piece will not be ripped brittly, because the toughness of thin sheets is greater than that of thick sheets, but as the load is increased, shear stresses may be produced that will cause slip at 45° in the *thickness* of the sheet rather than across

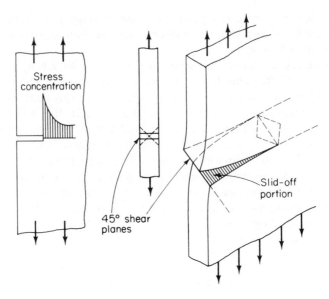

Figure 14-23 Out-of-plane tearing in thin sheets at stresses below the yield stress, caused by slip-off through the thickness of the sheet.

the width. It can be shown (Exercise 14-10) that the average stress required to do this in a thin sheet is less than σ_y. Then out-of-plane tearing can easily take place and produce complete fracture (Fig. 14-23). It is a locally ductile fracture, which could not take place in a thick sheet, because the area of the shear planes would be so big that the normal stress required would probably be sufficient instead to cause generalized yielding across the width.

Such fractures should be familiar to the reader in aluminum foil, chocolate wrappers, etc. They occur because the material is thin. Since aircraft skins and pipelines are also thin, this sort of fracture can be dangerous.

Although produced by slip, and thus locally ductile, out-of-plane tear fractures are comparatively low-energy fractures. Since they are not accompanied by generalized yielding throughout the sheet, such fractures should really be classified as brittle.

Exercise 14-10

Explain in terms of dislocation theory why out-of-plane tearing fractures in thin sheets are possible at stresses below the generalized yield stress. Why are such fractures not possible in thick sheets?

Solution.   The tearing fracture in thin sheets occurs because, literally, dislocations go through the (thin) thickness of the sheet out to the other side, and slip-off results. The situation is analogous to a screw dislocation; the original set of dislocations, sufficient to produce complete slide-off at the start of the fracture, remains sufficient to produce slide-off all the way along, merely by gliding forward. In thick sheets, however, incremental growth of the crack requires more edge dislocations to be injected and activated in the slip lines. The yield zones must extend over bigger and bigger proportions of the cross section for the crack to advance. This is much more difficult to achieve than slip-off. It is important to realize that tearing fractures succeed because the *plastic* displacement at the tip of the crack is accommodated by *elastic* displacements elsewhere. There is no need for plastic zones to span the entire cross section (see Fig. 14-23).

REFERENCES

14-1. Gordon, J. E., *The New Science of Strong Materials—or Why You Don't Fall Through The Floor.* London: Penguin Books, 1968.

14-2. Gurney, C. and coworkers, various papers in the *Proceedings of the Royal Society (London)*, e.g., 1974, vol. 340A, page 213, which gives other references.

14-3. Kelly, A., *Strong Solids.* Oxford: Clarendon Press, 1964.

14-4. Knott, J. F., *Fundamentals of Fracture Mechanics.* London: Butterworths, 1973; New York: John Wiley/Halstead.

14-5. Lawn, B. R. and T. R. Wilshaw, *Fracture of Brittle Solids.* Cambridge University Press, 1975.

14-6. Polakowski, N. H., and E. J. Ripling, *Strength and Structure of Engineering Materials.* Englewood Cliffs, N.J.: Prentice-Hall, 1966.

14-7. Tada, H., P. Paris, and G. Irwin, *The Stress Analysis of Cracks Handbook.* Hellertown, Pa.: Del Research Corporation, 1973.

PROBLEMS

14-1. Give examples of materials with the following combinations of properties:
(a) Stiff and weak.
(b) Stiff and strong.
(c) Flexible and strong.
(d) Flexible and weak.

14-2. When a glazier scratches glass with diamond prior to cutting, is it better to cut a deep rather than a shallow groove?

14-3. The material of the stressed element shown in Fig. 14-24 fails by cleavage at 190 MPa and by shear at 147 MPa. Determine the fracture mode and calculate the value of S when fracture occurs.

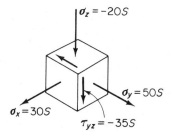

Figure 14-24

14-4. The stress concentration at the tip of a Griffith crack is approximately $\sqrt{2L/\rho}$, where L is the length of the crack and ρ is the crack tip radius. If fracture occurs when the "concentrated stress" is equal to the theoretical strength (i.e., $\sigma_{av}\sqrt{2L/\rho} = \sqrt{E\gamma_s/a_0}$), we have σ_{av} at fracture equal to $\sqrt{E\gamma_s\rho/2La_0}$. This suggests that sharply cracked bodies (where $\rho \rightarrow 0$) have *no strength* at all! This obviously is not true. Explain.

14-5. **(a)** Tough solids (e.g., hard steel) retain considerable strength when scratched; other solids (glass and stone) do not. Explain.

  (b) Why do shop windows, beer bottles, and drinking glasses that do have chips and scratches in them not break?

14-6. Explain why rubber is notch-insensitive.

14-7. How can a glasscutter further weaken glass after scratching but before the act of breaking?

14-8. Can "Griffith cracks" be seen under an optical microscope?

14-9. What are the peculiar colored patterns that sometimes are seen on toughened windshields of cars?

14-10. Explain the ideas behind "toughened glass."

14-11. Some materials that fail with no plastic extension when tested in tension are observed to exhibit some ductility when tested in compression or in pure shear. Why might this be so?

14-12. Coarse grain sizes in zinc and pure iron promote brittleness. Explain.

14-13. "Hardening without refinement of microstructure generally raises the transition temperature and promotes brittleness." Discuss this statement, illustrating your answer with reference to tempered martensite, blue-brittleness, temper-brittleness, hot-shortness, and strain-age embrittlement.

14-14. An asymmetrical double cantilever beam test piece has arms of unequal height h_1 and h_2 in contrast to the symmetrical specimen in Figure 14-11(a). If the cut sheet has thickness t, determine the expression for crack propagation by equal and opposite end loads P.

14-15. Calculate the fracture toughness, in J/m^2, of a plate 20 mm thick with the load-extension path $ABCD$ shown in Fig. 14-25. The crack is 3.4 mm long at A, 3.4 mm at B, 8.6 mm at C, and 13.8 mm at D.

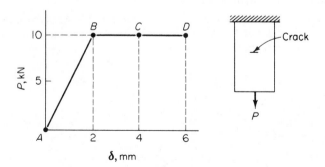

Figure 14-25

14-16. Calculate the fracture toughness of a flat plate specimen, 5 mm thick, of a polymer that yields load-deformation data as follows:

	Load, N	Crack Length, mm	Deflection at Load, μm
0	0	175.0	0
1	623	187.2	523.0
2	514	198.3	562.6

14-17. In a fracture-toughness test the load, and corresponding crosshead displacement, at which cracking commenced were P_1 and u_1. After the load had fallen to P_2 (with displacement u_2), by which time the crack had propagated some distance, the test piece was unloaded. The unloading line went back to the origin. Prove that the fracture toughness R is given by

$$R = \frac{\frac{1}{2}(P_1 u_2 - P_2 u_1)}{(A_2 - A_1)}$$

where A_1 is the initial crack area before cracking and A_2 the final crack area after unloading (Fig. 14-26). Assume that the R locus joining $P_1 u_1$ and $P_2 u_2$ is straight.

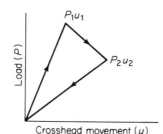

Figure 14-26   Crosshead movement (u)

14-18. A beam of material with initial height 76.4 mm is being torn apart by equal and parallel forces P, as shown in Fig. 14-27. The deflection δ is measured from a

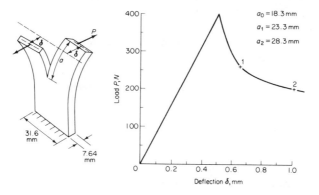

Figure 14-27

fixed point, as shown. From the data plotted in Fig. 14-27, calculate the fracture toughness of the solid. (Crack length a is measured along the crack.)

14-19. The following data for crack area, compliance, and cracking load applied to a certain structure. Determine R.

Crack Area, mm²	Compliance, nm/N	Cracking Force, kN
1300	73	24
1940	206	15
2580	447	12
3226	810	10

14-20. An elastic rod akin to King Arthur's sword Excalibur is embedded in a massive block, as shown in Fig. 14-28. Determine the load that will cause cracking along the interface between the rod and the block. Under what combination of properties and geometry might the rod itself break?

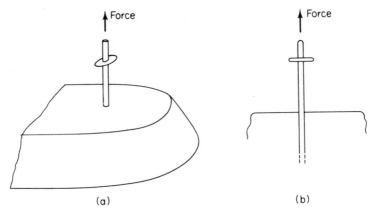

(a)   (b)

Figure 14-28

14-21. Select materials from Appendix 9 that would display notch brittleness in a comparison between plain and deeply notched round tensile bars 15 mm in diameter.

14-22. To overcome a failure problem, a designer decides to use thicker plates of a higher-strength steel as replacements for the broken components. Is this a good solution?

14-23. Rubbish often goes to grain boundaries when a metal crystallizes. What effect does this segregation have on strength properties?

14-24. Thin sheets are tougher than thick sheets (the tearing mode of failure being excluded). Explain.

14-25. Why is it difficult, in principle, to make strong materials?

14-26. Describe the likely modifications necessary in the Griffith theory to explain the mechanism of vacancy creep, described in Chapter 12.

15 | Irons

INTRODUCTION

Pig iron is the first crude stage in the manufacture of iron–carbon metal alloys from iron ore. *Cast irons* are somewhat purified forms of the crude iron that is tapped from the blast furnace; as the name implies, they are used for making castings. In general, the alloys known as cast irons contain greater amounts of carbon than typical steels, $2 \sim 4\%$ rather than less than 1%. Inspection of the iron–carbon equilibrium diagram (Fig. 8-6) shows a eutectic at 4.3% C, which has the lowest melting point ($1148°C$) of Fe–C alloys in that range of carbon contents; steels have melting points of about $1500°C$. The iron age in prehistory began when man was able for the first time to generate appropriate temperatures in fires to melt iron ores. Steels did not come into the picture until still higher temperatures could be generated.

Iron ores have various other elements in them, so that cast irons contain significant amounts of silicon, phosphorus, sulfur, and manganese. Such additional elements modify the simple equilibrium-phase diagram, in which the boundaries and temperatures are altered; thus it is possible to predict cast-iron microstructures from Fig. 8-6 *only* for the restricted group of "pure" irons, which contain few, if any, additional elements. Furthermore, not only do impurities alter the phase diagram, but silicon in particular also encourages the formation of *free graphite*. In steels and other cast irons, carbon is combined with iron as *cementite* (Fe_3C), which is the stable form (either free or in pearlite) in plain-carbon steels below the eutectoid transition temperature (we ignore the minute amounts of carbon dissolved in ferrite α-iron below the eutectic transition temperature). In cast irons containing silicon, free graphite becomes the characteristic stable form of carbon.

Free graphite makes the iron darkish, and dirty to handle; the irons containing no free graphite, on the other hand, have much cementite in their microstructure, which gives a white appearance to a fracture surface. Cast irons are therefore classified broadly as *gray* or *white*.

Cast irons as made are comparatively brittle, with little reduction in area to fracture in the tensile test. Nevertheless, they can have adequate mechanical properties for some applications, and there are plenty of uses for this cheap material (as will be discussed later); as an engineering material cast iron is second only to steel in annual tonnage produced. The extreme hardness of "chilled" cast iron was used to advantage in 1803 by Ransome in Ipswich, England, in the manufacture of plowshares, which up to that time had been made from *wrought iron*, which quickly became blunt; tremendous improvements in the speed of arable cultivation were thereby obtained. It was possible at that time to improve the ductility of white cast iron by special heat treatment, extra toughness in such *malleable* cast iron being obtained at the expense of strength. Indeed, before the advent in the 1850s of Bessemer's method of making steel on a tonnage basis, malleable cast iron and also wrought iron were the only strong ductile structural materials available. Small quantities of steel were being made at that time by the *cementation* and *crucible* processes, which were means of increasing the carbon content of wrought iron. These methods were known by the ancients, lost in medieval times, and rediscovered in the seventeenth century. Many steel artifacts from ancient times consisted of partially case-hardened wrought iron; *Damascus steel* was made by piling pieces of soft iron on pieces of high-carbon iron followed by repeated heating and hammering, which made the carbon distribution somewhat more uniform.

Nowadays there are many types of cast irons, including those alloyed with certain elements to produce particular characteristics such as heat, corrosion, and wear resistance. No other metal suitable for casting into intricate shapes possesses such a wide range of properties as the cast irons.

WHITE CAST IRONS

The microstructure of *white* irons, which have few impurities, can be interpreted more or less directly from the Fe–C equilibrium diagram (Fig. 8-6). At just below 1148°C there are two phases, γ austenite and θ cementite, the γ having 2.11% C. For carbon contents greater than 2.11% and less than 4.30% the microstructure is noneutectic γ and $(\gamma + \theta)$ eutectic; for carbon contents greater than 4.30% we have $(\gamma + \theta)$ eutectic plus noneutectic θ. The $(\gamma + \theta)$ eutectic is sometimes called *ledeburite* and, of course, exists only between 1148°C and 727°C. As the temperature drops, γ has a decreasing solubility for carbon (the line joining 2.11% back to 0.77%), so θ comes out of the γ, both from the free γ and from the eutectic. Thus the ledeburite eutectic changes composition and the free γ throws out θ precipitate into the grain boundaries (compare the Cu–Al system, Chap. 9). Therefore, just above 727°C the phases are still γ and θ, the γ now having 0.77% C, but the microstructure has changed from that at first solidification. For *hypoeutectic* irons (those with less than 4.30% C) the microstructure consists then of free austenite, ledeburite, and grain boundary cementite. For *hypereutectic* irons (more than 4.30% C) we have the ledeburite eutectic plus the original free ("massive") cementite. The $(\gamma + \theta)$ eutectic at 728°C is different from that at 1147°C because of the composition of the γ. As the temperature is reduced below 727°C, all the austenite of 0.77% C becomes pearlite.

Figure 15-1 Hypoeutectic white cast iron, etched in 4% picral solution, showing pearlite and cementite. (Courtesy BCIRA.)

Therefore an "ideal" hypoeutectic pure white iron would have a matrix at room temperature of pearlite and cemetite (Fig. 15-1). The pearlite has come from both the free austenite and the austenite in the $(\gamma + \theta)$ ledeburite eutectic; the cementite has come from both the $(\gamma + \theta)$ eutectic and the θ precipitate. Different pearlite grain sizes can be obtained by varying the cooling rate, as with steel. Martensitic structures can also be obtained by quenching.

Exercise 15-1

A white cast-iron casting of mass 15 kg is poured from a melt containing 2.7% carbon by mass. Determine the mass of eutectoid iron carbide present at room temperature.

Solution. From Fig. 8-6, at just less than 1148°C the mass of noneutectic γ is $(4.3 - 2.7)(15)/(4.3 - 2.11) = 10.96$ kg. At just above 727°C, $(2.11 - 0.77)/(6.69 - 0.77) = 22.6\%$ of the γ appears as θ_{ppt}, leaving 77.4% or 8.48 kg as free γ, all of which transforms to the eutectoid pearlite. Pearlite consists of $0.77/6.69 =$

11.51% of θ. Thus the mass of eutectoid iron carbide present at room temperature is $(0.1151)(8.48) = 0.98$ kg.

Although the foregoing description implies an equilibrium (slow) cooling rate, it would also be true only for a pure iron–carbon alloy. Even small amounts of silicon promote the dissociation of cementite θ into γ plus graphite (above 727°C), or α plus graphite (below 727°C). Thus to obtain a white-iron microstructure, the cooling rate in fact must be faster than one would think: obviously not fast enough to produce a martensite matrix, but fast enough to stop any silicon having time to catalyze the cementite dissociation to iron and graphite. Hence the term "chilled castings" is often used and the section size of castings is important: Because of differential cooling rates, large castings cannot generally be guaranteed to have white iron throughout, as the cooling rate in the center will be too slow.

White irons are extremely hard with little reduction in area to fracture in tension. Depending on the pearlite grain size, the tensile strength lies between perhaps 210 and 560 MPa. The major use for white irons (which are very difficult to machine) is in wear- and abrasion-resistant applications (railroad brakeshoes, pulverizing plant, shot blasting nozzles). Sometimes by appropriate cooling a white-iron microstructure is produced on a softer (but perhaps tougher) core of gray or "mottled" iron for wear-resistant applications.

MALLEABLE IRON

The ductility of white cast irons can be improved considerably by a reheating process whereby the iron becomes malleable. At temperatures somewhere between 800°C and 900°C, the room-temperature pearlite and cementite become a γ plus θ microstructure. If castings are left for several days in this temperature range, the θ breaks down into γ and free graphite (G). The graphite comes out as bumpy clusters (not in smooth, round blobs, as in *nodular* iron, described later in this chapter, nor the sharp-ended flakes characteristic of *gray* irons).* On cooling the castings to room temperature, the austenite either (1) transforms on air cooling or mild quenching simply to pearlite, as we would expect, giving a *pearlitic malleable* iron, with the graphite clusters sitting in a pearlite matrix, or (2) transforms on much slower furnace cooling down to ferrite and additional graphite, giving *ferritic malleable* iron (Fig. 15-2).

Air cooling is too quick to allow the θ in the pearlite (which comes from γ below 727°C) to transform, but furnace cooling is slow enough to allow this θ to dissociate (especially in the presence of some silicon), that is, P = $(\alpha + \theta) \rightarrow \alpha + (\alpha + G)$. These different microstructures are also influenced by whether an oxidizing or neutral atmosphere is used in the reheating furnace.

---

*The clusters are sometimes called *temper carbon*, as they moderate the brittleness of untreated white iron (compare with the meaning of tempered martensites, Chap. 10).

Figure 15-2   Ferritic (blackheart) malleable cast iron, etched in 5% nital solution, showing ferrite and bumpy clusters of graphite ("temper carbon"). (Courtesy BCIRA.)

The fracture surface of pearlitic malleable iron tends to be shiny in comparison with ferritic malleable iron (which has more temper carbon for the same total carbon content). Hence the terms *whiteheart* and *blackheart* malleable irons are applied to the pearlitic and ferritic forms.

Depending on the matrix, the tensile strength of malleable iron might be 400 to 700 MPa with 10% reduction in area to fracture. The improved ductility comes about from the absence of cementite in the microstructure and from the fact that the round shape of the free graphite "holes" in the microstructure is relatively harmless, at least in comparison with the sharp-ended flake geometry of gray cast iron. Malleable irons are used where toughness and ductility are required for parts cast to shape, for example, in automotive gear cases, suspensions, rocker arms, wheels and axles, chain links, hand tools, trailer hitches, boiler and manhole covers.

The malleabilizing process can be long and tedious; sometimes a week is required to heat, hold, and cool castings. Nevertheless, the effort (including the chilling precautions necessary to obtain a white-iron casting in the first place) was worth it—especially in the old days—since along with wrought iron, malleable cast iron was one of the few ductile predecessors of steel.

GRAY CAST IRONS

Gray irons have more carbon and silicon than white irons and are associated with relatively slow cooling rates. Strictly speaking, their microstructures are not predictable from the equilibrium diagram, because the large amount of silicon makes the cementite dissociate easily. Moreover, the percentage of carbon in the γ at the eutectic and eutectoid transformations is reduced by the large amount of silicon: For example, 2% Si reduces the 2.11% maximum solubility of C in γ to 1.5% and reduces the pearlite eutectoid carbon content to 0.7% or thereabouts instead of 0.77%. The percentage of carbon in the ledeburite eutectic is reduced also from 4.3% to 3.7% for 2% Si, and the transformations take place over a range of temperatures even under equilibrium conditions.

In as-cast gray irons cooling from the melt, the cementite that forms between 1148°C and 727°C (i.e., the θ in the ledeburite eutectic and the free θ precipitated as a result of the change in solubility of C in γ iron) is encouraged by silicon to decompose into iron (i.e., γ) and graphite. Of crucial importance, because of the effect on subsequent mechanical properties, is the fact that the graphite comes out as *sharp-ended flakes*. On cooling below 727°C the γ becomes pearlite, but the usual cooling rates in casting are not slow enough to allow the θ in this pearlite to dissociate and give yet more graphite, so at room temperature there results *pearlitic gray* cast iron.

With much slower cooling (or by reheating pearlitic gray iron back into the γ region followed by furnace cooling) the θ in the pearlite can also break down, so that at room temperature only α-ferrite and graphite flakes are seen, since pearlite $= (\alpha + \theta) \rightarrow \alpha + (\alpha + \text{graphite})$, plus all the graphite already there from the broken-down θ above 727°C. Hence the term *ferritic gray* cast iron.

The graphite in both these cases, on cooling from the melt, always appears as flakes that look like black worms under a microscope (Fig. 15-3). The flakes inhibit deformation of the normally ductile ferrite and produce a very

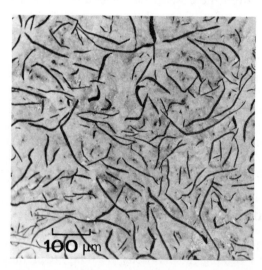

Figure 15-3   Flake gray cast iron, etched in 4% picral solution. (Courtesy BCIRA.)

brittle material. The flakes are essentially sharp cracks, as the graphite with which they are filled has negligible strength. The size and distribution of the flakes determine the particular strengths of gray irons, but in general terms they have practically zero reduction in area to fracture with tensile strengths ranging between 140 and 420 MPa, depending on the matrix. Gray cast irons are much stronger in compression than tension, the harmful cracks being closed up. Because of the wide range of flake sizes in different grades of gray iron and the consequent effect on the stiffness of the material, the Young's modulus can vary, in extreme cases, between about 70 GPa and 150 GPa, the smaller flake sizes of the stronger irons resulting in the higher moduli.

Exercise 15-2

Two tensile specimens of cast iron, both of initial diameter 12.8 mm, are tested to fracture. Specimen A had a maximum load of 58.0 kN and a fracture diameter of 12.4 mm. Specimen B had a maximum load of 18.6 kN and a fracture diameter of 12.8 mm. If these were pearlitic malleable iron and ferritic gray iron, respectively, describe any differences in (a) phases present, (b) shape of phases, and (c) spacing of phases that led to the measured differences.

Solution. (a) The phases present in Specimen A are ferrite α and cementite θ (combined as pearlite) and free cluster graphite; in Specimen B the phases are ferrite α and free flake graphite. (b) The principal difference in the shape of the phases will be the type of graphite. (c) There is likely to be no difference in the spacing of the phases. The measured differences in mechanical properties arise from the different characters of the free graphite, namely, sharp-ended flakes in the case of the gray iron, which promotes easy fracture, and bumpy clusters of graphite in the malleable iron, which are less deleterious and permit the pearlite matrix to display some of its ductility before the weakening effect of the graphite "holes" is felt.

The strengthening mechanisms that operate in the α and $(\alpha + \theta)$ matrices in irons are the same as those in steels, discussed in earlier chapters.

Exercise 15-3

Show that the relative strengths of different pearlitic gray cast irons, namely, 140 MPa for ASTM-20 with large flake size (flakes about 250 μm long) and 420 MPa for ASTM-60 with small flake size (about 30 μm long), follow the rules of brittle fracture (Chap. 14).

Solution. The mechanics of fracture would say that the strengths are inversely proportional to the square root of the flake size, whence

$$\frac{\sigma_{ASTM-20}}{\sigma_{ASTM-60}} = \sqrt{\frac{3}{25}} = 0.35$$

The given strength ratio is $140/420 = 0.33$, so the data seem consistent.

The relatively low strength and low ductility of gray irons make them exceedingly easy to machine. Applications include automobile engine blocks, flywheels, brake drums, piston rings, machine bases, and gears.

DUCTILE, NODULAR, OR SPHEROIDAL GRAY IRONS

The ductility of gray irons can be improved by adding *magnesium* (particularly) and/or *cerium* to the melt, which in gray irons containing only small amounts of sulfur makes the graphite come out as round blobs (not the flakes of untreated gray irons, nor the clusters of malleable white irons) directly from the melt. There is no reheating as with malleable white iron. The carbon content of such irons (variously called by all the names given in the heading above) is usually kept above 3% to ensure good castability.

With different cooling rates, *pearlitic* (Fig. 15-4) or *ferritic nodular* irons may be obtained. The absence of the sharp-ended flake form of the graphite, with its stress-raising problems, improves the strength and ductility of the gray irons immensely. The reduction in area to fracture in tension can be as great as 25%, and the tensile strength can range between 420 and 800 MPa. Precise values for both depend on the matrix and on any subsequent heat treatments, such as quenching, tempering, annealing, or surface hardening. Typical applications are for crankshafts, heavy-duty gears, and automobile door hinges.

Magnesium is above its boiling point (1110°C) at the temperatures of liquid cast iron, so certain precautions must be taken when adding this nodule-forming element; for example, it may be introduced as a nickel–17% magnesium alloy or in a pressure-tight ladle. (Similar problems occur in making brass from its elements, since zinc boils at 907°C, whereas copper does not melt until 1088°C.)

Nodular iron is a much newer material than malleable cast iron, having been developed during World War II. It is replacing malleable (white) iron for

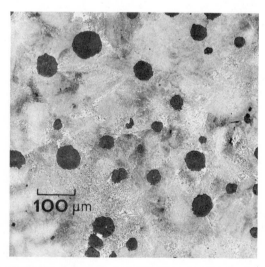

Figure 15-4 Pearlitic nodular cast iron, etched in 4% picral solution. (Courtesy BCIRA.)

many applications, since it is produced from the melt and does not require lengthy reheating procedures.

Exercise 15-4

Draw a schematic microstructure for each of the following, clearly labeling every phase: (a) pearlitic malleable cast iron, (b) ferritic nodular cast iron, (c) white cast iron, heated to 950°C for 24 hours and water-quenched.

Solution. Figure 15-5 illustrates the phases and microstructures.

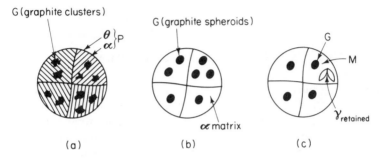

Figure 15-5 Solution to Exercise 15-4.

INOCULATED AND ALLOY IRONS

Irons that would otherwise be white irons of low silicon and carbon contents are given additions of graphitizer inoculants in the melt to cause graphite to

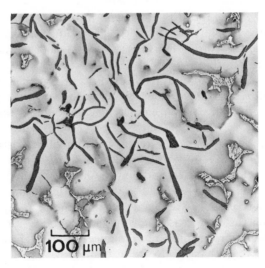

Figure 15-6 "Nomag" austenitic flake cast iron, etched in 4% picral solution. (Courtesy BCIRA.)

appear in the casting in a finely dispersed flake form rather than as clusters, with a consequent increase in strength (say, to 500 MPa). For example, calcium silicide is used in the *Meehanite* range of heavy duty irons.

Alloying elements, such as nickel, chromium, copper, or molybdenum, can be added to irons for special purposes. Nickel tends to produce a gray iron (see next section), acting like silicon in this regard. Large amounts of nickel (up to 25%) affect the Fe–C equilibrium diagram so much that the γ phase can exist at room temperature, giving *austenitic irons* (Fig. 15-6), which are nonmagnetic and have high electrical resistance and a high coefficient of thermal expansion.

COMPARISON AMONG DIFFERENT CAST IRONS

For engineering purposes different types of cast iron are not normally defined in terms of chemical composition. Rather, cast irons are specified according to mechanical properties and any other special properties required for particular purposes, with the chemistry being left to the foundry. This comes about because of the widely different compositions of the starting iron ores. Impurities other than silicon, such as sulfur, manganese, and phosphorus, can affect foundry practice and the final properties vary markedly. Deliberate additions of alloying elements likewise can be made to produce special cast irons. A range of heat treatments can be given to cast irons to produce matrix structures varying from very soft to very hard; for example, they can be reheated into the γ-region and then heat-treated as steels are to give *martensitic* or *tempered martensitic* structures. Appendix 10 gives an indication of the range of mechanical properties usually encountered.

Figure 15-7 shows the effect of silicon in promoting the dissociation of cementite into free graphite, and thus the transition between white and gray iron microstructures. *Mottled irons* are white/gray borderline cases, and the regions of the diagram labeled "mixture" correspond with a mixed pearlitic/ferritic microstructure. Cooling rate affects Fig. 15-7 in that as the percentage of silicon goes up, faster cooling rates would be be required to preserve a white iron microstructure, and at high enough silicon contents that is not always possible.

The inherent virtues, such as good machinability, corrosion resistance, and wear resistance, that are possessed by gray cast iron are shared by both malleable and nodular irons. In addition the latter have reasonable ductility, and either type could be substituted for steels in many applications. Service requirements and economics will dictate the final choice. Gray iron is lowest in cost and also the easiest to cast and obtain a sound, essentially nonporous structure. Nodular iron is similar but is subject to more shrinkage on solidification and hence requires special care in casting. Malleable iron is the most difficult to cast (as white iron), and the size of castings is more restricted (smaller) than nodular iron. If nodular iron does not require an anneal to

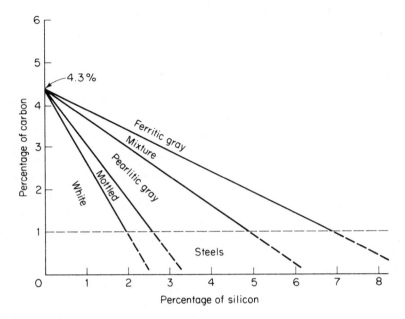

Figure 15-7 The effect of silicon in promoting the dissociation of cementite into free graphite. Mottled irons are borderline cases between white and gray iron.

ensure uniformity of properties and can thus be used directly on cooling from the melt, it is much cheaper than malleable iron.

Exercise 15-5

Describe the influence of phosphorus in cast irons.

Solution. In irons made from the common ores containing phosphorus, iron phosphide is probably the most important inclusion. The presence of phosphorus can offer some benefits, as during casting it increases the fluidity of the iron by lowering its melting point; this feature made possible the intricate casting found in Victorian ironwork, allowing fine detail and smooth finishes. However, it ·tends to separate during solidification and leave a liquid network, which can subsequently freeze with markedly different contraction from the parent phases. This results in weak porous regions, particularly in sections greater than, say, 10 mm across. In heavier-duty irons the phosphide must be eliminated, leaving graphite as the factor controlling strength.

WROUGHT IRON: ARMCO IRON

Wrought iron, as developed in the eighteenth and ninteenth centuries, was the result of "burning out" the excess carbon of pig iron in an attempt to produce pure iron, which would be comparatively ductile. Pig iron was remelted in a *puddling furnace* lined with a bed of iron oxide. It took about two hours of extremely arduous and skilled work on the part of the puddler, constantly stirring the iron with a rake or *rabble* introduced through a hole in the furnace door, before this amount was converted into a glowing plastic ball of wrought iron. (As the carbon content decreased, the melting point increased, and the furnace was incapable of reaching those higher temperatures.) The spongy mass of iron and slag had to be pounded or *shingled* under a heavy forging hammer to expel the cinder and to form it into a bloom suitable for conversion in the rolling mill into bars. Even so, bars rolled directly from the bloom in this way were of such poor quality as to be useless. To produce good-quality iron it was necessary to cut the bars into short lengths, bind them in bundles, forge them under the hammer, and roll them again at least once.

Rolling elongates the slag into fibers, and since these have comparatively weak interfaces, cracks become diverted along the slag fibers, producing a characteristic "woody" or fibrous fracture (Fig. 15-8). Wrought iron typically displays 25–35% elongation on a 50-mm gage length, with a yield strength of 200 MPa and tensile strength of 320 MPa; excessive phosphorus from the ore reduces these properties. Wrought iron has comparatively good shock resistance because of the fibrous fracture, and although rarely used nowadays, it was employed for chains, railway couplings, and crane hooks long after mild steel had become readily available. In fact, in the Aston process of manufacturing wrought iron, molten iron (already "purer" than wrought iron) from a

Figure 15-8 Fibrous, "woody" fracture of wrought iron nail.

Bessemer converter was poured into a synthetic slag to form pasty balls that could be rolled into bars.

The many thousands of miles of wrought iron rails that were laid during the great age of railroad construction, and all the wrought iron plates used for building bridges and ships' hulls and in boiler making in early Victorian times, were all produced by the extremely laborious process of puddling and repeated rolling. By the 1850s puddling had become a serious bottleneck, limiting production of the basic material on which the progress of the Industrial Revolution depended. This led to Nasmyth's idea of substituting a jet of steam in place of the rabbling bar, to increase the temperature of the furnace, and was the incentive for the introduction of Kelly's, Bessemer's, and Siemens' more modern methods of making tonnage steel.

Ingot (Armco) iron is almost pure ferrite (say, 0.03% C in Fe). It is made by fully oxidizing metal in the basic open-hearth furnace, which is capable of coping with the melting point of pure iron (1538°C). It is thus different from wrought iron, in that the slag is removed in the molten state, and there are no slag fibers in the ferrite matrix. The mechanical properties are lower than those of wrought iron, for example, yield strength 150 MPa and tensile strength 280 MPa. The Brinell hardness of Armco iron is about 80 Bhn; note that $S_u =$ 3.45 Bhn MPa here but that this relationship does *not* hold for cast irons in general.

Even purer iron can be made by electrolytic processes. The principal applications for these pure irons, such as in electric transformer cores, use their high magnetic properties.

REFERENCES

15-1. Brick, R. M., R. B. Gordon, and A. Phillips, *Structure and Properties of Alloys.* New York: McGraw-Hill Book Co., 1965.

15-2. Rollason, E. C., *Metallurgy for Engineers.* London: Edward Arnold, 1956.

15-3 Rolt, L. T. C., *Victorian Engineering.* London: Penguin Books, 1970.

PROBLEMS

15-1. Name the phases present in the following, and describe the shape of graphite whenever it is present (all held for 2 ks):

 (a) Ferritic gray iron, heated to 650°C.

 (b) Pearlitic nodular iron, heated to 950°C.

 (c) White cast iron of 2.67% carbon at 20°C.

 (d) Pearlitic malleable iron at 20°C.

15-2. Name the phases present in the following, and describe the shape of graphite whenever it is present (all held for 2 ks).:

 (a) White cast iron of 2.37% carbon, heated to 1047°C.

(b) Ferritic nodular iron at 20°C.

(c) Pearlitic malleable iron, heated to 950°C.

(d) Pearlitic gray iron at 20°C.

(e) 10100 steel, water-quenched from 810°C to 20°C.

15-3. Sketch and label the shape(s) of all nonmetallic phases (including all intermetallic compounds) present for each of the following:

(a) Pearlitic gray cast iron.

(b) Ferritic malleable cast iron.

(c) Pearlitic nodular iron.

15-4. Give reasons, in brief, why each of the three alloys in Problem 15-3 is manufactured.

15-5. A white cast-iron component with 2.40% carbon has a mass of 30 kg. Calculate the mass of noneutectoid iron carbide (θ, Fe_3C) present at room temperature.

15-6. A white-cast iron specimen of mass 1.5 kg contains 2.2% C. At room temperature find:

(a) The mass of eutectic iron carbide.

(b) The mass of eutectoid ferrite.

(c) The mass of noneutectic, noneutectoid iron carbide.

15-7. Calculate the mass of eutectic, eutectoid iron carbide present at 20°C in a 1.73-kg casting of white cast iron of 3.14% carbon.

15-8. **(a)** Calculate the mass of noneutectic eutectoid present in 7.23 kg of white cast iron of 2.67% carbon at 20°C.

(b) Calculate the mass of iron present in the total eutectoid in 7.23 kg of white cast iron of 2.67% carbon at 20°C.

15-9. Calculate the total mass of graphite at 20°C in a 100-kg mass of pearlitic malleable iron of 3.76% carbon.

15-10. Calculate the total mass of carbon in the noneutectic, noneutectoid iron carbide in a 7.38-kg bar of white cast iron of 2.69% carbon content.

15-11. **(a)** Calculate the mass of eutectic, noneutectoid iron carbide at 21°C in a white cast-iron casting of mass 1.3 kg and composition 3.4% C, 0.70% Si, 0.60% Mn, balance Fe.

(b) Calculate the mass of eutectoid, noneutectic ferrite at 21°C.

15-12. A ferritic ductile iron has a tensile strength of 400 MPa, whereas a pearlitic malleable iron has a strength of 670 MPa. Fully describe the heat treatments of each, show their final microstructures, label all phases, and explain why one is stronger. Which one would you expect to have the higher true strain at fracture, and why?

15-13. Consider the following specimens: Specimen G contains 3.5% C, 1.8% Si, 0.5% Ce, balance Fe; Specimen H contains 2.1% C, 0.7% Si, balance Fe; Specimen J contains 3.1% C, 1.9% Si, balance Fe. If all are air-cooled from the melt in the form of bars 10-mm in diameter, which has the greatest hardness?

15-14. The cast irons in the two tensile tests below were of identical composition. Cast iron A was an as-cast test piece 12.8 mm in diameter. Cast iron B was the same size and received the same treatment except that after casting it had been reheated to 950°C for 3 ks and furnaced-cooled.

	d_0, mm	P_{max}, kN	d_f, mm
Cast iron A	12.8	24.0	12.8
Cast iron B	12.8	17.3	12.8

(a) Explain, from a fundamental point of view, why the strengths are different.

(b) Name the phases present in each cast iron.

(c) Name the cast irons.

15-15. Describe the critical differences that would cause a designer to specify a pearlitic malleable cast iron for one application and a ferritic nodular cast iron for another application.

15-16. Although the chemical compositions of white and gray cast irons can be quite different, they are both brittle in the as-cast condition. Explain why.

15-17. Explain how the ductility of as-cast gray iron can be increased without altering the composition or reheating to the liquid state.

15-18. A relatively thick section is to be made from some type of cast iron and is to possess maximum ductility. Specify which type of cast iron you would use and explain you choice.

15-19. Typical properties of gray cast iron and fully annealed 1080 steel are given below. Sketch on log–log coordinates their two σ–ϵ curves, marking elastic and plastic regions.

C.I.: $S_u = 200$ MPa; $S_y = 200$ MPa; $E = 90$ GPa; $v = 0.3$; $\epsilon_f = 0.002$; %R.A. = 0%

1080: $S_u = 650$ MPa; $S_y = 400$ MPa; $E = 210$ GPa; $v = 0.3$; $\epsilon_y = 0.002$; %R.A. = 45%

16 Ceramics and Glasses

Ceramics are inorganic, nonmetallic solids whose microstructures in the broadest terms consist of mixtures of crystalline phases and glasses (see Fig. 4-4). The traditional family of ceramic materials incorporates clay products (china, bricks, tiles, refractories, abrasives), cement, enamels, and glasses. Many of the raw materials for these products are complex silicate and aluminum silicate minerals naturally occurring in the earth's crust. For example, ordinary clays are based on micrometer-size platelets of kaolinite crystals $[Al_2(Si_2O_5)(OH)_4]$; silica (SiO_2) is a principal component of common glass and also of various abrasive, whiteware, and refractory formulations. Other refractories, often used as linings of metallurgical furnaces, are manufactured from nonsilicate minerals such as the chromium and magnesium ores called chromite and magnesite.

The general strength characteristic of traditional ceramics is that they are brittle—chipped chinaware and drinking glasses, cracked plaster, and so on are not uncommon—so that ceramic bodies are often only lightly stressed in use, unless that loading is consistently compressive, as in masonry arch bridges or in a structural brick wall. Even so, stones such as flint, which could be broken into pieces having sharp edges, were used by early man as tools and weapons. Despite the characteristic of brittleness, ceramic raw materials are relatively inexpensive, and a multitude of applications have been found for ceramic products. Indeed, ceramics are favored for some uses because of particular combinations of chemical, electrical, and thermal properties not found in other materials. Many new ceramics have been developed that possess special properties.

There are two principal engineering applications for ceramic materials. The first concerns the traditional function of refractory solids for furnaces and linings of furnaces, whose origins are lost in antiquity. Refractory bricks are manufactured, in the main, from natural minerals. The metallurgical industries (particularly ferrous metals) are the major users of refractory materials—for melting, refining and casting, and hot forming and subsequent heat treatments. As strength is not of paramount concern, such applications are not considered in detail in this book. Indeed, since they must cope with liquid slags and metals, it is their chemical behavior that is often crucial and not their strength.

The second application *does* concern strength, particularly strength at high temperatures. As ceramics are degenerate metals in essence (i.e., metal carbides, oxides, and so on), they display great thermal stability. In the development of engines of high thermodynamic efficiency and of chemical plants, the requirements of designers for materials with adequate mechanical properties at high temperatures (among which creep resistance figures strongly) could be met with ferrous alloys as long as the operating temperatures did not exceed 500°C. Among such iron-base alloys are the 18 Cr or the 18 Cr–8 Ni austenitic steels (see Chap. 11), and also various tungsten and molybdenum alloys. High mechanical strength can be expected only at temperatures below the recrystallization range, and since it appears impossible to raise that temperature range in iron-base alloys significantly above 800°C, recourse was had to other materials when higher temperatures were specified. Thus were developed various "superalloys" of nickel, cobalt, and chromium such as the nimonic series. Even these alloys are eliminated for operating temperatures above 1000°C.

Nevertheless, materials are required in such applications as gas turbines, jet engines, nuclear plants, fuel burners, and high-temperature chemical plants for which a service temperature of 1000°C and higher is desirable. New types of ceramic are being developed for these applications. In addition to possessing adequate mechanical properties these solids must clearly be resistant to oxidation, so they also find use in the chemical industry. For example, alumina (Al_2O_3) ceramics have high strength and high wear resistance and are good insulators. They are used in high-speed machining of metals. The carbides, borides, and nitrides of the transition elements are variously used as cutting tools, dies, and for parts of aeroengines operating at high temperature. Traditional ceramics, although perhaps retaining their integrity at high temperatures, would not possess adequate strength (for example) in the uses here mentioned for the new ceramics.

In fact, a principal market for the new ceramics utilizes their hardness and wear resistance at ordinary temperatures rather than their high-temperature properties. For example, wire-drawing dies, seals, bearings, and thread guides in textile machinery are common applications. In the electrical industry they have wide use as insulators. Cutting tools require strength at high temperatures, and ceramics are used in high-temperature applications involving hostile environments along with high mechanical and thermal stresses. Ceramic fibers are now woven into blankets for insulation purposes and are replacing traditional firebricks on occasion.

Except for the glasses themselves, most ceramics consist of crystalline phases surrounded by glassy binders. Many of the new ceramics are purer than the traditional ceramics and furthermore aim to have much less glass-phase material. In some ceramics, particularly those carbides and borides used as cutting tools, the glassy phase is replaced by a metallic binding phase. For example, "cemented" titanium carbide, TiC, can have mixtures of cobalt, chromium, and nickel as the agent binding together the carbide, which is present in powder form. Not only is some small measure of ductility conferred on the ceramic (while retaining the high strength of the hard phase), but, more important, the thermal shock resistance is improved. If the ceramic content in such systems is reduced, we arrive at metallic alloys strengthened by refractory particles as opposed to a refractory held together by a metal. These hybrid materials are called *cermets*, among which are aluminum dispersion-hardened with particles of alumina (Al_2O_3), called *SAP alloys*, and molybdenum alloys containing 1–2% alumina. Other types of new ceramics are discussed later in this chapter.

SOLIDIFICATION AND SHAPING OF CERAMICS

In most ceramics other than the glasses themselves, the other hard crystalline phases solidify first, being surrounded by glassy binders. During solidification there is a large volume change associated with the formation of the solid phase (see Chap. 4). Moreover, the solid crystal regions often have anisotropic coefficients of expansion that are different from those of the vitreous matrix. The low and anisotropic thermal conductivity of these materials, when coupled with the brittle nature of the solid regions, gives rise to marked residual stress and cracking problems on solidification. Consequently the melting and casting techniques well known in metal production cannot be used with ceramics. Furthermore, because of their brittleness, ceramics cannot be shaped by plastic deformation methods as is common for many metals.

Manufacturing processes for ceramics usually start with the raw material ground into *powder* form. The method of shaping traditional ceramics is then to mix the powder into a liquid paste that has appropriate rheological (flow) properties to permit forming. Such shaping may be done by hand, as with potter's clay, or by machine, as in the extrusion of sewer pipe and some types of brick. The body is then dried and fired at a temperature appropriate for the ceramic between, say, 800 and 1800°C. Firing causes the powder particles to stick together, thus retaining the shape of the formed body. The mechanics of binding may relate to melting of parts of the ceramic mixture, to diffusion reactions across particle interfaces, or to processes of *sintering*, in which surface tension at high temperature consolidates the particles. Common cement reacts with water to form a cementative matrix and usually dries and hardens by chemical reaction at ambient, rather than furnace, temperatures; hydration may continue over a long period of time, the strength of concrete increasing in step.

There will be a reduction in the overall volume of ceramic bodies after firing as the voids between the particles are eliminated. This shrinkage must be allowed for in design. Voids are never completely eliminated, however, and it is inevitable that bodies made in this way from powders will be porous. For example, if powder particles of a single size are used, cubic packing considerations (Chap. 5) suggest that some 40% of the volume consists of voids between the particles. Practical powders, of course, have a range of sizes, and some small particles will fill some of the holes between the large particles. Nevertheless, the body will retain some porosity. As the size, shape, and distribution of pores markedly affect the properties of a material, *porosity* is as significant a microstructural feature in ceramics as the crystalline and glassy solid phases themselves (compare the size, shape, and distribution of free graphite in cast irons).

In order to improve density, *high pressures* are used in some processes of forming ceramic powders. The methods concern either *cold pressing* followed by *sintering* or the related method of *hot pressing*. Both techniques are used in "powder metallurgy," which concerns the manufacture of articles from metal powders. Cold pressing and sintering of dry ceramic powder mixtures is actually similar to the slurry and paste methods just described, except that high-capacity presses are used so that much higher green densities (i.e., densities before firing) may be achieved. In hot pressing the powder is heated and pressed at the same time so that highly dense, strong bodies result. Because of the combination of simultaneous pressure and heat, lower temperatures can be employed for hot pressing than for sintering after cold pressing. Hot pressing equipment is more expensive, however.

Exercise 16-1

Other things being equal, are ceramic powder compacts likely to be less dense or more dense than metal powder compacts after sintering or hot pressing?

Solution. Ceramic powders are harder than metal powders in general, so that they will not flow and compact as much as metal powders. For the same reason the areas of contact between particles will be smaller in ceramics than in metals; this, coupled with inherently lower rates of diffusion, means that sintering will not be as efficient in ceramics as in metals. All in all, this tends to produce more porous ceramic products.

There are special versions of sintering and hot-pressing methods. For example, *liquid-phase* sintering at temperatures just above the melting point of the metallic binder is carried out in the manufacture of cermets, in order that the metal phase may run into as many pores as possible to produce a dense body. *Reaction* sintering and hot pressing deliberately encourage solid-state transformations at the high temperature to obtain higher densities. In this way dense silicon nitride (Si_3N_4) is immediately sintered in the same process in which it is produced chemically from heating silicon powder compacts in nitrogen at about 1400°C. Another process is *sinter forging*, in which the article is forged into shape as sintering takes place.

The method of pressing determines the final density achievable. Friction between the powder particles and at the die walls means that some regions of a pressed shape may have poor compaction in relation to other regions. This means that some areas of a body are finally weaker than others. Dieless *isostatic* pressing (done cold before sintering or done hot) is coming into use; here the object is placed in a flexible bag surrounded by fluid under hydrostatic pressure. Oddly shaped compacts may be readily pressed to give quite uniform density. Many ceramic materials prepared in these ways have porosities of less than a few percent.

As the raw powder sizes are very small, bodies having acceptably smooth surface finish can be produced. In fact, as the fired, sintered or hot-pressed article is so hard and brittle, little can be done in the way of post-forming, other than by grinding, spark machining, or ultrasonic drilling, for example. In the case of Si_3N_4 mentioned earlier, it is possible first to form the silicon powder into the roughly dimensioned shape required and then to perform an intermediate nitriding at 1100°C in nitrogen, which converts perhaps 5% of the body to Si_3N_4. The mechanical properties of partially nitrided articles are such as to allow easy machining. Fully nitrided articles are then obtained as already described.

Fully dense ceramic solids can be obtained under special conditions. These include single crystals of artificial gemstones, such as sapphire (a variety of alumina), which are used as jeweled bearings for delicate mechanisms. Pore-free alumina can be made from powder by adding small amounts of Li and MgO before sintering. The trade name Lucalox is used by General Electric for such fully dense, translucent Al_2O_3. *Glass–ceramics* form an important group of dense ceramics based on Al_2O_3/SiO_2. It can be arranged that these materials are wholly glassy at first, in which state they can be readily shaped by conventional glass-forming techniques (as described in the next section). Subsequently nucleating agents, such as TiO_2 or ZrO_2, cause controlled devitrification (i.e., controlled removal from the character and appearance of glass) so that the final pore-free structure consists of perhaps 97% very small (micronsize) crystals, with a small residual glassy phase. Glass–ceramics are essentially free of the interparticle and grain boundary weaknesses commonly found in ceramic bodies made by the usual methods. Most particularly these solids have an almost zero expansion coefficient, so that they have important applications in thermal shock situations. *Pyroceram* is one trade name for glass–ceramics. Similar sorts of devitrified structures can be produced from metallurgical slags and can be used to make tiles.

SOLIDIFICATION AND SHAPING OF GLASSES

The difficulties of differential phase shrinkage that occur in ceramics during solidification are not present in predominantly glassy solids. Therefore, instead of shaping first and then using high-temperature "welding," the raw materials of glasses can be fused at high temperature and then shaped at lower temperatures (after cooling and reheating if required) much in the same manner as metals. Thus glass articles can be cast, sheet and plate glass can be made by rolling, glass tubes and glass fibers can be drawn, and, of course, glass blowing and blow molding are important techniques for necked containers.

Viscosity is the principal parameter determining the temperatures required for different forming processes. For mechanical working, the glass should not be too "runny" (otherwise, it will not retain its shape) yet must flow readily in order to give reasonably low working loads. The working range of viscosity is 1 to 10 000 kPa-s, which for common soda–lime glass means a temperature range between about 600 and 900°C. Such glass "melts" at, say, 1400°C. Relief of residual stresses arising from working processes can be performed at about 400°C, when the viscosity is perhaps 1 TPa-s; this is termed *annealing* in the glass trade, although *stress relieving* would be the metallurgical term.

The mechanical behavior of hot glass can reasonably be represented by a linear viscous relation between stress and strain rate. Following Chap. 12, we may write, in terms of tensile stress σ and tensile strain rate $\dot{\epsilon}$,

$$\sigma \propto \dot{\epsilon}$$

Now $\dot{\epsilon} = \dot{L}/L = -\dot{A}/A$, where L and A are the current length and cross-sectional area, respectively, of a bar under tension. Then, writing $\sigma = P/A$, where P is the tensile load, we have

$$\frac{P}{A} \propto \frac{-\dot{A}}{A}$$

That is, $\dot{A}$ must be constant (as P is constant along the bar). Thus, whatever the initial cross-sectional profiles, all cross sections lose area at the same rate, and originally narrow regions do not neck down faster than others. This feature of stability against necking is important in many glass-forming operations, not least of which is the production of glass rods (more than 1 mm in diameter) by pulling vertically upwards from hot glass. Thick fibers (0.1 $\sim$ 1 mm) are drawn from larger-diameter rods. Thin fibers, on the other hand, are drawn through nozzles in the base of electrically heated platinum crucibles.

An important development in manufacturing sheet glass is *Pilkington's float-glass* method, in which a layer of glass is cast on top of a bath of liquid tin in a protective atmosphere. The lower surface acquires the smooth finish of the tin, and the top surface is "fire-polished" by external radiation, so that subsequent mechanical polishing, found necessary in ordinary plate glass manufacture to improve the strength, is avoided. Surface damage, which markedly affects the strength of glass, as discussed in following sections, is thereby reduced.

STRUCTURE OF CERAMICS AND GLASSES

As described in Chap. 4, the changes in viscosity with temperature of glassy, amorphous substances are so gradual that no clear-cut melting or freezing temperature can be identified. The glass-transition temperature is defined as that at which the viscosity passes through the (arbitrarily chosen) value of 100 TPa-s. For this reason glass is often termed a supercooled liquid, X-ray results for the structure showing molecular arrangements normally associated with liquids.

Glass at room temperature has frozen into it a molecular arrangement appropriate to some higher temperature, called the *fictive temperature*. As a result of the rapid cooling that can occur in the production of glass fibers (which at 70 m/s can cool to room temperature from 1300°C within some 30 mm of the orifice of the crucible from which they are drawn), the glass may have a fictive temperature many hundreds of degrees above the glass-transition temperature T_g. If glass is annealed, the fictive and transformation temperatures become identical.

Common bottle and window glass is made from silica, SiO_2, with additions of Na_2O and CaO fluxes that, among other things, reduce the temperature of fusion. Optical glass (so-called flint glass) with high refractive indices contains PbO as the active flux. When boric oxide (B_2O_3) is used along with the principal ingredient silica, Pyrex-type glasses with a low coefficient of expansion result. In general, in these and other glasses, the interatomic bonding is mainly covalent in nature and the amorphous array is probably an irregular three-dimensional network lattice of oxygen polyhedra akin to cross-linked network and ladder polymers (see Fig. 4-3 and Chap. 13). Under an optical microscope glasses are generally featureless.

Exercise 16-2

Explain why the addition of metal oxides to silica produces softening.

Solution.   The metal atoms effectively reduce the number of cross links in the random glass structure, so that lower viscosities are attained at lower temperatures than otherwise. This behavior is analogous to plasticization in polymers, whereas the vulcanization of rubber aims to produce exactly the opposite effect (Chap. 13).

Ceramic microstructures can be determined by polishing and etching cross sections and viewing under an optical microscope with reflected light, as done with metals. In addition, very thin ceramic sections (about 20 μm thick) can be viewed by transmitted light; this technique is popular in geological investigations.

Equilibrium-phase diagrams have been constructed for many ceramic systems; in these, mixtures of compounds are plotted instead of metallic elemental mixtures as in Chap. 8. For example, the equilibrium diagram for the MgO–CaO binary system has the general shape of the Ag–Cu system (Fig. 8-2) with a eutectic composed of about 66% CaO and 34% MgO at 2370°C, and end regions of limiting solid solubility of 8% MgO in CaO and 17% CaO in MgO at the eutectic temperature, both decreasing to very small values at lower temperatures. Other systems have more complex diagrams to a greater or lesser extent; some multicomponent ceramic systems have very complicated diagrams. It must be remembered that these diagrams describe the crystalline phases present and not any glassy binding phase.

Exercise 16-3

Are there likely to be difficulties associated with the construction of phase diagrams for ceramics, in comparison with metal alloys?

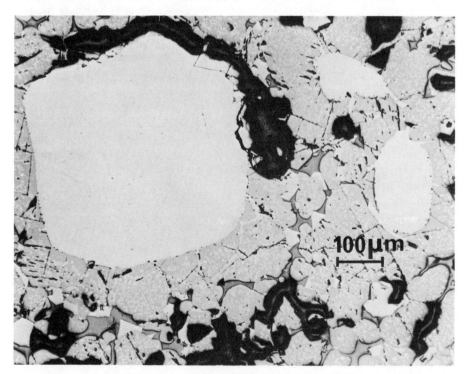

Figure 16-1 Magnesite–chrome refractory (hard fired); large, white grains are chrome, smaller, light gray grains are periclase (native magnesium oxide) together with exsolved magnesia–ferrite particles, medium-gray grains are silicate particles, and black regions are pores. (Courtesy British Ceramic Research Association Ltd.)

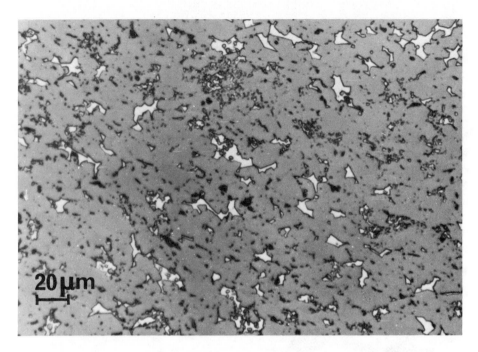

Figure 16-2 Self-bonded SiC, in which SiC forms on original SiC grains. White regions are excess Si. (Courtesy British Ceramic Research Association Ltd.)

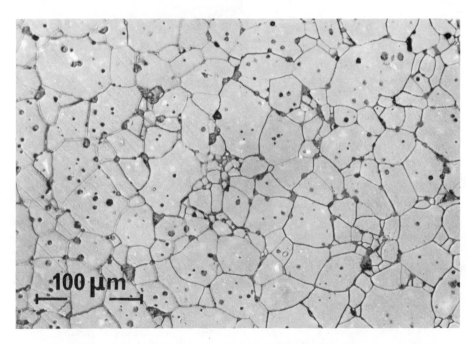

Figure 16-3 Pure Al_2O_3, thermally etched to show grains. Rounded pores appear black. (Courtesy British Ceramic Research Association Ltd.)

Figure 16-4 Hot-pressed Si_3N_4 (MgO used as a pressing aid). Small pores appear black. (Courtesy British Ceramic Research Association Ltd.)

Solution. Both cooling curve methods and quenching methods (akin to the construction of TTT curves in alloys) are used, among others. The low thermal conductivity and high viscosity of the liquid phases in many ceramic systems means that equilibrium is attained only after exceedingly long times. Indeed, it may be said that for some systems equilibrium is never attained, and the phases produced on cooling are "metastable" phases. It can be shown on thermodynamic grounds that such phases are not the phases of lowest energy, which equilibrium demands, so that it is not always clear that the observed phases represent the long-term structure. (See Exercise 4-1 for an example involving glass.)

A typical ceramic microstructure will consist of diverse crystalline phases in a glassy matrix, together perhaps with inclusions of impurities and, most particularly, pores. The character of pores significantly affects the properties of ceramics. Some pores remain isolated, being those gaps not joined by lenses between the original powder particles. Other ceramics may have regions where the pores are interconnected. Such features affect strength, thermal conductivity, and so on. Figure 16-1 shows a polished section of magnesite–chrome refractory such as used in steel-making furnaces. The black regions are connected pores, which are featured prominently in this complicated microstructure. Figure 16-2 displays self-bonded SiC, formed by firing a mixture of carbon and

SiC in an atmosphere of silicon; new silicon carbide grows on the original SiC grains. Excess silicon is shown by the white regions. Pure Al_2O_3, thermally etched to show grains, appears in Fig. 16-3; the rounded pores appear black. Much finer pore size is shown in Fig. 16-4, which is a micrograph of hot-pressed Si_3N_4.

STRENGTHS

Extreme brittleness is the pervading characteristic of most glasses and ceramics in bulk. The room-temperature fracture toughness of typical glass is only about 20 J/m^2 (which with a Young's modulus of 70 GPa gives a critical stress-intensity factor K_c of about 1 MPa$\sqrt{m}$). Furthermore, ordinary glass contains many surface microflaws, which act as starter cracks. As described in Chap. 14, Griffith performed his classical fracture experiments on glass rods, and the microflaws in glass have come to be called Griffith cracks. They have been shown to be a few micrometers in size, with a numerical density of perhaps 10^9 m^{-2}, which is consistent with the observed tensile strength of bulk glass of some 70 MPa and toughness of some 20 J/m^2. Dense ceramics, whose inherent resistance to crack propagation is also comparatively small, display low practical strengths (10–100 MPa) because of similar sorts of surface flaws. In addition, porous ceramics have low engineering strengths as a result of internal voids.

Glass in the form of thin filaments (less than 50 μm in diameter) and other ceramics such as Al_2O_3 and Si_3N_4 in fiber or whisker form usually have significantly higher tensile strengths, which can approach 4 GPa (already noted in Chap. 14). Such strengths are almost equal to the theoretical strength of approximately $E/10$ and represent in the case of glass a 50 : 1 increase over the strength of the bulk solid. Although a similar trend of increasing strength with decreasing size is not unknown for some metals and high polymers, the improvements are not so marked in these solids. The strength–diameter relation is explained on a *statistical* basis: There is less chance of a flaw or void being present in a small volume of material (i.e., in thin fibers rather than thick rods), so there must be a trend of higher strengths in thinner filaments and whiskers. Statistical approaches to breaking strength are discussed in a later section. The high strength of commercial glass filaments is used to advantage in glass-fiber composites (Chap. 17), and other composites exploit increases in both strength and modulus of fibers and whiskers of graphite, boron, silicon nitride, and so on.

It seems clear that if the starter-crack flaws in glass and ceramic solids can be reduced in size or otherwise rendered harmless, the strength of the solids in bulk should be improved. Where do the damaging flaws come from? They arise in the main from handling, which can produce extremely high local contact stresses in rubbing, for example. Thus freshly made glass rods (1–3 mm in diameter) produced by vertical drawing from molten glass can display the sort of strengths associated with thin filaments (0.7 ~ 4 GPa), if tested *immediately* after manufacture and if care is taken to *avoid handling* the gage section. Even so, the customary bulk strength of 70 MPa will be displayed by such rods as soon as they are subjected to even the simplest handling (the scratching caused by touching with a coat sleeve on a test piece gage length, for example).

Figure 16-5 contrasts the usual strength–diameter behavior, as first observed by Griffith, with the data of Thomas for freshly drawn fibers and rods. Clearly, if the microcrack flaws can be eliminated, bulk glass should be as strong as thin filaments. In commercial practice there will still be strength–size variations, but appropriate coatings (*sizes*) are used on filaments in order to protect them in transit between the glass plant and the glass-fiber manufacturing industry; terrific scatter in strengths would result if uncoated filaments were allowed to rub over one another on a winding drum.

Sometimes it is possible to reduce the number of internal voids during or after manufacture, as in the hot isostatic pressing of ceramics. Moreover, commercial glass rods sometimes contain bubbles, and experiments show that it is possible to improve the strength to above some 1.8 GPa by floating out the bubbles by holding the bulk glass at 1400°C or so for a week prior to the rods' being made.

Methods are available to increase the strength of surface-damaged glass. If the sharp-ended flaws in the glass can be healed up or put into a state of compressive stress, the strength should revert to the high values of undamaged glass. One method is to *acid-polish* the glass in a hydrofluoric/sulfuric acid mixture, which etches away the sharp flaws. Commercial glass rods can be made to show strengths of over 1 GPa in this way; because of internal micro-

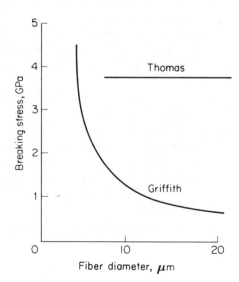

Figure 16-5 Strength–fiber-diameter behavior for glass filaments. Griffith results are contrasted with data on freshly drawn fibers and rods obtained by Thomas.

bubbles, higher strengths are rarely achieved. Mechanical polishing of sheet glass can remove some of the worst flaws to give strengths of perhaps 100 MPa. The inverse relation between flaw size and strength (Chap. 14) is well established by experiments in which mechanically polished glass is deliberately sandblasted; the strength is reduced right down to 14 MPa. Glass and ceramics such as alumina are readily "flame-polished" and their strengths much improved thereby, provided that no chemical degradation or change occurs in the process.

Exercise 16-4

Why is pottery glazed?

Solution. Glazes used for pottery are glasses of low thermal expansion which, because of the difference in expansion coefficient of the base ceramic, end up in a state of residual compressive stress on cooling down. Microflaws are thereby squeezed together, and higher surface tensile stresses are required to cause fracture. In a similar manner so-called *toughened* glass is rapidly quenched during manufacture from the plastic state so as to leave the free surfaces in a state of residual compressive stress.

Chemical effects play their part in determining the strengths of ceramics and glasses. Although acid polishing, as just described, smooths out the microflaws that arise from scratching, some chemicals produce the opposite effect and can intensify the action of the flaws by enlarging them through dissolution. Again, the fracture toughness can be altered by the environment. Glass is thus susceptible to the effects of water layers adsorbed on the surface, so that atmospheric humidity affects the strength. Exposure of 120 h in 100% relative humidity reduces the tensile strength of freshly drawn rods from 3.5 GPa to 2.1 GPa; lower humidities produce smaller effects. Likewise the phenomenon of *delayed fracture* (also wrongly called *static fatigue*) in glass, where sudden fracture at low stress occurs after a long time (see Chap. 18), is caused by small flaws slowly growing to the larger critical size necessary at that low stress to satisfy the particular fracture mechanics formula for the test piece geometry involved.

As a result of the limited size of samples, brittleness in handling, and alignment difficulties on the testing machine, there is often not the freedom of choice of test piece geometry with ceramics and glasses as with other solids. For these reasons, three-point bend testing is popular, with the maximum outside tensile breaking stress being calculated from some beam formula. This stress is sometimes called the *modulus of rupture* of the solid. Appendix 11 gives typical room-temperature quasi-static strength values for a selection of ceramics and glasses.

The mechanical properties of concrete materials are discussed further in Chap. 17.

THERMAL STRESSES AND THERMAL SHOCK

Materials usually expand when heated and contract when cooled. Allowance for such movements must be made in the design of engineering components. If such changes in dimensions are restrained as the temperature is changed, thermal stresses will be set up in the body. For example, a rod of length L, heated uniformly through a temperature interval ΔT, will extend by $L\alpha\Delta T$, where α is the coefficient of thermal expansion. If that movement is completely resisted, it is equivalent to imposing a strain of magnitude $\epsilon = L\alpha\Delta T/L = \alpha\Delta T$ on the rod. If the strain is within the elastic limit (as is usually the case), a compressive stress of magnitude $\sigma = E\alpha\Delta T$ is generated in the rod; if the strain is great enough to go over into the plastic region (which is uncommon), the associated stress would be given by some expression such as $\sigma = K\epsilon^n$. Tensile stresses of similar magnitudes are set up when a body is fully restrained from contracting during cooling.

Brittle solids, such as ceramics and glasses, which fail at bulk elastic stresses σ_f (given by some fracture-mechanics formula involving K_c and inherent flaw size), can therefore break under thermal stress alone when contraction is prevented. In the case of simple bars, uniformly heated, fracture occurs when the temperature change from the starting condition satisfies $\sigma_f = E\alpha\Delta T$. The coefficients of thermal expansion of many ceramics are about $(4 \sim 10) \times 10^{-6}(°C)^{-1}$; by using the bend strengths in Appendix 11 as appropriate indicators of the thermal failure tensile stress σ_f, it can be shown that the value of ΔT ($= \sigma_f/E\alpha$) necessary to cause fracture in many ceramics is some 100–200°C. That is, a bar in thermal and mechanical equilibrium, if fully restrained from movement, requires a hundred or so degrees Celsius slow change in (uniform) temperature to cause fracture. In comparison, brittle metals that fail with little if any plastic flow (such as tool steels) require greater ΔT (say, 400°C), since their σ_f/E is larger than that for ceramics, even though α is also greater.

In practice, components are not fully restrained. Even so, thermal stresses may develop in a body that is free to move if its temperature is not uniform. The foregoing treatment assumed the same temperature level throughout the body. However, if there are temperature gradients in the body (caused, for example, by nonuniform heating compounded by bad thermal conductivity in the body), internal stresses will be set up as different parts of the solid, which would expand or contract by different amounts at the different temperature levels, are constrained in their movement by adjacent regions; the restraint is now internal, not external. The final displacement and strain field is a compromise, wherein regions that would have expanded much are restricted in that movement and regions that would not have expanded much are forced to suffer greater displacements than usual for the same temperature change. For example, a piece of material at room temperature when placed in a furnace goes through a heating stage when the outsides are hotter than the inside, the magnitude of the differences in temperature depending on the conditions of heat transfer and the thermal conductivity of the solid. As the outsides are restrained in their expansion by the insides, thermal stresses are set up whereby the outsides are in compression and the inside goes into tension. Of course, the overall stress distribution is such that the net load and moment are zero in the free body. The thermal conductivity of metals is quite high, so that the associated stresses are low and few problems are caused, particularly if the heating rate is low. In the case of ceramics and glasses, however, where the thermal conductivity is generally lower, such stresses can sometimes cause problems—not only for ovenware and so on, but during manufacture when bodies can crack during firing.

The reverse process of cooling produces similar effects, and severe quenching can produce much more rapid changes in temperature, much more severe thermal gradients, and hence much more severe thermal stresses. Time is required to transfer the heat out of the hot, thick sections, so that cooling rates are different at different locations in the body. The so-called "mass effect," Jominy curves, and related problems of hardenability applying to metals are discussed in Chap. 10. Thermal gradients (which are the driving forces of heat transfer) are more severe when the thermal conductivity of the solid is low, so that, other things being equal, the possibility of thermal cracking during quenching is more likely with ceramics and glasses than metals. Even so, quench cracks do sometimes occur in martensite, as a result of critical combinations of high thermal stress in thick sections and low ϵ_f for martensite.

Resistance to thermal shock is the name given to the ability of a solid to withstand sudden changes in temperature. Thermal shock, as evinced by cracking of glasses into which hot liquids have been poured or by cracking of ovenware, is more of a problem with ceramics and glasses than with metals, although it must not be neglected in the design of turbine blades manufactured from refractory metals. Sometimes thermal cracking breaks a component in one go; at other times, myriad small cracks are initiated and progressively enlarge with successive thermal shocks, thereby degrading the retained strength in a cumulative fashion. For example, the starter-crack pores in some ceramics grow slowly with increasing number of shocks until the retained strength is incapable of withstanding the severity of the shock, when the body breaks apart. The appearance of surface cracking on metals is called *thermal checking*; similar patterns occur in mud when ponds dry out and in varnish coatings on old furniture and paintings.

One traditional type of thermal shock test consists of successively plunging a small specimen a number of times from some high temperature into a cold fluid and then looking for surface cracks. The temperature from which the quenchings take place is increased and the process repeated until cracks are observed. The final temperature at which cracks are initiated is used to rank materials for given geometries and environments (i.e., quenching media). The type and size of specimen and the heating and quenching schedules vary from worker to worker and are arbitrary. Such tests are sometimes continued at increasing temperature levels until the already initiated cracks propagate in shock so that the specimens break completely. Then, in addition to the merit-ranking based on crack initiation, some idea of the life of the part is available by adding up the number of heating and quenching cycles to failure. Insofar as the heating and quenching schedules are arbitrary, the results are quite empirical. Nevertheless, useful data can be generated for thermal cracking situations such as those encountered in interrupted machining cuts where the tool repeatedly enters and leaves the workpiece.

For a given type of specimen and given conditions of heat transfer at the interface between the initially hot body and cold quenching medium (or initially cold body and hot medium), it is possible to perform calculations for the thermal stresses induced in a body in terms of the thermal conductivity of the body. Then instead of the simple thermal stress of magnitude $E\alpha\Delta T$ produced by the same change in *uniform* temperature throughout a fully restrained body, we have

$$\sigma = ME\alpha\Delta T$$

where M depends on the thermal conductivity, heat-transfer coefficient, and geometry of the body and ΔT is now understood to occur suddenly at the boundaries of the body, rather than uniformly and slowly as before.

In general, M varies inversely with k, the thermal conductivity, in a complicated fashion. Thus the critical shock temperature difference to cause cracking is now dependent on $k\sigma_f/(E\alpha)$. This parameter is sometimes known as the *spalling resistance index* and may be used to rank materials with respect to likely resistance to thermal shock. Large values of $k\sigma_f/(E\alpha)$ are favorable, as they mean that large thermal gradients are required before shock occurs. Materials satisfying this condition have high strength, good thermal conductivity, low stiffness, and low coefficients of thermal expansion. Appendix 11 gives values of $k\sigma_f/(E\alpha)$. Glasses have values less than 1 kW/m, and ceramics less than 10 kW/m. For comparison, refractory metals such as molybdenum or the nickel-based superalloys might have $k\sigma_f/(E\alpha)$ around 100 kW/m and high-strength steel even larger values.

Exercise 16-5

Describe the characteristics required of an ideal cutting tool material, and relate the properties of coated, cemented solids to this ideal.

Solution.   The final properties of engineering solids are compromises between the ideal and the attainable. Ideal tools would have: (1) high elastic moduli to minimize deflection, (2) high hardness at operating temperature to resist deformation and to provide wear resistance, (3) high fracture strength, (4) high thermal conductivity and a low thermal expansion coefficient to resist thermal shock, (5) high chemical stability to resist attack by workpiece and environment, (6) high stability at high temperatures.

Some of these properties are mutually irreconcilable. The chemical stability and hardness of Al_2O_3 and TiC, especially at the temperatures encountered in cutting, make them excellent candidates as materials for cutting tools. However, they are not as strong as WC-based materials, which has limited their application. On the other hand, plain WC-alloys do not have the wear resistance. Thus TiC/WC alloys are manufactured in an attempt to combine the best features of both or to tailor the tool to the application (e.g., wear resistance is sometimes sacrificed for shock resistance in tools suitable for roughing and milling operations).

The development of disposable, indexable tool inserts has allowed even better materials, having coated layers, to be developed. The part of the tool in contact with the workpiece can be made of good wearing material and bonded to a tough substrate with high thermal conductivity and thermal mass (which provides efficient heat removal from the cutting edge) together with high elastic modulus to provide support to the cutting edge.

Care should be taken about exactly what σ_f means. It could be the stress that would initiate a crack and catastrophically break the body. However, there are situations, already mentioned, particularly with porous bodies, in which the multitude of small flaws are enlarged by a thermal shock but complete rupture does not occur. That is, *crack arrest* occurs after some propagation, where excess strain energy at crack initiation is converted into crack kinetic energy but is ultimately absorbed by the body. It can be shown that the damage, or extent of crack propagation, caused in these circumstances is inversely proportional to $(ER/\sigma_f^2) = (K_c/\sigma_f)^2$, where R is the dynamic fracture toughness in the shock environment (K_c is the dynamic critical stress-intensity factor). This type of subcritical crack growth under shock conditions is therefore less likely to occur in solids with high toughness-to-strength ratios. It can also be shown that the extent of damage in bodies of a given porosity is less for a large number of small pores than for a few large pores. Appendix 11 gives values of K_c/σ_f.

The two parameters $k\sigma_f/(E\alpha)$ and K_c/σ_f should correlate with the various empirical tests mentioned earlier, which establish (1) critical quenching temperatures at which surface cracks are initiated and (2) the number of further shocks that can be withstood before complete rupture.

Subsequent room-temperature testing of ceramic bodies that have suffered thermal shock but have not been destroyed shows a degraded strength, as the fracture loads are those to break test pieces containing larger flaws (extent of damage) rather than the original small pore size. Figure 16-6 shows retained room-temperature bend strengths after ten thermal cycles as a function of shock temperature interval for (a) a type of titanium carbide cutting tool material and (b) a cordierite (lithium–aluminum–silicate) ceramic used in turbines. In Fig. 16-6(a) a critical quenching temperature interval of about 400°C is shown; beyond this there is a marked reduction in residual bend strength, which remains relatively constant thereafter. The unshocked high-temperature strength of the carbide (open points) remains constant at the

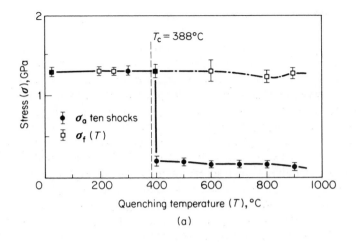

(a)

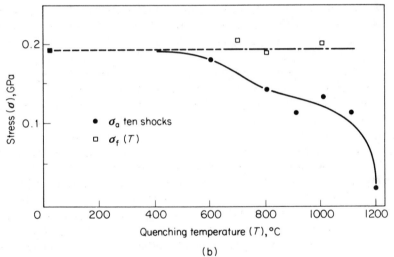

(b)

Figure 16-6 Retained room-temperature bend strengths after ten thermal cycles as
a function of shock temperature interval for (a) a type of TiC cutting tool material,
(b) a cordierite (lithium–aluminum–silicate) turbine ceramic. [From Y. W. Mai and
A. G. Atkins, *J. Mater. Sci.*, **10** (1975), p. 1904.]

room-temperature value to temperatures beyond the critical temperature. The
behavior shown in Fig. 16-6(b) is different: The after-shock strength gradually
decreases as the shock-temperature interval increases. The contrasting pattern
is bound up in differences in the ease of crack initiation and growth at high
temperatures; that is, the parameters $k\sigma_f/(E\alpha)$ and K_c/σ_f change their relative
ranking with temperature, since k, σ_f, E, and α alter much more in some
materials than in others.

STATISTICAL STRENGTHS OF SOLIDS

Even in a given "state of damage" (as-drawn, acid-etched, ground and polished, shot-blasted, and so on in the case of glass), the microflaws in ceramic
bodies will actually vary in size and number, so it is inevitable that there will
be a statistical variation in the fracture loads of test pieces (and hence the
commonly quoted strengths) even when the samples are taken from the same
large piece. We might anticipate that, under conditions of hydrostatic tensile
stress, final fracture would result from propagation of the severest crack. However, when the stress varies throughout a brittle ceramic or glass body containing a distribution of flaws, final fracture might result from propagation of a
smaller crack whose orientation to the stress field was appropriate. In this way
the breaking loads of brittle solids are likely to depend on the size of the
regions under stress—particularly, perhaps, those under tensile stress.

Again, as the breaking load of a test piece reflects the load necessary for
propagation of the microflaws and as, according to the mechanics of fracture,
that load depends on the geometry of the test piece and the mode of deformation (Chap. 14), we should not expect the same "tensile strength" from tests
in, say, simple tension and bending on specimens of different sizes. Each type
of specimen is likely to have different distributions of strength about different
mean values. These variations in strength are in addition to any variations to
be expected on the basis of accuracy of the experiments.

A strength value quoted from a given series of experiments is the arithmetic mean of all the results. Other sets of experiments on the same brittle
material are unlikely to give the same result. It is not really possible to indicate
an exact value for the fracture load or stress, but by using methods of statistics
it is possible to quote a *probability of rupture* p at a given stress σ. We have
$p = \phi(\sigma)$, which is a distribution curve of probability against strength. At low
values of applied stress, $p \to 0$; when a body is highly stressed, $p \to 1$. When the
$p = \phi(\sigma)$ relation is known for a test piece of a given material and size, the
(different) $p = \phi'(\sigma)$ relation for test pieces of other sizes can be related to it as
follows:

$$(1 - p_V) = (1 - p_0)^V$$

where p_0 (a function of σ) is the distribution curve for a body of unit volume
and p_V (a different function of σ) is the distribution curve for a body of volume
V times unit volume.

It is convenient to define a new parameter B, called the *risk of rupture*, as
$B = -\log(1 - p_V) = -V \log(1 - p_0)$. In an isotropic, homogeneous solid
where the chance of failing is the same everywhere throughout the volume,
p is simply some function $f(\sigma)$ of the magnitude of the stress. Then $B = \int_V f(\sigma)\, dV$; that is, the risk of rupture depends on the stress and volume of the
body.

The arithmetic mean (σ_{av}) from a series of experimental results for breaking stress or load (which come from the wide statistical variation of strengths

that are possible in the test piece) can be shown to be related to B by

$$\sigma_{av} = \int_0^\infty \exp(-B) \, d\sigma$$

As put forward by Weibull in Sweden in 1939 there is good agreement with experimental data on the strength of brittle solids if $f(\sigma)$ is set equal to $k\sigma^m$ (where k and m are constants for a particular material of a specific size) and, additionally, if the volume integration for B is taken *only* over those regions under *tensile* stress. If k is written as $(1/\sigma_0^m)$, $f(\sigma) = (\sigma/\sigma_0)^m$, where σ_0 is that stress, on the curve best fitting the data, such that the cumulative probability of fracture is 0.633 (see Exercise 16-6). (The use of the symbol m in this context should not be confused with its use as the index of strain-rate sensitivity in Chap. 12 and elsewhere.)

Exercise 16-6

Determine the meaning of σ_0 in the foregoing theory.

Solution. If $B = -\log(1 - p_V)$, then $p_V = 1 - \exp(-B) = 1 - \exp[-\int_V f(\sigma) \, dV]$, which becomes, if $f(\sigma)$ is independent of volume (as stated),

$$p_V = 1 - \exp[-f(\sigma) \, V]$$

For unit volume

$$p_0 = 1 - \exp[-f(\sigma)] = 1 - \exp\left[-\left(\frac{\sigma}{\sigma_0}\right)^m\right]$$

When $\sigma = \sigma_0$, $p_0 \to 1 - \exp(-1) = 0.633$; that is, σ_0 is that stress such that the cumulative probability of fracture of a specimen of given size (up to and including that stress level) is 63.3%.

Thus m is a material property known as the *Weibull modulus* and is a measure of material variability; the larger the value of m, the more homogeneous the solid. For $m \to \infty$, $f(\sigma)$ is zero for all values of σ less than the characteristic strength σ_0 and p is then 1 for $\sigma = \sigma_0$; that is, all specimens break at the same stress—there is no variability. For $m \to 0$, $f(\sigma)$ is 1 and the probability of failure is the same for any value of σ; that is, fracture is likely to occur at any stress. In practice m ranges between 5 and 20 for most ceramics. Figure 16-7 gives an example of a Weibull fit to strength data for reaction-bonded Si_3N_4; the theoretical curve corresponds with $m = 17.55$ and $\sigma_0 = 1.025\sigma_{av}$. Particular values for m and σ_0 for other tests on other materials can be obtained by methods of best-fit. Moreover, the Weibull modulus m can be shown to be related to σ_{av} and the statistical standard deviation of a given set of data by

$$m \approx \frac{1.2\sigma_{av}}{\text{standard deviation}}$$

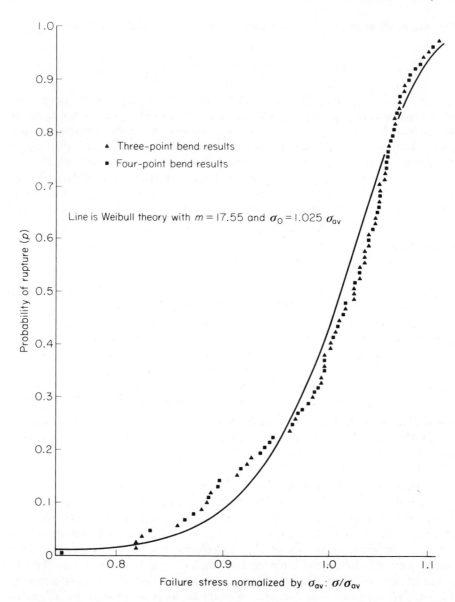

Figure 16-7 Three- and four-point bending failure data for reaction-bonded Si_3N_4. Solid line is the Weibull distribution using $m = 17.55$ and $\sigma_0 = 1.025\sigma_{av}$. (Courtesy Fulmer Research Institute, U.K.)

Weibull's statistical approach suggests that nominally identical test pieces of the same material have different risks of rupture depending on whether they are tested in simple tension, three- or four-point bending, torsion, and so on. The reasons for the differences concern the different-size regions of the test piece subjected to tensile stresses by the various modes of deformation. For example, a bar tested in simple tension has the same nominal tensile stress over all cross sections, but the same bar in bending has only those regions on one side of the neutral axis under tensile stress. Flaws in the compressive regions do not influence the breaking load. This means that the statistically anticipated average failure loads will vary with the mode of deformation, as shown by the following example.

Consider rectangular bars of some brittle solid, which are tested (1) in simple tension, (2) in pure bending (i.e., under the action of simple moments only, without shear forces, such as obtained in four-point bending). In case (1) the risk of rupture $B = \int_V f(\sigma)\,dV = f(\sigma)V$, since the stress (σ) is independent of elemental volume position. Therefore, using the Weibull function $f(\sigma) = k\sigma^m$, we have $B = kV\sigma^m$.

In case (2) the tensile stress varies linearly with distance y from the neutral axis according to $(y/h)\sigma$, where σ is the tensile stress on the outside of the bar and $2h$ is the depth of the cross section, which also has width w. Therefore $f(\sigma) = k(y/h)^m\sigma^m$. This varies with position, and therefore elemental volume slices $dV = Lw\,dy$ along the length L of the bar experience different $f(\sigma)$ depending on y. As $f(\sigma)$ is clearly related to dV, the integration for B becomes

$$B = \int_V f(\sigma)\,dV = \int_0^h k\left(\frac{y}{h}\right)^m \sigma^m Lw\,dy$$

where the limits are 0 to h because we are considering only those regions suffering tensile stresses. Therefore

$$B = \frac{k\sigma^m Lwh^{(m+1)}}{h^m(m+1)}$$

$$= \frac{kV\sigma^m}{2(m+1)}$$

We see that the risk of rupture at stress level σ is $2(m+1)$ times as great in case (1) as in case (2). This means that the statistical mean breaking stress of nominally identical samples in pure bending will be greater, on the average, than the breaking stress in simple tension.

We have that $\sigma_{av} = \int_0^\infty \exp(-B)\,d\sigma$, so in case (1)

$$\sigma_{av} = \int_0^\infty \exp(-kV\sigma^m)\,d\sigma$$

In case (2) $\sigma_{av} = \int_0^\infty \exp[-kV\sigma^m/2(m+1)]\,d\sigma$. If the variable in the integration of case (2) is changed to $\lambda = [\sigma/2(m+)]^{1/m}$, it follows that

$$\sigma_{av} = [2(m+1)]^{1/m} \int_0^\infty \exp(-kV\lambda^m)\,d\lambda$$

The integral expressed in terms of λ is the same as the integral expressed in terms of σ in case (1). Thus

$$\frac{(\sigma_{av})_{bending}}{(\sigma_{av})_{tension}} = [2(m + 1)]^{1/m}$$

As mentioned before, only when m takes on very large values do the two σ_{av} become similar. In other instances, if m has the value 3, for example, $(2 \times 3 + 2)^{1/3} = 2$, so bending samples are "twice as strong" on the average as tension samples.

It can be shown that three-point bending, where only a small region at the center span is subjected to the greatest tensile stress, gives higher σ_{av} for brittle substances that have wide scatter in their strengths (i.e., m small). One reason for the popularity of modulus of rupture bending data over simple tension data is immediately apparent!

Exercise 16-7

In round bars of some solid for which $f(\sigma) = k\sigma^3$, show that the risk of rupture in torsion is nearly 5 times as great as in bending.

Solution. An elemental volume, distant y from the neutral axis of a circular bar in bending, is given by $dV = 2L\sqrt{(r^2 - y^2)}\, dy$ (Fig. 16-8). Using $f(\sigma) = k(\sigma y/r)^3$, we can readily show that $B = [4/(15\pi)]Vk\sigma^3$.

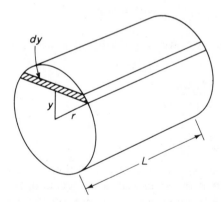

Figure 16-8 Solution to Exercise 16-7, showing geometry of round bar stressed in bending.

In torsion of a circular rod, $f(\sigma)$ relates to the torsional stresses τ at some radius y acting on the annular volume $dV = 2\pi Ly\, dy$. Thus using $f(\sigma) = k(\tau y/r)^3$, we find that the value of B is $\frac{2}{5}Vk\tau^3$. Pure shear is equivalent to mutually perpendicular tensile and compressive stresses of equal magnitude; then if fracture is associated with a critical stress level, $\sigma \equiv \tau$ in the foregoing. Thus the risk of rupture in torsion is $(15\pi/4)(\frac{2}{5})$, or nearly 5 times as great in torsion as in bending. The reason is clear: All of the outside skin in torsion is likely to spawn a fracture, and there is more chance of cracking than in bending, where only part of the surface is subject to tensile stress.

Given the laboratory data for the failure strength of a certain material, Weibull statistics can be used in design calculations to find the appropriate working stress in a component made from that material for which only a certain limited number of failures can be tolerated. For example, let us assume that three-point bending results gave $\sigma_0 = 250$ MPa and $m = 15$ for some brittle solid. A component, whose volume is 1000 times greater than the test pieces from which the σ_{av} result was obtained, is to be designed such that only one failure in 1000 components can be tolerated. What should be the working stress? We have $p_V = \frac{1}{1000}$, $\sigma_0 = 250$ MPa, and $V = 1000$. Consequently, in $p_V = 1 - \exp[-(\sigma/\sigma_0)^m V]$ (Exercise 16-6), we have

$$10^{-3} = 1 - \exp\left[-\left(\frac{\sigma}{250}\right)^{15} \cdot 1000\right]$$

That is, $\ln 1.001 = 1000(\sigma/250)^{15}$, or $10^{-6} = (\sigma/250)^{15}$, which gives $\sigma = 100$ MPa. This, of course, is smaller than the stress found from the laboratory (small) test piece.

REFERENCES

16-1. Bradt, R. C., D. P. H. Hasselman, and F. F. Lange (eds.), *Fracture Mechanics of Ceramics*. Procceedings of Symposium at Pennsylvania State University, 1973. New York: Plenum Press, 1974.

16-2. Evans, A. G., and T. G. Langdon, *Progress in Materials*. Science V., vol. 21, no. 314 (eds. B. Chalmers, J. W. Christian, and T. B. Massalski). Oxford: Pergamon, 1976.

16-3. Gordon, J. E., *Structures—or Why Things Don't Fall Down*. London: Penguin Books, 1978.

16-4. Kingery, W. D., H. K. Bowens, and D. R. Uhlmann. *Introduction to Ceramics*, 2nd ed. New York: John Wiley & Sons, 1975.

16-5. Pomeroy, C. D. "Concrete, An Alternative Material." Fourteenth John Player Memorial Lecture, Institution of Mechanical Engineers, London. *Proceedings*, **192** (1978), 14, p. 135.

PROBLEMS

16-1. Explain why the strength of nearly all glass articles can be increased by etching off a thin surface layer with hydrofluoric acid.

16-2. Some glass articles can be strengthened by annealing to remove residual stresses, whereas other glass articles are strengthened by tempering, an operation that deliberately introduces residual stresses. Discuss the seeming contradiction between these two facts. (It may be of assistance to know that tempering introduces compressive residual stresses at the surface of glass.)

16-3. Why are Al_2O_3 cutting tools used with negative rakes? (The blades of woodworking planes have "positive" rakes; a negative-rake blade is one set with the blade trailing the direction of cut instead of attacking the direction of cut.)

16-4. Explain why a ceramic component subjected to a nonuniform stress field may not fail at the point of maximum nominal stress.

16-5. Explain why the room-temperature physical properties of glass fibers such as density, modulus, and index of refraction correspond with properties of bulk glass at higher temperatures.

16-6. Explain why, in the manufacture of glass rods by pulling upwards from hot glass, the rods do not neck and fail.

16-7. Why is it said that a "silver" spoon should be put in a drinking glass before a hot liquid is poured in?

16-8. Distinguish between those physical properties of a solid which determine the resistance during thermal shock (1) to crack initiation and (2) to crack propagation. Using the data in Appendix 11, rank the various ceramic and glass solids in terms of resistance to cracking in thermal shock. Where in your ranking list would common engineering metals lie?

16-9. Relate the method of ranking employed in Problem 16-8 to the two types of thermal shock behavior observed in practice, examples of which are shown in Fig. 16-6.

16-10. Under what circumstances is a relatively low thermal conductivity required in ceramics?

16-11. Show that the average strength of a series of tensile tests (which correspond to $B = \frac{1}{2}$) is $(\frac{1}{2})^m \sigma_0$, where the probability of rupture is related to the applied tensile stress by $(\sigma/\sigma_0)^m$. Show how m reflects the spread of strength values.

16-12. A body is to be designed so that failure will not occur more than once per 10 000 components. If the working stress of the components cannot be less than 150 MPa, what stress level must be attained in the laboratory on model components whose volume is $\frac{1}{100}$ of the full size component? Your answer will depend on m, which is not known a priori.

17 Composites

In the broadest sense of the term, structural composites include any mixtures of separate material components combined to give a useful material. Such a composite often possesses the best qualities of the individual constituents and sometimes displays composite properties that none of the components possess. Composite materials may be classified as follows:

1. *Fibrous composites*, having continuous or discontinuous *filaments* or *whiskers* in a matrix. Examples are wood, bone, plastics reinforced with glass fiber or carbon fiber, and metals reinforced with boron fiber.
2. *Laminated composites*, having layers of various solids (such as clad metals, laminated glass, plastic-impregnated cloths and papers, laminated fibrous composites).
3. *Particulate composites*, having particles in a matrix. Examples are concrete (stones in mortar), ABS polymers (see Chap. 13), and some rocket propellants (inorganic particles in a rubber matrix). Particulate composites having metal matrices may have either nonmetallic particles (such as ceramic particles giving cermets; see Chap. 16) or particles of other metals (which encompasses inclusions in metals).

This chapter concentrates (but not exclusively) on the engineering properties of fibrous and laminated composites.

The discovery of very high strengths in ultrafine filaments of many solids provided the impetus for development of man-made fibrous composite materials. Mixtures of two or more distinct materials on a very fine scale (of the order 100 μm or less) now produce solids with a wide range of properties impossible to achieve with conventional materials. In many cases very strong, lightweight solids result. Structures manufactured from such composites can therefore be lighter for the same working stresses or stronger for the same mass. Clearly, they find application in aeronautical and space designs, as well as in many other fields.

HIGH STRENGTH

Nearly flawless crystalline solids in which dislocations cannot move may be expected to exhibit maximum strengths of the order of E/10 (where E is the modulus of elasticity; see Chaps. 14 and 16). The only cases where strengths approaching such values have actually been observed are for very small samples of small-diameter filaments (with diameters in the range of 1 μm), usually called *whiskers*. For example, fracture strengths of iron whiskers have been measured up to 14 GPa (about 0.07E). Measurements on whiskers of other materials likewise give high strengths: Cu, 0.02E; Ni, 0.02E; Al_2O_3, 0.05E; SiC, 0.04E; graphite, 0.03E. Whiskers are discontinuous, but continuous fila-

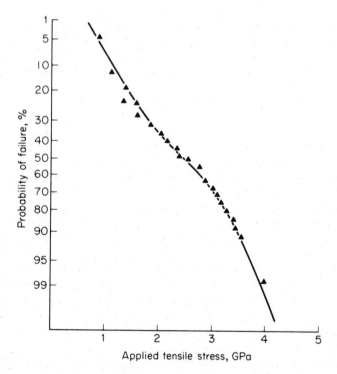

Figure 17-1 Cumulative strength distribution for boron-on-tungsten filaments of diameter approximately 100 μm.

ments displaying respectably high strengths of materials such as glass, boron, and carbon are available. For example, glass fibers can have strengths of $0.014E$; boron filaments have strengths of $0.009E$ in fibers 100 μm diameter; and carbon fibers have strengths of $0.004E$ in fibers 8 μm in diameter. The mechanical properties of several filamentary materials are given in Appendix 12.

Chapters 14 and 16 discuss the role of flaws and surface quality in the strength of brittle solids and the origins of these size effects on strength, whereby small samples are likely to display values approaching the theoretical strength. Large lumps of the same solids (which in many cases cannot be manufactured anyway) display much lower strengths because there is greater likelihood of finding imperfections in the form of microcracks or dislocations in large volumes of materials. The associated statistical variation in strength values is also described in Chap. 16. Figure 17-1 shows a strength distribution for boron-on-tungsten filaments. When numbers of filaments with such characteristics are used in composites, there will be a statistical variation in the composite strength, but it will not be so far-ranging as that for separate filaments, as the strong filaments present help to hold up the composite structure, even when some low-strength components have failed.

MODULI

In a quest for strength, we might simply look for materials with high yield strengths. We know that the strengths of solids can often be improved by cold work, by heat treatment of some sort, or by alloying to introduce extra phases (Chaps. 2, 9, 10, and 11). However, the usefulness of ultrastrong materials is limited by their *stiffnesses* or *elastic moduli*. The reason should be clear from Fig. 17-2. If the working stress level is much increased but the modulus remains unchanged, the working strains increase in the same proportion as the

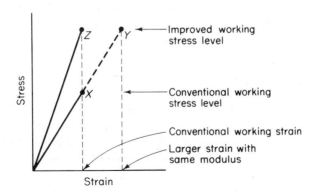

Figure 17-2 The need for increased stiffness as well as increased strength to keep working deflections small.

strength (point X to point Y). Translated into deflections, this means that although objects designed with the strong materials can bear heavier loads (or equivalently require less material to take the old loads) they do so at much greater deformations. For compact items with small deflections this may not matter, but for structures such as bridges and airplane wings the use of structural material of low elastic modulus would lead to the unfortunate situation shown in Fig. 17-3.

If, with these potentially ultrastrong fibrous materials, we wish to operate at the same elastic deformations as we use at present with traditional engineering materials, then we must look for *increased E* as well as increased working stress (point X to point Z in Fig. 17-2). Moreover, those parts of a structure in compression at high working stresses require higher moduli to prevent Euler buckling. It is possible, of course, as J. E. Gordon has pointed out, for the situation in Fig. 17-3 to be attained by materials with high E if their breaking strain is great enough.

Figure 17-3 The consequence of increased strength with no improvement in stiffness; 1.6% strain in aircraft wing spar booms (bend radius of beam = thickness/2 × strain). (After J. E. Gordon.)

Since it has been shown that the theoretical strength of a solid is proportional to E, it seems that high strength and high stiffness will be simultaneously satisfied in the whisker and filamentary materials being discussed in this chapter. This should be contrasted with improvements in strength of conventional materials where E is not altered by cold working, heat treating, or other methods.

DENSITY

Another factor of interest in many applications is the mass of a given part. As we shall see, many strong composite materials have low mass, which enhances their attractiveness. As a measure of strength for a given mass of structure, the *specific strength* is defined as (strength/density). For those very strong steels with $S_u = 2$ GPa and $\rho = 7.8$ Mg/m^3, the specific strength is some 0.25 (km/s)^2. On the other hand, for graphite whiskers of strength 17 GPa and $\rho = 2.25$ Mg/m^3, it is some 7.6 (km/s)^2. These figures demonstrate the potential of such high strengths: If a cable 6 mm in diameter could be made wholly of continuous carbon filaments of 1.6 GPa strength, it could support a tensile load of 45 kN, and a 3-m length of the cable would weigh about 200 g. The same mass of wire rope 3 m long would support 6.5 kN.

In addition to the concept of specific strength, the elastic modulus can be divided by density to give *specific elastic modulus* or *specific stiffness*. Appendix 12 lists typical values of these quantities for some common engineering materials, and Appendix 13 gives these values for some high-strength filamentary materials. Note the remarkable fact that the specific moduli of most of the common engineering solids are all about 27 $(km/s)^2$. Most of their specific strengths are between 0.02 and 0.2 $(km/s)^2$. This strongly suggests that we must look at new materials for efficient improvements in strong and stiff structures of low mass. The use of glass-filament composites represents a sort of halfway point in the transition from traditional to new materials. The specific modulus of glass filaments is the same as for traditional engineering materials, but the specific strength is perhaps six times that of bulk glass. The growth of the glass-fiber industry shows to what extent such composite materials can be used, despite the disadvantage of large working strains. There is, in fact, an additional factor: *In bending*, specific strengths and specific moduli are more properly represented by the ratios (strength/density^2) and (modulus/density^3) than by the simple ratios given in Appendix 13 for plain tension (see Problem 17-12). Then, because the density of plastics reinforced by glass fibers is so much lower than that of common engineering metals, the bending stiffness of unidirectional glass-fiber composites can be about 20 times the bending stiffness of the same mass of steel and 2 times that of aluminum alloy beams (see Exercise 17-4). Even so, Fig. 17-3 emphasizes that materials of higher specific stiffness than glass are required for many applications.

MATRICES AND THE MAKEUP OF COMPOSITES

It is all very well to have identified high-strength, high-modulus solids, but if most of them are obtainable only in whisker or filament form, how can they be used? How can we "put together" the whiskers or filaments to make structurally useful shapes? Perhaps we can make bundles like rope, or perhaps hold together the separate high-strength components in some *matrix*. This latter is the route normally chosen. By using an epoxy or polyester resin as a glue, filaments and whiskers can be made into useful structural materials. Sometimes ropes, mats, and cloths are made from filaments before these components are stacked or combined to form the final structure (a process called *laying up*) but an adhesive matrix is always necessary to hold things together. The matrix can be a ductile metal too on occasion. By analogy with reinforced concrete (which has ductile bars or meshes of steel running through a brittle stone/cement matrix) we often refer to filamentary composites as *fiber-reinforced* plastics or metals.

The matrix plays an important and complex role in composites. Apart from binding fibers together, the matrix acts as a medium to transfer external loads on the composite to the filaments, because it is rare that the fibers themselves are in direct contact with the external loading system. As we shall see, the interfacial bond between filament and matrix is most important. It is the load-transfer agent and, as we shall also see, is important in determining the toughness of a composite. Among other functions of the matrix are the following:

1. It can be made to align filaments in important stress directions and to permit "composite action" in tensile, compressive, and shear modes.
2. It separates fibers, thus preventing cracks from passing catastrophically from one fiber to another through the composite.
3. It protects the fibers from mechanical damage and perhaps also from environmental damage (e.g., moisture or high-temperature oxidation).
4. If ductile, the matrix provides a mechanism for slowing down cracks.
5. If brittle, the matrix may depend on the *fibers* to act as crack stoppers.

The density of most laminating resins is some 1.0 Mg/m^3. Because of the fairly large difference between this and the density of many reinforcing filaments $(2.0 \sim 2.5 \text{ Mg/m}^3)$ there are important increases in the average density of fiber-reinforced composites as the amount of fibers packed into the composite increases. Consequently for structures of fixed mass there exists an optimum number of fibers per unit volume (an optimum *volume fraction*; see next section) to give, for example, maximum bending resistance.

VOLUME FRACTION

The amount of reinforcement in a composite is normally given in terms of its *volume fraction*, the percentage or decimal fraction of the total composite volume taken up by the filaments. Usually we wish to pack in as much reinforcement as possible, but, as in crystal packing (Chap. 4), there are geometrical limits. Practical composites have 50–60% of filamentary reinforcement.

Exercise 17-1

(a) Whiskers with specific gravity 3 are bonded with a resin of specific gravity 1.2. What percentage mass of the composite is taken up by the whiskers if their volume fraction is 0.3?

(b) If a metal matrix is used instead, with a specific gravity of 8.9 (nickel), what is the mass content of the whiskers in the composite?

Solution.   (a) The relative mass of whiskers is

$$\frac{3(0.3)}{3(0.3) + 1.2(0.7)} = \frac{0.9}{1.74} = 52\%$$

(b) In this case we have

$$\frac{3(0.3)}{3(0.3) + 8.9(0.7)} = \frac{0.9}{7.13} \approx 13\%$$

It is seen that there are good arguments for the use of lightweight resin matrices where possible. If metal matrices are used (in order that the composite may operate at higher temperatures, for example), a penalty is paid in mass.

Exercise 17-2

What is the maximum volume fraction attainable with aligned round fibers?

Solution. The hexagonal cross section will produce the maximum volume fraction with aligned fibers, as shown in Fig. 17-4. If there are m rows of filaments with n fibers in each row, the total cross-sectional area of the fibers is $mn\pi d^2/4$. The circumscribing matrix area is given by

$$d\left[1 + (m - 1)\frac{\sqrt{3}}{2}\right]d\left(n + \frac{1}{2}\right)$$

Thus the volume fraction of fibers aligned in the manner shown is

$$v_f = \frac{mn\pi d^2}{4d^2[1 + (m - 1)\sqrt{3}/2](n + \frac{1}{2})}$$

which for very large m and n becomes

$$v_f = \frac{\pi}{2\sqrt{3}} \quad \text{or} \quad 90.7\%$$

Practical fiber-reinforced composites rarely exceed two thirds of this value.

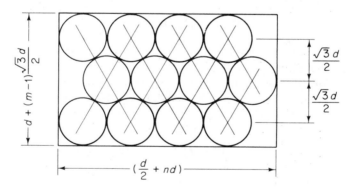

Figure 17-4 Solution to Exercise 17-2, showing hexagonal array of round fibers, with m rows of fibers having n fibers in each row.

COMPOSITE STRESS AND MODULUS

Consider a unidirectional array of continuous filaments embedded in some matrix. Although the behavior of such a composite is far from being fully understood, a reasonable approximation for both the composite modulus and strength is given by the *rule of mixtures*.

Let a tensile load W be applied in the direction of the aligned filaments, and assume that all components extend equally, that is, that they have equal strains $\epsilon_f = \epsilon_m = \epsilon_c$, where the subscripts f, m, and c mean filament, matrix, and composite, respectively (Fig. 17-5).

If the stresses induced in the components by this strain level are σ_f and σ_m, then

$$W = \sigma_f A_f + \sigma_m A_m \qquad (17\text{-}1)$$

where A_f and A_m are the individual cross-sectional areas of the filaments and matrix.

The average composite stress is $\sigma_c = W/A_c$, where A_c is the total cross-sectional area, given by $A_c = (A_f + A_m)$. Thus

$$\sigma_c = \frac{\sigma_f A_f}{A_f + A_m} + \frac{\sigma_m A_m}{A_f + A_m}$$

The quantities $A_f/(A_f + A_m)$ and $A_m/(A_f + A_m)$ represent the fractions of the total cross section taken up by the filaments and matrix. If we multiply the numerator and denominator of these area fractions by the total length of the composite, we have volume fractions v_f and v_m. Thus

$$\sigma_c = v_f \sigma_f + v_m \sigma_m \qquad (17\text{-}2)$$

or, since $v_f + v_m = 1$,

$$\sigma_c = v_f \sigma_f + (1 - v_f)\sigma_m \qquad (17\text{-}3)$$

This equation is the basic rule of mixtures and is clearly analogous to the lever rule in phase diagrams (Chap. 8).

Again, for the effective modulus E_c of the composite, we use the fact that the stresses σ_f and σ_m are given by $E_f \epsilon_f$ and $E_m \epsilon_m$. So in Eq. (17-1)

$$W = E_f \epsilon_f A_f + E_m \epsilon_m A_m \qquad (17\text{-}4)$$

Now in terms of E_c and ϵ_c, $W = E_c \epsilon_c A_c$. Assuming that $\epsilon_f = \epsilon_m = \epsilon_c$, we have

$$E_c A_c = E_f A_f + E_m A_m \qquad (17\text{-}5)$$

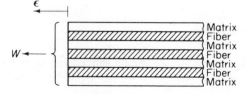

Figure 17-5 Load applied in the direction of aligned filaments.

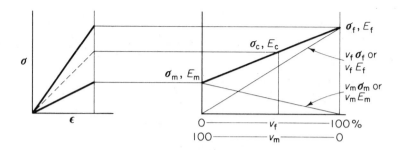

Figure 17-6 Meaning of Eqs. (17-3) and (17-6) for stress and stiffness in unidirectional composites.

which transforms as above into a volume-fraction rule of mixtures for the composite modulus:

$$E_c = v_f E_f + v_m E_m = v_f E_f + (1 - v_f)E_m \tag{17-6}$$

Graphically, Eqs. (17-3) and (17-6) are represented by Fig. 17-6.

Thus the Young's modulus slope for a given mixture lies proportionally between those of its components, and the strength points lie similarly between those of the components.

Exercise 17-3

Boron filaments of $E_f = 380$ GPa are made into a unidirectional composite with an epoxy matrix of $E_m = 2.5$ GPa. What are the moduli, parallel to the fibers, (a) for $v_f = 0.05$ (i.e., 5%) and (b) for $v_f = 0.50$?

Solution. $E_c = (1 - v_f)E_m + v_f E_f$.
(a) $E_c = (1 - 0.05)2.5 + (0.05)380 = 2.38 + 19 = 21.4$ GPa
(b) $E_c = (1 - 0.5)2.5 + (0.5)(380) = 1.25 + 190 = 191$ GPa. Note that the contribution from the matrix may often be neglected.

It is clear that if the strains through the stressed composite are assumed to be the same everywhere, then the stresses in the components will be different ($\sigma = E\epsilon$, and the E's are different). We could postulate, in fact, that the stresses were equal all over and thus allow unequal strains in the components. We then would obtain for the composite strength and modulus "harmonic" (i.e., reciprocal) rules of mixtures:

$$\frac{1}{\sigma_c} = \frac{v_f}{\sigma_f} + \frac{1 - v_f}{\sigma_m} \tag{17-7}$$

$$\frac{1}{E_f} = \frac{v_f}{E_f} + \frac{1 - v_f}{E_m} \tag{17-8}$$

Figure 17-7 displays the two different models of composite action. Equation (17-6) was first used by Voigt early this century in calculating the moduli

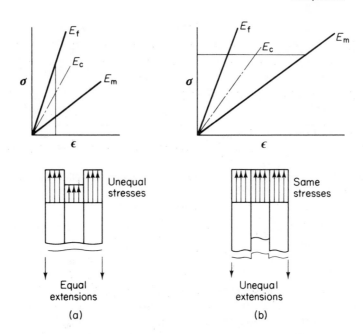

Figure 17-7 Different assumptions for composite action by Voigt (equal component extensions) and by Reuss (equal component stresses).

of polycrystalline solids, and Eq. (17-8) by Reuss a little later. In contrasting the two types of behavior (equal strains with unequal stresses or unequal strains with equal stresses), one is reminded of the mechanics of springs or resistors in parallel and in series. Real composite behavior is somewhere between the two (they are mathematical bounds to σ and E) but tends toward the uniform strain case. This is because the uniform stress model suggests slippage at the interface between fiber and matrix. If the interfacial bonding between the components is good (i.e., the gluing is good), we should not expect much relative movement. The final displacement of the composite under load is a compromise movement. The parts with low E's want to stretch more than those with high E's and vice versa (contrast quench cracks and thermal shock in Chap. 16, toughened glass, and so on).

Exercise 17-4

Show that in a composite for which $E_f \gg E_m$, the condition for maximum flexural rigidity of a beam of given mass is that the density of the matrix should be two thirds of the overall density of the composite, that is, the mass of reinforcing fiber should be one third of the total mass of the beam.

Solution. The mass m of the composite beam is given by

$$m = \rho_f t_f + \rho_m t_m = \rho t$$

where t is the overall thickness of a beam of given plan area, t_f and t_m the equivalent thicknesses of fiber and matrix, respectively, and ρ (with similar subscripts) is density. Then

$$\rho = v_f \rho_f + (1 - v_f)\rho_m$$

The resistance of the beam to bending is proportional to EI, where for a rectangular beam $I = wt^3/12$, w being the planar width of the beam. For a beam of fixed mass and plan area, maximum flexural rigidity (maximum EI) therefore comes from maximum Et^3, or maximum

$$[v_f E_f + (1 - v_f)E_m]\left(\frac{m}{\rho}\right)^3 = \frac{[v_f E_f + (1 - v_f)E_m]m^3}{[v_f \rho_f + (1 - v_f)\rho_m]^3}$$

Differentiating with respect to v_f and letting $E_m \approx 0$ gives the condition for a maximum as

$$v_f = \frac{\rho_m}{2(\rho_f - \rho_m)}, \quad \text{or} \quad \rho_m = \frac{2\rho}{3}$$

or equivalently that the mass of reinforcing fiber is one third of the total mass of the composite beam. (This answer is the same as the "skin mass criterion" for maximum flexural rigidity in honeycomb sandwich laminates when the core modulus is neglected.)

Exercise 17-5

Explain why copper-coated steel wire is easier to cold-work than plain wire but the maximum attainable drawing reduction of the coated wire is less than for plain steel wire.

Solution. The overall yield strength of copper-coated steel wire is smaller than that of steel alone. This follows from the rule of mixtures applied to the component yield strengths, that is, from

$$(\sigma_y)_{coated} = (1 - v_f)(\sigma_y)_{Cu} + v_f(\sigma_y)_{steel}$$

where v_f refers to the volume fraction of steel. Cold working is thus easier because $(\sigma_y)_{Cu} < (\sigma_y)_{steel}$ (say, 120 MPa compared with 250 MPa).

During deformation of the copper–steel composite at common uniform strain, compressive stresses are set up in the lower-strength copper coating and tensile stresses are set up in the higher-strength steel (Fig. 17-8), since the copper wants to flow at a lower stress than the steel. A simple Mohr's circle diagram illustrates the origin of those stresses during *plastic* flow. (The earlier derivations in this chapter have been cast in terms of elastic deformations.)

If the copper and steel components were being deformed alone, they would have individual Mohr's circles as in Fig. 17-8(a) and (b). When combined in a sandwich and compressed (as when passing through a wire-drawing die), the applied compressive load and stress must be the same in both, so effectively the small copper circle shifts to the left to match the steel circle applied stress [Fig. 17-8(c)]. However, this leaves an unbalanced horizontal stress (AB) in the weaker component. A horizontal tensile force therefore must exist in the stronger component to balance this out, so that the final Mohr's circle shifts to the right, [Fig. 17-8(d)]. The tensile forces generated by stresses OD acting on the particular area of steel balance those compressive forces generated in the cross section of copper acted on by stresses OC.

It is seen that if the fracture of the steel core of the wire can be described in terms of a tensile failure stress (not strictly true, but an adequate hypothesis), the presence of

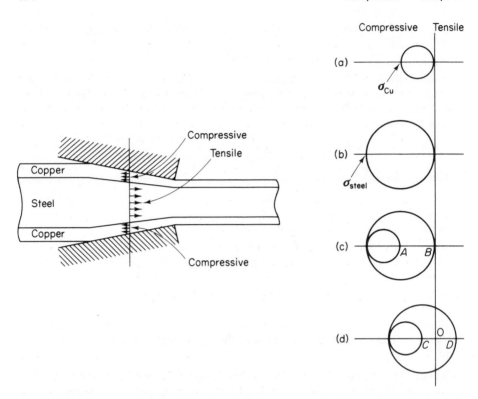

Figure 17-8 Solution to Exercise 17-5, showing stress distributions in the mechanical working of clad materials (in this case copper-coated steel wire).

an additional induced tensile stress (OD) in the steel core produces breakages at lower total strains than in plain wire.

The presence of compressive stresses helps reinforced concrete in a similar way. Another way of putting the result is to say that the strains in the weaker part of composites are less than they would be if only the weak solid were being stressed, and that in the strong component greater strains are attained for a given load in a composite than if tested alone. Use is made of this in the cold working of difficult-to-form materials such as stainless steel and titanium; their yield strengths are large and they work-harden appreciably, but "pack-rolling" between layers of soft metals reduces the necessary working loads.

CRITICAL TRANSFER LENGTH;
CONTINUOUS AND DISCONTINUOUS FILAMENTS;
ASPECT RATIOS; TENSILE STRENGTH

When a fiber-reinforced composite is loaded, it is rare for individual filaments to be acted on by the external load; we do not grab the individual filaments, which in many cases are smaller than 100 μm in diameter. Rather, the matrix

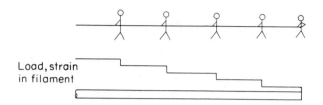

Load, strain
in filament

Figure 17-9   Buildup of load and strain in a filament over a transfer length analogous to buildup of load in a tug of war.

acts as a medium to transfer loads to the filaments. The bond between the
filament and matrix is the load-transfer agent, making the matrix and filament
"move" together. The nature of the bond (chemical or mechanical) varies in
different composite systems, but for convenience we can represent the interfacial bond strength by a constant shear stress τ.

   The existence of a bond strength builds up stresses in the filaments. We
can liken the buildup of load in the filaments to the buildup of load in the rope
in a tug of war (Fig. 17-9).

   The load is transferred from the matrix to the fiber at the end regions
only, because as the load builds up so does the strain, and when that strain
reaches the overall strain in the fiber there is no relative slip between matrix
and fiber. The adjacent matrix then becomes virtually stress-free in the region
corresponding to the middle of the fiber. In this way high-strength filaments
effectively take the major portion of the external loads, as may be seen by
investigating the force equilibrium on an element of length dL near the end of
a filament embedded in a lump of matrix [Fig. 17-10(a)].

$$\frac{\sigma_f \, \pi d^2}{4} + \tau \pi d \; dL = (\sigma_f + d\sigma_f) \, \frac{\pi d^2}{4}$$

Therefore

$$\frac{d\sigma_f}{dL} = \frac{4\tau}{d}$$

so

$$\sigma_f = \frac{4\tau L}{d} \tag{17-9}$$

since $\sigma_f = 0$ at the end of the filament where L is zero. Thus the stress in the
filament builds up linearly from the ends at the rate $\tan^{-1}(4\tau/d)$, and the
tensile stress in the adjacent matrix decreases similarly. The stress buildup in
the filament does not go on indefinitely, merely over a length sufficient to
transfer the particular external load into the fiber [Fig. 17-10(b)]. It is at this
point that the stress in the filament becomes sufficiently high and the stress in
the matrix becomes sufficiently low that the strains in the filament and in the
matrix are the same. There is then no shear stress across the interface.

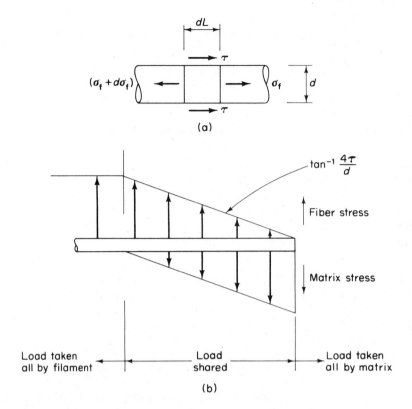

Figure 17-10 Equilibrium of stresses in an element of the transfer length; σ_f is the stress in the filament, τ the interfacial stress between filament and matrix.

Exercise 17-6

A single fiber of some material is embedded in a matrix that is subjected to a force of 40 mN. If the diameter of the filament is 10 μm and $\tau = 20$ MPa, over what length does load transfer take place?

Solution. When the load is fully transferred,

$$\sigma_f = [4 \times 10^{-3}]/[\pi(10 \times 10^{-6})^2/4] = 510 \text{ MPa}$$

Then $L = (510 \times 10^6)(10 \times 10^{-6})/(4 \times 20 \times 10^6) = 64$ μm at one end of the filament. The load is transferred back to the matrix at the other end of the same filament, so that $64 + 64 = 128$ μm of the filament would be used up in load transfer. For the same τ, thicker filaments require longer L's to build up to the same stress. Were $d = 100$ μm in the foregoing example, L would be 10 times as long, that is, 640 μm at each end of the fiber.

There must be a limit to the load buildup, and in brittle filaments that occurs when the breaking stress of the fiber is reached. The middle length of the filament is conventionally described as all being subjected to this maximum stress at fracture. Obviously, it will normally break only in one spot (at some diameter variation or where the material is locally weak), so that a constant fracture stress along the length is strictly unrealistic. The real stress distribution must peak at the fracture point and blend into that value elsewhere. Nevertheless, the simple distribution, with linear buildup at the ends and with the central section of the filaments capable of sustaining a stress at or near the fracture stress, is an adequate model for describing the mechanics of composites. It has already been mentioned that filaments display a statistical variation in failure strengths (Fig. 17-1). Consequently, in what follows, where a value is given for the breaking strength of fibers, it must be construed as a representative average value for the family of filaments being employed.

If the average filament breaking stress is $(\sigma_f)_{max}$, the corresponding load transfer length is $(\sigma_f)_{max}(d/4\tau)$ at each end of a filament. Unless the filament length is greater than twice this value, that is, $(\sigma_f)_{max}(d/2\tau)$, the breaking strength is never reached [Fig. 17-11(a)]. The length $(\sigma_f)_{max}(d/2\tau)$ is called the *critical transfer length* (L_c) of the filament/matrix combination. The average working stress in filaments of length L (which is here assumed to be longer than L_c) is given, as shown in Fig. 17-11(b), by

$$(\sigma_f)_{average} = \frac{1}{L} \int_0^L \sigma_f \, dL = \sigma_f\left(1 - \frac{L_c}{2L}\right) \qquad (17\text{-}10)$$

If a filament is loaded up to its breaking stress, the average stress in the fiber is $(\sigma_f)_{max}(1 - L_c/2L)$. Depending on the relative values of L_c and L, the mean stress may be within, say, 95% of $(\sigma_f)_{max}$, (i.e., when $L_c/2L < 0.05$ or $L > 10L_c$) or perhaps only 50% of $(\sigma_f)_{max}$ [when $L_c/2L = 0.5$ or $L = L_c$; see Fig. 17-11(c)].

Clearly, account should be taken of these differences when the rule-of-mixtures expressions for stress, Eqs. (17-2) and (17-3), are applied at failure. Thus instead of

$$(\sigma_c)_{max} = (1 - v_f)(\sigma_m)_{max} + v_f(\sigma_f)_{max} \qquad (17\text{-}11)$$

for breaking strength of composites (the derivation of which implicitly assumes long, continuous fibers running through the composite), it is preferable to write

$$(\sigma_c)_{max} = (1 - v_f)(\sigma_m)_{max} + v_f(\sigma_f)_{max}\left[1 - \frac{L_c}{2L}\right] \qquad (17\text{-}12)$$

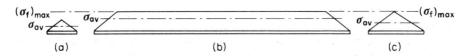

Figure 17-11   The influence of fiber length L on the stress level transferred from the matrix: (a) fibers shorter than the critical transfer length L_c; (b) $L \gg L_c$; (c) $L = L_c$.

The particular lengths of fibers therefore affect the value of composite breaking strengths, as the following example shows.

Exercise 17-7

A composite is made up of 40% of filaments 100 μm in diameter of strength 3 GPa in a matrix of strength 80 MPa. The interfacial bond strength is 60 MPa. Determine the composite breaking strength if the filaments have average lengths of (a) 100 μm and (b) 2 mm.

Solution.

$$L_c = (\sigma_f)_{max} \frac{d}{2\tau} = \frac{3 \times 10^9 \times 100 \times 10^{-6}}{2 \times 60 \times 10^6} = 2.5 \text{ mm}$$

Therefore (a) for $L = 100$ mm, $L_c/2L = 2.5/200 = 0.0125$ and

$$\sigma_c = (1 - 0.4)80 \times 10^6 + 0.4 \times 3 \times 10^9 \times (1 - 0.0125)$$

$$= 48 \times 10^6 + 1.2 \times 10^9 \times 0.9875 = 10^9(0.048 + 1.19) = 1.24 \text{ GPa}$$

(b) for $L = 2$ mm, $L_c/2L = 2.5/4 = 0.625$. So

$$\sigma_c = (1 - 0.4)80 \times 10^6 + 0.4 \times 3 \times 10^9(1 - 0.625)$$

$$= 48 \times 10^6 + 1.2 \times 10^9(0.375) = 10^9(0.048 + 0.45) = 500 \text{ MPa}$$

Note the significant difference in σ_c, the reason for which lies in the values of average stress distribution in the fibers. Note that when the filaments are long, the contribution from the fibers ($\approx v_f \sigma_f$) swamps the contribution from the matrix ($\approx v_m \sigma_m$). Often in the rule-of-mixtures calculations for both σ and E, we neglect the contribution from the matrix.

$L_c/2L$ can be written $(\sigma_f)_{max} d/(4\tau L)$, which shows that for a given τ, the average stress in stressed filaments is close to $(\sigma_f)_{max}$ when L/d is large. L/d is called the *aspect ratio* of the fibers.

We see that, even if a composite is made up of lengths of filament shorter than the total size of the article (which therefore are not continuous through the composite), the composite strength is the same as if they were continuous, for all practical purposes, provided that L/d is large enough. As $(\sigma_f)_{max}/\tau \approx 50$ for many brittle filament/brittle-matrix systems, large composite strengths are attainable if the aspect ratio is greater than, say, 100. For filaments 10 μm in diameter, this means that they need only 1 mm or longer to come within 90% of the "ideal" composite strength given by the simple rule of mixtures, Eqs. (17-2) and (17-3). Thus short whiskers can be used to good effect, and in practice, filaments are long and have favorable aspect ratios. Therefore, although there is a strict distinction between continuous and discontinuous filamentary composites, it is somewhat academic unless L/d is small or for some reason the interfacial bond is very weak.

The simple rule of mixtures diagram (Fig. 17-6) is modified to take account of varying L/d as shown in Fig. 17-12. Low L/d prevents large $(\sigma_c)_{max}$

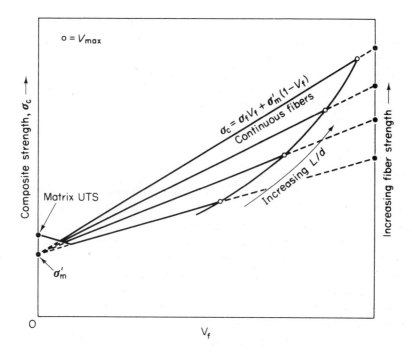

Figure 17-12 Modifications to the simple rule-of-mixtures diagram (Fig. 17-6) to account for finite L/d.

from being attained both because of the low average stress in the fibers and because the maximum attainable volume fraction is restricted.

Figure 17-12 also shows a modification to the simple diagram in Fig. 17-6 at the very low volume fraction end, where paradoxically it seems that the addition of fibers *reduces* the composite strength *below* that of the plain matrix. This comes about as follows: In Eq. (17-11) or (17-12) it is implicit that $(\sigma_m)_{max}$ and $(\sigma_f)_{max}$ occur at the same strain. But the failure strain of the matrix in the composite could be less than or greater than that of the filaments (Fig. 17-13). In fact, glass filaments do tend to have roughly the same fracture

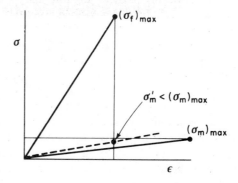

Figure 17-13 When the failure strain of the fiber is smaller than that of the matrix, the matrix stress is smaller than its own fracture stress. (Voigt equal strain model for composite behavior is assumed.)

strains as the common glass-fiber matrices (0.02 ~ 0.05), but boron and graphite filaments have fracture strains that are smaller (less than 0.01). When the
failure strain of the fiber is smaller than that of the matrix, the matrix σ_m to be
used in Eqs. (17-11) and (17-12) will not be its own breaking stress $(\sigma_m)_{max}$ but
rather a lower value appropriate to the failure strain of the filament. In this
case it turns out that at low filament-volume fractions, $(\sigma_c)_{max}$ from Eqs. (17-
11) and (17-12) can be *less* than $(\sigma_m)_{max}$. Thus until a *critical volume fraction* of
filament is achieved producing $(\sigma_c)_{max} > (\sigma_m)_{max}$, there is no gain to reinforcement.

Exercise 17-8

What is the critical volume fraction for a brittle-fiber/brittle-matrix composite made of
fibers with $(\sigma_f)_{max} = 3$ GPa, $(\sigma_m)_{max} = 100$ MPa, and $E_m = 2$ GPa with $(\epsilon_f)_{max} = 0.01$?

Solution. At $\epsilon = 0.01$, σ_m is not 100 MPa but rather is given by $(E_m)(\epsilon_f)_{max} =
(2 \times 10^9)(0.01) = 20$ MPa. Then if $\sigma_c \leq 100$ MPa at volume fractions less than the
critical fraction, using Eq. (17-11) for simplicity, we have

$$100 \leq (1 - v_f)20 + v_f(3000) \text{ MPa}$$

Therefore

$$80 \leq 2800 v_f$$

So $(v_f)_{critical} = 0.0285$, or roughly 3%. Of course, practical composites have much
greater v_f.

COMPRESSION STRENGTH

When a unidirectional composite is loaded in compression parallel to the
fibers, the mode of failure depends on the strength of the filament/matrix bond.
If the bond is weak, the fibers and matrix detach themselves (debond) at low
applied loads, the filaments buckle elastically, and the compressive strength
never reaches the composite tensile strength. Even if the interfacial bond is
strong, failure of the matrix in shear keeps the compressive strength lower.
This is particularly true with discontinuous fiber composites, as matrix shear
can occur without the necessity for elastic distortion of the filaments. Composites with high filament-volume fractions fail by *creasing* in compression,
which is a shear instability (Problem 17-22).

An approximate expression for compressive strength that applies to
many, but certainly not all, systems is

$$(\sigma_c)_{max} = \frac{0.63 G_m}{1 - v_f} \tag{17-13}$$

where G_m is the matrix shear modulus. For many epoxy resins G_m is about
1 GPa, so the compressive strength of a $0.5 v_f$ composite might be some
1.2 GPa. The tensile strength of a $0.5 v_f$ boron–epoxy composite along the
direction of the filaments would be some $(0.5)(3.5) = 1.7$ GPa.

ANISOTROPY IN COMPOSITES:
ALIGNED AND RANDOM FIBER DISTRIBUTIONS

The strength and stiffness of composites reinforced with aligned strong fibers is highly anisotropic. The variation of tensile strength with orientation angle for two unidirectional composites is shown in Fig. 17-14. The mode of failure in an aligned continuous composite depends very much on the orientation of the fibers with respect to the applied stress. When the loading is parallel to the filaments, failure is governed by fiber tensile breakage. If the loading is shifted from the fiber direction, the matrix or the interface (whichever is weaker) will be increasingly loaded in shear until shear failure occurs even though the fiber tensile stresses are still low. When the applied stress is perpendicular to the fiber direction, failure occurs by tensile fracture of the matrix or interface, the filaments playing comparatively little part in the process. By determining the composite tensile strengths along and normal to the aligned filaments, together with the failure stress in shear parallel to the fibers, it is possible to arrive at the failure stress at any arbitrary angle using an empirical theory, akin to the Tresca condition for yielding (see Chap. 2). Failure is predicted in three distinct modes, over which one of the three stress parameters predominates (Fig. 17-14). Failure criteria for other composites based on the Maxwell–von Mises combined-stress condition have been proposed.

LAMINATES

Maximum use is made of the reinforcement when the composite is stressed parallel to the fibers, as with simple struts or the caps of panel stiffeners. For other applications it is more useful to arrange the filaments in a composite in more than one direction. Thus *laminates* are employed, which have layers with regular orientations such as the cross-ply arrangement (alternate layers at right angles, as in plywood sheets) and angle-ply schemes ($\pm\theta$ where $30° < \theta < 60°$); most applications have multiple plies with three or four angles.

The highly anisotropic character of single layers is ironed out when composites are made up from layers with different orientations. For example, Fig. 17-15 shows the variation of tensile strength with orientation for composites made with unidirectional filaments, cross-plied filaments, and angle-plied fibers. The latter composite is said to be *pseudo-isotropic* in behavior, as the variation of strength with angle is not marked; note in comparison that the unidirectionally aligned system has a transverse strength of only some 8% of its strength parallel to the filaments.

Much of the difference in elastic anisotropy that determines stiffness also becomes less marked in multilayer sheets, and it often happens that well-known isotropic "strength-of-materials" theories give reasonably adequate descriptions of composite behavior. That having been said, however, the *scissoring* (or free edge problem) between layers must not be forgotten. There

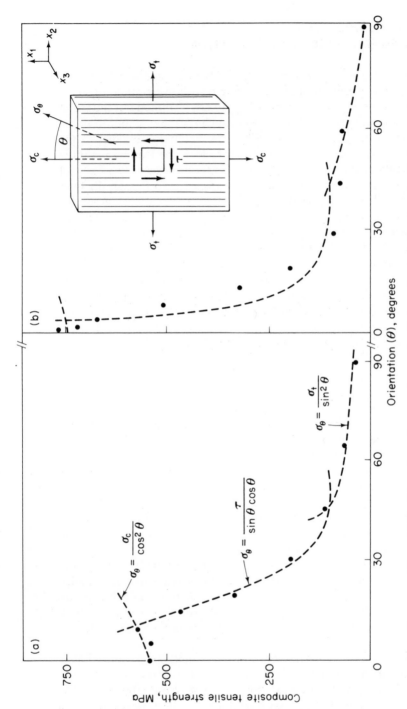

Figure 17-14 Comparison of measured strengths of composites, as a function of angle between stress axis and fiber direction, with predictions of a maximum-stress model based on strength along the filaments (σ_c), strength perpendicular to the filaments (σ_t), and shear strength parallel to the filaments (τ). (Compare the Mohr's circle constructions in Chap. 2.) (a) 0.57 v_f type I carbon fibers in epoxide resin; (b) 0.50 v_f SiO$_2$ fibers in aluminum. [From B. Harris, *Composites*, **3** (1972), p. 152, after J. Dimmock and M. Abrahams, and P. W. Jackson and D. Cratchley, by permission.]

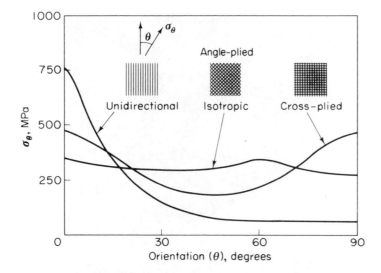

Figure 17-15 Variation of laminate composite tensile-strength with orientation for unidirectional, cross-plied, and angle-plied laminates of glass-fiber-reinforced plastics. Angle-plied arrangement gives quasi-isotropic behavior. [After J. W. Davis and N. R. Zurkowski, 3M Corporation, by permission.]

interlayer and edge shear stresses are set up in the composite to counteract the rotation of one layer (individually anisotropic) relative to a differently oriented, adjacent layer when subjected to external loads. Edge delamination and loss of structural integrity can result.

If, instead of the laminae being aligned, the fibers are arranged randomly in two dimensions, the average breaking strength will be about one third that of unidirectional composites, and if the filaments are three-dimensionally random, about one sixth. Therefore only the strongest reinforcement fibers can be considered for use in nonaligned composites. Of course, limits are also set on v_f in random reinforcement by the difficulties of close packing.

FRACTURE TOUGHNESS

The components of many practical filament-reinforced composites are often brittle. Consequently, although they have high fracture strengths and are stiff, such composites are usually not very tough. We have seen (Chap. 14) that for reliable use in design, a material must have an adequate resistance to crack propagation. It is essential therefore to ensure toughness in composites; otherwise, we shall be using expensive constituent materials to produce a composite whose potential strength we can never use, because we are afraid of it breaking catastrophically if there is some local imperfection that causes a crack to start.

One peculiarity of fracture in filamentary composites is that the fracture

Figure 17-16 Stubblelike fracture surface of an aligned graphite/epoxy composite. (Photograph courtesy W. H. Durrant.)

surfaces are rarely simply planar and have bits of fiber sticking out of the matrix (Fig. 17-16). As we shall see, this phenomenon helps to produce some extra toughness in composites. The cause of the stubblelike feature is as follows: If the fibers are not continuous and a matrix crack (originating perhaps at a bubble, dust particles, or notch in the matrix) approaches a fiber in such a way that the end of the fiber is within a length $L_c/2$ of the fracture plane, we expect that filament end to be *pulled out* of the matrix rather than broken; this comes about simply because within $L_c/2$ of the ends of the filaments, the breaking stress is not reached [Fig. 17-10(b)]. On the other hand, continuous filaments of uniform strength should fail in the plane of the matrix crack, with no pull-out; this applies also to discontinuous filaments whose ends are embedded at distances *greater* than $L_c/2$ below the main fracture plane. Now real fibers have weak points from place to place along their length and vary statistically in strength from fiber to fiber, so that a fiber in the path of the crack will be broken either where it crosses the plane of the crack (i.e., where the stress is greatest) or at a flaw in the fiber that is near, but not necessarily on, the plane. Thus fracture surfaces appear "whiskery," and real composites (having continuous filaments but weak points) possess properties between the two extremes shown by those with full pull-out and no pull-out.

On statistical grounds it is argued that the average pull-out length for random filament fracture is $L_c/4$ because the shortest is 0 and the longest $L_c/2$. This provides a means of approximately estimating the interfacial stress that existed before the composite broke.

Exercise 17-9

Pull-out lengths average 1 mm on a fracture surface of a filamentary composite whose fibers have $(\sigma_f)_{max} = 2.5$ GPa and $d = 50$ μm. Estimate the interfacial bond strength.

Solution. $L_c/4 = 1$ mm, so $L_c = 4$ mm. $L_c = (\sigma_f)_{max}\, d/(2\tau)$, so

$$\tau = (\sigma_f)_{max}\, d/(2L_c) = \frac{2.5 \times 10^9 \times 50 \times 10^{-6}}{2 \times 4 \times 10^{-3}} = 15.7 \text{ MPa}$$

When a filament fractures off the main fracture plane, debonding must take place above, below, and around the filament fracture to allow the main crack to propagate (Fig. 17-17). After the interfacial bond has been broken, there is an interfacial frictional stress opposing the fibers coming out of their holes. The friction stress can be as large as the interfacial stress was before debonding (if the interfacial break occurs by slip in the matrix adjacent to the fibers), but usually it is rather less than τ. Even so, work must be done pulling fibers out as the original main crack front propagates. This work contributes *significantly* to the toughness of composites.

The work of pulling out an isolated fiber can be determined as follows. In Fig. 17-17 a fiber of diameter d has broken a distance h below the main crack plane, where $0 < h < L_c/2$. It is shown as having been pulled out a distance x. If the interfacial frictional stress is τ', the total force on the cylindrical debonded area opposing pull-out at that instant is $\tau'\pi d(h - x)$. When pulled out a further distance dx, the work done is $\tau'\pi d(h - x)\, dx$. The total work of pull-out for the isolated fiber over the whole distance h is therefore

$$\int_0^h \tau'\pi d(h - x)\, dx = \frac{\tau'\pi dh^2}{2}$$

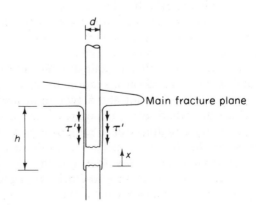

Figure 17-17 Debonding between filament and matrix necessary to permit main crack to propagate when fiber fractures off the principal plane of fracture.

As already explained, the shortest pull-out length is 0 and the longest is $(L_c/2)$; this sets limits on values taken by h over a number of filaments. Thus the average pull-out work per filament is given by

$$\frac{1}{L_c/2} \int_0^{L_c/2} \frac{\tau'\pi dh^2}{2}\, dh = \frac{\tau'\pi dL_c^2}{24} \tag{17-14}$$

If there are N filaments in the crack advance area A,

$$v_f = \frac{N\pi d^2}{4A}$$

The pull-out work of these N filaments divided by the advancing crack area represents the pull-out contribution $R_{\text{pull-out}}$ to the composite fracture toughness. Therefore

$$R_{\text{pull-out}} = \frac{N(\tau'\pi dL_c^2/24)}{A} = \frac{v_f \tau' L_c^2}{6d} \tag{17-15}$$

$$= \frac{v_f \tau' d(\sigma_f)_{\max}^2}{24\tau^2} \tag{17-16}$$

since $L_c = (\sigma_f)_{\max} d/(2\tau)$.

Exercise 17-10

Determine the pull-out contribution to toughness in a carbon-fiber/polyester-matrix composite with the following properties: $v_f = 0.4$, $(\sigma_f)_{\max} = 1.6$ GPa, $d = 8$ μm, $\tau = 7$ MPa, $\tau' = 1$ MPa.

Solution. $R = (0.4)(8 \times 10^{-6})(1.6 \times 10^9)^2(1 \times 10^6)/[24(7 \times 10^6)^2] = 6.97$ kJ/m^2. Notice that $L_c = (\sigma_f)_{\max} d/(2\tau) = 1.6 \times 10^9 \times 8 \times 10^6/(2 \times 7 \times 10^6) = 0.91$ mm, so that the longest pulled-out fiber on a stubbly fracture surface would be some 0.5 mm in length and the shortest 0 mm.

Some other contributions to composite toughness come from the work of debonding and the creation of new surfaces, both in the main fracture plane and also in the cylindrical areas around the pulled-out filaments.

Composite fracture toughness is the sum of all the various dissipative work components per unit cross-sectional area of fracture *referenced to the area of the main crack*. As cylindrical debond areas and pull-out lengths are related to events off the main fracture plane, a consequence of dividing the total work of crack propagation by merely the projected area of the main matrix crack—and not the actual total area of new surfaces, which would include the cylindrical debond areas—is that a *synergism* in toughness occurs. That is, the toughness of a composite is greater than the sum of the toughnesses of the components. Thus, even though many filamentary composites are made from components that are individually brittle, the composites can have respectable resistance to crack propagation. The total toughness of typically laid-up composites with about 50% volume fraction of fibers may be

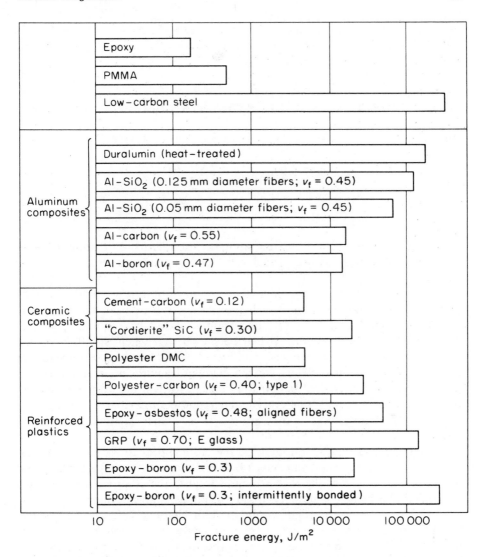

Figure 17-18 The fracture toughness of various types of materials. [Adapted from B. Harris, *Composites*, **3** (1972), p. 152.]

50 kJ/m². This should be contrasted with the fracture toughness of various other solids, shown in Fig. 17-18.

Another aspect of the cracking behavior of filamentary composites is that partially pulled-out fibers bridge the crack opening in the wake of the advancing matrix crack. The effect is enhanced the longer the pull-out lengths, as discussed in the next section. Cracked composites thus have an ability to hang together in situations in which other solids of similar toughness would break apart. Note that the full contribution to toughness given by pull-out is achieved only when the filaments have pulled all the way out, at which time they no longer bridge the crack faces.

HIGH TOUGHNESS AND HIGH STRENGTH COMBINED

Recall the rule of mixtures for composite strength, Eq. (17-12):

$$(\sigma_c)_{max} = (1 - v_f)(\sigma_m)_{max} + v_f(\sigma_f)_{max}\left(1 - \frac{L_c}{2L}\right) \qquad (17\text{-}12)$$

$$= (1 - v_f)(\sigma_m)_{max} + v_f(\sigma_f)_{max}\left[1 - \frac{(\sigma_f)_{max}\,d}{4\tau L}\right] \qquad (17\text{-}17)$$

This demonstrates that for greatest strengths the interfacial bonding must be strong, that is, τ must be large. Much research has gone into the production of strong interfaces to that end.

However, Eq. (17-16) for the pull-out contribution to toughness suggests that τ must be small to obtain tough composites. When τ is small, L_c is long, so that the fibers pull out over long lengths with a correspondingly large dissipation of toughness work. It can be shown that when the other contributions to total toughness are also considered, high-toughness composites are produced when τ is small. Thus, when a filament fractures in a composite that displays a strong interfacial bonding, a crack is formed in the matrix around the fracture, and usually this crack is energetic enough to run through the composite, breaking fibers as they are met in a "zipper" action. The work of fracture is very low.

High strength and high toughness therefore appear to be mutually exclusive. For reasons explained in Chap. 14, this is not a satisfactory situation, and use cannot be made of the high strength because the composite is relatively brittle. Various means of getting around this impasse have been proposed.

In one method the high tensile strength characteristic of strong interfacial filament matrix bonding can be combined with the high fracture toughness of weak interfacial bonding, when the filaments are arranged to have random alternating sections of high and low shear stress (and locally low and high interfacial toughness). Such weak and strong areas can be achieved by intermittent coating of the fibers with appropriate substances (Fig. 17-19). The strong regions (especially if longer than $L_c/2$) ensure that the filament strength is picked up; weak areas randomly in the path of running cracks provide the source of toughness and pull-out work.

Extra-long pull-out lengths may also be achieved when a special mechanism of debonding at the filament–matrix interface takes place ahead of the running crack. This *Cook-Gordon mechanism* relates to tensile stresses that are generated *parallel* to the crack, which can produce debonding and crack-tip blunting when sufficiently weak interfaces are present ahead in the path of the

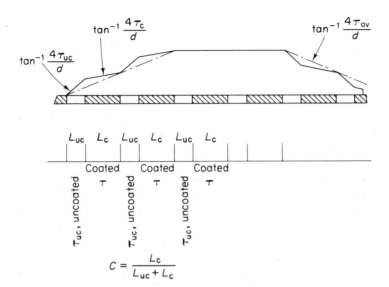

Figure 17-19 Intermittent bonding of filaments (alternating bands of high and low interfacial stress and low and high toughness) in order to achieve simultaneous high strength and high toughness. Subscript c means coated with low shear strength material; subscript uc means uncoated. [From A. G. Atkins, *J. Mater. Sci.*, **10** (1975), p. 819.]

crack (Fig. 17-20). Such an effect is normally absent in conventionally made composites because they have continuously strong bonds. Were the interfacial conditions appropriate to allow Cook–Gordon debonding in conventional composites, the interfacial τ would probably be too low to give adequate tensile strength. Some results for intermittently bonded boron–epoxy composites are shown in Fig. 17-21. Two types of coating materials were used, one (polyurethane) that encouraged Cook–Gordon debonding and another (sili-

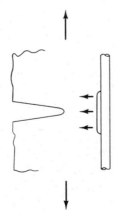

Figure 17-20 The Cook–Gordon debonding and crack-blunting mechanism caused by tensile stresses *parallel* to the crack, which can debond sufficiently weak interfaces in the path of the crack.

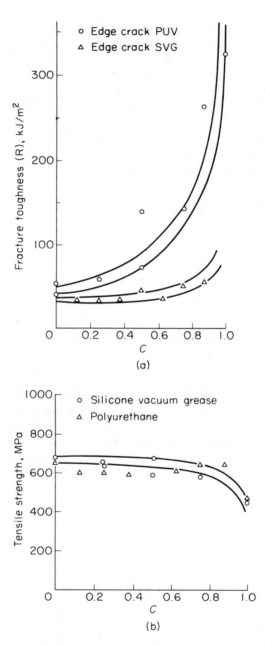

Figure 17-21 (a) Fracture toughness and (b) tensile strengths of 0.25 v_f intermittently bonded boron–epoxy composites (filament diameter 140 μm) as a function of C, which is the ratio of coated length to pattern repeat distance along the fiber (repeat distances variously 19, 25, and 51 mm). Much improved toughness with polyurethane coatings, which permit Cook–Gordon debonding. [From A. G. Atkins, *J. Mater. Sci.*, **10** (1975), p. 819.]

cone grease) that did not. The differences in toughness with, and without, the enhanced pull-out lengths are clearly shown. As there are no significant differences in tensile strength, it suggests that a distinction should be made between the property of interfacial shear strength and the property of interfacial toughness in the coatings.

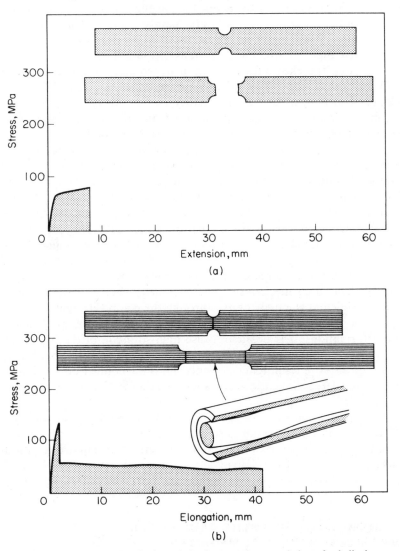

Figure 17-22 The Morley duplex reinforcing member, consisting of a helical core and surrounding sheath, which gives rise to a stress-controlled interfacial debonding mechanism. A semicircular notch localizes failure in even a strongly work-hardening austenitic stainless steel (a), whereas a brittle epoxy matrix, reinforced with duplex members, continues to carry load at high elongations (b). [From J. G. Morley, *Chemistry in Britain* (Nov./Dec. 1974), by permission.]

Related to the foregoing is the study by Gordon and Jeronimidis on wood, the toughness of which (about 9 kJ/m^2) is exceptionally high in relation to its other mechanical properties and to its density. The specific work of fracture is greater than the energy likely to be absorbed by fiber pull-out, so that a special energy-absorbing mechanism seems to exist for trees and presumably for other plants. It turns out that in the cell walls of the timber, the cellulose fibrillae of which the cells are mainly composed are disposed in a preponderantly helical manner. When pulled in tension, the cell walls buckle in such a way that longitudinal extensions of the cell up to 20% are possible. Thus, although the elastic fracture strain in wood as a whole seldom much exceeds 1%, cells close to the fracture surface typically extend 15–20% before breaking, absorbing much energy as they do so. Arrays of model cellulose fibers, made by winding glass and carbon filaments into hollow helices with resin, exhibit experimental fracture toughnesses up to 400 kJ/m^2.

Another approach by Morley and coworkers employs a new type of reinforcing member, consisting of a core and a surrounding sheath. In the laboratory hypodermic tubes have been used which are filled with an inner core of helical wire, the unstressed helix diameter of which is greater than the internal bore of the tube so that there is a frictional bond between the core and sheath. Under normal loading conditions the reinforcing members stiffen and strengthen the matrix in the usual way as with any other fiber-reinforced solid. The shear strength of the interface is reduced, however, as tensile loads on the core member increase; this gives rise to a simple stress-controlled interfacial debonding mechanism. The core helix wire is therefore prevented from fracturing and is merely drawn through the composite against the friction at the walls of the sheath. As the decoupling mechanism is a result of small elastic deformations, it is possible to use high-strength, brittle metal core elements, or even bundles of polymer-bonded glass or carbon fibers, in the construction of the core-reinforcing elements. Thus the conflict between high yield strengths and large elongations to failure encountered particularly in metals can be avoided.

The simple geometric stress concentration in the test piece shown in Fig. 17-22(a) localizes failure even in a strongly work-hardening austenitic stainless steel. A brittle epoxy matrix of the same geometry, but reinforced with duplex members, continues to carry load at very high elongations [Fig. 17-22(b)]. This sort of behavior is important in the design of energy-absorbing structures.

DESIGNING WITH FILAMENT-REINFORCED
COMPOSITES: HYBRID COMPOSITES

Laminates of thermosetting resins (mainly polyesters and epoxies) and glass fibers (GRP) are extensively used as medium-cost materials for a variety of products ranging from car and boat shells to large tanks and pipes. For many structures where the design parameters are deflection limits of buckling stability rather than ultimate strength, the relatively low elastic modulus of GRP is a limiting factor. Composites based on higher-modulus fibers, such as boron, carbon, and kevlar, have greatly superior specific strengths, but their use is restricted by their high price. Even so, production aerospace structural members and some automotive components are now being manufactured (at least in part) with these stiffer filaments, particularly carbon fibers. Moreover, sporting goods such as shafts for golf clubs are available in carbon-fiber-reinforced plastics (CFRP).

In an attempt to bridge the cost/performance gap between GRP and, say, CFRP, small amounts of stiff fibers can be added to GRP. Such *hybrid* composites containing more than one material may be of various types and different construction, depending on the desired properties and design objectives. Stiffness and strength of laminated structures can be increased substantially by the addition of appropriate surface layers, such as carbon filaments on glass-fiber-mat-reinforced plastics.

Appendix 13 gives representative densities and mechanical properties for a range of composite materials. Designers should remember that practical composites do not always achieve the moduli and strengths predicted by theory, or the same values obtained in the laboratory, as a result, for example, of difficulties in manufacturing large homogeneous structures (whether the fibers are aligned, in cloths, or random). Reasonable agreement with ideal behavior occurs when the loading is predominantly tensile and filaments regularly aligned (as in helically wound reinforced tubes). It is when major loadings are compressive that problems arise, as already mentioned, and then the performance of practical composites falls well below ideal behavior. Similar comments relating to tensile or compressive stress fields apply in fatigue (Chap. 18).

CONCRETE MATERIALS

Conventional concrete contains Portland cement, aggregate, and water. A wide range of properties can be produced depending on the proportions of the mixture and the type of aggregate. The cement reacts with the water to form a matrix for the aggregate, the hydration sometimes continuing over many years if plenty of water is present. An adequately strong material can usually be produced in a few hours at outdoor temperatures above freezing; factory-made concrete products may be heat-cured in order to speed up production. The strength of the cementative matrix is governed mainly by porosity, which depends on the initial water/cement ratio and the extent of hydration at the time in question; consequently strengths increase with time and it is customary to proof-test concretes for reference purposes 28 days after manufacture. Subsequently the strength of concretes may increase by 25% a year later.

Conventional concretes have relatively low fracture toughness and hence low tensile strengths (about 3 MPa), but they do have reasonable compressive strengths (greater than 10 MPa). It is customary in civil engineering to use steel reinforcing bars to help concrete structures carry tensile loads, and if furthermore these bars are tensioned, it is possible to put the concrete into compression and effectively increase the loads that can be carried without cracks forming in the concrete. In this way the flexural strength of tensioned reinforced concrete can be doubled over that of simple concrete.

Traditional solid reinforcing bars are comparatively large ($10 \sim 50$ mm in diameter), and even if steel wire is used in place of solid rods, individual strands are usually not less than a few millimeters in diameter; similarly, steel wire mesh reinforcement is not particularly small. In recent years, however, the types of filament already described in this chapter have been successfully incorporated into concrete and cement. Fibers of steel, glass, polypropylene, and so on can increase the tensile strength of cements to perhaps 20 MPa. An added bonus is that it becomes possible to fabricate thin sheets of fiber-reinforced cements that maintain their integrity during hardening; such techniques have long been known in the asbestos cement industry. The toughness of an asbestos cement is some 700 J/m^2; high-alumina cement reinforced with glass fiber can have toughnesses of some 20 kJ/m^2.

Concrete impregnated with polymers can also be made with compressive and tensile strengths much improved over conventional concrete. Polymers such as PMMA, PS, and PVC may be incorporated either as aqueous emulsions during the mixing process (when the water in the emulsion reacts with the unhydrated cement to develop strength and stiffness in the usual way) or impregnated as monomers to already hardened conventional concrete, with subsequent polymerization being effected by gamma radiation or by a thermal catalytic process. Concrete made in these ways possesses improved resistance to attack from acids.

In comparison with other engineering solids, concretes require quite small amounts of energy for their production—on the various bases of energy per unit mass, per unit stiffness, per unit strength, and so on. Appendix 14 gives a broad indication of the range of values. The achievement of specified performance criteria using low-energy materials is likely to be of growing importance in the future.

REFERENCES

17-1. Cottrell, A. H., *The Mechanical Properties of Matter*. New York: John Wiley and Sons, 1964.

17-2. Gordon, J. E., *The New Science of Strong Materials—or Why You Don't Fall Through the Floor*. London: Penguin Books, 1968.

17-3. Kelly, A., *Strong Solids*. Oxford: Clarendon Press, 1964.

17-4. National Engineering Laboratory, Glasgow, *Designing Hybrid Composite Materials*. Research Summary No. 633, 1977.

17-5. Pomeroy, C. D., "Concrete, An Alternative Material." Fourteenth John Player

Memorial Lecture. Institution of Mechanical Engineers, London. *Proceedings*, 192 (1978), 14, p. 135.

PROBLEMS

17-1. Describe those properties of high-performance composites that make them attractive in spite of their high cost.

17-2. It is proposed to hang long specimens in a vertical hole 1 m in diameter in the ground, loaded only by their own mass. Find the maximum lengths that can be suspended, for each of the following materials (neglect changes in gravity with the depth of the hole):

   (a) 1020 steel.

   (b) Glass-reinforced tape.

   (c) Boron–epoxy.

   (d) Graphite whiskers.

Material	Average Density, Mg/m^3	S_u, MPa	E, GPa
1020 steel	7.85	500	210
glass-reinforced tape	1.90	770	27
boron–epoxy	2.34	1 800	250
graphite	1.65	20 000	500

17-3. Calculate the stress that can be withstood by a boron–epoxy composite having $v_f = 0.4$ at some strain level where, separately, the born filaments require a stress of 1.4 GPa and the matrix 81 MPa.

17-4. Determine the critical lengths of carbon fibers 8 μm in diameter possessing $(\sigma_f)_{max} = 1.6$ GPa in a polyester matrix with $\tau = 20$ MPa. Likewise determine L_c for boron–epoxy with fibers 100 μm in diameter with $(\sigma_f)_{max} = 3$ GPa and $\tau = 70$ MPa.

17-5. In a tensile test of monofilament of boron embedded in an epoxy matrix, several breaks occurred in the filament before the specimen fractured. Explain clearly how it is possible to develop a stress high enough to cause a second fracture in the broken fiber after it breaks the first time. Draw a sketch of the broken filament in order to make your explanation clear.

17-6. Mylar tape of 34.2 MPa strength is reinforced with continuous glass filaments to

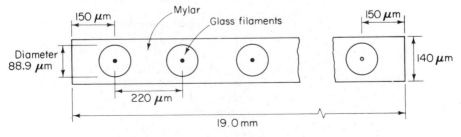

Figure 17-23

produce a composite strength of 356 MPa. If the cross section of the tape has the dimensions given in Fig. 17-23, calculate the strength of the glass filaments.

17-7. A composite specimen contains parallel graphite filament bundles 135 μm in diameter aligned in a square pattern (shown in Fig. 17-24 in cross section) in a matrix of epoxy.

  (a) Calculate the volume fraction of filament in this specimen.

  (b) The graphite strength is 2.93 GPa, and the epoxy stress at the fracture strain of the graphite is 83.7 MPa. Estimate the strength of the specimen.

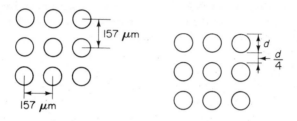

| **Figure 17-24** | **Figure 17-25** |

17-8. Graphite filaments 7.00 μm in diameter and having fracture strength 850 MPa (at a failure strain of 0.0140) are to be cast in a linear elastic epoxy matrix of strength 112 MPa (with failure strain of 0.0670). If the filaments are to be aligned in a square matrix with the distance between filaments at least one fourth the filament diameter (Fig. 17-25), calculate the maximum strength the composite will have.

17-9. You wish to manufacture a composite tape made of a single layer of aligned continuous glass filaments 125 μm in diameter, which have a fracture strength of 700 MPa, embedded in a matrix of resin that has a strength of 35 MPa at the same fracture strain as the glass filaments. If the tape is to be 208 μm thick and 50 mm wide and to have a strength of 200 MPa, how many strands of glass filament must it contain?

17-10. Tensile data on a continuous-filament glass-fiber-reinforced tape are as follows:

Test	Fracture Stress, MPa	Orientation of Tensile Load
A	120	Longitudinal to fibers
B	20	45° to fibers
C	32	Transverse to fibers

  (a) Explain the difference between the strengths of test A and test C.

  (b) Explain the difference between the strengths of test B and test C.

17-11. The world record for pole vaulting stood at 15 ft (4.6 m) in 1940. Fiberglass poles replaced bamboo poles in 1962, and by 1970 the world record stood at 18 ft (5.5 m). Other things being equal, explain how the difference in moduli between bamboo poles (about 45 GPa) and glass-fiber poles (about 40 GPa) contributes to the improved vaulting record.

17-12. Show that flexural stiffness is proportional to (beam thickness)^3, and that consequently for a given length, width, and mass of sheet material the specific stiffness in bending varies as E/ρ^3 rather than as E/ρ, which is used in Appendices 12 and 13 for simple tension. (ρ is density.)

17-13. Show that in a unidirectional glass-fiber composite the maximum stiffness per unit mass is obtained at 25% volume fraction fibers, using $E_f/E_m = 20$ and 2.5 Mg/m³ for the density of glass filaments with 1.0 Mg/m³ for the density of the laminating resin.

17-14. Show that in a composite for which $(\sigma_f)_{max}$ of the filaments is much greater than $(\sigma_f)_{max}$ of the matrix, the maximum bending strength is obtained when the mass of reinforcement is one half of the total mass. (This corresponds with the optimum skin mass fraction for maximum bending strengths of sandwich beams.) Explain why an optimum volume fraction exists and why in bending it is not useful to aim for the highest fiber volume fraction possible. (Your answer will be relevant to Problem 17-13 as well.)

17-15. A test for fracture toughness of a cross-ply boron–epoxy composite yields the load–extension curve shown in Fig. 17-26. Calculate its fracture toughness.

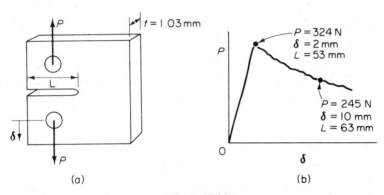

Figure 17-26

17-16. Data from a compact tension specimen of molded fiberglass of thickness 2.63 mm and crack growth increment between points 1 and 2 of 6.35 mm are given in Fig. 17-27. Calculate the fracture toughness.

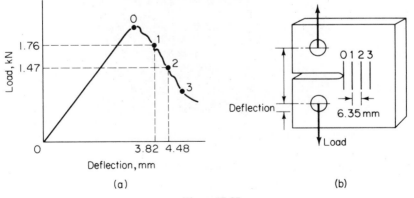

Figure 17-27

17-17. Compacts of dry cement powder that are subsequently hydrated can be made with compressive strengths of over 370 MPa. Explain why they are not used commercially, and suggest modifications which would permit their wider use.

17-18. Hardened cement paste is so alkaline that rusting of steel reinforcing material is inhibited in the absence of ingress of surface water down cracks. What is likely to happen to ordinary glass filaments if used for fiber reinforcement?

17-19. "Holes are enormously cheaper, both in money and in energy, than any conceivable form of high-stiffness material." (J. E. Gordon.) Discuss.

17-20. The stress–strain characteristics of linear elastic graphite filaments 7 μm in diameter and of a linear elastic epoxy are given as follows:

Material	Fracture Stress, GPa	Fracture Strain
Graphite filaments	3.00	0.010
Epoxy	0.50	0.050

If you are limited to a filament volume fraction of 0.35, calculate the maximum load that can be supported by a rod 10 mm in diameter of a composite of aligned graphite and epoxy.

17-21. Graphite filaments 7.00 μm in diameter and having fracture strength 850 MPa (at $e_f = 0.0140$) are to be cast in a linear elastic epoxy matrix of strength 112 MPa (at $e_f = 0.0670$). If the filaments are to be aligned in a hexagonal matrix with the distance between filaments at least one fourth the filament diameter, calculate the maximum strength the composite will have.

17-22. Figure 17-28 shows a crease compression failure in 40% v_f carbon-fiber-reinforced epoxy. Discuss how this mode of failure arises.

Figure 17-28 Specimen width is 2.5 mm. (Courtesy C. R. Chaplin.)

17-23. Show that a volumetric strain of magnitude

$$\frac{\cos(\alpha - \gamma)}{\cos \alpha} - 1$$

occurs in the sheared slice in Fig. 17-28, where γ is the shear strain in the slice and α is the inclination of the slice to the horizontal in the photograph. Discuss how such a volume increase may be accommodated in a fiber-reinforced composite and show that a limit is set on the axial displacement associated with this type of failure.

18

Fracture by Gradual Crack Growth: Fatigue and Stress-Corrosion Cracking

Under certain circumstances, cracks can grow gradually in a structure at average stresses below the yield stress. If this crack growth progresses far enough, the crack reaches a critical size that will lead to catastrophic growth (see Chap. 14), or the crack so reduces the net cross section that the tensile strength is exceeded (see Chap. 2). In either case the structure fails suddenly, without warning. A large proportion of service fractures occur in this fashion.

The circumstances that can cause gradual growth of a crack are: (1) *fatigue*, when the stress in a part varies with time, and (2) environmentally promoted crack growth, usually called *stress-corrosion cracking* or *hydrogen embrittlement*, which can occur under constant tensile stress over a period of time. Fatigue seldom occurs without acting in concert with some environmental effects, except under carefully controlled laboratory conditions, for example, testing in a vacuum, so the two modes of gradual crack growth are often combined.

This chapter describes the fundamentals of these fracture modes and the conditions that promote or restrict crack growth, with the objective of assisting the engineer-designer in avoiding or minimizing gradual crack growth in his structures.

CONDITIONS FOR FATIGUE

Fatigue fracture occurs when a part is subjected to many cycles of varying stress, in which at least part of each cycle is tension. The magnitude of stress required to cause fatigue is less when the number of cycles is larger. This

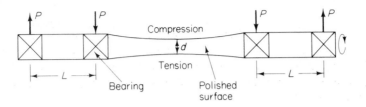

Figure 18-1   Rotating–bending test for fatigue.

interdependence of stress and number of cycles can be measured in a simple *rotating–bending test.* Figure 18-1 shows how a cylindrical specimen with a ground and polished cross section in its midsection is loaded in four-point bending, with four equal forces P. The bending moment M in the midsection is thus a constant, PL, over the entire polished region, and the stress σ_t at the bottom (tensile) side at the section of minimum diameter d is thus

$$\sigma_t = \frac{32M}{\pi d^3} = \frac{32PL}{\pi d^3}$$

After the specimen has rotated one half turn, so that this same point is now on the top side, the stress becomes compressive:

$$\sigma_c = \frac{-32PL}{\pi d^3}$$

When this same point rotates through the positions midway between top and bottom, its stress is zero. Thus a single point on the surface of the specimen goes through a complete sinusoidal stress cycle from tension to compression to tension with each full rotation. Although it is not necessary for the cyclic variation in stress to be either as smooth or as repetitive as this, or as symmetrical from tension to compression, this test is commonly used as a standard to measure the maximum fatigue life under a given stress. We will examine later in this chapter those variables for a real part that serve to lower this life. For the present we must note that surface finish is very important in fatigue, so that the standard specimens described above have ground and polished cross sections as shown in Fig. 18-1. In other tests with which we have dealt (e.g., simple monotonic tension tests) surface condition is not so important for ductile materials.

Several identical specimens are required to establish the fatigue life. For each, a value of the load P is selected, and the specimen is rotated until the specimen fractures. The testing machine has a cycle counter that records the number of cycles at the instant of fracture. Most tests are of relatively short duration because a machine rotating at 1800 revolutions per minute (30 revolutions per second) will expose the specimens to about 10^5 cycles per hour and 2.50×10^6 cycles per day. After the first specimen fractures, a new specimen is tested at a different load P, and so on.

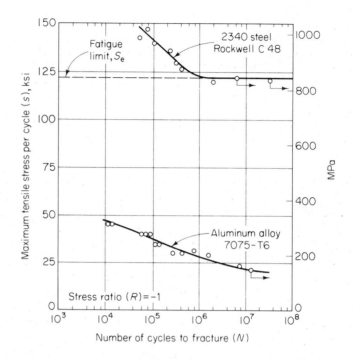

Figure 18-2 Results of rotating–bending fatigue tests, in the form of s–N curves. [From ASM Committee on Analysis of Fatigue Failures, "Fatigue Failures," *Metals Handbook: Failure Analysis and Prevention*, 8th ed., vol. 10, ed. Howard E. Boyer. (American Society for Metals, 1975), p. 96.]

Figure 18-2 is the plot of maximum stress (s) vs. number of cycles (N) for fatigue fracture of a steel of hardness 48 R_C and of a high-strength aluminum alloy. For the steel, stresses below about 825 MPa do not cause fracture for any number of cycles tested, and presumably not for an infinite number of cycles. The stress below which fatigue does not occur in steel is called the *fatigue limit* or *endurance limit*, S_e.

In the absence of tests to determine the fatigue limit for a steel, the value of the mean fatigue limit $\bar{S}_e$ for a polished rotating beam specimen can be estimated from the following:

$$\bar{S}_e = 0.5 S_u \quad \text{for } S_u \leq 1.4 \text{ GPa}$$

$$\bar{S}_e = 700 \text{ MPa} \quad \text{for } S_u > 1.4 \text{ GPa}$$

Because there is a wide statistical spread in the fatigue limit, the above values may not provide for conservatively safe design. Actual measured values must therefore be used for design. Furthermore, the above fatigue limit will be reduced by numerous service factors, which will be discussed later in this chapter.

Aluminum alloys do not exhibit a fatigue limit. In Fig. 18-2 the slope of the curve for the aluminum alloy indicates that for higher numbers of cycles the stress for fatigue fracture will be even less than the lowest value shown in this curve. Thus the "fatigue limit" for aluminum is often quoted as the stress for 500×10^6 cycles. The absence of a true fatigue limit in aluminum alloys leads to a mandatory requirement for periodic inspection of highly stressed aluminum structures to locate fatigue cracks and replace such parts before the cracks become dangerously large. (See Chap. 14 for a discussion of the influence of crack size on fracture.)

Fatigue in engineering structures exposed to a wide and possibly random spectrum of loads will depend on the variable range and absolute magnitude of the stresses produced, so exact simulation of each case is impossible. However, tests are possible with a sinusoidal variation of stress that is different from the equal tension-to-compression cycle of the rotating–bending test. All that is essential is that some portion of each cycle be tensile. For any test that varies from a maximum stress s_{max} to a minimum stress s_{min}, the *stress ratio R* is defined as

$$R = \frac{s_{min}}{s_{max}}$$

(The symbol R is also used elsewhere in this book to denote fracture toughness; since both these meanings are in common use, we will use R for both. The meaning should always be clear from the context.) Thus for the rotating–bending test in Fig. 18-1, $R = -1$. A test with $R > 0$, for example, will always be in tension. The number of cycles required for fracture will thus depend on the stress ratio as well as the magnitude of s_{max}.

Other factors include the type of loading (bending, tension, or torsion, for example), the shape of the loading curve (of which sinusoidal is the simplest), the frequency of cycling, the environment, temperature, and specimen size and surface condition (notched, rough, smooth, etc.). Although it is not appropriate here to present the complete formal procedures used for designing to resist fatigue fracture (see Ref. 18-10 for such discussion), we will describe later in this chapter the influence of some of these variables on fatigue life. In order to understand how these factors can operate, it is first necessary to present a brief discussion of the mechanisms of fatigue crack initiation and growth.

Exercise 18-1

How do airlines prevent fatigue cracks in their aluminum aircraft from growing to catastrophic size?

Solution.   Regular and very careful inspection is required to locate and remove potentially dangerous fatigue cracks. Failure to locate cracks has led to many aircraft disasters; in most cases these were in relatively new designs in which the regions susceptible to fatigue were not completely anticipated.

FATIGUE CRACK INITIATION

Fatigue would probably not occur in a perfectly homogeneous material that is absolutely smooth on its external surfaces and uniformly loaded. Under elastic stresses the material would uniformly deform elastically, and when the stress is removed the elastic deformation would be completely reversed and thus be returned to the initial configuration. No matter how many times this stress cycle is repeated, there would be no permanent change in the structure.

It is the nonuniform characteristics of real materials that lead to fatigue under cyclic stress. Consider a polycrystalline metal that contains nonmetallic impurities or inclusions, natural dislocations that arise from solidification processes, and an imperfectly smooth exterior surface. When a part made of this metal is stressed in the nominally elastic range, portions of many dislocations will move and can contribute to very small, irreversible changes in the configuration of the part. These changes are initially much smaller than can be detected in a stress–strain test, so we would consider such a load–unload cycle to be elastic, that is, "fully reversible." In fact, even the first cycle has caused some local irreversible changes.

An example of such an irreversible change under a stress ratio $R = -1$ (recall that $R = -1$ means fully reversed tension–compression) is the *Cottrell–Hull mechanism*, which we have here modified and altered for simplicity. Consider a crystal located at a free surface with a stable internal dislocation configuration such as shown in cross section in Fig. 18-3(a). The dislocations on slip plane A are blocked by a barrier (for example, dislocations on an intersecting plane) at the inner end of the plane, as are the dislocations (of opposite sign) on parallel slip plane B. (As this is a very special configuration, many stress cycles may be required before this kind of mechanism even begins to operate.) When a tensile stress is applied to the ends of the specimen, the local resulting shear stress τ causes dislocation 1 on slip plane A to move to the free surface and to leave a step, as shown in Fig. 18-3(b). The dislocations on slip plane B cannot move because of the barrier.

When the applied stress is compressive, as in Fig. 18-3(c), dislocation 2 on slip plane B can now move freely to the surface to produce a second (and opposite) step, as in Fig. 18-3(d). Thus we have the genesis of a fatigue crack, but since its depth is roughly the Burgers vector, or of order 300 pm, the crack and the change in shape of the specimen are still undetectable. Thus we see an example of a permanent change in a part that we have previously considered elastic and fully reversible. If there are no further tension–compression cycles, the effects of the emergence of dislocations 1 and 2 in Fig. 18-3 are negligible. The infinitesimal crack formed becomes important only if it grows through cyclic stressing.

The next stress cycle can cause dislocations 3 and 4 to move in a similar manner to increase the crack length. It is also possible that dislocation 3 is slightly blocked by an impurity atom, and it is not until many cycles later that the local stresses are of sufficient magnitude to move dislocation 3, at which

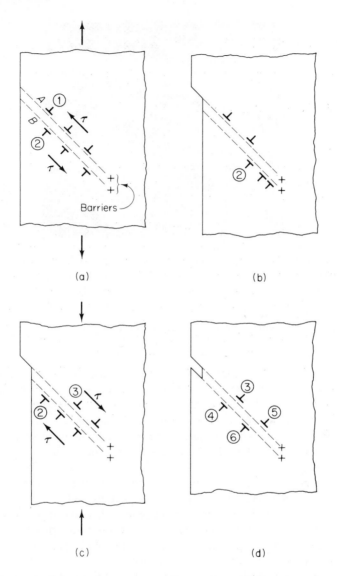

Figure 18-3 Modified Cottrell–Hull mechanism for fatigue crack initiation.

time the process can continue. If this process continues for enough cycles, it will produce a detectable crack. If the minimum detectable crack length is, for example, 10 μm, then for this one crack in this one grain to be detected would require 10 μm/300 pm $\approx$ 30 000 cycles at the very least. (It would also require a dislocation-generation mechanism to create the large number of dislocations required.) Thus in order to create a small fatigue crack, many stress cycles are required.

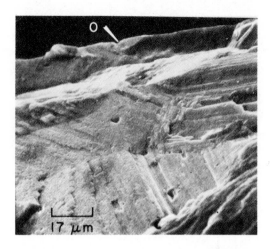

Figure 18-4 Stage I fatigue fracture surface in aluminum alloy 2024-T3 (UNS A92024). Initiation at free surface, pointer at 0, with {111} of each grain as fracture plane. Stage II begins at lower right. [From ASM Committee on Fractography by Electron Microscopy, "Interpretation of Scanning-Electron-Microscope Fractographs," *Metals Handbook: Fractography and Atlas of Fractographs*, 8th ed., vol. 9, ed. Howard E. Boyer. (American Society for Metals, 1974), p. 68.]

This initial crack formation during fatigue is called *Stage I fatigue*. The fracture surface, which is the new surface created by the slip on two parallel planes, shown in Fig. 18-3(d), is therefore the plane of the slip process and is very flat when confined to a single grain. Stage I fatigue usually extends 2 to 5 grains from the origin, which requires slip in several adjacent grains. We have already seen (Chap. 5) that slip in fcc aluminum occurs on {111} in a ⟨110⟩ direction, so we would expect to find Stage I fatigue fracture surfaces in aluminum to be {111} planes. Indeed they are, but Stage I fracture surfaces in aluminum have also been observed to be {110} and {100}; these observations are not inconsistent with our simple Stage I model, as the slip direction ⟨110⟩ also lies in the {110} and {100}. Thus fatigue initiation processes are more complex than simple slip processes. One example of the appearance of a Stage I fatigue fracture surface is shown in Fig. 18-4.

The foregoing mechanism considers only one of several possible ways that dislocations can lead to fatigue cracking. A foreign impurity particle, a surface imperfection, or a plastically deforming grain surrounded by elastic grains can act to cause local stress concentrations that will in similar ways lead to creation and growth of small fatigue cracks. The more severe the initial imperfection, the more rapidly the fatigue cracks are formed. Because of the difficulty of observing the initiation and growth of small cracks in the earliest stages of fatigue, and probably also because of the large number of variables involved, the dominant processes at this stage are not yet fully understood.

FATIGUE CRACK GROWTH

Once a fatigue crack is created, its growth to finite size is controlled by progressively larger scale effects than operated in its earliest stages. It is in this realm that the importance of a tensile component to the stress cycle becomes apparent. This period of crack growth is called *Stage II fatigue*.

Because metals and alloys exhibit a wide range of properties, no one mechanism completely describes the events of Stage II fatigue, but the essential elements are included in the discussion that follows, for a moderately ductile, medium-strength alloy. Figure 18-5(a) shows the condition at the end of Stage

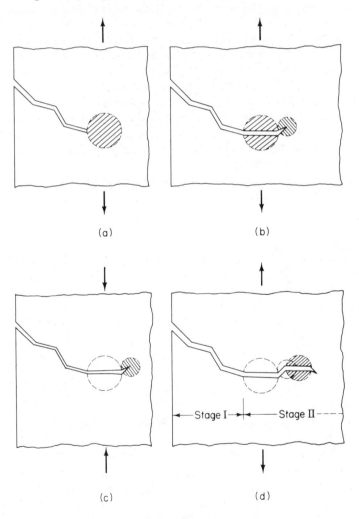

(a)

(b)

(c)

(d)

Figure 18-5 Sequence of Stage II fatigue crack growth.

I, where four grains have generated slip-induced cracks; each crack passes through a different grain and thus has a different orientation and produces a zigzag path. At this point the initial crack has become long enough to produce on the next tensile cycle a stress concentration at the crack tip that is high enough to cause local yielding, over the shaded region shown as circular here. The local strain hardening that results then causes this region suddenly to crack transversely under the tensile stress, as in Fig. 18-5(b). The growth of this crack is limited to the width of the damaged region during this cycle, so the crack progresses only across the width of the shaded region. In the context of Chap. 14, here denoting K_c as the initial stress-intensity factor required for crack growth and K as the stress-intensity factor resulting from the applied load, we can say that in the shaded region in Fig. 18-5(a), $K \geq K_c$, so that the crack grows, but it stops when it reaches the undamaged region of higher K_c when $K < K_c$. It is likely then that when the transverse crack runs out of the strain-damaged region, it will revert for a short distance to the shear mode and will create a new, small strain-hardened region (shaded) and then stop.

Under either tensile stress reduction or stress reversal, the crack partially or completely closes up [Fig. 18-5(c)]. Because the region in the vicinity of the new tip has been locally strain-hardened, it may not close up entirely, as a result of local plastic flow. The process is then repeated on the next application of a tensile stress [Fig. 18-5(d)].

Stage II fracture is thus very different in operation from Stage I fracture and produces a very different fracture appearance. The terminus of crack

Figure 18-6 Fatigue striations in TEM fractograph of Type 302 stainless steel (UNS S30200) at 427°C, $R = 0.2$, $\sigma_{max} = 124$ MPa, which failed after 1.6 million cycles. Propagation direction is downward. [From John A. Fellows, "Transmission-Electron-Microscope (TEM) Fractographs From Replicas," *Metals Handbook: Fractography and Atlas of Fractographs*, 8th ed., vol. 9, ed. Howard E. Boyer. (American Society for Metals, 1974), p. 333.]

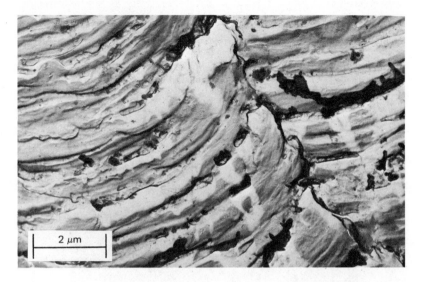

Figure 18-7   Stage II service fracture in a 0.06% C steel pin in reversed bending. This TEM micrograph exhibits several features frequently found in service fractures. The striations are irregularly spaced as a result of varying stress amplitude. Post-fracture squashing of the surface, resulting from the compression portion of the stress cycle, can be seen in several locations. The squashing also has partially closed up some surface cracks, which causes the replica to be torn when removed, to be seen in the micrograph as large, feathery, dark blotches. (See Exercise 18-2 for a discussion of fracture growth direction.)

advance for each cycle in Stage II is usually a slight change in direction of the crack, as shown in Fig. 18-5(d), when the crack extends into the undamaged region. This change in crack contour extends along the full length of the crack, perpendicular to the plane of the sketch, and appears as a narrow band, or *striation*, on the fracture surface. Figure 18-6 shows the appearance of such striations when the fluctuating stress amplitude is constant; as expected, the advance of the crack is approximately the same distance for each cycle. Figure 18-7 exhibits the typical variation in striation width that results from the variable stress amplitude of a part that fatigued in service. Additional examples of service-fatigue fractures are given in Chap. 19.

Exercise 18-2

In which direction is the crack growing in Fig. 18-7?

Solution.   Toward the lower right. A discontinuity of some sort, probably a grain boundary, runs from upper left to lower right and acts to retard the growth of the crack. This leads to a backward curving of the striations in the vicinity of the discontinuity. This effect, if it occurs, can also be useful in discriminating between postfracture damage marks, or scrapes, and true striations, since scrapes (as well as many striations) appear virtually straight and parallel at this high magnification.

Figure 18-8   Fatigue fracture in a tractor axle, showing progression marks over upper two thirds of fracture surface. Crack growth direction is down in photograph, culminating in sudden fracture over the rough oval-shaped region at the bottom. (Axle courtesy of Richard B. Felbeck. Photograph courtesy of W. H. Durrant.)

Service-fatigue fracture often occurs in parts that are not constantly in use, so that the Stage II fatigue crack remains at a fixed length for some time. Ordinary rust of the fracture tip that can occur during these periods leads to a feature clearly visible to the unaided eye, called *progression marks, clamshell marks,* or *beach marks* (because of their similarity to the marks left in the sand at the ocean shore by the receding tide). Figure 18-8 shows classic beach marks. A marked change in maximum stress range extending for a period of time can also cause beach marks, representing the line of demarcation between the regions of high- and low-stress Stage II fracture. Beach marks are not a positive indication of fatigue, only of gradual and interrupted crack growth, as beach marks can also result from interrupted stress-corrosion cracking. This will be discussed later in this chapter.

FINAL FRACTURE

The last stage of fracture, Stage III, is not fatigue at all. When the Stage II fatigue crack becomes long enough so that $K = K_c$ under the stress amplitude applied, where K_c is the critical stress-intensity factor of the undamaged metal,

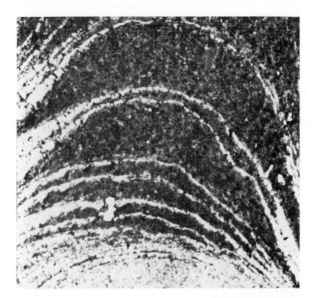

Figure 18-9 Alternate fatigue (light, narrow bands) and ductile fracture in forged 7075-T6 aluminum (UNS A97075). This is a transmitted light photograph of a plastic replica. [From ASM Committee on Use of Fractography for Failure Analysis, "Use of Fractography for Failure Analysis," *Metals Handbook: Fractography and Atlas of Fractographs*, 8th ed., vol. 9, ed. Howard E. Boyer. (American Society for Metals, 1974), p. 120.]

the part fractures completely and suddenly. Alternatively, for ductile metals, when the Stage II fatigue crack becomes so large that the remaining cross section of metal can fail by tensile overload, the part fails suddenly by a ductile mode. Thus, in either case, the final Stage III fracture occurs without prior warning. The sudden fracture of the tractor axle in Fig. 18-8 thus occurred without prior warning and left the rough fracture surface visible in the lower portion of the photograph.

It is quite common to find incomplete Stage III fracture interspersed with Stage II fracture, as in Fig. 18-9. This can occur because insufficient energy is present in the stressed structure to drive the Stage III fracture completely through the part. Figure 18-9 shows the propensity of the Stage II fatigue fracture to favor growth at the free surfaces of the part (the left and right edges of the figure) and thus to tend to straighten the contour of the crack tip, while the Stage III fracture favors growth of the central, internal region of the crack and thus tends to bend the contour of the crack tip.

FACTORS INFLUENCING FATIGUE

For a fatigue crack to initiate and grow, the following are essential:

1. The stress must vary.

2. At least a portion of each stress cycle must be tensile.

3. The crack must initiate at or very near a free surface or a large discontinuity.

It is obvious that an infinity of possible combinations of stress–time history can lead to fatigue in a given part, and that the introduction of additional variables even further complicates the problem of predicting fatigue fracture. These variables have been simplified for design purposes, and some of the references (Refs. 18-9 and 18-10) cited at the end of this chapter describe the procedures used in design. We will therefore discuss here the practical service factors that can alter fatigue life, from the point of view of how these factors influence the initiation and propagation of fatigue cracks.

The *stress level* has already been discussed in terms of the *s–N* curve (Fig. 18-2), for fully reversed tension–compression. Another way to demonstrate sensitivity of fatigue to stress is to show the rate of growth in fatigue of preexisting cracks of length *a*. Figure 18-10 shows the effect of stress range

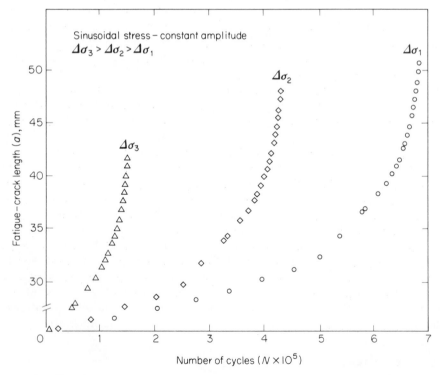

Figure 18-10 Effect of cyclic stress range on crack growth. [From Stanley T. Rolfe and John M. Barsom, *Fracture and Fatigue Control in Structures: Applications of Fracture Mechanics*, © 1977, p. 233. Reprinted by permission of Prentice-Hall, Inc., (Englewood Cliffs, N. J.).]

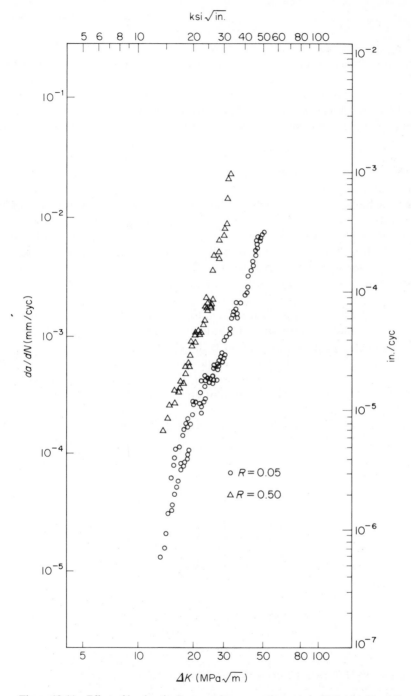

Figure 18-11 Effect of load ratio R on crack propagation rate in Ti-8 Al-1 Mo-1 V in an argon environment. [From Richard W. Hertzberg, *Deformation and Fracture Mechanics of Engineering Materials* (New York: John Wiley & Sons, 1976), p. 509, Fig. 13.32.]

$\Delta\sigma = \sigma_{\max} - \sigma_{\min}$ on the rate of growth of a crack in identical specimens, presumably for the same stress ratio R. Thus any increase in $\Delta\sigma$ will increase crack growth rate and decrease fatigue life. Stress concentrations that result from details such as notches, fillets, holes, keyways, splines, gear teeth, and screw threads all lead to higher fatigue crack growth rates. When these factors are unavoidable, the designer must accordingly take account of their weakening effect.

Note in Fig. 18-10 that the slope of each curve at any point is da/dN and that this increases as N increases, which means that as the crack grows longer it grows at a faster rate. This should be expected, since a longer crack usually leads to a higher value of applied stress-intensity factor K. Since the applied average stress σ determines the applied stress-intensity factor K, via the appropriate fracture-mechanics formula for the crack geometry and type of loading (see Chap. 14), changes in stress intensity at the crack tip (ΔK) are determined by the changes in remote stress ($\Delta\sigma$) in the same formula. The rate of crack growth da/dN is thus a function of ΔK. Figure 18-11 (discussed in the next paragraph) is an example of this now-common form of presenting crack growth data: The rate of growth of the fatigue crack of length a for each cycle N, da/dN, is the ordinate, and the range of stress-intensity factor ΔK that results from the cyclic variation in $\Delta\sigma$ is the abscissa.

The stress ratio R influences fatigue life in a predictable manner. If the stress range $\Delta\sigma$ is held constant, increasing R means that the maximum tensile stress will be increased, while the minimum stress becomes more tensile. Figure 18-11 shows this effect in a titanium alloy. For a given ΔK, say 20 MPa$\sqrt{\text{m}}$, da/dN for $R = 0.50$ is roughly 5 times greater than da/dN for $R = 0.05$. Another way of expressing this relationship is to say that when R is increased, the

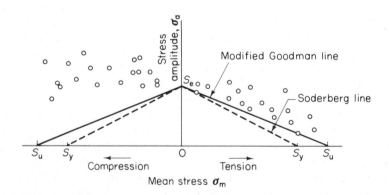

Figure 18-12 Typical high-cycle fatigue failures (circles) over a range of stress values. Note that these failures all occur at values of stress amplitude σ_a that exceed the limits set by the modified Goodman line. The Soderberg line, which is similar to the modified Goodman line except that it runs to the yield stress, is shown for comparison. [From Joseph E. Shigley, *Mechanical Engineering Design*, 3rd ed. (New York: McGraw-Hill, 1977), p. 202, Fig. 5-24b.]

magnitude of permissible ΔK and $\Delta \sigma$ for infinite life will decrease; thus for higher mean stress σ_m, where

$$\sigma_m = \frac{\sigma_{max} + \sigma_{min}}{2}$$

the allowable $\Delta \sigma = \sigma_{max} - \sigma_{min}$ must be reduced if fatigue is to be avoided. The designer uses this relationship as the *modified Goodman line*, shown in Fig. 18-12. As the mean stress σ_m increases, the stress range for high-cycle fatigue fracture decreases. Note that σ_m may exceed the yield stress S_y, so a typical design calculation needs to be further limited to hold $\sigma_{max} < S_y$ to avoid yield failure.

Exercise 18-3

If the form of the graph of da/dN vs. ΔK is known (e.g., Fig. 18-11), can we calculate the number of cycles that a broken part was subjected to in its lifetime?

Solution. We can estimate its life but cannot precisely determine the number of crack-growth cycles that it experienced except by actually counting all the Stage II striations (see Fig. 18-7) across the fracture. Note that because the da/dN data must of necessity record only the growth of the crack after it is long enough to be detected, the number of cycles for initiation of the crack and the rate of growth of the crack prior to its being of detectable length are unknown. Thus we cannot integrate the crack growth from the condition $a = 0$. The best we can do is make some estimate of the early crack-growth conditions from tests and predict the remaining life of the part once an observable crack length has been reached. Techniques for estimating fatigue life are discussed in more detail in Chap. 19.

The total number of service stress cycles in many structures is much less than 10^6, so the designer can design for higher stress than the fatigue limit S_e (see Fig. 18-2). The smaller the number of cycles, the higher the allowable stress. When the local maximum tensile stresses in a perceptible region about the crack tip are above the yield stress, fatigue usually occurs with less than about 10^4 or 10^5 cycles; this behavior is called *low-cycle fatigue*. A simple and extreme example is the fracture of a low-carbon steel nail after 100 or so cycles of plastic reversed bending. In this case the successive striations that define each step of crack growth are visible at relatively low magnification, at least in the latter stages of fracture when each striation is rather wide, as in Fig. 18-13. Design procedures have been established (see Ref. 18-10) for estimating the service stresses that parts can tolerate under low-cycle fatigue loading, usually in the range 10^3 to 10^6 cycles.

Surface finish, and surface conditions in general, are critical to fatigue. A "smooth surface" is never really without imperfections of some kind to provide a value for surface crack length a, no matter how small. Thus the smoother a surface is, the smaller is the value of a and the greater the value of $\Delta \sigma$ must be in order to reach a given value of ΔK. Alternatively, if the value of $\Delta \sigma$ remains constant, a smoother surface experiences a reduced ΔK and conse-

Figure 18-13 Fracture surface of a common low-carbon steel nail of diameter 2.5 mm, fractured in low-cycle fatigue by reversed bending in 222 cycles. Fatigue fractures initiated at the lower right, where the tensile bending stress was higher, and at the upper left, with the final fracture at the region of broad striations in the upper left center of the SEM micrograph.

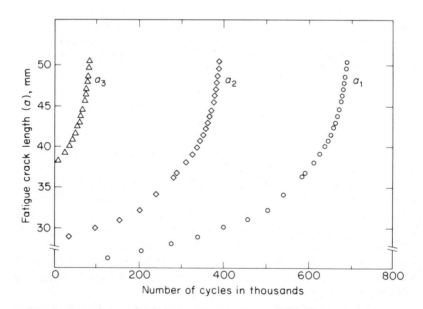

Figure 18-14 Effect of initial crack length on crack growth; a_3, a_2, and a_1 are initial crack lengths; $a_3 > a_2 > a_1$. [From Stanley T. Rolfe and John M. Barsom, *Fracture and Fatigue Control in Structures: Applications of Fracture Mechanics,* © 1977, p. 234. Reprinted by permission of Prentice-Hall, Inc. (Englewood Cliffs, N. J.).]

quently will sustain more cycles to fracture than will a rougher surface. Figure 18-14 shows the effect of increasing initial crack length on the number of cycles required to reach a given (critical) crack length.

The longest fatigue life for a given material therefore requires the smoothest possible surface. Grinding and polishing provide the best commercially attainable surface quality in metals, and such treatment is usually required for parts subjected to severe fatigue conditions. Axles and shafts subjected to bending stresses and high numbers of cycles are therefore ground and polished. Surface finish of poorer quality, such as an ordinary machined surface, an as-cast or as-forged surface, a rusty or corroded surface, or, even worse, a gouged or undercut surface, will substantially reduce fatigue life below the S_e measured in the polished rotating–bending test.

Exercise 18-4

Which is more sensitive to surface damage, a polished steel surface or an as-forged steel surface?

Solution. The polished surface will suffer greater reduction of fatigue life than the as-forged surface as a result of surface damage. The as-forged surface already has numerous crack-initiation regions because of its roughness, so it is less sensitive to further surface damage. The polished specimen, because it is relatively free of surface flaws at the outset, will experience a substantial loss in fatigue life when its surface is damaged.

Figure 18-15 shows fatigue fractures that initiated at an improperly machined undercut between two teeth of an automotive steering gear. This has the effect of producing a notch in a region that is already under higher stresses because of the shape of the teeth; under such circumstances fatigue life can be markedly reduced. A polished part must be carefully protected from damage before and during its service life if it is to fulfill the designer's expectations. Small nonmetallic inclusions located at or very close to a free surface can likewise act as initiating cracks. Large, macroscopic nonmetallic inclusions, seams, cracks, holes, and voids located inside a part can act as a free surface and can allow initiation of fatigue cracks if the local stress is sufficiently high.

It is possible to improve fatigue life through surface hardening. The usual methods are carburizing, nitriding, and shot-peening. Carburizing is usually applied to steels of low (about 0.20%) carbon content, which are selected to give good toughness. The part is heated in an atmosphere containing carbon so as to increase the carbon content by diffusion into a very thin surface layer to, say, 0.60–0.80% C; then the part is quenched to transform this outer layer to martensite. Nitriding similarly introduces enough nitrogen into the surface to produce hard nitrides. In both cases this very hard layer inhibits the slip processes that lead to early crack initiation (Fig. 18-3) and thus contributes substantially to increased fatigue life, far out of proportion to its small fraction of the volume of the whole part. At the same time the part retains its tough-

(a)

(b)

Figure 18-15 (a) Teeth of a fractured automotive steering gear, with fractured parts held together. Fatigue fracture initiated along an improperly machined slot that produced an undercut between the first two teeth on the right. (Note that the undercut appears even worse at the left side of the slot between the middle two teeth.) (b) Fracture surface of the left (larger) portion of the gear shown in (a). The three white markers indicate possible fatigue initiation sources; the left marker shows a partial crack that is not visible here but is located exactly along the undercut.

ness because it consists almost completely of a relatively soft low-carbon steel.

Shot-peening not only hardens a thin layer at the surface, but when properly done it also leads to compressive *residual stresses* at the surface. Under service loads the net stresses in this surface thus become more compressive than they would otherwise be, with consequent increase in fatigue life.

The environment surrounding a part during fatigue can profoundly affect the rate of crack growth. Because this is a special case involving different mechanisms from those discussed so far, corrosion fatigue will be discussed at the end of this chapter, after stress-corrosion cracking.

STRESS-CORROSION CRACKING

Sometimes a crack can propagate slowly when a part is under a constant tensile stress, with no cyclic stresses applied. This acutely hazardous mode of fracture is called *stress-corrosion cracking*. In contrast to ordinary corrosion (such as rust), which leads to a more or less uniform and gradual loss of metal at a free surface exposed to a corrosive environment, stress-corrosion cracking leads to crack growth in the presence of an appropriate environment (see below) that occurs fastest where the tensile stresses are highest. Furthermore, the cracks that result are perpendicular to the tensile stress. Thus stress-corrosion cracking leads to the worst possible situation: the growth of cracks in regions of highest tensile stress, oriented in the most severely weakening manner.

The essential features of stress-corrosion cracking can be demonstrated in a simple experiment. A cast flat bar of polystyrene, 3.2 mm thick by 13 mm wide, is forced between fixed blocks so as to cause it to bend to a predeter-

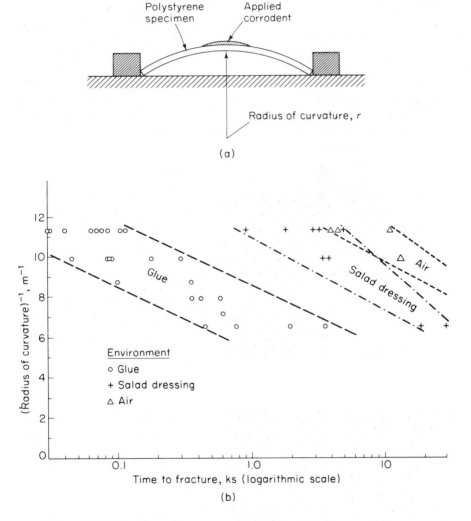

Figure 18-16 Experimental stress-corrosion cracking in polystyrene: (a) test set-up; (b) results.

mined radius of curvature r [Fig. 18-16(a)]. To a first approximation, the tensile stress on the top center face of the bar is related to $1/r$. (Of course, since relaxation occurs in a viscoelastic material such as polystyrene, as described in Chap. 12, the stress is dropping throughout this experiment. But the behavior of all specimens will be similar, so the results will still be comparable.) As soon as the specimen is in place, a stress-corroding environment suitable for polystyrene is applied to the top center surface of the specimen. Acetone is such a substance; in fact, pure acetone is so effective as to be useless for test purposes, for a drop of acetone on the top surface leads to almost instantaneous fracture. We have found that acetone-based glue and ordinary salad dressing provide suitably slow stress-corrosion cracking in these tests. The substance is placed on the top center surface, and this is time zero. A series of parallel flaws grows normal to the top surface in the region of the corrodent until the longest flaw provides, in conjunction with the stress, the condition $K = K_c$ (see Chap. 14), and the part breaks. This fracture occurs "suddenly and without warning," in much the same manner as service fractures. Figure 18-16(b) shows that the rate of crack growth is sensitive to the corrodent and is slower for lower stress. The fact that fracture also occurs (very slowly) in air, in the absence of any deliberately applied stress-corrosion cracking environment, suggests that air itself, or probably the moisture in the air, also acts as a mild corrodent.

This same effect of moisture sometimes occurs in glass and is often called (improperly) "static fatigue." As no cycling of stress is involved, the phenomenon is better called *delayed fracture*: A crack in a bottle grows gradually under the combined effects of residual tensile stress present because of faulty heat treatment and any applied tensile stress such as from internal pressure, until after an elapsed time of perhaps years, the crack reaches critical dimensions and the bottle explodes. This behavior is another example of stress-corrosion cracking, and anyone who has ever lost an unopened bottle of expensive Scotch to this phenomenon (as one of the authors did) will never forget it.

Many metals and alloys of commercial importance are susceptible to stress-corrosion cracking in certain special environments. (Fortunately, the number of corrodents for each alloy is small, or otherwise metals would be of limited usefulness.) For example, the austenitic stainless steels (see Chap. 11) are generally sensitive to stress-corrosion cracking in chloride environments. Figure 18-17 shows the rather wide range of cracking sensitivity in boiling magnesium chloride of austenitic stainless steels of different compositions. Most alloys exhibit some sensitivity to stress-corrosion cracking in certain environments. One example is the so-called "season cracking" of moderately cold-worked 70 Cu–30 Zn brass in the presence of ammonia; the tensile stress is provided by the residual stress that remains after cold working, but if the brass is severely cold-worked, the residual stress is reduced and the brass becomes less susceptible to stress-corrosion cracking.

The *threshold stress* for stress-corrosion cracking is that tensile stress below which stress-corrosion cracking does not occur, even for long times. The corresponding *critical threshold stress-intensity factor for stress-corrosion*

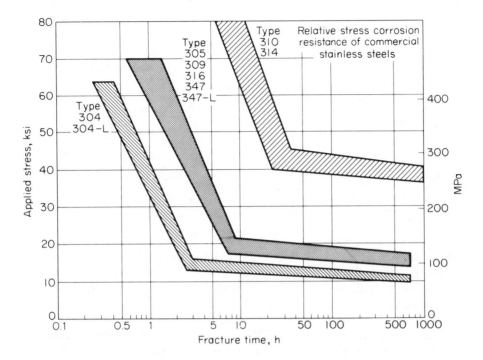

Figure 18-17 Stress-corrosion cracking of austenitic stainless steels in boiling magnesium chloride. See Appendix 7 for composition of these stainless steels. [From Denhard, E., *Corrosion*, **16** (1960), No. 7, p. 131.]

cracking, K_{scc}, thus represents the stress intensity below which cracking will not occur. The shape of the curves in Fig. 18-17 is somewhat similar to the s–N curves in Fig. 18-2, where the fatigue limit is analogous to the threshold stress of Fig. 18-17. Typical values of threshold stresses for the onset of stress-corrosion cracking are from 10 to 70% of the yield stress. Stress-corrosion cracking is thus a potentially serious problem because fracture can occur under static stress that is well below the yield stress.

The mechanisms that lead to stress-corrosion cracking are associated with changes in the bulk solid in the region of the tip of a crack. (Note the similarity to the mechanisms of fatigue, where cyclic stress also alters the properties of the bulk solid in the region of the crack tip.) If we assume the presence of some surface flaws, cracks, or irregularities, which are present in all solids, the corrodent, either liquid or gaseous, diffuses to the tip of the flaw, where tensile stresses are highest [Fig. 18-18(a)]. The actual mechanism by which the corrodent causes cracking may be either (1) the physical dissolution of the interatomic bonds in the region of highest tensile stress, or (2) a local reduction in energy to create new fracture surface, with a consequent local reduction in the K_c required for cracking. In either case the effect is the same,

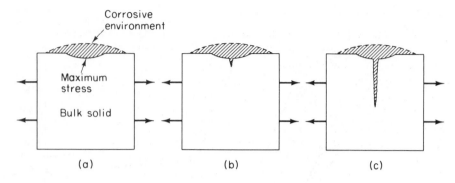

Figure 18-18 Macro-scale mechanism of stress-corrosion cracking: (a) initial condition; (b) after short crack growth; (c) after more crack growth.

and a short length of new crack is formed at the tip of the prior flaw [Fig. 18-18(b)]. The increment of new crack is short because only a very limited amount of the bulk solid has been altered by the corrodent, which has only limited access. Thus we see that the crack grows gradually and continuously [Fig. 18-18(c)], not by short jumps as in fatigue. Figure 18-19 shows the early branched cracks that develop in 304 stainless steel (UNS S30400) exposed to boiling magnesium chloride, for which data were given in Fig. 18-17. Such branching of cracks is characteristic of many stress-corrosion cracks. The rate

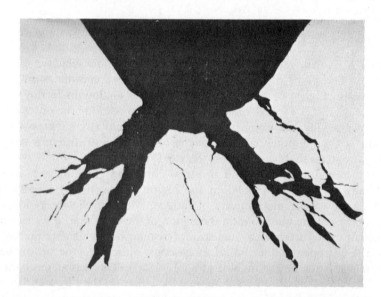

Figure 18-19 Stress-corrosion cracks at the root of a notch in 304 stainless steel exposed to boiling 42% magnesium chloride solution. [From H. L. Logan, *The Stress Corrosion of Metals* (New York: John Wiley & Sons, 1966), p. 120, Fig. 4.9.]

of crack growth is limited by (1) the rate of diffusion of fresh corrodent to the new tip of the crack, (2) the effectiveness of the corrodent in producing cracking, and (3) the local stress level. This process continues until the crack reaches critical dimensions and the entire part fails.

The gross rate of crack growth can vary as the enviromental stress-corrosion cracking conditions vary in concentration, temperature, or velocity, and as the average stress varies. Thus the fracture surface may exhibit *progression marks* or *beach marks* that appear similar to the appearance of a fatigue fracture. Figure 18-20 shows the appearance of a stress-corrosion crack (the small concentric crescents) and the final sudden fracture (the balance of the fracture surface) in a high-strength steel. Thus beach marks indicate only that a crack has grown gradually, over some time. Furthermore, the absence of beach marks does not indicate that the fracture is not fatigue or stress-corrosion cracking.

On a microscopic scale, however, the presence of striations is positive evidence of fatigue, never of stress-corrosion cracking alone. The fracture mode of stress-corrosion cracking can be either intergranular or transgranular, so discrimination between stress-corrosion cracking and single-load fracture is often difficult. However, when a local region of intergranular cracking leads to gross fracture by a quasi-cleavage mode, this evidence points strongly to

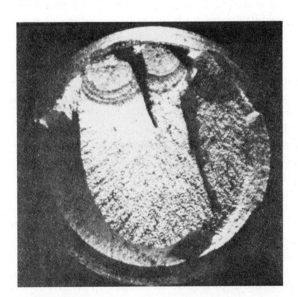

Figure 18-20 Progression marks from two origins of stress-corrosion cracking, at top. Steel of hardness 50 R_C under static load in distilled water. [From ASM Committee on Fractography by Light Microscopy, "Interpretation of Light-Microscope Fractographs," *Metals Handbook: Fractography and Atlas of Fractographs*, 8th ed., vol. 9, ed. Howard E. Boyer. (American Society for Metals, 1974), p. 42.]

Figure 18-21   Intergranular fracture from hydrogen embrittlement in high-carbon steel.

stress-corrosion cracking as the cause of the first crack. Figure 18-21 shows the intergranular fracture surface located near the fracture origin that resulted from exposure of a 0.62% C steel rivet under residual stresses to a dilute solution of hydrochloric acid with an applied voltage of 6 V for 1.2 ks. (If a system itself constitutes an electrolytic cell, hydrogen embrittlement, discussed in the next section, can occur without an applied voltage; for the example in question, the voltage was applied in order to accelerate and control the process.) Although this fracture is the result of hydrogen embrittlement, the appearance is typical of the intergranular fracture often associated with stress-corrosion cracking.

Since each alloy is susceptible to its own very limited set of environments, a full description of all possible hazardous combinations is beyond the scope of this book. Some of the references listed at the end of this chapter give more detailed information. Appendix 15 lists just a few of these known combinations. Among the common metals that are relatively more susceptible to stress-corrosion cracking are low-carbon steel, high-strength steels, austenitic stainless steels, high-strength aluminum alloys, and brasses and other copper alloys. Since the effective concentration of the corrodent can be quite low, it is essential that the service environment be carefully studied by the designer. The designer has a responsibility either to avoid using a material susceptible to the anticipated service environment or to protect the material from exposure to the environment.

Exercise 18-5

What should a designer do if he has no choice but to use a material susceptible to stress-corrosion cracking, and without adequate protection?

Solution. If, after careful testing, the designer has determined that the part can operate for a reasonable time before serious cracks develop, he must warn the user to inspect the part for cracks at prescribed intervals, in the same way that aluminum alloy structures are inspected for fatigue cracks. Such a procedure would be taken only after thorough study of all the variables of the service environment.

HYDROGEN EMBRITTLEMENT

Atomic hydrogen can diffuse through solid metal because its effective atomic radius is relatively small. Once inside the bulk metal, the hydrogen can accumulate at internal inclusions, voids, and discontinuities in the form of gaseous hydrogen. These small bubbles can achieve a pressure as high as 1.3 GPa for maximum hydrogen concentration and are thus able to produce small local fractures. Under tensile stress as low as 40% of yield stress sustained for a period of time, these fractures may join up and eventually lead to sudden complete fracture of the part in the same fashion as stress-corrosion cracking. Furthermore, there is evidence that hydrogen acts to reduce the surface energy at the tip of a crack under tensile stress, and to this extent hydrogen is just another environment that can lead to stress-corrosion cracking. For this reason we shall usually treat hydrogen embrittlement as a special case of stress-corrosion cracking. But because of the problem of the irreversible small cracks produced by gaseous hydrogen, the subject of hydrogen embrittlement warrants separate discussion.

Hydrogen embrittlement causes trouble in many high-strength metal alloys, chiefly in high-strength steels, such as AISI-SAE 4340 (UNS G43400) quenched and tempered to above 1.4 GPa, or high-carbon plain-carbon steels, quenched and tempered to above 1.4 GPa. The hydrogen can come from any of several sources during manufacture and service: pickling, electroplating, stray currents, and service environments. Hydrogen can be charged into a part when the part is the cathode in an aqueous conducting solution, formed by the dissociation of water to H^+ and 0^{--}, where the H^+ picks up an electron from the cathode and is free to diffuse into the cathode as atomic hydrogen. The embrittlement in Fig. 18-21 was produced in this way.

If the part is not under tensile stress during and after the charging, the hydrogen can usually be driven off by heating the part to 150 to 200°C for an hour or two. If not later exposed to hydrogen again, there will be no damage. If the part has been plated, the plating will hinder the desorption of hydrogen, so longer baking times (e.g., 190°C for 3 to 4 h) are employed, with even longer times for cadmium plating.

If the part is under tensile stress (from applied service loads or from residual stresses) after being charged with hydrogen, and the hydrogen concentration and time are sufficient to cause local cracking, then the part has been permanently damaged. No amount of baking will eliminate the cracking.

Steps to mitigate hydrogen embrittlement are thus:

1. Use a lower-strength alloy. For example, steels of strengths below 700 MPa are generally resistant to hydrogen embrittlement, steels from 700 to 1000 MPa are slightly susceptible, and steels over 1000 MPa are very susceptible. Within each group of alloys, certain classes have been found to have high resistance to stress-corrosion cracking. Reference 18-3 provides detailed rankings of alloys according to their susceptibility to stress-corrosion cracking.
2. Avoid treatments and service conditions that can lead to hydrogen absorption.
3. Bake to remove hydrogen as described above.
4. Reduce residual stresses.
5. Reduce severity of notches under tensile stress.

Martensitic and precipitation-hardening stainless steels can experience hydrogen embrittlement, especially in their higher-strength forms (see Chap. 11). Baking can eliminate hydrogen picked up during manufacture, but only isolation from hydrogen-producing environments can protect these alloys from cracking in service. In the absence of other practical protection, these stainless steels must be heat-treated to substantially reduced strength levels to provide protection from hydrogen damage.

Fractographic analysis of hydrogen embrittlement is substantially the same as for stress-corrosion cracking. Although it is known that the initial cracking of quenched and tempered steel is always intergranular and on prior austenite grain boundaries for hydrogen embrittlement, and that hydrogen damage may produce some other subtle differences in the fracture, there is no unambiguous method to separate general hydrogen-embrittlement fracture from stress-corrosion cracking. Thus hydrogen embrittlement should be treated as a special case of stress-corrosion cracking.

CORROSION FATIGUE

In nearly every instance, fatigue in service occurs in conjunction with the possibility of corrosion and/or stress-corrosion cracking; this is called *corrosion fatigue*. In its simplest form ordinary corrosion can lead to surface pitting that will substantially reduce the fatigue limit of, for example, a polished part. In addition, the presence of either residual tensile stresses or the tensile component of the fatigue stress cycle (especially for the continuous tension in tests with stress ratio $R > 0$) can lead to simultaneous crack growth by fatigue and by stress-corrosion cracking. Figure 18-22 shows the marked increase in crack growth rate da/dN that results when a 17-4 PH stainless steel (UNS S17400) is tested in sea water. Thus, in addition to the basic mechanisms of fatigue and stress-corrosion cracking, there exists the additional possibility

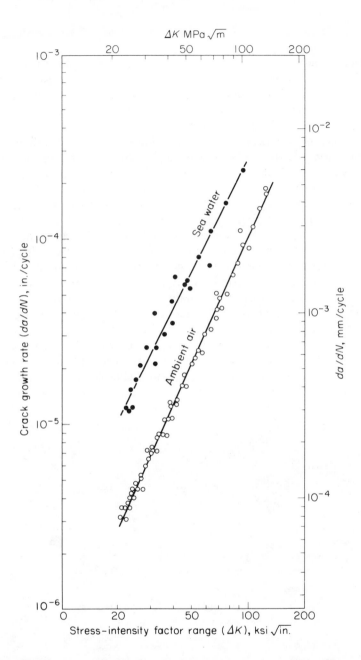

Figure 18-22 Corrosion-fatigue crack growth data for 17-4 PH stainless steel (UNS S17400). [From T. W. Crooker, F. D. Bogar, and W. R. Cares, "Effects of Flowing Natural Seawater and Electrochemical Potential on Fatigue-Crack Growth in Several High-Strength Marine Alloys," *Corrosion-Fatigue Technology*, ASTM STP 642, H. L. Craig, Jr., T. W. Crooker, and D. W. Hoeppner, eds. (Philadelphia: American Society for Testing and Materials, 1978), pp. 189–201. Copyright American Society for Testing and Materials, 1916 Race Street, Philadelphia PA 19103. Adapted with permission.]

of interaction between the two fracture modes. For example, it is possible as a result of the local strain hardening at the crack tip (see Fig. 18-5) that during periods of rest from fatigue cycling (or when the service cyclic stresses are low) stress-corrosion cracking can lead to more rapid crack growth than would occur in the absence of the fatigue damage.

We may thus expect some of the characteristics of both fatigue and stress-corrosion cracking to be associated with corrosion-fatigue fracture. Corrosion pits are often seen on the fracture surface, presumably forming after the local fracture occurs. In fact, the entire fracture surface may be coated with corrosion products that are difficult to remove; these nonconducting products appear white in a scanning electron microscope because they build up an electron charge. Figure 18-23 shows such a service fracture, in which the harmful environment may have been calcium chloride (road salt) and water. Other characteristics of corrosion-fatigue fracture may include beach marks visible to the naked eye, striations visible in an electron microscope, and secondary cracking parallel to the main crack in the initiation region. Corrosion fatigue exhibits no single unique characteristic that sets it apart from other fracture modes.

In contrast to stress-corrosion cracking, corrosion fatigue does not require a narrow range of specific environments. Any effective corrodent may contribute to corrosion-fatigue fracture, and higher concentrations are generally more effective. These observations suggest the possibility of a fairly simple

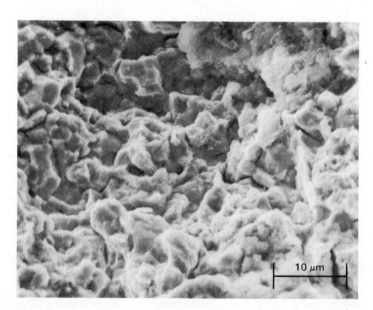

Figure 18-23 SEM micrograph of corrosion-fatigue service fracture of case carburized AISI 8650 (UNS H86500) steel track pin from a battle tank that had traveled 2521 miles (4056 km). Intergranular fracture and surface corrosion appear to dominate and thus mask any possible fatigue striations.

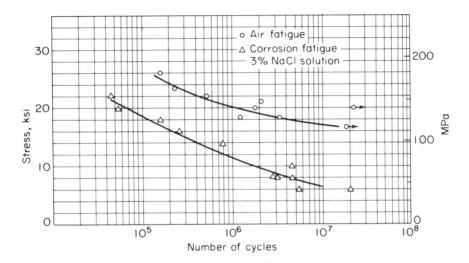

Figure 18-24 Reduction of fatigue strength in an Al–Zn–Mg alloy tested in 3% NaCl solution. [From P. J. E. Forsyth, *The Physical Basis of Metal Fatigue* (New York: American Elsevier Pub. Co., 1969), p. 116, Fig. 6.13.]

mechanism for corrosion fatigue, as already mentioned, in which crack growth by corrosive attack in the region of the crack tip is accelerated by the high local plastic strain in this region that results from the fatigue cycling. At the same time, this picture may be complicated somewhat by the fact that the corrodent may remove the hard surface film from the metal, allow dislocations to escape more easily, and thus promote the growth of the fatigue crack by a shear mode of the sort described by Fig. 18-3. The reduction of fatigue life by such a mode is shown in Fig. 18-24 for an Al–Zn–Mg alloy, where the length of initial crack growth by shear is greater under the corrosive condition.

REFERENCES

18-1. American Society for Metals, *Metals Handbook*, 8th ed., vol. 10, "Failure Analysis and Prevention." Metals Park, Ohio: American Society for Metals, 1975.

18-2. Beachem, Cedric D., ed., *Hydrogen Damage*, Metals Park, Ohio: American Society for Metals, 1977.

18-3. Brown, B. F., *Stress Corrosion Cracking Control Measures*. Washington, D. C.: U.S. Dept. of Commerce, Nat. Bur. of Standards, Mono. 156, June 1977.

18-4. Brown, B. F., ed., *Stress Corrosion Cracking in High Strength Steels and in Titanium and Aluminum Alloys*. Washington, D. C.: Naval Research Lab., U.S. Govt. Printing Office, 1972.

18-5. Craig, H. L., Jr., T. W. Crooker, and D. W. Hoeppner, eds., *Corrosion-Fatigue Technology*, ASTM STP-642 Philadelphia: American Society for Test. and Mat., 1978.

18-6. Forsyth, P. J. E., *The Physical Basis of Metal Fatigue*. New York: Amer. Elsevier Pub. Co., 1969.

18-7. Logan, Hugh L., *The Stress Corrosion of Metals*. New York: John Wiley & Sons, 1966.

18-8. Mann, J. Y., *Fatigue of Materials*. London and New York: Melbourne Univ. Press and Cambridge Univ. Press, 1967.

18-9. Rolfe, Stanley T., and John M. Barsom, *Fracture and Fatigue Control in Structures*. Englewood Cliffs, N. J.: Prentice-Hall, 1977.

18-10. Shigley, Joseph E., *Mechanical Engineering Design*, 3rd ed. New York: McGraw-Hill Book Company, 1977.

PROBLEMS

18-1. Why is it not possible to estimate the prior number of fatigue cycles that were applied to a broken part by dividing the observed width of the fatique-fracture surface by the observed width of a single striation?

18-2. Why might fatigue striations be difficult to observe in a brittle metal?

18-3. Why might fatigue striations be difficult to observe in a very ductile metal?

18-4. What fractographic evidence can indicate whether a fracture is solely the result of fatigue or if stress-corrosion cracking has occurred?

18-5. What are the minimum essential requirements for stress-corrosion cracking to occur in a part in service?

18-6. Why do most parts in service with alternating loads not survive as long as would be indicated by the results of a standard rotating–bending test?

18-7. What are the differences between Stage I and Stage II fatigue crack growth?

18-8. What do beach marks or progression marks indicate?

18-9. What characteristics or properties of a fabricated metal part can be altered (or differently specified by the designer) in order to increase the fatigue life of the part?

18-10. What measures can be taken to minimize the effects of stress-corrosion cracking?

18-11. How can hydrogen embrittlement be avoided?

18-12. What is the difference between fatigue striations and progression marks (also called beach marks or clamshell marks)?

18-13. What does acetone-based glue do to polystyrene under tensile stress? Under compressive stress? Explain why in each case.

19 Failure Analysis

When an engineering structure fails to perform its intended function, a *failure analysis* should be made. The impact of such an analysis can range from the potential economic crippling of a nation's economy, down to searching for a replacement for a broken pencil lead or rubber band (the latter failures are usually not even analyzed). When an engine pylon of a DC-10 jumbo jet fractured on takeoff from O'Hare airport in Chicago in mid-1979, with the loss of 273 lives, the resulting grounding of all DC-10's for many weeks while a failure analysis was being performed led to severe problems for airlines and passengers alike. Even if no lives are lost and no damage occurs as a result of a failure, the design engineer and the user have a responsibility for determining the cause of the failure so that similar events can be avoided in the future. And when injury, death, and property damage occur, the question of responsibility for the failure is raised because insurance and other contractual obligations are usually involved. In cases of dispute, these matters are taken to courts of law for ultimate resolution under statutes of *product liability*, which are designed to aid in establishing responsibility.

This chapter describes the process of analysis of failures that involve deformation and fracture. Excluded from discussion here are such matters as design deficiencies, the necessity of adequate guarding devices to protect users of equipment, and other engineering failures such as inadequate flow rates or heat-transfer rates. The job of the *failure analyst* is to determine the original, proximal cause of a failure. In this task he has two big advantages over the designer, for he knows (1) that a failure did occur and (2) where it occurred. As we shall see, failure analysis may also reveal the mode of failure. For example, the DC-10 designers stated during the investigation following the crash that they had never considered the possibility of the mode of failure that actually occurred (fatigue fracture of the pylon). The designer has a much broader and in some ways more difficult job than the failure analyst, for the designer must consider absolutely all possible modes of failure, and then design his structure to avoid them.

It is our aim here to describe the procedures of failure analysis so that the reader will be able, if necessary, to determine for himself the reasons for a failure. Because this subject is still rather new as a formal area of inquiry (although it has, of course, been carried on for thousands of years), there is very little published literature on the subject (see the references at the end of the chapter). We shall therefore draw on our own experience for examples to illustrate the range of solutions that one might employ in the course of analyzing failures.

EXAMINATION OF PARTS

Visual inspection. Usually the first information available to a failure analyst is that obtained from broken parts. (Some failures may involve only plastic deformation and no fracture, but even so we shall include these also when we speak of "broken parts" or "failed parts.") Sometimes a brief visual examination may reveal the *fracture mode* and the *proximal cause* of the fracture within a few seconds. However, an investigator must be careful not to become firmly attached to any one failure theory until after the investigation is completed. *Working hypotheses* for the failure should be established, with the purpose of the investigation eventually to exclude all hypotheses except one, if possible. In many cases the initial hypothesis will stand up throughout the entire investigation. Figure 19-1 shows two photographs of a void 16 mm in diameter that initiated fracture in the neck of a cast-iron gooseneck used to charge molten zinc alloy under pressure into a die-casting mold. Preliminary examination showed that the surface of the void was markedly different from the fracture surface, and this hypothesis ultimately proved to explain the fracture.

3X inspection. Once the part is visually examined, closer inspection with a low-power magnifying glass of about 3X magnification will reveal details of areas of interest. Not only should the fracture surface be carefully examined, but the entire part should be searched for scrapes, gouges, upset areas, machining marks, and other evidence of forces applied to the part. A 3X glass is often a more useful device in failure analysis than an expensive electron microscope. Figure 19-2 shows an indentation in the lip of a ball-joint socket of the steering linkage of an automobile that was involved in a crash. The essential question in such a situation is, did this indentation occur before the accident, or is it a result of the accident? Calculations (see Exercise 19-5 later and Reference 19-4) showed that forces sufficiently large (in this case,

(a)

Figure 19-1 Casting void in the fracture surface of a cast-iron gooseneck. (a) A portion of the tip of the fractured gooseneck with void (arrow) near the region of maximum circumferential stress. (b) Closeup of the void.

(b)

29.5 kN) to lead to such an indentation could only have arisen during a crash, indicating that the steering mechanism was intact at the instant of the crash. Obviously, then, a fracture found elsewhere in the steering mechanism could not have been responsible for the crash, which led to the conclusion in this case that the car crashed because of driver error.

Figure 19-2   Indentation in the lip of a ball-joint socket.

Exercise 19-1

How can we be certain that the indentation shown by the arrow in Fig. 19-2 came from the ball-joint stud and not from some other source?

Solution.   We cannot be absolutely certain, but we can be reasonably sure. If the indentation dimensions and curvature match the portion of the ball-joint stud that contact the indentation when the ball-joint stud is rocked into contacting position, then it is reasonably certain that this caused the indentation. Such reasoning is usually sufficient for both engineering analysis and courtroom testimony.

Indentations and scratch marks can indicate prior points of contact and thus reveal something of the history of a part. Figure 19-3 shows marks in the inside of the base of an automobile bumper jack that tipped over the only time it was used. These marks indicate that the jack column was improperly inserted into this base, vertically rather than at an angle as it was designed to be.

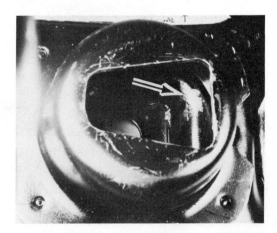

Figure 19-3   Rub marks (arrow) indicating incorrect seating of column in the base of an automobile bumper jack.

Such accidental misuse could have been prevented in this case by a simple guide installed in the base to prevent incorrect insertion of the column.

The *absence* of indentations and scratches is sometimes as important as their presence. A tractor being towed, shortly after a steering system overhaul, veered suddenly off the road and severely injured the driver. The only possible explanations were either (1) that the driver was careless and was going too fast, smashed the vehicle, and sheared the (missing) pin that connected the upper end of the steering shaft to the lower end through a collar, or (2) that the critical pin fell out and caused loss of steering. Because the steering system was operating just prior to the accident, the pin could not have been missing following the overhaul. Figure 19-4 shows the completely undamaged hole for the pin in the lower end of the steering column; calculations showed that the pin could not have sheared without also indenting the side of the hole in the shaft. The repair mechanic admitted on his second deposition that he may have neglected to install a retaining clip on the pin, thus allowing the close-fitting pin to fall out after a short period of service.

Figure 19-4   Undamaged hole that held pin connecting upper and lower portions of tractor steering column.

Optical microscope inspection.   A low-power (30–70X) binocular optical microscope provides the next level of magnification useful for fractography. Damage that may have occurred to the fracture surface for other reasons after the time of fracture is usually visible in such a microscope. Many instances of such *postfracture damage* result from complex events that occur during and after a collision or crash, so the observed damage often cannot be related to contact with a particular part. But when mating fracture surfaces are forced into contact after the initial fracture is complete, the postfracture damage may extend over large fractions of the fracture area because the surfaces match so well. Furthermore, one would expect to find the damaged regions visible on both portions of the mating surfaces, although not necessarily in perfectly symmetrical locations when relative motion of the fractures is

(a)

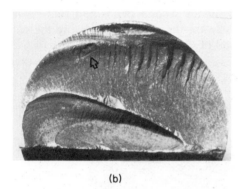

(b)

Figure 19-5 Punch press drive pin. (a) Small fractured end of pin that caused malfunction, showing extensive postfracture damage. (b) Larger portion of pin, showing fatigue fracture and postfracture damage in concave region (arrow) of final cleavage fracture. (One or more small pieces between these two parts were lost through secondary fractures, as these surfaces do not match over their entire area.)

possible. Such a damaged region can be seen on a portion of the fracture surface of a punch press drive pin in Fig. 19-5(a). This end of the pin broke off as a result of fatigue arising from a poorly machined fillet radius, was mashed at least once against the pin, then got caught in the small gap between the driving flywheel and driven wheel of the punch press, caused a "double trip," and injured the operator. The concave fracture surface of the larger part of the

Figure 19-6 Fatigue fracture of an intramedullary rod that initiated at a prior surface indentation (arrow).

pin, shown in Fig. 19-5(b), could have been damaged only by the matching convex fracture surface in Fig. 19-5(a).

Prior indentation damage can be the cause of a fracture. The fracture surface of a reinforcing rod used on the upper leg bone is shown in Fig. 19-6. This rod had been installed in a woman's broken femur following an automobile accident, to hold the bones in place during healing. After 11.5 months (30 Ms) the rod broke, the result of fatigue failure. At the fatigue initiation site there is a surface indentation (arrow); a similar indentation exists on the matching fracture surface, indicating that the two indentations are two sides of a single dent that existed before the fracture. Because of the sensitivity of fatigue to poor surface condition (Chap. 18), this dent led to the initiation of fatigue at that point.

Exercise 19-2

Why would a dent promote fatigue?

Solution. To repeat the arguments given in Chap. 18, a surface dent represents a region of stress concentration where local slip can induce cracks before general slip. Furthermore, the plastic deformation that has occurred under the dent will restrict slip in this region and thus will lead to earlier fracture than in an undamaged region. Good surface quality is thus essential to long fatigue life.

SEM/TEM. The scanning electron microscope (SEM) and transmission electron microscope (TEM) can often reveal the fracture mode more clearly and explicitly than optical microscopy. In the magnification range of 100 to 10 000X, most of the important fractographic discriminations are possible.

Because of the possibility that high magnification of the surface contours may be required to determine the fracture mode, the fracture surface should be kept in the condition that it had immediately following fracture, if at all possible. The matching fracture surfaces should not be brought together to see how well they fit; this matter can be established with the surfaces 5 mm apart, if there is any question. The fracture surfaces should not be touched with the fingers or any tools, or cleaned or wire-brushed, and the surfaces should be kept free of moisture. A thin layer of light rust-preventive oil is effective for protection of ferrous metals if they must be stored before SEM/TEM examination.

The TEM was described briefly in Chap. 5 for use in examination of thin foils of metal to observe dislocations that become visible because of Bragg diffraction of the electron beam. When used for fractography, the TEM sends an electron beam through a very small (diameter about 2 mm) *replica* of the fracture surface to be examined. This polymer replica of the fracture surface is first shadowed at an angle with a dense metal, often chromium, to accent the hills and valleys of the surface contours. Then a thin layer of carbon, which is relatively transparent to electrons, is coated over the entire surface to provide some strength and the plastic is then dissolved away so that it will not be vaporized in the electron beam. The entire replica is usually supported at the outset by a very fine wire grid of about 125 μm spacing, so that when the replica is viewed in the TEM the fracture region is visible between the lines of the wire grid. The fracture contours are revealed because the electron beam is blocked off in those regions where the chromium has been deposited.

One replica 2 mm in diameter can require several hours of study at high magnification for a complete examination. For example, at 10 000X the area to be studied through the fluorescent viewing screen, which is about 100 mm wide, would be about 20 m in diameter if viewed at one time. However, it is seldom necessary to examine the entire replica, because in many instances a number of traverses will indicate the characteristic fracture modes present. But when searching for isolated patches of fatigue striations or intergranular fracture surface, more thorough examination may be required.

Exercise 19-3

How does the investigator know where on the fracture surface to take replicas?

Solution. He bases his decision on the observations from his earlier optical examinations and an SEM examination (see below) if possible. From experience it is possible to select regions that may exhibit fatigue, stress-corrosion cracking, or brittle cleavage, for example. If a preliminary study of replicas is inconclusive, more replicas may be required.

The SEM allows the direct examination of any electrically conducting surface that can fit into the specimen chamber (usually less than 100 mm in diameter). The image is formed by the secondary electrons reemitted at an angle from the surface that is being bombarded with a narrow electron beam. This incident beam is moved over a line-by-line raster (where the electron beam sweeps along a horizontal line, then along a second horizontal line just under the first, and so on, until it has covered a selected rectangular area, and then it starts over again at the top line) covering the very narrow width of the region being examined, and the subsequent enlarged image is synchronized with the raster on a television screen. Since the texture and orientation of each point on a surface control the intensity of the secondary electron emission from that point, this procedure provides an image of the fracture surface. The viewer must educate himself and his eyes, particularly regarding intensity, because secondary electrons do not behave the same as reflected light: Nearly vertical surfaces emit *more* electrons, and flat, horizontal surfaces emit *fewer* electrons. In addition, the tops of ridges are bright lines because these are favored regions for electron emission.

The TEM and the SEM each have advantages for fractography. Obviously, the SEM cannot be used in many instances when the fracture surface cannot be cut, since many fractured parts are more than 100 mm in diameter or are too long to fit into an ordinary specimen chamber. If such destruction of the part is not permitted, then the SEM usually cannot be used. Since the TEM replica does not alter the fracture surface except by stripping off some of the corrosion and other contaminants clinging to the surface, any number of TEM replicas can be made without damage to the surface. Furthermore, the replicas provide a permanent and easily stored record of the surface. A lifetime of TEM replicas for a failure analyst can be stored in a volume of 0.1 L. When the SEM can be used, its greatest assets are the following: (1) At lower powers, in the range 20 to 100X, the SEM provides much greater depth of focus than is possible with optical microscopy; (2) at higher powers, up to perhaps 5000X, the SEM reveals the contours of rough fracture surfaces, holes, tears, cracks, and fissures, with great accuracy; and (3) specimen preparation for the SEM is very simple, consisting only of cleaning the sample and fastening it to a mount with conducting adhesive.

The TEM is very poor in replicating rough surfaces, as the plastic of the replica that flows into cracks either tears when it is stripped or forms a smooth surface across rough fractures and thus does not accurately represent the actual surface. However, for relatively smooth, subtle contours the TEM is very powerful because the shadowing of the replica accentuates small contour differences, which the SEM cannot do. Because of the inherently greater complexity of the SEM, it does not provide as good resolution as the TEM at higher magnifications, for example, above 10 000X.

Numerous examples of fracture-mode discrimination by SEM/TEM examinations are illustrated in Refs. 19-2 and 19-5. In brief, *cleavage fracture* appears as relatively flat regions joined by vertical cliffs called *river markings*. The typical fracture appearance of ductile fracture is that of *microvoids*, where small voids form and grow by slip processes until the walls between the voids tear apart in the final stages of fracture. The fracture surface appears as a series of craters of a wide range of possible sizes. Fatigue can appear either with or without striations (see Chap. 18), and the striations can be regular, irregular, more-or-less straight, very wavy, or even discontinuous and interspersed with cleavage or microvoid fracture. Stress-corrosion cracking may appear as intergranular fracture, where the corrosion-related cracking has preferentially followed grain boundaries, and these appear as clear facets at distinct angles to each other (see Fig. 18-21).

But stress-corrosion cracking and fatigue may also occur with no such characteristic features, and the SEM/TEM examination may not provide unambiguous discrimination of fracture mode. Thus analysis by SEM/TEM should be considered just one more technique available to the failure analyst, which may provide additional information to be used in conjunction with information obtained through other means.

An example of the discrimination made possible by SEM/TEM is shown in Fig. 19-7. A pin [manufactured from AISI 4320 (UNS G43200) and case-hardened to 60 R_C] of one link of the lifting chain of a fork-lift truck exhibits very faint beach marks at the initiation site indicated by the arrow in Fig. 19-7(a). But one would have expected the designer to have considered fatigue as the chief fracture mode against which to design. Figure 19-7(b) shows the intergranular fracture surface observed in the TEM at the initiation site; in the

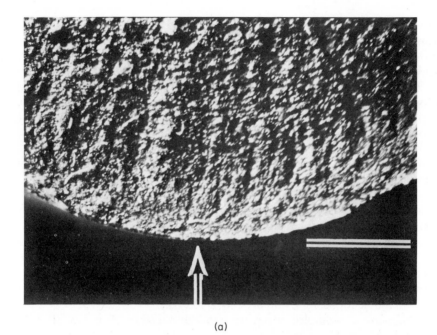

(a)

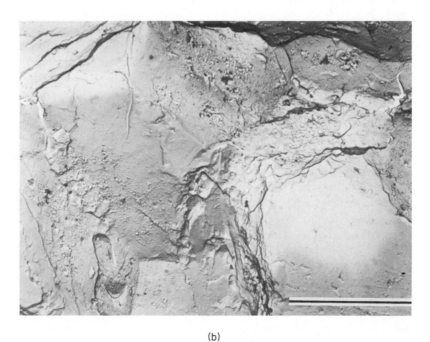

(b)

Figure 19-7   Chain pin of fork-lift truck. (a) Initiation site at bottom center (bar is 1 mm long). (b) Intergranular fracture at initiation site (bar is 5 μm long).

absence of fatigue striations (which are often not observed in fatigue of high-hardness steel) it is not clear whether fatigue was significant, but stress-corrosion cracking was certainly an essential element. We would not have known this without the TEM.

Sometimes unexpected modes are found with a TEM. Figure 19-8(a) shows the fracture surface of one column of a stepladder 3 m high; this

(a)

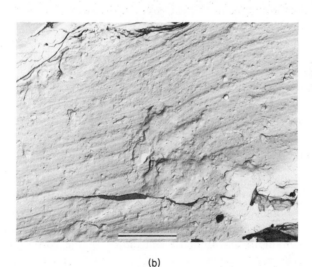

(b)

Figure 19-8 Fracture surface of a column of an aluminum alloy (UNS A96005) ladder. (a) Full cross section of column, with flat, featureless fracture in short flange region at right, changing to ductile fracture and tearing in long flange region at left. (b) TEM micrograph of short flange region fracture, showing fatigue striations; most of the short flange fracture region exhibited similar striations (bar is 5 μm long).

fracture might be expected to be the characteristic microvoid fracture of a high-strength aluminum alloy (UNS A96005). However, a routine TEM examination showed extensive fatigue striations [Fig. 19-8(b)] throughout the short flange region at the right in Fig. 19-8(a). None of the experts involved in this case had ever before seen fatigue fracture in an aluminum ladder, but subsequent stress analysis showed how it could occur. The evidence held up through the trial; when true striations are present, fatigue has definitely occurred.

Exercise 19-4

How do we know that the pattern in Fig. 19-8(b) represents true fatigue striations?

Solution. This TEM micrograph has several features that point to fatigue. (1) The striations are not perfectly straight; sometimes postfracture rubbing of a fracture surface will leave very straight parallel lines, and this possibility is thus excluded here. (2) The striations are unevenly spaced, and this is characteristic of service fatigue resulting from a range of stress levels. (3) Bending of the striations around the barrier in the center of the micrograph indicates a crack-growth direction toward the top of the micrograph; this direction agrees with the apparent crack-propagation direction at the macroscopic level.

Many of the published SEM/TEM fractographs are of fracture surfaces made under controlled laboratory conditions where the fracture is carefully protected from the elements. The fracture analyst may be discouraged when the quality of his fractographs resembles, for example, Fig. 18-23, the result of weeks or months of exposure to rain, mud, and road salt. Nevertheless, under close inspection the essential features of the fracture mode can usually be determined and an accurate analysis made.

Sometimes it becomes important to determine the chemical composition of a portion of a fracture surface. Such was the case in the fracture of the tungsten carbide cutting tip of a wood-cutting circular saw blade. The plaintiff's expert claimed that a contaminant visible (optically) on the fracture surface was brazing alloy that had entered a crack that was present in the tungsten carbide tip at the time it was brazed to the steel blade. An attachment for the SEM provides a means for measuring the characteristic energy of the secondary X-rays emitted by each element on a surface bombarded with an electron beam. By measuring the strength of the secondary beam at each energy level, it is possible to determine semiquantitatively the relative abundance of each element present on the surface. In the case of the saw blade tip, none of the elements of the brazing alloy was found to be present in the surface contaminant in question. In fact, some of the contaminant was vaporized under the highly concentrated beam required for local analysis, so it was concluded that chiefly nonmetallic material, not brazing alloy, had collected on the surface.

There are numerous other techniques for analysis of surfaces, but they are generally slow and very expensive. In most cases the failure analyst can obtain all the high-magnification information that he needs from a SEM and/or TEM.

Photographs.   Before any changes or cuts are made in failed parts, detailed photographs should be taken. Probably not one photograph in twenty will prove useful, but at the outset it is impossible to know which ones will be. For future reference the analyst should include in some photographs a scale of length. He should also keep a detailed log of all photographs taken, noting exact orientations of all parts.

The informed failure analyst must take his own pictures, for no commercial photographer, no matter how skilled, can discriminate the important from the trivial. Figure 19-9(a) shows an overall view of one half of a failed crank 1.35 m in diameter from a 71-MN forging press, and Fig. 19-9(b) shows six of the 12 separate fatigue fractures that precipitated this failure. Without the

(a)

(b)

Figure 19-9   Failed forging press crank. (a) One half of crank, with fractured end at left. (b) Six fatigue fracture sites along periphery of fracture; the largest fatigue fracture, at left, is about 50 mm wide.

second photograph the analysis would have been incomplete. Likewise, Fig. 19-10(a) shows an interesting but irrelevant fracture of the end of the tank of a railroad tank car; this fracture *resulted* from a collision that was *caused* by brittle fracture of the main support frame [Fig. 19-10(b)], resulting from a poor design detail and very low temperature.

If the failure analyst is fortunate enough to be able to visit the site of a failure, observations and photographs taken there can be of great help in

(a)

(b)

Figure 19-10   Railroad tank car. (a) The most conspicuous fracture, the result of a collision that followed the original fracture. (b) Brittle fracture of the main support frame that occurred at very low air temperature. Arrow indicates fracture origin.

(a)

Figure 19-11 Railroad siding. (a) Wavy track. (b) Old, partially fractured track reinstalled at siding. The spalled side of the track, at top right of photograph, is here installed so as to be on the outside of the track, the side that is not worn by wheel flanges.

(b)

understanding the reasons for a fracture. For example, the general condition of the track of a railroad siding (Fig. 19-11) provided valuable support to the allegation of poor installation and/or maintenance that caused the fracture of the track, derailment, and the death of a bystander.

Dimensions. The physical dimensions of key parts involved in a failure should be measured and recorded. Because at this stage the investigator usually does not know the cause of the failure, he must record more information that he probably will need. If a subsequent stress analysis (see below) becomes necessary, the critical dimensions of the part can be very important. He can later obtain the detail drawings of the part and can compare these with the observed dimensions. Such comparison will show whether the part is worn, bent, or damaged.

Figure 19-12(a) shows the apparently normal left rear axle flange of an automobile that lost its left rear wheel while traveling on a highway at 30 m/s,

(a)

(b)

Figure 19-12 Automobile wheel failure. (a) Outer face of left rear axle flange, which had been improperly manufactured with a slightly convex surface rather than absolutely flat. (b) Inner face of the left rear wheel, which failed in fatigue from transfer of the radial wheel forces by the lug bolts instead of by friction with the wheel face. Note the small cracks at the periphery of the holes shaped by the hexagonal lug nuts. (c) Fatigue striations typical of many such regions observed during TEM examination of the fracture surfaces of these cracks. Scale bar is 1 μm long.

rolled over, and caused two deaths and five severe injuries. Careful measurements showed this outer surface to have a convexity of about 0.8 mm across its face, while the manufacturing drawings required that it be at least absolutely flat or have a maximum *concavity* of about 0.1 mm across its face. Such a subtle discrepancy in manufacturing tolerance led to improper contact be-

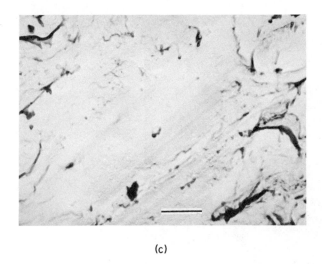

(c)

Figure 19-12 (cont.)

tween axle flange, brake drum, and wheel so that the wheel lug bolts rather than the face of the wheel supported the radial forces between the wheel and axle flange. The resulting fatigue fracture of the region surrounding the lug bolts allowed the wheel to come off over the lug nuts [Fig. 19-12(b)] and caused the accident. An example of the widespread fatigue striations found in the cracks in the wheel is shown in Fig. 19-12(c). Without careful measurement of the axle flange we would not have known that the observed fatigue fractures originated from a manufacturing error.

Sometimes careful observation of the configuration of a failed part can be very revealing. A car drove into a canal and the driver was drowned, but an autopsy showed that the driver had a high concentration of carbon monoxide in his blood. The cast-iron exhaust manifold was found to be fractured, and it was uncertain whether (1) the impact and/or thermal shock of the water caused it to crack; or (2) the exhaust manifold was previously cracked and had leaked the exhaust gas that asphyxiated the driver. Figure 19-13 shows one portion of the mismatching fracture surfaces of the exhaust manifold pressed tightly together (such practice, even though it violates the stricture against bringing fracture surfaces together, is sometimes necessary), indicating that some wear had occurred after the fracture and prior to the accident. This observation, coupled with the absence of any other possible source of carbon monoxide in this case, demonstrated that the cracked manifold and leaky cowling seal that allowed the gas to enter the heater duct to the passenger compartment were responsible for the accident.

Measurement of the dimensions of indentations present on parts can be used to estimate the magnitude of the force required to produce the indenta-

Figure 19-13 Fractured sections of a cast-iron exhaust manifold pressed tightly together. Prior wear at several points, such as at arrow, show that the fracture had occurred prior to and had caused the asphyxiation of the driver.

tion (Ref. 19-4). Although this knowledge alone seldom explains the reasons for the fracture, it often limits the possible time of occurrence of the indentation, as discussed below and with Fig. 19-2.

HARDNESS TESTS

Of all the mechanical tests, hardness measurement is the fastest and the least destructive (see Chap. 2). The hardness of all but the most brittle steels (see Fig. 2-20) and of many nonferrous metals can be used to determine the approximate strength. Appendix 4 gives approximate conversions between the more common hardness scales, and an estimate of strength. For lower-hardness alloys, such as aluminum, the hardness (usually Rockwell E) can be used in conjunction with the chemical analysis (see below) to determine the probable heat treatment. From this knowledge, the yield and ultimate tensile strength can be obtained from handbooks such as Ref. 19-3.

The wide variety of hardness scales allows the failure analyst to select the one that measures the hardness with the least damage to the part. For this reason the 3000-kg$_f$ (29.4-kN) Brinell test is seldom used except for cast irons where the point-to-point variation in hardness requires the averaging effect of the relatively large Brinell indentation. For steels, Rockwell C (for hard steels) and Rockwell B (for soft steels) are fairly standard. Often a complex part cannot be properly supported to withstand the 100 kg_f (981 N) or 150 kg_f (1.47 kN) of the Rockwell B or C tests, respectively. The Rockwell A scale, using 60 kg_f (588 N), provides an alternative. Even better, the superficial Rockwell scales, which require a separate testing machine, provide hardness measurements using forces of 15, 30, and 45 kg_f (147, 294, and 441 N) and either a diamond brale (the "N" scales) or a ball $\frac{1}{16}$ in. (1.6 mm) in diameter (the "T" scales). The superficial scales must be used with caution, however, because they measure hardness in a very thin layer of the surface, and over a very small region. If bulk hardness is needed, either the surface must be ground off or a different hardness test must be used. For very thin parts, the superficial test is often the only choice available.

The simplest application of hardness is as a test against the hardness specified by the manufacturer. A more specific use of hardness in failure analysis is in utilizing the derived strength values in a stress analysis of the structure. When a brass safety latch of an 8.9-kN steel hook failed to prevent a large steel panel from falling off the hook, comparison of the hardnesses and yield strengths of the brass latch with the cold-worked steel latch for which the unit was originally designed showed that the steel latch would have been about 2.5 times as strong as the substitute brass and probably would have protected against the failure. (It was never clear why the manufacturer would use an expensive metal such as brass instead of low-carbon steel in a noncorrosive application.)

A rare use of hardness is to infer the temperature–time history of a part that has been through a fire. If a heat-treated part is reheated, the hardness can change, and this information may be useful in inferring the conditions in a fire. Such a case is discussed under operational tests in a later section of this chapter.

MUTUAL INDENTATION HARDNESS ANALYSIS

Following a failure, parts often exhibit indented regions where they were in contact with other parts. It is possible to estimate the magnitude of the forces required to produced observed indentations (Ref. 19-4). Once the forces are known, it is often possible to establish unequivocally whether these forces arose prior to or at the instant of the failure. Such knowledge can thus be of immense aid to failure analysis.

As described in Chap. 2, hardness H in general is the force F required to produce an indentation divided by the projected area A of the indentation:

$$H = \frac{F}{A} \tag{19-1}$$

In the Brinell test, for example, the numerical value of Bhn is the force F (in

kg_f, unfortunately) divided by the actual area of the spherically indented surface (in mm^2); for present purposes this area is roughly equal to the projected area A.

For a real work-hardening material, the hardness is related to the yield stress Y at the completion of a conventional ball indentation (such as Brinell):

$$H = cY \tag{19-2}$$

where c is a function of geometry. For example, $c = 2.8$ in the Brinell test (H_B), where the indenting ball is always much harder than the material being tested. For other very hard indenters of different geometries, and also when the two contacting bodies have hardnesses comparable with one another, c will change. Experiments show that the geometric constant c is a function of the shape and relative hardnesses of the two indenting surfaces, up to the point, for example, where a ball is about 2.5 times the hardness of a flat; beyond this there is no further change. These results are given in Fig. 19-14 for a number of representative geometries of contacting surfaces such as ball on flat (hard-

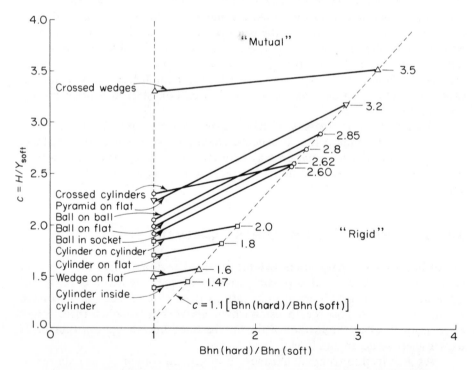

Figure 19-14 Geometric factors ($c = H/Y_{soft}$) for many geometries, for use in mutual indentation hardness analysis. [From A. G. Atkins and D. K. Felbeck, "Applying Mutual Indentation Hardness Phenomena to Service Failures," *Metals Eng. Quarterly*, **14**, 2 (May 1974), p. 55. Copyright American Society for Metals. Reprinted with permission. Also *Chartered Mechanical Engineer*, **21**, 6 (June 1974), p. 78.]

ness test), wedge on flat, and crossed cylinders. The hardnesses of both soft and hard contacting surfaces must be known or estimated.

Combining Eqs. (19-2) and (19-1) gives

$$F = cYA \tag{19-3}$$

Since the softer material usually is indented more, the value $c = H/Y_{soft}$ from Fig. 19-14 can be used in Eq. (19-3) to determine the force F that caused an indentation in the soft surface of known Bhn $= H_B = c_B \cdot Y_{soft} = 2.8 \cdot Y_{soft}$.

$$F = c \cdot \left(\frac{H_B}{c_B}\right) \cdot A \tag{19-4}$$

For consistent units, since 9.807 N $= 1$ kg$_f$, the useful form of Eq. (19-4) is

$$F\ [N] = c \cdot \frac{\text{Bhn}\ [\text{kg}_f/\text{mm}^2]}{2.8} \cdot A\ [\text{mm}^2] \cdot \frac{9.807\ [N]}{1\ [\text{kg}_f]}$$

$$F = 3.50 \cdot c \cdot \text{Bhn} \cdot A \tag{19-5}$$

F is the force [N] to cause an indentation A [mm^2] in a material of hardness Bhn [kg$_f$/mm^2] under geometric and relative hardness conditions that give the constant c.

Exercise 19-5

If the indentation in Fig. 19-2 has an area of 34.8 mm^2, the hardness of the indented ball-joint socket is measured as 85 R$_B$, and the hardness of the stud is estimated to be about 300 Bhn, show how the required indentation force of 29.5 kN, already mentioned earlier, was derived.

Solution. From Appendix 4, 85 R$_B$ is equivalent to 165 Bhn (3000 kg$_f$). Thus Bhn(hard)/Bhn(soft) $= 300/165 = 1.82$. From Fig. 19-14, for "cylinder inside cylinder," c has reached the constant value of 1.47 [for all Bhn(hard)/Bhn(soft) above 1.34]. Then, from Eq. (19-5), $F = (3.50)(1.47)(165)(34.8) = 29.5$ kN. Note that the hardness of the stud has no effect on the answer if it is greater than $1.34 \cdot 165$ Bhn $= 221$ Bhn, because the stud is not indented.

METALLOGRAPHIC TESTS

If we are permitted to cut out a portion of the part at a location removed from the fracture site, we usually perform a routine metallographic examination. A thin sliver of the part cut out is mounted in plastic, polished, and directly examined with the metallographic microscope in the unetched condition (for porosity, cracks, seams, etc., and for graphite size and distribution in cast irons) and/or in the etched condition to determine the microstructure. Seldom does such examination alone lead to an explanation for a failure, but it does supplement and support other tests such as the measurement of hardness and observations of the fracture surface.

In cases not involving lawsuits it is often desirable to cut out a sample containing one cut perpendicular to the fracture surface. (Such destruction of evidence is almost never allowed in a lawsuit.) Such a specimen, mounted and polished, can provide information that is unavailable from surface fractography, concerning the paths of secondary cracks, for example as in Fig. 18-19, or whether the main crack is intergranular or transgranular.

SPECTROGRAPHIC ANALYSIS

After the sliver has been removed for metallographic examination, the small sample (about 5 to 10 g; for steel this is a cube about 10 mm on each side) removed from the part can be quantitatively analyzed for chemical composition. (We often make an additional hardness test on this sample at this stage before it is analyzed, because it is of convenient size and can easily be ground off. Since it will be consumed in the analysis, no additional damage is done to the parent part.) Most failure analysts do not maintain the necessary specialized and expensive equipment and staff, so a commercial laboratory is used for the analysis.

Knowing the chemical composition of a failed part usually provides no new information and is therefore just a routine check against specifications. However, discrepancies occur just often enough to warrant the test. In a case of fatigue failure of a pin of a paper-shearing machine that led to injury of the operator, the pin was analyzed and found to be a plain-carbon steel of 0.06% C, whereas the specification was for 0.45% C. This discrepancy explained the fatigue fracture, in conjunction with a stress analysis of the conditions in the fractured region. In this case a replacement pin, of the same dimensions and part number as the failed pin, ordered from the manufacturer's distributor, had a composition of 0.17% C, still not close to the 0.45% C in the specification. It was clear that the manufacturer was not maintaining close control over the steel used for this pin; one may wonder how many other failures have resulted from other similar pins not manufactured according to specification.

A spectrographic analysis can sometimes save a lot of investigative time. A stainless steel boat-hoist strap, plastic-coated to prevent damage when lifting boats, fractured and dropped a 12-m-long boat on its owner and killed him. Early in the study a sample of the strap was sent out for chemical analysis, in the expectation that the grade of stainless steel used might have had some special problems because the fracture appeared rusty. When the analysis showed that it was 0.18% C plain-carbon steel, the case was over.

ADDITIONAL INFORMATION

Although examination of the parts usually provides the chief activity of the failure analyst, he should make full use of all other available information that is pertinent to the failure.

Witnesses. People who were present at the time of a failure are usually questioned about what they saw. Their reports, in the form of statements, interviews, or as formal depositions, often provide useful supplementary information to the analyst. However, all such information should be considered to be of secondary value, as compared with the hard facts obtained from examination of the parts. People's memories are unreliable, especially at a time of crisis, and often much time has passed before they are deposed by the lawyers involved.

The statements of completely disinterested people should be weighed most heavily of all witnesses because they are free of apparent bias. Associates of an injured person often act so at to protect themselves and their reputations. In most states in the United States an employer cannot be sued by an injured employee because the employer is automatically obligated to provide workmen's compensation insurance to pay any injured employee. But the employer is still motivated to provide a public image of doing no wrong, so statements, test results, and other information provided by the employer should be weighed as if coming from a biased source. Injured parties to a lawsuit, or employees of a defendant, who stand to gain from the successful outcome of the suit may or may not provide all the facts in the course of an inquiry; their potential gain from their side's winning the lawsuit may lead them to distort or alter the facts of a case in their favor. These cautions to the independent failure analyst apply no matter who his client is, for plaintiff and defendant alike have been known to lie to their own experts in an attempt to win their case. The expert should therefore listen, but cautiously, to all he is told and carefully weigh what he hears against what he sees.

Exercise 19-6

What should a failure analyst do when he becomes convinced that his client's product liability case is without any foundation in the facts?

Solution. He should tell his client as quickly as possible. Society is served by engineers who accept reality, not by those who would distort it to suit their own ends.

Police and fire department reports. In many instances such formal reports are available to the analyst. Their value usually lies in the indisputable, hard facts of the investigation: date, time, physical locations of parts, injured people, photographs, and compass orientation. Conclusions as to cause should be largely discounted. Most police officers simply are not qualified to say that an accident resulted from the broken steering component observed after the accident, even though they would like to believe that they are so qualified. And most firefighters are not very good at determining the source of a fire; expert fire investigators are specially trained for this purpose.

Other experts. A factual report of a qualified accident investigator can be very useful, if he narrowly limits himself to his on-site observations and avoids reaching conclusions beyond his qualifications. For studies involving areas outside his own field, the failure analyst should not hesitate to recommend that other experts be brought into a case: experts in, for example, hydraulics, structures, dynamics, or specialists in the specific equipment involved, such as automobiles, cranes, or aircraft. We work regularly with people in such areas and are thus able to complete an analysis much more quickly than without them. An expert who tries to deal in a very broad technical area becomes so general that he does not serve his clients well.

Drawings and specifications.   As has already been mentioned, comparison of the dimensions, hardness, and chemical composition of a part with the specified values may be of great assistance in establishing the cause of failure. Even if the part meets all specifications, such comparison should be made in order to eliminate these deficiencies as possible causes of failure.

Exercise 19-7

If a part is found not to meet its design specifications, such as in dimensions or hardness, is that sufficient evidence to identify the cause of a failure?

Solution.   No. A *causal relationship* must be established between any design deficiency and a resulting failure before a failure can be explained. A substandard windshield wiper blade on an automobile does not explain an alleged steering linkage failure. Much litigation time is wasted in such allegations, in what might be termed "smoke-screen" failure analysis, where an expert attempts to hide a nonexistent causal relationship behind a cloud of irrelevancies.

The design capacities of a structure should be determined from the manufacturer. Use of a device beyond its capabilities has been the cause of many failures. Figure 19-15 shows the fatigue fracture of the yoke of a universal joint of a large-capacity end loader. (An earlier investigator had already butchered the evidence by cutting out the center of the fractured area.) While the machine was driving down a public highway, the yoke fractured and the drive shaft flew out and severely injured a bystander. The history of this

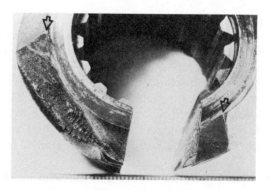

Figure 19-15   Yoke of universal joint of an end loader (having a section cut out earlier), showing two fatigue fracture origins (arrows).

machine indicated that after manufacture it had been adapted for heavy-duty use in a landfill operation, which involved the addition of heavy protective underplating armor and installation of a heavy-duty torque converter that increased the torque transmitted by this joint. Since small increases in stress

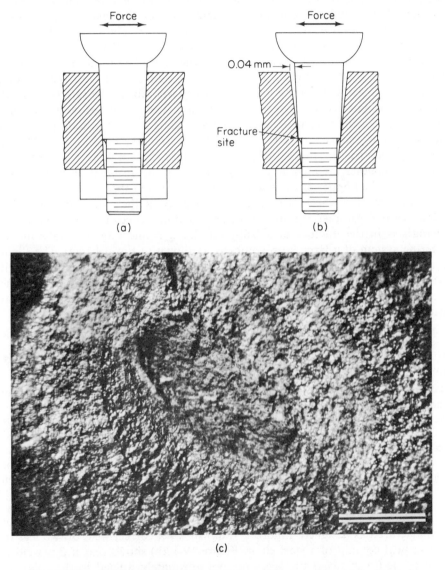

(a)   (b)

(c)

Figure 19-16   Ball-joint shank of steering linkage of farm combine. (a) As designed, shank taper fits taper of supporting plate. (b) As measured after the failure, shank has ±0.04 mm possible lateral motion, with consequent high stress at fracture site (the dimensions are exaggerated in the sketch). (c) Fatigue fracture; scale bar is 2 mm long.

above the fatigue limit can drastically reduce the life of a part, this failure was attributed to the excessive loads on the part, which were not considered in the original design.

Sometimes a second-year engineering student could have determined the reasons for a failure. In one such case, a sling made of wire rope 9.5 mm in diameter failed while supporting a weighing hopper that was being installed inside a grain elevator; one man fell 10 m to his death. The safe rated capacity of the wire rope alone (for a factor of safety of 5) was 10.8 kN, and this was confirmed by subsequent tensile testing of the rope. The total known gravity force exerted by the hopper, the two men standing on it, and the hoisting block was 32.7 kN. Thus the load was clearly well above the safe rated load, but this does not necessarily explain the failure. The important point was that the strength of the wire rope when used as a sling around a hook (as in this accident) was reduced to 60% of the rated strength, or 33.1 kN. This is virtually identical to the actual load at failure, 32.7 kN. It was never established whether the men involved in setting up this installation knew either the mass of the hopper or the fracture load of the wire rope.

Failure of a wheel as a result of convexity of the axle flange was discussed earlier. Another example of the profound influence of small dimensional changes is in the fracture of a ball-joint shank connecting two links in the steering system of a large farm combine, as shown in Fig. 19-16(a). The slight mismatch between the taper of the shank and the taper of the plate to which it was attached [Fig. 19-16(b)] led to an unexpected bending moment at the bottom end of the taper, with consequent fatigue fracture [Fig. 19-16(c)] that caused the combine to run off the road and be seriously damaged.

Exercise 19-8

How is it possible to discriminate in Fig. 19-16(c) between fatigue and, for example, excessive tensile stress from overtorquing the nut on the shank?

Solution. (1) The simplest discrimination results from the macroscopic appearance of the fracture. Fracture progression lines are clearly visible in Fig. 19-16(c). (2) This pattern is typical of bending fatigue (Ref. 19-2), not overload tension. (3) Measurements on the parts showed them to correspond to Fig. 19-16(b) rather than Fig. 19-16(a) as was intended by the designer. (4) Overload tension from overtorquing usually occurs at the time of overtorquing or not at all, as subsequent relaxation in service of the torsional moment on the bolt (shank) lowers the maximum principal tensile stress from its initial value.

The best efforts of the designer may be worthless if the part is not manufactured as designed. Figure 19-17(a) shows a correct weld of two legs to the vertical column of a steel chair. Figure 19-17(b) shows that if this weld is incomplete (in this case the legs were not adequately welded to the column), the chair can fail under normal service loads. It caused a 120-kg man to fall and be injured.

If the design permits excessive wear or accidental abuse to occur during normal service by a reasonable and careful operator, failure can occur. A

(a)

Figure 19-17 Improperly welded joint between vertical column of a steel chair (horizontal in photos) and the legs. (a) Correct weld joining two legs and column. (b) Incorrect weld, only joining two legs, did not penetrate to column (arrow).

(b)

motor crane [Fig. 19-18(a)] had its wire rope fail while lifting a very light load of two men in a man-bucket, injuring them both. Lack of guides for the wire rope as originally installed had allowed the rope when slack to be caught on the lower corner of the boom. Figure 19-18(b) shows the wear at this point that had occurred prior to the accident. This location is out of the view of the operator. This led to either wear or damage so severe that the wire rope could fail under a very light load. The problem was solved after this accident by addition of guards to prevent the misguiding.

Applicable standards. No part ever failed in service solely because it did not meet the requirements of standards written to serve as guidelines for safe practice. But the failure analyst should consult standards, especially if a failed component had been purchased to comply with a particular standard. Standards for metals, materials, dimensions, tests, and safe practice have been established by such organizations as American Society of Mechanical Engineers (ASME), Society of Automotive Engineers (SAE), American Society for

(a)

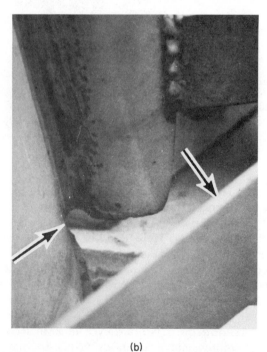

(b)

Figure 19-18 Wire rope failure because of misguiding. (a) Right side of motor crane with drum and wire rope at upper left. (b) Lower left end of boom (boom extends upward to the left in photo) worn from wire rope that came out of U-shaped trough of boom (left arrow). Guard to prevent such misguiding (right arrow) was added after accident.

Testing and Materials (ASTM), American Iron and Steel Institute (AISI), and American National Standards Institute (ANSI), to name but a few. Superimposed on these are state and federal governmental requirements such as by the Occupational Safety and Health Administration (OSHA).

For the failure analyst these standards often give a good view of industry practice, and they are often very useful in describing functions and designs of parts. But the standards are certainly not absolute, since each serves a different purpose. For example, the people who manufacture wire rope want their wire rope to perform as well as possible in service, so they recommend that the ratio of the diameter of the sheave (pulley) to the diameter of the wire rope be (for example, for 6 × 25 FW*) 36 : 1, with a minimum of 24 : 1. (Larger sheave diameter causes less bending and consequently leads to longer life for wire rope.) The SAE, on the other hand, wants hoisting equipment to be kept to moderate size and cost, so it recommends a minimum ratio of sheave diameter to wire rope diameter of 12 : 1.

The failure analyst must carefully examine any deviations from applicable standards to determine their possible influence on the failure in question. Even if the part was incorrectly made or does not meet some standard, if this error or deviation did not cause this failure, then it is irrelevant. Efforts to suggest incompetence through elaborate presentation of such irrelevant information may help win cases by obfuscation before juries, but they waste a lot of time and do not further the cause of impartial failure analysis.

MISAPPLICATION

Sometimes a part or assembly is simply used in a manner for which it was never intended. Figure 19-19 shows seven of about 100 gear teeth (made of UNS G51400 steel) that failed by fatigue in three separate but identical 233-kW screw pump installations, designed to pump purified water from a sewage treatment plant up a 5-m head in order to dump it into an adjacent river. When identical failures occur in separate installations, one suspects similar causes. In this case the gear reduction system specified by the designer was not rated high enough for the capacity required for this installation, and the consequent excessive stress led to the fatigue failures.

When an adjustable aluminum alloy staging 5 m long used for lowering workers from the roof of a building failed, it dropped two men 25 m to the ground, killing one and injuring the other. Investigation showed that the structural frame of the staging was not overloaded, but it had been overloaded at an earlier time, damaged, and repaired by riveting. The person who chose to rivet this had of course not performed a stress analysis on the structure; a subsequent stress analysis revealed that, although the riveted joint was vir-

---

*This is a wire rope made of 6 strands, each strand made of 25 wires [19 larger wires plus 6 much smaller *filler wires* (FW)].

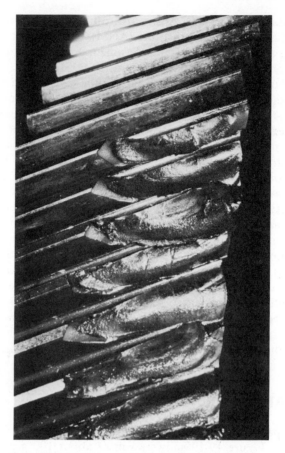

Figure 19-19   Fatigue fractures of teeth of an overloaded reduction gear.

tually as strong in tension as the continuous beam it replaced, it had only about one-third its bending moment capacity. Thus a misapplication consisting of repair by an unqualified person led to injury and death.

Exercise 19-9

How could the manufacturer of a device of such sophisticated design as this aluminum alloy staging (see above) prevent such an accident?

Solution.   Since the purpose of the choice of design and material is to produce a very light but strong staging, any lowering of the design stresses would increase the mass. In such a case the simplest alternative is to place conspicuous and permanent warnings on the structure instructing the user to scrap it if it is damaged and stating that it cannot be safely repaired.

ANALYSIS

Many failures can be fully explained without further analysis, as we have seen, for reasons such as improper strength, dimensions, or chemical composition, or obvious abuse or overload. But in some instances an analysis of the conditions that existed at the time of the failure can lead to an order-of-magnitude answer that may either exclude a potential explanation for the failure or include it as possible. We mentioned in the preceding section a case where stress analysis of a repaired staging led to full explanation for a failure. In this section we will show how fracture analysis and dynamics can be used together to evaluate the validity of a proposed explanation for an accident.

An unusual case involved a *question* of fracture, that is, whether a fracture had occurred at all. A man fell while skiing and flew into a large tree, so injuring himself as to be paralyzed from the chest down. Witnesses and friends said he was an excellent skier, and observed that his tracks during the turn just prior to the accident crossed over and had apparently cut the top off the stub of a small tree of about 20 mm diameter. The fracture surface of the tree appeared fresh and was flush with the surface of the snow. In the litigation that followed (about 4 y later), the plaintiff skier claimed that he had struck the stump of this small tree and severed the top off it, leading to his fall and injury. He could not prove the existence of the stump except by the statements of some of the witnesses and one photograph that showed the tree stump as little more than a dot in the snow, for the area of the accident site had been completely altered by a bulldozer during the spring following his accident. The owner of the ski area had removed not only the stump but even the large tree that the plaintiff had struck. The defendant ski slope owner claimed there was never any stump, that the plaintiff "caught an edge" of his ski in the snow, flipped over, and struck the large tree.

As the plaintiff had struck the large tree with the back of his head after completing about $\frac{3}{4}$ of a forward revolution in the air, the truth of this matter hinged on the dynamics of the situation: (1) Could the change in angular momentum from fracturing a tree lead to the $\frac{3}{4}$ turn, and/or (2) could "catching an edge" in the snow lead to the $\frac{3}{4}$ turn? Distances, velocities, and elevations were all known, so all that were needed were estimates of stresses for fracturing green wood (handbook values) and for "fracturing" Michigan snow (determined by experiment). In brief, question (1) turned out to be well within the range of possibility, while question (2) was impossible, by a factor of at least 4. The defendant settled during trial.

A question regarding whether one possible explanation for fracture was valid was likewise settled by a simple dynamic analysis. A horizontal platform suspended near the top and inside a 200-m-high chimney tipped and dropped three men to their deaths. One possibility proposed was that the crane operator had dropped a loaded bucket on the platform and caused one of the platform's welded support lugs to fracture. A simple analysis of the forces that would result from such an elastic impact showed that if the weld had been sound, the bucket could have been dropped from 4 m above the platform and would not have produced forces that exceeded the strength of the weld. Since there was no possibility that the bucket could have been dropped from such a height, this possibility was thereby completely excluded.

OPERATIONAL TESTS

Sometimes a test that to a limited extent simulates the conditions of failure can help in understanding what is possible and what is impossible. Such a test is seldom used to simulate fully the circumstances of a failure because there are usually unknown variables in an actual failure that can make a difference in the result of a test.

Even a simple test can be very informative. Two boys were "jogging" down a hall in a high school, and when they tried to turn right to enter a doorway, one of them skidded, he said, past the doorway and 5.2 m further to the end of the hall, crashed through a wire-reinforced window in an exit door, and severely cut his arm. He sued the school for having a floor that was too slippery. These facts were subject to testing, which involved measurement of the coefficient of friction between the actual floor and the kind of shoes worn by the plaintiff, using a student of about the same mass. The resulting friction force could then be used to calculate the initial speed (8.2 m/s) that would be required for the plaintiff to slide the 5.2-m distance from the doorway and hit the glass door; this speed approaches the Olympic record for the 400-m dash, 9.1 m/s, which would not be considered "jogging." Furthermore, it would be nearly impossible to remain upright during such a slide because the coefficient of friction was 0.35. From such a simple test it was concluded that the plaintiff and his friend were just running too fast and did not slow in time to avoid hitting the door; sliding on the floor for 5.2 m did not occur and thus had no influence on the accident.

Exercise 19-10

Why did the wire-reinforced window break? If a completely unbreakable window were in place, the plaintiff would not have been cut.

Solution. The wire-reinforced window met the building safety code (see discussion of standards above) in effect at the time of construction, and it is capable of withstanding a higher force than a nonreinforced window. Since no window can be completely unbreakable or completely "safe," we must compromise between cost and safety, trying to achieve the most increase in safety for the least cost. Other solutions, such as reduction of the window size, moving its location, or completely eliminating it, should also be considered by the designer.

Simple laboratory experiments can test a hypothesis for a failure. To establish why a blade had fractured in a power lawn mower, we simply ran the unit with a new blade. Speed measurement revealed that the speed governor did not function, and the excessive speed had contributed to the fracture. Another simple test was made in a case involving an injury sustained after a car fell off an automobile jack, and the injured mechanic claimed that the gripping portion of the jack had slipped off the bumper because the jack was poorly designed. To test this, a car of the same model and year was lifted several times with this jack, under the most unstable conditions possible, and the car never fell. This led to the conclusion that some other factor, not revealed in the plaintiff's testimony (see section on witnesses above), had caused the jack to fall.

A knowledge of the effect of heat treatment on microstructure can define limits on the possible temperature–time history of a heat-treated part that has been through a fire. In one instance a plaintiff claimed that a fire on his premises was caused by the absence of a synthetic rubber disk in a valve seat that allowed a gas regulator (pressure reducer) to pass gas of too high a pressure to a gas-fired space heater, leading to the fire. He had reached this conclusion because after the fire the disk was missing from the aluminum alloy disk holder to which it ought to have been glued when manufactured. The aluminum alloy was 2011-T3 (UNS A92011), specified to be quenched and aged at room temperature, which would give 98 R_E (121 Bhn) hardness as heat-treated; production samples of this part were tested and found to be about 98 $\pm$ 3 R_E. The hardness of the disk holder in question was 48 R_E, so clearly it had been through a fire. Although the exact duration of the fire could not be established, an assumption of 1 h at temperature appeared to be reasonable. New aluminum disk holders were therefore exposed to 1 h at temperatures ranging from 316°C to 538°C (it begins to melt at about 538°C), then half were air-cooled and half were furnace-cooled. A reasonable assumption was that the actual cooling rate fell somewhere between these two extremes. The lowest exposure temperature that gave a final hardness of 48 R_E was 360°C, and the highest was 538°C. Since the synthetic rubber of the missing disk degrades into ashes at about 480°C, these tests showed that it was consistent to have the final hardness of the disk holder 48 R_E and at the same time to have achieved high enough temperatures completely to destroy all evidence of the synthetic rubber disk. As is often the case, these test results did not *prove* one single hypothesis but only demonstrated its consistency. Thus such tests are used in conjunction with other evidence and observations to identify the cause of a failure.

Exercise 19-11

How do we know that the disk was not improperly heat-treated by the manufacturer to an initial hardness of 48 R_E?

Solution. We do not, and no test can answer the question. A completed failure analysis seldom excludes all other explanations, but a rational person will tend to accept the most probable explanation. In the above question we would have to accept the two manufacturing errors, in sequence, of incorrect heat-treatment and omission of the

(a)

(b)

Figure 19-20 Test of acceleration of dummy on small end loader. (a) Prior to silage drop; note instrumentation wires. (b) After silage drop. (c) Prior to reverse drop of end loader to ground.

rubber disk, in order to explain the facts. Such a situation seems much less probable than that the disk was heated to 480–538°C in the fire.

Sometimes a full-scale field test, under carefully controlled conditions, can provide useful data. A man operating a very small end loader [Fig. 19-20(a)] broke his neck while unloading compacted silage from a trench silo in the winter, when some of the silage fell from about 5 m high. He claimed that the silage fell on the bucket of the end loader and caused the vehicle to tip suddenly forward. Either the resulting upward acceleration of the back end of

(c)

Figure 19-20 (cont.)

the vehicle or the deceleration when the back end of the vehicle hit the ground broke his neck in compression, he claimed, and he sued the manufacturer for not having designed the shock-absorbing system of the seat so as to protect him from such an injury. (He did not sue for lack of a guard cage around the driver because he had declined this option at time of purchase.) We made some preliminary calculations of the angular acceleration of the vehicle that would result from the maximum possible amount of silage (95 kg) falling on the bucket from a 5-m height. The maximum acceleration of a rigid seat, obtained from this calculation, was insufficient to break his neck. But the calculation involved many simplifying assumptions, notably the estimate of the dynamic behavior of the tires under such an impact. We therefore recommended a full-scale test, complete with an instrumented dummy having mechanical properties similar to those of the human body. An enormous mass of silage (5 Mg), far greater than the bucket could hold, was dropped on the bucket of the small end loader from a 5-m height [Fig. 19-20(b)], and it barely lifted the rear wheels off the ground. The resulting measured maximum acceleration (7.8 m/s^2, or 0.8 g's) of the dummy was the deceleration that occurred when the vehicle returned to the ground; this is well below that required to break a man's neck. The other extreme possibility was the deceleration from the vehicle's falling on its rear wheels after a full forward tip. Figure 19-20(c) shows the unit tipped fully forward, held by a jack, at a 42° angle. When the jack was suddenly pulled out, the maximum vertical deceleration of the seat was 27.5 m/s^2 (2.8 g's) and of the dummy 39.2 m/s^2 (4.0 g's) when the wheels struck the ground. Such deceleration could indeed cause injury, but on the basis of the first test there was no evidence that dropping any amount of silage on the bucket could cause the vehicle to tip forward 42°. It was concluded that the plaintiff's neck had been broken when some silage fell directly on his head; in winter the moist silage freezes and is dug out in clumps, and it is altogether possible that a large clump fell directly on the plaintiff. There was no other consistent explanation possible after the test results were analyzed.

ASSESSMENT

Most of the failures described in this chapter were explained rather clearly at some point in the investigation by simple measurements or tests. The formal procedure of evaluation is thus usually so direct that we are hardly aware of the process. But in fact we consider all tenable hypotheses for the failure at the outset of the investigation, then eliminate each in turn as possibilities following completion of the tests and analyses. If we have done everything correctly, if all the facts are known accurately, and we have considered all the possibilities, we should be left with only one hypothesis at the end of the study. In this way we are often able to establish whether damage caused an accident or was the result of the accident, as has been discussed above. But usually we lack all the facts, and we are even given incorrect, misleading information, so the failure analyst must work his way through his study rather like a detective trying to solve a crime.

Sometimes a complete understanding of a failure comes rather late. In a failure of a wire rope of an overhead hoist, operated by a person standing on the floor using a drop pedestal control, the block and hook of a hoist, *carrying no load*, dropped on the operator–plaintiff's head and caused severe injury. This was a real puzzle, as the wire rope could safely support many times the mass of the block and hook. The entire wire rope involved, which was about 50 m long, had been examined but yielded no explanation. During the trial the plaintiff's expert described the physical surroundings in the building housing the hoist, and showed photographs of the several large floor-mounted machines in this area. When the expert launched into an irrelevant discussion about the ratio of sheave diameter to wire rope diameter being too small, it was clear that this certainly had no bearing on this failure. Something had to have applied a large force to break the wire rope, but what? That evening the entire wire rope was examined again, stretched out in a parking lot. This led to evidence that had been overlooked during the earlier examination: a sharp kink about 10 m from the end that had been noticed before but forgotten because it made no sense at the time. Just prior to the failure, the height of the hoist and position of the block were such as to place this kinked region of the wire rope where it could have caught on a heavy machine bolted to the floor. So instead of lifting just the block and hook, the plaintiff was unknowingly attempting to lift this large bolted-down machine. When the wire rope broke at the point of contact with a sheave, it dropped the block and hook. If the witnesses to the accident were aware that the wire rope had caught on the bolted-down machine, they never mentioned it.

REFERENCES

19-1. American Society for Metals, *Case Histories in Failure Analysis*. Metals Park, Ohio: American Society for Metals, 1979.

19-2. American Society for Metals, *Metals Handbook*, 8th ed., vol. 10, "Failure Analysis and Prevention." Metals Park, Ohio: American Society for Metals, 1975.

19-3. American Society for Metals, *Metals Handbook*, 9th ed., vol. 2, "Properties and Selection: Nonferrous Alloys and Pure Metals." Metals Park, Ohio: American Society for Metals, 1979.

19-4. Atkins, A. G., and D. K. Felbeck, "Applying Mutual Indentation Hardness Phenomena to Service Failures." American Society for Metals, *Source Book in Failure Analysis*. Metals Park, Ohio: American Society for Metals, 1974, p. 364.

19-5. Engel, Lothar, and Hermann Klingele, *An Atlas of Metal Damage*. Englewood Cliffs, N.J.: Prentice-Hall, 1981. Techniques of failure analysis using the SEM are described, and it includes an excellent collection of SEM micrographs covering a wide range of fracture conditions.

19-6. Guy, Henry, *Fractography and Microfractography*. Philadelphia: Heydon & Son, 1979.

19-7. Hertzberg, Richard W., *Deformation and Fracture Mechanics of Engineering Materials*. New York: John Wiley & Sons, 1976.

19-8. Rolfe, Stanley, T., and John M. Barsom, *Fracture and Fatigue Control in Structures*. Englewood Cliffs, N.J.: Prentice-Hall, 1977.

19-9. Shank, M. E., *A Critical Survey of Brittle Failure in Carbon Plate Steel Structures Other Than Ships*. Ship Structure Committee Report, Serial No. SSC-65. Washington, D.C.: Ship Structure Committee, National Academy of Sciences, 1953. (Also reprinted as Welding Research Council Bulletin No. 17.)

19-10. Smith, Charles O., *The Science of Engineering Materials*, 2nd ed. Englewood Cliffs, N.J.: Prentice-Hall, 1977.

PROBLEMS

19-1. Describe all the possible fracture mode(s) that can produce each of the following characteristics:
(a) Cleavage fracture.
(b) Beach marks.
(c) Intergranular fracture.
(d) Striations.
(e) Microvoids

19-2. Name four kinds of fracture appearance that might be observed through examination of fracture surfaces with a magnifying glass. Tell what inferences can be drawn from each.

19-3. Why is an optical microscope not used to observe and identify fracture surface features such as intergranular fracture, microvoids, and striations?

19-4. In which ways is the transmission electron microscope better than the scanning electron microscope? In which ways is the SEM better than the TEM?

19-5. What information can be deduced from a plastically indented region observed after an accident?

19-6. What purposes are served by a chemical analysis of a failed part?

19-7. What does a hardness test reveal to the failure analyst?

19-8. Under what circumstances should a metallographic examination be made, following a failure?

19-9. If a full-scale test is being considered as an aid to analyzing a failure, what cautions should be observed before the test is performed, if it is to have significance for the actual failure?

19-10. Name two observable fracture characteristics which when found in combination constitute a *unique* identification of fracture mode for each of the following. (Note that the combination of the two constitutes a positive identification of the fracture mode, although either of the characteristics may also be observed in other fracture modes.)

  (a) Fatigue in aluminum alloy.

  (b) Hydrogen damage in high-strength AISI-SAE 4340 steel.

  (c) Ductile fracture of AISI-SAE 1020 steel.

  (d) Stress-corrosion cracking of high-strength aluminum alloy.

19-11. (a) Without emphasis on the derivation of the theory, explain the main features of engineering significance that occur when a sphere is loaded against a flat surface. In particular, emphasize the changes in the character of the conditions that occur as the load is increased.

  (b) An engineering joint, represented in simplified form in Fig. 19-21, is part of a mechanism that has broken at another place. It has been argued that the failure was caused by excessive loads. Examination shows that the socket has been indented with a circular area of indentation of approximately 4.1 mm diameter; the hemispherical surface has a similar, if somewhat shallow-

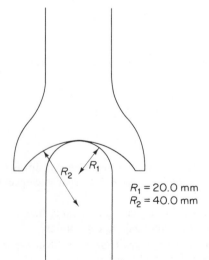

$R_1 = 20.0$ mm
$R_2 = 40.0$ mm

Figure 19-21 Courtesy of Dr. J. F. Archard.

er, indentation. Both parts are of steel, the socket has a hardness of 150 Bhn, and the spherical component has a hardness of 225 Bhn.

Make an estimate of the load to which the component has been subjected. Discuss the way in which the spherical surface might have been indented, despite the fact that it has a higher hardness than the socket; thus suggest an approximate value for the hardness required to avoid any indentation when in contact with the socket. (Courtesy of J. F. Archard.)

Appendices

APPENDIX 1. PREFIXES FOR USE WITH SI UNITS

Multiplication Factor	Prefix	SI Symbol
10^{18}	exa	E
10^{15}	peta	P
10^{12}	tera	T
10^{9}	giga	G
10^{6}	mega	M
10^{3}	kilo	k
10^{-3}	milli	m
10^{-6}	micro	μ
10^{-9}	nano	n
10^{-12}	pico	p
10^{-15}	femto	f
10^{-18}	atto	a

APPENDIX 2. FACTORS FOR CONVERSION TO SI UNITS

Conversion factors in the following table are given as a number between 1 and 10, followed by the letter E (for exponent), a plus or minus symbol, and two digits that indicate the power of 10 by which the number must be multiplied to obtain the value in the appropriate SI unit. For example, to convert from pound-force/inch^2 (psi) to pascal (N/m^2), the factor listed is 6.894 757 E+03. Thus 1 psi is 6.894 757 kPa, and 1000 psi is 6.894 757 MPa.

To convert from	To	Multiply by
Area		
foot^2	m^2	9.290 304 E−02
inch^2	m^2	6.451 600 E−04
Energy and Work		
Btu (Int. Table)	J	1.055 056 E+03
calorie (Int. Table)	J	4.186 800 E+00
erg	J	1.000 000 E−07
foot-pound-force	J	1.355 818 E+00
watt-second	J	1.000 000 E+00
Energy/Area (Toughness)		
erg/cm^2	J/m^2	1.000 000 E−03
foot-pound-force/inch^2	J/m^2	2.101 522 E+03
inch-pound-force/inch^2	J/m^2	1.751 268 E+02
Force		
dyne	N	1.000 000 E−05
kilogram-force	N	9.806 650 E+00
pound-force	N	4.448 222 E+00
poundal	N	1.382 550 E−01
Length		
angstrom	m	1.000 000 E−10
foot	m	3.048 000 E−01
inch	m	2.540 000 E−02
mile (U.S. nautical)	m	1.852 000 E+03
mile (U.S. statute)	m	1.609 344 E+03
Mass		
grain	kg	6.479 891 E−05
pound-mass	kg	4.535 924 E−01
slug	kg	1.459 390 E+01
Mass/Volume (Density)		
g/cm^3	kg/m^3	1.000 000 E+03
pound-mass/inch^3	kg/m^3	2.767 990 E+04
slug/foot^3	kg/m^3	5.153 788 E+02

To convert from	To	Multiply by
Power		
Btu (Int. Table)/hour	W	2.930 711 E−01
foot-pound-force/s	W	1.355 818 E+00
horsepower	W	7.456 999 E+02
Pressure or Stress		
atm. (760 torr)	Pa	1.013 250 E+05
bar	Pa	1.000 000 E+05
cm of mercury (0°C)	Pa	1.333 22 E+03
cm of water (4°C)	Pa	9.806 38 E+01
dyne/cm^2	Pa	1.000 000 E−01
kg-force/cm^2	Pa	9.806 650 E+04
kg-force/mm^2	Pa	9.806 650 E+06
newton/m^2	Pa	1.000 000 E+00
pound-force/inch^2 (psi)	Pa	6.894 757 E+03
torr (mm Hg, 0°C)	Pa	1.333 22 E+02
Temperature		
degree Celsius (°C)	K	$t_K = t_C + 273.15$
degree Fahrenheit (F)	K	$t_K = \dfrac{t_F + 459.67}{1.8}$
Time		
day (mean solar) (d)	s	8.640 000 E+04
hour (mean solar) (h)	s	3.600 000 E+03
year (365-day) (y)	s	3.153 600 E+07
year (tropical)	s	3.155 693 E+07
Velocity		
foot/minute	m/s	5.080 000 E−03
km/hour (km/h)	m/s	2.777 778 E−01
mile/hour (U.S. statute)	m/s	4.470 400 E−01
Viscosity		
centipoise	Pa-s	1.000 000 E−03
poise	Pa-s	1.000 000 E−01
pound-force-s/ft^2	Pa-s	4.788 026 E+01
Volume		
barrel (oil, 42 U.S. gal)	m^3	1.589 873 E−01
fluid ounce (U.S.)	m^3	2.957 353 E−05
foot^3	m^3	2.831 685 E−02
gallon (Imperial liquid)	m^3	4.546 122 E−03
gallon (U.S. liquid)	m^3	3.785 412 E−03
inch^3	m^3	1.638 706 E−05
liter (L)	m^3	1.000 000 E−03
Miscellaneous Special Conversions		
ksi $\sqrt{\text{in.}}$	MPa $\sqrt{m}$	1.098 843 E+00
mile/gallon (U.S. liquid)	km/L	4.251 437 E−01

APPENDIX 3. PLASTIC STRESS–STRAIN CONSTANTS FOR METALS

| | | | $\sigma = K(\epsilon_{cw} + \epsilon)^n$ | |
| | | | K, MPa | n |
UNS	Metal	Condition		
A91100	1100 Al	Annealed	182	0.20
A92024	2024 Al	T-4	700	0.15
C11000	Cu	Annealed 1 h at 540°C	550	0.55
C11000	Cu	Annealed 1 h at 820°C	480	0.48
C26000	70–30 brass	Annealed 1 h at 540°C	770	0.56
C33100	70–30 leaded brass	Annealed 1 h at 680°C	735	0.50
—	Armco iron	Normalized	560	0.32
G10200	1020 steel	As hot-rolled	800	0.22
G10450	1045 steel	As hot-rolled	980	0.14
H43400	4340 steel	As hot-rolled	1500	0.10
G52986	52100 steel	Annealed 1 h at 820°C	1470	0.07
		Spheroidized	1150	0.18
S30400	18–8 stainless steel	Annealed 1 h at 870°C	1500	0.51

Note: ϵ_{cw} is the prior true prestrain from cold work.

APPENDIX 4. APPROXIMATE HARDNESS CONVERSIONS FOR METALS

A. Hardness conversions: hardened steel and hard alloys*

| Rockwell Scale | | | | | Dph, 10 kg_f | Bhn, 3000 kg_f | Tensile Strength | |
| | | | | | | | 10^3 psi (approx.) | MPa (approx.) |
C	A	15N	30N	45N				
80	92.0	96.5	92.0	87.0	1865	—	—	—
79	91.5	—	91.5	86.5	1787	—	—	—
78	91.0	96.0	91.0	85.5	1710	—	—	—
77	90.5	—	90.5	84.5	1633	—	—	—
76	90.0	95.5	90.0	83.5	1556	—	—	—
75	89.5	—	89.0	82.5	1478	—	—	—
74	89.0	95.0	88.5	81.5	1400	—	—	—
73	88.5	—	88.0	80.5	1323	—	—	—
72	88.0	94.5	87.0	79.5	1245	—	—	—
71	87.0	—	86.5	78.5	1160	—	—	—
70	86.5	94.0	86.0	77.5	1076	—	—	—
69	86.0	93.5	85.0	76.5	1004	—	—	—
68	85.5	—	84.5	75.5	942	—	—	—
67	85.0	93.0	83.5	74.5	894	—	—	—
66	84.5	92.5	83.0	73.0	854	—	—	—
65	84.0	92.0	82.0	72.0	820	—	—	—
64	83.5	—	81.0	71.0	789	—	—	—
63	83.0	91.5	80.0	70.0	763	—	—	—

APPENDIX 4A. (cont.)

	Rockwell Scale				Dph, 10 kg$_f$	Bhn, 3000 kg$_f$	Tensile Strength	
							10^3 psi (approx.)	MPa (approx.)
C	A	15N	30N	45N				
62	82.5	91.0	79.0	69.0	739	—	—	—
61	81.5	90.5	78.5	67.5	716	—	—	—
60	81.0	90.0	77.5	66.5	695	614	—	—
59	80.5	89.5	76.5	65.5	675	600	—	—
58	80.0	—	75.5	64.0	655	587	—	—
57	79.5	89.0	75.0	63.0	636	573	—	—
56	79.0	88.5	74.0	62.0	617	560	—	—
55	78.5	88.0	73.0	61.0	598	547	301	2080
54	78.0	87.5	72.0	59.5	580	534	291	2010
53	77.5	87.0	71.0	58.5	562	522	282	1940
52	77.0	86.5	70.5	57.5	545	509	273	1880
51	76.5	86.0	69.5	56.0	528	496	264	1820
50	76.0	85.5	68.5	55.0	513	484	255	1760
49	75.5	85.0	67.5	54.0	498	472	246	1700
48	74.5	84.5	66.5	52.5	485	460	237	1630
47	74.0	84.0	66.0	51.5	471	448	229	1580
46	73.5	83.5	65.0	50.0	458	437	221	1520
45	73.0	83.0	64.0	49.0	446	426	214	1480
44	72.5	82.5	63.0	48.0	435	415	207	1430
43	72.0	82.0	62.0	46.5	424	404	200	1380
42	71.5	81.5	61.5	45.5	413	393	194	1340
41	71.0	81.0	60.5	44.5	403	382	188	1300
40	70.5	80.5	59.5	43.0	393	372	182	1250
39	70.0	80.0	58.5	42.0	383	362	177	1220
38	69.5	79.5	57.5	41.0	373	352	171	1180
37	69.0	79.0	56.5	39.5	363	342	166	1140
36	68.5	78.5	56.0	38.5	353	332	162	1120
35	68.0	78.0	55.0	37.0	343	322	157	1080
34	67.5	77.0	54.0	36.0	334	313	153	1050
33	67.0	76.5	53.0	35.0	325	305	148	1020
32	66.5	76.0	52.0	33.5	317	297	144	993
31	66.0	75.5	51.5	32.5	309	290	140	965
30	65.5	75.0	50.5	31.5	301	283	136	938
29	65.0	74.5	49.5	30.0	293	276	132	910
28	64.5	74.0	48.5	29.0	285	270	129	889
27	64.0	73.5	47.5	28.0	278	265	126	869
26	63.5	72.5	47.0	26.5	271	260	123	848
25	63.0	72.0	46.0	25.5	264	255	120	827
24	62.5	71.5	45.0	24.0	257	250	117	807
23	62.0	71.0	44.0	23.0	251	245	115	793
22	61.5	70.5	43.0	22.0	246	240	112	772
21	61.0	70.0	42.5	20.5	241	235	110	758
20	60.5	69.5	41.5	19.5	236	230	108	745

* From Vincent E. Lysaght and Anthony DeBellis, *Hardness Testing Handbook* (American Chain & Cable Co.). Reprinted with permission of Page-Wilson Corporation, which holds the copyright.

B. Hardness conversions: soft steel, gray and malleable cast iron, and most nonferrous metals*

		Rockwell Scale				Bhn, 500 kg$_f$ (10-mm ball)	Bhn, 3000 kg$_f$ and Dph, 10 kg$_f$	Tensile Strength	
B	15T	30T	45T	E	A			10^3 psi (approx.)	MPa (approx.)
100	93.0	82.0	72.0	—	61.5	201	240	116	800
99	92.5	81.5	71.0	—	61.0	195	234	112	772
98	—	81.0	70.0	—	60.0	189	228	109	752
97	92.0	80.5	69.0	—	59.5	184	222	106	731
96	—	80.0	68.0	—	59.0	179	216	103	710
95	91.5	79.0	67.0	—	58.0	175	210	101	696
94	—	78.5	66.0	—	57.5	171	205	98	676
93	91.0	78.0	65.5	—	57.0	167	200	96	662
92	90.5	77.5	64.5	—	56.5	163	195	93	641
91	—	77.0	63.5	—	56.0	160	190	91	627
90	90.0	76.0	62.5	—	55.5	157	185	89	614
89	89.5	75.5	61.5	—	55.0	154	180	87	600
88	—	75.0	60.5	—	54.0	151	176	85	586
87	89.0	74.5	59.5	—	53.5	148	172	83	572
86	88.5	74.0	58.5	—	53.0	145	169	81	558
85	—	73.5	58.0	—	52.5	142	165	80	552
84	88.0	73.0	57.0	—	52.0	140	162	78	538
83	87.5	72.0	56.0	—	51.0	137	159	77	531
82	—	71.5	55.0	—	50.5	135	156	75	517
81	87.0	71.0	54.0	—	50.0	133	153	74	510
80	86.5	70.0	53.0	—	49.5	130	150	72	496
79	—	69.5	52.0	—	49.0	128	147	—	—
78	86.0	69.0	51.0	—	48.5	126	144	—	—
77	85.5	68.0	50.0	—	48.0	124	141	—	—
76	—	67.5	49.0	—	47.0	122	139	—	—
75	85.0	67.0	48.5	—	46.5	120	137	—	—
74	—	66.0	47.5	—	46.0	118	135	—	—
73	84.5	65.5	46.5	—	45.5	116	132	—	—
72	84.0	65.0	45.5	—	45.0	114	130	—	—
71	—	64.0	44.5	100	44.5	112	127	—	—
70	83.5	63.5	43.5	99.5	44.0	110	125	—	—
69	83.0	62.5	42.5	99.0	43.5	109	123	—	—
68	—	62.0	41.5	98.0	43.0	107	121	—	—
67	82.5	61.5	40.5	97.5	42.5	106	119	—	—
66	82.0	60.5	39.5	97.0	42.0	104	117	—	—
65	—	60.0	38.5	96.0	—	102	116	—	—
64	81.5	59.5	37.5	95.5	41.5	101	114	—	—
63	81.0	58.5	36.5	95.0	41.0	99	112	—	—
62	—	58.0	35.5	94.5	40.5	98	110	—	—
61	80.5	57.0	34.5	93.5	40.0	96	108	—	—
60	—	56.5	33.5	93.0	39.5	95	107	—	—
59	80.0	56.0	32.0	92.5	39.0	94	106	—	—
58	79.5	55.0	31.0	92.0	38.5	92	104	—	—
57	—	54.5	30.0	91.0	38.0	91	103	—	—

APPENDIX 4B. (cont.)

		Rockwell Scale				Bhn, 500 kg$_f$ (10-mm ball)	Bhn, 3000 kg$_f$ and Dph, 10 kg$_f$	Tensile Strength	
B	15T	30T	45T	E	A			10^3 psi (approx.)	MPa (approx.)
56	79.0	54.0	29.0	90.5	—	90	101	—	—
55	78.5	53.0	28.0	90.0	37.5	89	100	—	—
54	—	52.5	27.0	89.5	37.0	87	—	—	—
53	78.0	51.5	26.0	89.0	36.5	86	—	—	—
52	77.5	51.0	25.0	88.0	36.0	85	—	—	—
51	—	50.5	24.0	87.5	35.5	84	—	—	—
50	77.0	49.5	23.0	87.0	35.0	83	—	—	—
49	76.5	49.0	22.0	86.5	—	82	—	—	—
48	—	48.5	20.5	85.5	34.5	81	—	—	—
47	76.0	47.5	19.5	85.0	34.0	80	—	—	—
46	75.5	47.0	18.5	84.5	33.5	—	—	—	—
45	—	46.0	17.5	84.0	33.0	79	—	—	—
44	75.0	45.5	16.5	83.5	32.5	78	—	—	—
43	74.5	45.0	15.5	82.5	32.0	77	—	—	—
42	—	44.0	14.5	82.0	31.5	76	—	—	—
41	74.0	43.5	13.5	81.5	31.0	75	—	—	—
40	73.5	43.0	12.5	81.0	—	—	—	—	—
39	—	42.0	11.0	80.0	30.5	74	—	—	—
38	73.0	41.5	10.0	79.5	30.0	73	—	—	—
37	72.5	40.5	9.0	79.0	29.5	72	—	—	—
36	—	40.0	8.0	78.5	29.0	—	—	—	—
35	72.0	39.5	7.0	78.0	28.5	71	—	—	—
34	71.5	38.5	6.0	77.0	28.0	70	—	—	—
33	—	38.0	5.0	76.5	—	69	—	—	—
32	71.0	37.5	4.0	76.0	27.5	—	—	—	—
31	—	36.5	3.0	75.5	27.0	68	—	—	—
30	70.5	36.0	2.0	75.0	26.5	67	—	—	—
29	70.0	35.5	1.0	74.0	26.0	—	—	—	—
28	—	34.5	—	73.5	25.5	66	—	—	—
27	69.5	34.0	—	73.0	25.0	—	—	—	—
26	69.0	33.0	—	72.5	24.5	65	—	—	—
25	—	32.5	—	72.0	—	64	—	—	—
24	68.5	32.0	—	71.0	24.0	—	—	—	—
23	68.0	31.0	—	70.5	23.5	63	—	—	—
22	—	30.5	—	70.0	23.0	—	—	—	—
21	67.5	29.5	—	69.5	22.5	62	—	—	—
20	—	29.0	—	68.5	22.0	—	—	—	—
19	67.0	28.5	—	68.0	21.5	61	—	—	—
18	66.5	27.5	—	67.5	—	—	—	—	—
17	—	27.0	—	67.0	21.0	60	—	—	—
16	66.0	26.0	—	66.5	20.5	—	—	—	—
15	65.5	25.5	—	65.5	20.0	59	—	—	—
14	—	25.0	—	65.0	—	—	—	—	—
13	65.0	24.0	—	64.5	—	58	—	—	—
12	64.5	23.5	—	64.0	—	—	—	—	—

APPENDIX 4B. (cont.)

Rockwell Scale						Bhn, 500 kg$_f$ (10-mm ball)	Bhn, 3000 kg$_f$ and Dph, 10 kg$_f$	Tensile Strength	
								10^3 psi (approx.)	MPa (approx.)
B	15T	30T	45T	E	A				
11	—	23.0	—	63.5	—	—	—	—	—
10	64.0	22.0	—	62.5	—	57	—	—	—
9	—	21.5	—	62.0	—	—	—	—	—
8	63.5	20.5	—	61.5	—	—	—	—	—
7	63.0	20.0	—	61.0	—	56	—	—	—
6	—	19.5	—	60.5	—	—	—	—	—
5	62.5	18.5	—	60.0	—	55	—	—	—
4	62.0	18.0	—	59.0	—	—	—	—	—
3	—	17.0	—	58.5	—	—	—	—	—
2	61.5	16.5	—	58.0	—	54	—	—	—
1	61.0	16.0	—	57.5	—	—	—	—	—
0	—	15.0	—	57.0	—	53	—	—	—

* From Vincent E. Lysaght and Anthony DeBellis, *Hardness Testing Handbook* (American Chain & Cable Co.). Reprinted with permission of Page-Wilson Corporation, which holds the copyright.

C.   Hardness conversions: steels*

BRINELL 10-mm steel ball Load 3000 kg$_f$		VICKERS diamond pyramid hardness, kg$_f$/mm^2	ROCKWELL	
Diam. mm	Hardness, kg$_f$/mm^2		C 150-kg$_f$ load 120° diamond cone	B 100-kg$_f$ load $\frac{1}{16}$-in. steel ball
2.20	780[a]	1,150	70	—
.25	745[a]	1,050	68	—
.30	712[a]	960	66	—
.35	682[a]	885	64	—
.40	653[a]	820	62	—
.45	627[a]	765	60	—
.50	601[a]	717	58	—
.55	578[a]	675	57	—
.60	555[a]	633	55	~120
.65	534[a]	598	53	—
.70	514[a]	567	52	—
.75	495[a]	540	50	—
.80	477[a]	515	49	—
.85	461[a]	494	47	—
.90	444	472	46	—
.95	420	454	45	~115

APPENDIX 4C. (cont.)

BRINELL 10-mm steel ball Load 3000 kg$_f$		VICKERS diamond pyramid hardness, kg$_f$/mm^2	ROCKWELL	
			C 150-kg$_f$ load 120° diamond cone	B 100-kg$_f$ load $\frac{1}{16}$-in. steel ball
Diam. mm	Hardness, kg$_f$/mm^2			
3.00	415	437	44	—
.05	401	420	42	—
.10	388	404	41	—
.15	376	389	40	—
.20	363	375	38	~110
.25	352	363	37	—
.30	341	350	36	—
.35	331	339	35	—
.40	321	327	34	—
.45	311	316	33	—
.50	302	305	32	—
.55	293	296	31	—
.60	285	287	30	105
.65	277	279	29	104
.70	269	270	28	104
.75	262	263	26	103
.80	255	256	25	102
.85	248	248	24	102
.90	241	241	23	100
.95	235	235	22	99
4.00	229	229	21	98
.05	223	223	20	97
.10	217	217	18	96
.15	212	212	17	96
.20	207	207	16	95
.25	201	201	15	94
.30	197	197	13	93
.35	192	192	12	92
.40	187	187	10	91
.45	183	183	9	90
.50	179	179	8	89
.55	174	174	7	88
.60	170	170	6	87
.65	167	167	4	86
.70	163	163	3	85
.75	159	159	2	84
.80	156	156	1	83
.85	152	152	—	82
.90	149	149	—	81
.95	146	146	—	80
5.00	143	143	—	79
.05	140	140	—	78
.10	137	137	—	77

APPENDIX 4C.   (cont.)

BRINELL 10-mm steel ball Load 3000 kg_f		VICKERS diamond pyramid hardness, kg_f/mm^2	ROCKWELL	
Diam. mm	Hardness, kg_f/mm^2		C 150-kg_f load 120° diamond cone	B 100-kg_f load $\frac{1}{16}$-in. steel ball
.15	134	134	—	76
.20	131	131	—	74
.25	128	128	—	73
.30	126	126	—	72
.35	123	123	—	71
.40	121	121	—	70
.45	118	118	—	69
.50	116	116	—	68
.55	114	114	—	67
.60	111	111	—	66
.65	109	109	—	65
.70	107	107	—	64
.75	105	105	—	62
.80	103	103	—	61
.35	101	101	—	60
.90	99	99	—	59
.95	97	97	—	57
6.00	96	96	—	56

*From D. Tabor, *The Hardness of Metals.* (Oxford: Clarendon Press, 1951).

[a]With the standard steel ball, Brinell hardness values above about 400 are not considered reliable.

HARDNESS CONVERSION: The values given here are approximate only and apply only to steels of uniform chemical composition and uniform heat treatment. The values are less reliable for nonferrous materials and are not recommended for case-hardened steel.

APPENDIX 5. MECHANICAL PROPERTY RANGES
FOR COMMON ENGINEERING MATERIALS AT ~20°C

Material	Density, kg/m^3	Elastic Modulus, GPa	Strength, MPa	Critical Stress Intensity Factor, MPa $\sqrt{m}$	Nom. Fract. Strain —	Maximum Strength/ Density, (km/s)^2
Ductile steel	7850–7870	200	350–800	170	0.2–0.5	0.1
Cast iron	6950–7700	55–140	140–650	5–20	0.0–0.2	0.09
Cast Al alloys	2570–2950	65–72	130–300	—	0.01–0.14	0.1
Wrought Al alloys	2620–2820	68–72	55–650	15–100	0.05–0.30	0.2
Polymers	900–2200	0.1–21	5–190	0.5–3	0.0–8.0	0.1
Low-alloy steel	7800–7900	200	1000–2000	15–170	0.05–0.25	0.25
Glasses, ceramics	2200–4000	40–140	10–140	0.2–5	0.0	0.05
Stainless steels	7600–8000	200	300–1400	55–120	0.10–0.30	0.2
Copper alloys	8200–8900	100–117	300–1400	10–100	0.02–0.65	0.17
Titanium alloys	4400–4800	116	400–1600	50–140	0.05–0.30	0.35
Mg alloys	1740–1840	45	80–370	—	0.01–0.25	0.2
Moldable glass fiber resin	1700–2100	11–17	55–440	20–50	0.003–0.015	0.2
Graphite–epoxy	1500	200	1000	45–>120	0–0.02	0.65
Tool steels: Hardness to 68 R_C; stable up to 600°C; strength to 2000 MPa				30–100		
Sintered carbides: Hardness at 20°C, 86 to 93 R_A (69 to well above 70 R_C equivalent); to 85 R_A max. at 760°C				3–21		
Superalloys: Strength is approximately 140 MPa at 650°C, 50 MPa at 750°C, 25 MPa at 900°C				—		

APPENDIX 6. PROPERTIES OF PURE METALS AT ~20°C

Metal	Symbol	Structure	Lattice Const., pm		Atomic Rad., pm	Density, kg/m³	Melting Point,		Elastic Modulus, GPa	Shear Modulus, GPa	Symbol
			a	c			°C	°F			
Aluminum	Al	fcc	405	—	143	2 700	660	1220	62	24	Al
Chromium	Cr	bcc	288	—	125	7 190	1875	3407	248	95	Cr
Copper	Cu	fcc	362	—	128	8 960	1083	1981	110	42	Cu
Gold	Au	fcc	408	—	144	19 300	1063	1945	80	31	Au
Iron	Fe	bcc	287	—	124	7 870	1538	2800	196	76	Fe
Lead	Pb	fcc	495	—	175	11 400	327	621	14	5	Pb
Magnesium	Mg	hcp	321	521	160	1 740	650	1202	44	17	Mg
Molybdenum	Mo	bcc	315	—	136	10 200	2610	4730	324	125	Mo
Nickel	Ni	fcc	352	—	125	8 900	1453	2647	207	80	Ni
Platinum	Pt	fcc	393	—	139	21 400	1769	3217	73	28	Pt
Silver	Ag	fcc	409	—	144	10 500	961	1761	76	29	Ag
Tin	Sn	bct	583	318	—	7 300	232	449	43	17	Sn
Titanium	Ti	hcp	295	468	—	4 510	1668	3035	116	45	Ti
Tungsten	W	bcc	316	—	137	19 300	3410	6170	345	133	W
Vanadium	V	bcc	304	—	132	6 100	1900	3450	131	50	V
Zinc	Zn	hcp	266	495	133	7 130	420	787	—	—	Zn

APPENDIX 7. COMPOSITIONS OF SOME STAINLESS STEELS*

Type Number	C	Mn Max.	P Max.	S Max.	Si Max.	Cr	Ni	Mo	UNS
17-4 PH	0.07 Max.	1.00	0.040	0.030	1.00	15.50/ 17.50	3.00/ 5.00		S17400
				Nb 0.15/0.45, Cu 3.00/5.00					
17-7 PH	0.09 Max.	1.00	0.040	0.040	1.00	16.00/ 18.00	6.50/ 7.75		S17700
				Al 0.75/1.50					
301	0.15 Max.	2.00	0.045	0.030	1.00	16.00/ 18.00	6.00/ 8.00		S30100
302	0.15 Max.	2.00	0.045	0.030	1.00	17.00/ 19.00	8.00/ 10.00		S30200
303	0.15 Max.	2.00	0.20	0.15 Min.	1.00	17.00/ 19.00	8.00/ 10.00	0.60[a] Max.	S30300
304	0.08 Max.	2.00	0.045	0.030	1.00	18.00/ 20.00	8.00/ 10.50		S30400
304L	0.030 Max.	2.00	0.045	0.030	1.00	18.00/ 20.00	8.00/ 12.00		S30403
305	0.08 Max.	2.00	0.045	0.030	1.00	19.00/ 21.00	10.50/ 13.00		S30500
309	0.20 Max.	2.00	0.045	0.030	1.00	22.00/ 24.00	12.00/ 15.00		S30900
309S	0.08 Max.	2.00	0.045	0.030	1.00	22.00/ 24.00	12.00/ 15.00		S30908
310	0.25 Max.	2.00	0.045	0.030	1.50	24.00/ 26.00	19.00/ 22.00		S31000
310S	0.08 Max.	2.00	0.045	0.030	1.50	24.00/ 26.00	19.00/ 22.00		S31008
314	0.25 Max.	2.00	0.045	0.030	1.50/ 3.00	23.00/ 32.00	19.00/ 22.00		S31400
316	0.08 Max.	2.00	0.045	0.030	1.00	16.00/ 18.00	10.00/ 14.00	2.00/ 3.00	S31600
316L	0.030 Max.	2.00	0.045	0.030	1.00	16.00/ 18.00	10.00/ 14.00	2.00/ 3.00	S31603
317	0.08 Max.	2.00	0.045	0.030	1.00	18.00/ 20.00	11.00/ 15.00	3.00/ 4.00	S31700
321	0.08 Max.	2.00	0.045	0.030	1.00	17.00/ 19.00	9.00/ 12.00		S32100
				Ti 5 × C Min.					
347	0.08 Max.	2.00	0.045	0.030	1.00	17.00/ 19.00	9.00/ 13.00		S34700
				Nb 10 × C Min.					
348	0.08 Max.	2.00	0.045	0.030	1.00	17.00/ 19.00	9.00/ 13.00		S34800

APPENDIX 7. (cont.)

Type Number	C	Mn Max.	P Max.	S Max.	Si Max.	Cr	Ni	Mo	UNS
			Nb 10 × C Min., Co 0.20 Max, Ta 0.10 Max.						
AM-350	0.07/ 0.11	0.50/ 1.25	0.040	0.030	0.50	16.00/ 17.00	4.00/ 5.00	2.50/ 3.25	S35000
				N 0.07/0.13					
403	0.15 Max.	1.00	0.040	0.030	0.50	11.50/ 13.00			S40300
405	0.08 Max.	1.00	0.040	0.030	1.00	11.50/ 14.50			S40500
				Al 0.10/0.30					
410	0.15 Max.	1.00	0.040	0.030	1.00	11.50/ 13.50			S41000
416	0.15 Max.	1.25	0.060	0.15 Min.	1.00	12.00/ 14.00		0.60[a] Max.	S41600
416 Se	0.15 Max.	1.25	0.060	0.060	1.00	12.00/ 14.00			S41623
				Se 0.15 Min.					
420	Over 0.15	1.00	0.040	0.030	1.00	12.00/ 14.00			S42000
430	0.12 Max.	1.00	0.040	0.030	1.00	16.00/ 18.00			S43000
430F	0.12 Max.	1.25	0.060	0.15 Min.	1.00	16.00/ 18.00		0.60[a] Max.	S43020
430F Se	0.12 Max.	1.25	0.060	0.060	1.00	16.00/ 18.00			S43023
				Se 0.15 Min.					
431	0.20 Max.	1.00	0.040	0.030	1.00	15.00/ 17.00	1.25/ 2.50		S43100
440A	0.60/ 0.75	1.00	0.040	0.030	1.00	16.00/ 18.00		0.75 Max.	S44002
440B	0.75/ 0.95	1.00	0.040	0.030	1.00	16.00/ 18.00		0.75 Max.	S44003
440C	0.95/ 1.20	1.00	0.040	0.030	1.00	16.00/ 18.00		0.75 Max.	S44004
442	0.20 Max.	1.00	0.040	0.030	1.00	18.00/ 23.00			S44200
446	0.20 Max.	1.50	0.040	0.030	1.00	23.00/ 27.00			S44600
				N 0.25 Max.					

* Society of Automotive Engineers, Inc., *Unified Numbering System for Metals and Alloys*, 2nd ed., Warrendale, PA 15096: Soc. of Auto. Engineers, 1977. Copyright Society of Automotive Engineers, Warrendale, PA 15096. Reprinted with permission.

[a] Optional.

APPENDIX 8. REPRESENTATIVE MECHANICAL
PROPERTIES OF POLYMERS

Thermoplastics	T_g °C	E, GPa	σ_y, MPa	Izod Impact J/m of Notch	Hardness Rockwell	Acronym or Trade Name
Acrylonitrile–Butadiene–Styrene	80	2.5	50	500	70 R_R	⎰ ABS,
Glass-filled		7.0	110	200	—	⎱ Cyclolac
Polyamides	60	2.8	70	53	119 R_R	⎰ Nylon 6,
30% glass-filled		7.0	150	145	121 R_R	⎱ Durethan
Polycarbonate	150	2.5	68	800	70 R_M	PC, Lexan
30% glass-filled		6.0	90	>1000	86 R_M	
Polyethylene						
Low-density	−20	0.18	11	—	43 R_D	PE
High-density		0.56	30	300	65 R_D	
Polyethylene Terephthalate	67	8	135	—	—	⎧ PETP, Dacron Terylene Fibers; Melinex Film
Polymethylmethacrylate	100	2.8	70	35	99 R_M	PMMA, Perspex, Plexiglas
Poly-4-Methylpentene-1	55	1.4	28	350	70 R_L	TPX
Polyoxymethylene	−13	3.6	70	74	94 R_M	POM, Acetal Delrin, Hostaform
Polyphenyleneoxide	180	2.4	62	210	90 R_R	PPO, Noryl
Glass-filled		8.4	120	60		
Polyphenylenesulphide	150	3.3	75	20	123 R_R	PPS, Ryton
40% glass-filled		7.8	150	20	123 R_R	
Polypropylene	0	1.3	35	70	90 R_R	PP
Glass-filled		7.0	55	100	100 R_R	
Polystyrene	100	2.7	50	200	70 R_M	PS
Polysulphone	200	2.5	70	700	120 R_R	
30% glass-filled		8.3	117	960	84 R_M	
Polytetrafluoroethylene	120	0.4	25	160	52 R_D	PTFE, Teflon
Polyvinylchloride	80	2.0	25	100	110 R_R	PVC

Thermosets	T_g, °C	E, GPa	σ_y, MPa
Alkyds (glass-filled)	(200)	15	45
Epoxides	(150–250)	3	60
Epoxides (glass-filled)	(150–250)	21	140
Melamines (glass-filled)	(200)	17	50
Phenolics (glass-filled)	(200–300)	18	80
Polybutadienes (glass-filled)	(200)	11	65
Polyesters (glass-filled)	(200)	14	70
Polyimides (glass-filled)	(350)	21	190
Silicones (glass-filled)	(300)	8	40
Ureas	(80)	7	60
Urethanes (solid)	(100)	7	70

APPENDIX 9. TYPICAL VALUES OF QUASI-STATIC TOUGHNESS AND YIELD STRESS

Material Type	E, GPa	σ_y, MPa	R, kJ/m^2	K_c, MPa$\sqrt{m}$	K_c/σ_y, $\sqrt{m}$
Glasses	70	138	0.01	1	0.01
Ceramics	300	280	0.1	5	0.02
PMMA	2.8	70	0.5	1	0.02
Cast iron	100	300	1	10	0.03
Boron-fiber epoxy composites	170	2000	30	70	0.04
Tough polymers (Polycarbonate)	2.5	63	4	3	0.05
High-strength aluminum-base alloys	70	600	28	44	0.07
High-strength steels (4340)	210	1400	140	170	0.12
High-strength titanium-base alloys	110	1100	170	136	0.12
Maraging steel	210	960	150	180	0.20
Low-carbon steels	210	280	140	170	0.60

APPENDIX 10. TYPICAL COMPOSITIONS AND MECHANICAL PROPERTIES OF CAST IRONS

Category	Class	Typical Composition, %			Tensile Strength, MPa	Bhn	Percentage Elongation at Fracture; 50 mm Gage Length
		Carbon	Silicon	Manganese			
Gray	ASTM-20 large flakes	3.20	2.30	0.65	140	155	~0
	ASTM-60 small flakes	2.65	1.35	0.65	430	230	~0
Ductile (nodular)	60-45-10	3.2–4.1 plus	1.8–2.8 0.1% Mg + Ce	0.80	420	150	>10%
	120-90-02		same range		840	260	>2%
White	Cupola	3.5	0.7	0.6	315	400	~0
Malleable	32510 (Ferritic)	2.5	1.2	0.4	385	140	18%
	80002 (Pearlitic)	2.3	1.2	0.8	700	250	5%

Young's modulus E is about 170 GPa for nodular and malleable irons. For gray irons E varies inversely with flake size (compliance of body); for example, for ASTM-20 class (250-μm flakes), $E = 80$ GPa, and for ASTM-60 (30-μm flakes), $E = 140$ GPa. See Exercise 15-3.

APPENDIX 11. REPRESENTATIVE ROOM-TEMPERATURE MECHANICAL PROPERTIES OF CERAMICS AND GLASSES

	Melting Point, °C	Tensile or Bend Strength (σ_f), MPa	Young's Modulus (E), GPa	Fracture Toughness (R), J/m²	Coefficient of Thermal Expansion (α), °C^{-1} × 10^{-6}	Thermal Conductivity (k), W/mK	Spalling Index ($k\sigma_f/E\alpha$), kW/m	Toughness/ Strength Ratio (K_c/σ_f) = $\sqrt{ER}/\sigma_f$, $\sqrt{m}$
Typical porous insulating firebrick	?	~15	20	? 30	7	1	0.1	0.05
Alumina (Al$_2$O$_3$)	2050	300 (700 whisker)	380	80	8	38	3.8	5.5
Magnesia (MgO)	2850	100	315	70	14.8	44	1	0.5
Silicon carbide (SiC)	2300 (decomposes)	200	350	50	4.5	209	26	0.02
Silicon nitride (Si$_3$N$_4$)	1900 (sublimes)	600 (1400 whisker)	320	80	2.5	8	6	0.01
Titanium carbide (TiC), Ni/Mo binder	3250	200	350	400	7	17	1.4	0.6
Tungsten carbide (WC), Co binder	2620	350	700	1000	1	34	1.4	0.8
Common bulk glass	Fuses at 1600	70 (up to 4000 filaments)	70	20	10	1	0.1	0.2
Pyroceram glass ceramic	Fuses at 1600	190	126	? 40	10	3	0.5	0.1
Concrete, reinforced	?	20	35	30	7	0.5	0.1	0.1
Carbon graphite	3600 (sublimes)	30	6	100	3	~100	170	0.3

NOTE: There is much more variation in properties of these types of solid than metals, variations occurring with production methods. Properties also vary with temperature and rate, for example, k for Al$_2$O$_3$ is 4 W/mK at 1000°C and is 13 W/mK at 1000°C for SiC.

APPENDIX 12. DENSITY AND MECHANICAL PROPERTIES OF BULK MATERIALS AND FILAMENTS

Bulk Material	ρ, kg/m^3	S_u, MPa	S_u/ρ, (km/s)^2	E, GPa	E/ρ, (km/s)^2
Low-carbon steel	7850	400	0.051	210	27
High-tensile steel	7850	2000	0.25	210	27
Aluminum alloy	2700	550	0.20	74	27
Magnesium alloy	1800	350	0.19	45	25
Nickel	8860	490	0.055	210	24
Copper	8940	240	0.027	120	13
Titanium alloy	4500	1000	0.22	120	27
Wood	600	60	0.1	14	23
Glass	2500	60	0.024	70	28
Filaments					
Glass	2500	1000	0.4	70	28
Iron	7850	14000	1.8	210	27
Nickel	8860	3800	0.43	210	24
Copper	8940	2900	0.32	120	13
Alumina (Al$_2$O$_3$)	4000	18700	4.7	390	98
Silicon carbide	3200	24100	7.5	560	175
Boron	2500	3450	1.4	380	152
Graphite	2250	21000	9.3	750	334
Carbon	2250	1600	0.71	360	160
Kevlar 49	1440	2700	1.9	130	90
Nylon	1100	400	0.36	4	3.6
Polypropylene	900	900	1.0	8	9

APPENDIX 13. DENSITY AND MECHANICAL PROPERTIES OF COMPOSITES

Material	ρ, kg/m^3	S_u, MPa	S_u/ρ, (km/s)^2	E, GPa	E/ρ, (km/s)^2
Boron–5505 epoxy	1990	1590	0.80	207	104
HTS graphite–5208 epoxy	1550	1480	0.95	172	111
Kevlar 49–resin	1380	1380	1.00	76	55
E glass–1002 epoxy	1800	1100	0.61	39	22
Boron–6061 Al	2600	1490	0.57	214	82
T50 graphite–2011 Al	2580	566	0.22	160	62

APPENDIX 14. ENERGY CONSUMPTION IN MANUFACTURE OF VARIOUS ENGINEERING SOLIDS

	Energy/Mass, GJ/1000 kg	Energy/Mass-Stiffness, m³/1000 kg	Energy/Mass-Strength, m³/1000 kg
Aluminum	200	2.8	870
Cast iron	45	0.3	300
Low-carbon steel	50	0.24	240
Glass	20	0.3	333
Brick	6	0.1	30
Concrete	2	0.05	80
Wood	1	0.07	10
Polyethylene	45	90	1500
Carbon-fiber-reinforced composite	4000	27	6000

APPENDIX 15. SOME COMMON KNOWN STRESS-CORROSION CRACKING (AND HYDROGEN-EMBRITTLEMENT) ENVIRONMENT–MATERIAL COMBINATIONS

	Low-C Steel	High-Strength Steel	Austenitic Stainless Steel	Al + Alloys	Cu + Alloys	Ni + Alloys	Ti + Alloys
Ammonia	X				X		
Ammonium chloride			X				
Ammonium hydroxide					X		
Calcium chloride			X				
Carbon dioxide or monoxide (wet)	X	X					
Carbon tetrachloride	X	X	X				
Chlorides (wet)		X	X	X			X
Chlorine		X	X				
Chromic acid	X				X	X	
Ethanol							X
Ferric chloride	X				X		
Hydrogen		X	X				
Hydrogen sulfide (wet)			X	X			
Magnesium chloride	X		X		X	X	
Manganese chloride			X				
Mercuric chloride			X	X	X	X	
Mercury			X	X	X	X	
Methanol							X
Nitric acid	X				X		
Oxygen			X		X		
Potassium chloride			X		X		
Potassium hydroxide	X		X			X	

APPENDIX 15.   (cont.)

	Low-C Steel	High-Strength Steel	Austenitic Stainless Steel	Al + Alloys	Cu + Alloys	Ni + Alloys	Ti + Alloys
Sodium carbonate			X				
Sodium hydroxide	X		X			X	
Steam				X		X	
Sulfides (wet)		X					
Sulfur dioxide					X		
Toluene					X		

NOTE: Not all alloys within a group are necessarily susceptible; see references for specific alloys. [ASM v. 10; *Corrosion Data Survey* (Metals Section), 5th ed., (Houston, Texas: National Assoc. of Corrosion Engineers). Copyright Nat. Assoc. of Corrosion Engineers and reprinted with permission.]

Index